Autonomous Underwater Vehicles

Related titles on radar:

Advances in Bistatic Radar Willis and Griffiths
Airborne Early Warning System Concepts, 3rd Edition Long
Bistatic Radar, 2nd Edition Willis
Design of Multi-Frequency CW Radars Jankiraman
Digital Techniques for Wideband Receivers, 2nd Edition Tsui
Electronic Warfare Pocket Guide Adamy
Foliage Penetration Radar: Detection and characterisation of objects under trees Davis
Fundamentals of Ground Radar for ATC Engineers and Technicians Bouwman
Fundamentals of Systems Engineering and Defense Systems Applications Jeffrey
Introduction to Electronic Warfare Modeling and Simulation Adamy
Introduction to Electronic Defense Systems Neri
Introduction to Sensors for Ranging and Imaging Brooker
Microwave Passive Direction Finding Lipsky
Microwave Receivers with Electronic Warfare Applications Tsui
Phased-Array Radar Design: Application of radar fundamentals Jeffrey
Pocket Radar Guide: Key facts, equations, and data Curry
Principles of Modern Radar, Volume 1: Basic principles Richards, Scheer, and Holm
Principles of Modern Radar, Volume 2: Advanced techniques Melvin and Scheer
Principles of Modern Radar, Volume 3: Applications Scheer and Melvin
Principles of Waveform Diversity and Design Wicks *et al.*
Pulse Doppler Radar Alabaster
Radar Cross Section Measurements Knott
Radar Cross Section, 2nd Edition Knott *et al.*
Radar Design Principles: Signal processing and the environment, 2nd Edition Nathanson *et al.*
Radar Detection DiFranco and Rubin
Radar Essentials: A concise handbook for radar design and performance Curry
Radar Foundations for Imaging and Advanced Concepts Sullivan
Radar Principles for the Non-Specialist, 3rd Edition Toomay and Hannan
Test and Evaluation of Aircraft Avionics and Weapons Systems McShea
Understanding Radar Systems Kingsley and Quegan
Understanding Synthetic Aperture Radar Images Oliver and Quegan
Radar and Electronic Warfare Principles for the Non-Specialist, 4th Edition Hannen
Inverse Synthetic Aperture Radar Imaging: Principles, algorithms and applications Chen and Martorella
Stimson's Introduction to Airborne Radar, 3rd Edition Baker, Griffiths, and Adamy
Test and Evaluation of Avionics and Weapon Systems, 2nd Edition McShea
Angle-of-arrival Estimation Using Radar Interferometry: Methods and applications Holder
Biologically-Inspired Radar and Sonar: Lessons from nature Balleri, Griffiths, and Baker
The Impact of Cognition on Radar Technology Farina, De Maio, and Haykin
Novel Radar Techniques and Applications, Volume 1: Real aperture array radar, imaging radar, and passive and multistatic radar Klemm, Nickel, Gierull, Lombardo, Griffiths, and Koch
Novel Radar Techniques and Applications, Volume 2: Waveform diversity and cognitive radar, and target tracking and data fusion Klemm, Nickel, Gierull, Lombardo, Griffiths, and Koch
Radar and Communication Spectrum Sharing Blunt and Perrins
Systems Engineering for Ethical Autonomous Systems Gillespie
Shadowing Function from Randomly Rough Surfaces: Derivation and applications Bourlier and Li
Photo for Radar Networks and Electronic Warfare Systems Bogoni, Laghezza, and Ghelfi
Multidimensional Radar Imaging Martorella

Autonomous Underwater Vehicles

Design and practice

Edited by
Frank Ehlers

The Institution of Engineering and Technology

Published by SciTech Publishing, an imprint of The Institution of Engineering and Technology, London, United Kingdom

The Institution of Engineering and Technology is registered as a Charity in England & Wales (no. 211014) and Scotland (no. SC038698).

First published 2020

The Institution of Engineering and Technology
Michael Faraday House
Six Hills Way, Stevenage
Herts, SG1 2AY, United Kingdom

www.theiet.org

British Library Cataloguing in Publication Data
A catalogue record for this product is available from the British Library

ISBN 978-1-78561-703-4 (hardback)
ISBN 978-1-78561-704-1 (PDF)

Typeset in India by MPS Limited

Contents

About the editor

Dr. Frank Ehlers obtained his Diploma degree in physics (1995) and his Dr. rer. nat. degree in theoretical physics (1998) from the Christian-Albrechts-University Kiel, Germany. He wrote his Ph.D. thesis on "Nonlinear blind source separation." For more than 20 years he worked as a scientist, project manager, program manager, or principal scientist in German and international research institutes dealing with maritime technology. He has conducted both application-oriented research and more fundamental research in the fields of signal processing, detection, classification, data fusion, coordinated distributed sensor systems, and autonomy for maritime scenarios.

He has become an associate editor of the *IEEE Transactions on Automation Science and Engineering* (*T-ASE*) with the primary areas: sensing and networks, networked robotics, many robot systems, and maritime robotic systems.

Editing this book on the design and practice of autonomous underwater vehicles continues a series of activities as a member of co-chair and guest editor teams for various conferences and journals: Blind Source Separation (2007), Track Before Detect Algorithms (2008), Robots and Autonomy (2009), Signal Processing for Maritime Applications (2010), Distributed Mobile Sensor Networks for Hazardous Applications (2012), Maritime Robotics (2013), Crossing the Reality Gap for Many Robot Systems (2012–2014), and Networked Cooperative Autonomous Systems (2015).

The topic of autonomy and mobile sensors is the focus of his recent activities as chair, member-at-large, and external evaluator in international research groups, panels, and organizations.

Foreword

Robotics and maritime research are two exciting fields of work which are to be combined when unmanned underwater vehicles enter the scene. On top of this, autonomy, as a disruptive technology, offers the implementation of effective, robust, and efficient systems, even in the harsh, highly variable, and still surprising maritime environment.

Is there a normative force of the factual also for this fascinating economic and scientific setup? To answer this question, first of all, we need to generate facts, by practicing the use of these maritime robots which we design as autonomous underwater vehicles (AUVs).

Written by a team of top-class international contributors, this book gives a state-of-the-art overview of the hot topic of "AUV Design and Practice." I am proud to state this: This book is the work of my colleagues and friends. I would like to thank all authors for their excellent work.

Many thanks to Nicki Dennis who proposed this editorial work to me. In the process of generating a book from a collection of chapters, as an initial step, the critical comments from the reviewers have been essential. I am grateful for their efforts. Furthermore, the process continued and concluded with the excellent cooperation with Olivia Watkins and Paul Deards from IET Books. Thank you for your help, support, guidelines, and for letting this book become reality.

Despite the fact that a lot of literature about AUVs is already available, compiling a book from the viewpoint of the interplay between design and practice and therefore containing all elements from control and navigation, communication and cooperation, to methodologies and applications has been an effort worthwhile. I spent many long evenings curiously reading through the submissions, mapping the insights gathered onto my own research project, thereby generating a more complete (holistic) picture. This is learning and exactly what the "Design and Practice" subtitle of the book promises. I hope many readers will join this experience.

Frank Ehlers
02.08.2019

List of authors

Alves, João	NATO Centre for Maritime Research and Experimentation (CMRE)	Italy
Ånonsen, Kjetil Bergh	Norwegian Defence Research Establishment (FFI)	Norway
Antonelli, Gianluca	Università degli Studi di Cassino e del Lazio	Italy
Arrichiello, Filippo	Università degli Studi di Cassino e del Lazio	Italy
Bays, Matthew J.	Naval Surface Warfare Center	USA
Bechlioulis, Charalampos P.	National Technical University of Athens	Greece
Behrje, Ulrich	University of Lübeck	Germany
Brink, Gunnar	Fraunhofer IOSB	Germany
Caiti, Andrea	Università di Pisa	Italy
Casalino, Giuseppe	Università di Genova	Italy
De Palma, Daniela	Università del Salento	Italy
Dias, Paulo Sousa	University of Porto	Portugal
Djapic, Vladimir	American Haval Motor Technology	USA
Fenucci, Davide	National Oceanography Centre	UK
Forgoston, Eric	Montclair State University	USA
Gade, Kenneth	Norwegian Defence Research Establishment (FFI)	Norway
Hagen, Ove Kent	Norwegian Defence Research Establishment (FFI)	Norway
Harris, Catherine A.	National Oceanography Centre	UK
Heshmati-alamdari, Shahab	KTH Royal Institute of Technology	Sweden
Hsieh, M. Ani	University of Pennsylvania	USA
Indiveri, Giovanni	Università di Genova	Italy
Isokeit, Cedric	University of Lübeck	Germany
Kar, Indra Narayan	Indian Institute of Technology Delhi	India
Karras, George C.	University of Thessaly	Greece
	National Technical University of Athens	Greece
Körner, Gerhard	Meelve	Germany
Kularatne, Dhanushka	University of Pennsylvania	USA
Kyriakopoulos, Kostas J.	National Technical University of Athens	Greece

Lennon, Jamie	Naval Research Laboratory	USA
Leonard, Elizabeth I.	Naval Research Laboratory	USA
Maehle, Erik	University of Lübeck	Germany
Marantos, Panos	National Technical University of Athens	Greece
McPhail, Stephen D.	National Oceanography Centre	UK
Meyer, Benjamin	University of Lübeck	Germany
Munafò, Andrea	SeeByte Ltd. Edinburgh	UK
Phillips, Alexander B.	National Oceanography Centre	UK
Pebody, Miles	National Oceanography Centre	UK
Pelekanakis, Konstantinos	NATO Centre for Maritime Research and Experimentation (CMRE)	Italy
Petroccia, Roberto	NATO Centre for Maritime Research and Experimentation (CMRE)	Italy
Pinto, José	University of Porto	Portugal
Pollini, Lorenzo	Università di Pisa	Italy
Prampart, Thomas	National Oceanography Centre	UK
Redfield, Signe A.	Naval Research Laboratory	USA
Rogers, Eric	University of Southampton	UK
Roper, Daniel T.	National Oceanography Centre	UK
Roy, Spandan	International Institute of Information Technology	India
Roy, Sayan Basu	Indraprastha Institute of Information Technology	India
Razzanelli, Matteo	Università di Pisa	Italy
Salam, Tahiya	University of Pennsylvania	USA
Salavasidis, Georgios	National Oceanography Centre	UK
	University of Southampton	UK
Sarma, Ashwin	BAE Systems	USA
Seto, Mae	Dalhousie University	Canada
Simetti, Enrico	Università di Genova	Italy
Smart, Micheal	National Oceanography Centre	UK
Sousa, João Borges	University of Porto	Portugal
Svendsen, Kathleen	Dalhousie University	Canada
Templeton, Robert	National Oceanography Centre	UK
Wettergren, Thomas A.	Naval Undersea Warfare Center	USA

Chapter 1
Introduction

Frank Ehlers[1]

Autonomous underwater vehicles (AUVs) generate two streams of interest, one in the foreground (robotics and the maritime environment) and the other in the background (autonomy), as will be explained in the following two sections (1.1 and 1.2). In Section 1.3, the organization of the chapters, in line with the discussion regarding robotics, maritime environment and autonomy, is explained. Section 1.3 contains also abstracts of all chapters. Their summaries, conclusions and suggestions for future work are collected in Chapter 19, where we try a 'localization and mapping' of the chapters' contents in the design space of AUVs, utilizing the coordinate system developed in this Introduction.

1.1 Robotics and the maritime environment

The foreground interest in AUVs stems from the combination of two extremely important fields of work: using robots in the maritime domain. Maritime robotics is growing in importance. Behaviours for many applications have been taken from concept to reality, delivering results and data that have started to change our models and our understanding of the maritime environment. Maritime robots are significantly changing the speed and accuracy with which undersea tasks (also as part of industrial applications) can be accomplished.

In the following paragraphs, three views on this combination are described: the roboticist's view, the statistical signal processing view and the system-of-systems view.

1.1.1 The roboticist's view

From the viewpoint of a roboticist, in general, the advances in embedded processors, architectures and modularity, sensors, middleware, communication and networking technology in the last few decades have accelerated interest in networked cooperative

[1]Bundeswehr Technical Center for Ships and Naval Weapons, Maritime Technology and Research (WTD 71), Kiel, Germany

autonomous systems, multi-robot systems and distributed sensor networks for applications such as manufacturing, logistics, process monitoring and enhanced situational awareness, plant safety, inspection, security and rescue operations.

Modern AUV designs, exploiting this general robotics trend in particular, become extremely suitable for the maritime environment because their components have better performance, but need less power and are powered from energy sources with higher storage capacities.

1.1.2 The statistical signal processing view

From the viewpoint of a statistical signal processing engineer, in general, a key element of the success story of robotic sensor systems is the way uncertainty is managed: uncertainty originating from insufficient sensor data, uncertainty about effects of future actions and, in the case of distributed sensors and actuators (like for a team of robots), uncertainty about communication lines.

The maritime domain continues to be important for our society. Significant investments continue to be made to increase our knowledge about what 'happens' underwater, whether at or near the sea surface, within the water column or at the seabed. The latest geophysical, archaeological and oceanographical surveys deliver more accurate global knowledge at increased resolution. Surveillance applications allow dynamic systems to be characterized. Underwater exploration is fundamentally reliant on the effective processing of sensor signal data. All maritime applications face the same difficult operating environment: fading channels, rapidly changing environmental conditions, high noise levels at sensors, sparse coverage of the measurement area and limited reliability of communication channels. These are the special uncertainties that AUVs have to handle.

1.1.3 The system-of-systems view

From the viewpoint of a system-of-systems engineer, in general, the advances that have made individual robots and autonomous systems more practical have enabled the research on and the development of cooperative systems, where capabilities are expressed by the team rather than by a super-capable individual. This is especially relevant in complex tasks that require capabilities that are varied in both quantity and difficulty. Moreover, one of the main advantages of having a team of robots, instead of a super-capable individual, is in the increased reliability due to redundancy. Performance metrics have to be used to evaluate the increases in quality (including reliability, efficiency and reduced costs) achievable through the use of networked cooperative systems.

For AUVs, a low-cost and rapidly deployable design is possible such that the AUV can remain operational for a very long time. The application of such systems for commercial, scientific and military purposes has already become reality.

1.2 Autonomy, or why the interplay between design and practice is so important

In addition to the foreground high interest in this topic (explained above: robotics and the maritime environment), there is another (maybe a bit hidden) extremely interesting aspect, autonomy.

Elements of autonomy are the adaptation of the system without human intervention but within the human-defined scope of the operation. There are, of course, various semi-autonomous designs possible, inserting human-to-system interaction, e.g. at the planning level and the decision level.

AUVs, in this book, are rather designed for sensing than manipulation tasks. In the future, there might be an increased interest in AUVs with manipulation capacity. As far as autonomous sensing is concerned, an implementation of AUVs might be described as a distributed multisensor multi-robot system with the ability to make decisions regarding sensor settings (i.e. an internal feedback from data analysis back to acquisition of data) and to act regarding own platform actuators on the basis of its own calculations, but within the limits of guaranteed effectiveness and robustness. The overall objective of autonomy for autonomous sensing, following this description, is maintaining, gathering and exploitation of process knowledge.

As such, autonomy is a disruptive technology, which means that the natural cycles consisting of 'design, practice, learn, re-design, practice again, learn more' run into a phase transition. The term phase transition is adopted from the description of critical phenomena where, although the atomic components of a system remain the same, the system behaves completely differently at external conditions belonging to different system phases. The interesting part here with autonomy is that, under certain external conditions that we have to provide, the autonomy internally performs the cycle of design, practice, learn, which we as engineers and scientists are used to perform, but the autonomy can perform this at machine speed and within a huge data/learning space. The question remaining is where to find these certain external conditions which allow such phase transition.

This is the scenario in which this book fits: For supporting the autonomy in the design space (including the design of behaviours), data from practices with current designs have to be provided as integrated (with the help of validated models) information to build up the critical mass for the autonomous optimization steps to reach their fixed point.

1.3 The organization of this book: from design to practice – there and back again

The order of chapters in this book follows the 'aviate, navigate, communicate' paradigm, which a pilot in trouble follows, and which applies to an AUV from the very first moment it dives. We add 'cooperate, reflect on the methods used so far and apply your capacities' to the sequence. In other words, we are adopting a 'roboticist's view' in terms of stacking behaviours.

In the remainder of this subsection, each paper will be presented. Conclusions from the chapters and from the (holistic) overview of all chapters will be presented in the Conclusion (Chapter 19) at the end of this book.

Part I: Control

Chapter 2

In the chapter 'Sensor-based motion control of autonomous underwater vehicles, part I: modeling and low-complexity state estimation' by George C. Karras, Charalampos P. Bechlioulis, Panos Marantos, Shahab Heshmati-Alamdari and Kostas J. Kyriakopoulos, the authors describe the kinematic and dynamic equations of underwater robotic vehicles including standard assumptions and simplifications that may lead to low-complexity but accurate and operational models. Furthermore, they describe a recursive-least-squares method for the online identification of the vehicle dynamic parameters. In addition, they present a state estimation technique which, according to the available sensors, can be adopted for either shallow or deep water operation. Keeping in mind the low-complexity requirements, the proposed state estimation method is based on the complementary filters notion. The goal of the chapter is to provide the reader with the appropriate analytical methods, in order to derive a simple yet accurate vehicle model, design a state estimation algorithm that could be easily integrated into the embedded system framework of an underwater robotic vehicle and perform a fast online dynamic parameter identification.

Chapter 3

In the chapter 'Sensor-based motion control of autonomous underwater vehicles, part II: robust motion control strategies' by Charalampos P. Bechlioulis, Shahab Heshmati-alamdari, George C. Karras, Panos Marantos and Kostas J. Kyriakopoulos, the authors present, in the first part, a non-linear model predictive control strategy for underwater robotic vehicles operating under various constraints. The purpose of the controller is to guide the vehicle towards specific waypoints. Various constraints such as: obstacles, workspace boundaries, control input saturation as well as predefined upper bound of the vehicle velocity (requirements for several underwater tasks such as seabed inspection scenario and mosaicking) are considered during the control design. The proposed scheme incorporates the full dynamics of the vehicle in which the ocean currents are also involved. The controller is designed in order to find the optimal thrusts required for minimizing the waypoint tracking error. Moreover, the control inputs calculated by the proposed approach are formulated in a way that the vehicle will exploit the ocean currents, when they are in favour of the waypoint tracking mission, which results in reduced energy consumption by the thrusters. In the second part of the chapter, novel position and trajectory tracking control schemes for AUVs are presented. The proposed controllers do not utilize the vehicle's dynamic model parameters and guarantee prescribed transient and steady-state performance despite the presence of external disturbances and kinematic constraints for the case of underactuated vehicles. Moreover, through the appropriate selection of certain performance functions, the proposed scheme can also guarantee the satisfaction of motion and performance constraints imposed by the desired task.

Chapter 4

The chapter 'Adaptive–robust control of autonomous, underwater vehicle with unknown system dynamics' by Spandan Roy, Sayan Basu Roy and Indra Narayan Kar proposes a new adaptive–robust control (ARC) architecture for AUVs in the face of uncertain as well as completely unknown system dynamics parameters. The conventional ARC mechanisms either presume that the overall uncertainties or its time derivative is norm bounded by a constant or require structural knowledge of the system. However, for an AUV, the upper bound of the uncertainty explicitly contains the system states. Thus, assigning constant upper bound to the uncertainty (or its time derivative) is restrictive if one considers a system like AUV. Moreover, due to the highly non-linear and coupled dynamics, true structural knowledge as well as parametric values of the AUV system is not always available. Considering such prevailing issues, Adaptive Switching gain-based Robust Control (ASRC) is conceptualized here for AUVs which avoids any presumption of constant upper bound on the overall uncertainties. The proposed control law does not utilize any prior knowledge of the parameters of the system dynamics, while negotiating the uncertainties regardless of the fact whether the uncertainties appear linearly or non-linearly in parameters. Besides, the proposed adaptive law for switching gain alleviates the over- and under-estimation issues commonly observed in the literatures of ARC. Extensive simulation studies also validate the effectiveness of the proposed ASRC law in uncertain scenarios compared to the conventional adaptive sliding mode design.

Part II: Navigation

Chapter 5

In the chapter 'AUV navigation: challenges and solutions' by Kjetil Bergh Ånonsen, Kenneth Gade and Ove Kent Hagen, the authors explain that the term navigation is commonly used to describe several different, but related concepts. Some use the term for the process of determining the position (in some reference frame) and the orientation of the vehicle only, whereas others also include the problem of how to get from a point A to a point B in the navigation concept. In their chapter, the authors focus solely on the former meaning of the term. In other words, they define navigation to mean estimation of the six degrees of freedom of a rigid body (any vehicle or device), i.e. the position and orientation of the body. The position can be expressed in various reference frames (local or global), and the orientation is defined as the spatial orientation of the body frame relative to a local reference tangential to the Earth ellipsoid. It is often customary to include the estimation of linear and angular velocity in the navigation problem. The uncertainties of the estimates are often also part of the navigation output, and the navigation can be performed in real time, or in post-processing. Throughout the chapter, a reference to the positioning part of the navigation problem is made by the term positioning. This is to be understood as the passive process of determining a position, not the more active process of physically placing an object in a certain position. The term localization is used, e.g., in conjunction with the simultaneous localization and mapping (SLAM) problem, and it is in this chapter to be understood as the same as positioning. The chapter provides a precise notation system for navigation purposes and the basic principles of a general inertial navigation system (INS), which are important for understanding the applicability and limitations of such systems for AUVs. Furthermore, important properties of a basic INS in an AUV and various sensors and methods that can be used for aiding the INS, including geophysical navigation and SLAM, are discussed. The focus is on single vehicle navigation only, but some of the methods discussed, especially those involving acoustic ranging, can be adapted to multi-vehicle scenarios by putting transponders on each vehicle and develop a communication protocol to enable the vehicles to exchange navigation information.

Chapter 6

The chapter 'Ultra-endurance AUVs: energy requirements and terrain-aided navigation' by Georgios Salavasidis, Andrea Munafo, Davide Fenucci, Catherine A. Harris, Thomas Prampart, Robert Templeton, Micheal Smart, Daniel T. Roper, Miles Pebody, Stephen D. McPhail, Eric Rogers and Alexander B. Phillips deals with a new class of ultra-endurance (of the order of months and thousands of kilometres) propeller-driven AUVs, the Autosub Long Range (ALR6000) and the Autosub Long Range 1500 (ALR1500). These platforms aim to combine the extreme range and endurance already achievable using subsea gliders, with the manoeuvrability, position control and payload capacity of a larger flight-style AUV. The long endurance is achieved through the use of high-energy-density lithium primary cells combined with optimization of speed and energy consumption within all systems. One possible method to face the challenge of unaided navigation is by exploiting geophysical information. In contrast to acoustic-aided methods, geophysical navigation can be fully implemented on board the vehicle. Hence, these methods are particularly appealing for long-range AUV missions. Terrain-aided navigation is a form of geophysical localization where water-depth observations are matched against an a-priori bathymetric reference map. Despite the challenges in such a mission, simulation results show that it may be possible to cross the Arctic Ocean even using a simple magnetic compass for heading estimation, given that the AUV avoids critical latitudes.

Chapter 7

In the chapter 'A simplified version of SLAM: data compression and data association for common UUV missions' by Ashwin Sarma, the author states that in the transiting stage of an unmanned undersea vehicle (UUV) mission, it is of interest to minimize platform localization error with minimal processing. Earlier work has derived a

SLAM-inspired estimator of platform location and velocity, dubbed 'velocity-over-ground' SLAM, that provides virtually identical performance in transit scenarios as conventional SLAM. The method lends itself to simple real-time operation as map building is not required. The 'velocity-over-ground' simplification was devised based on (a) the observation that the second measurement of a persistent contact was required for potential performance improvement in SLAM and (b) the intuitive idea that SLAM is providing velocity information since contact measurements can only be relative to the platform. We provide here a direct argument by arguing its optimality properties via connection to the maximum likelihood estimator. In addition, sonar data processing, measurement generation and data association methodologies to determine proper assignments between measurements and persistent bottom features are described in detail. The process can be currently completed before the next ping arrives suggesting near real-time SLAM performance in complex undersea environments.

Part III: Toolset

Chapter 8

In the chapter 'Toolchain for autonomous systems and its use in real-life scenarios: from science to defence' by Paulo Sousa Dias, José Pinto and Joao Borges Sousa, the authors describe their successful effort to build a toolchain that is able to interface with the vast variety of systems on the market. The LSTS Toolchain for Autonomous Systems is a modular toolchain that can be used as a whole or just parts. It contains an on-board control software to fully control the vehicle or to plug-in as a back-seat driver. Its message interface definition has flexible serialization options, and it contains a command and control software that can be extended through plug-ins (visual, data processing or communications), displays vehicles and other assets on a geo-referenced map, is used to plan a mission for a vehicle, with feasible manoeuvres and payload, and has a display for telemetry, data arriving from vehicles, allowing also to add extendable data export functions.

Chapter 9

In the chapter 'Telemetry for AUVs' by Joao Alves, Konstantinos Pelekanakis and Roberto Petroccia, the topic of telemetry for AUVs is introduced. The authors draw considerations on different techniques and approaches to be used when establishing underwater communications. In broad terms, telemetry refers to automated measurement and transmission of such measurements from a remote location. This wide definition tends then to be tailored to specific areas of use. In the context of AUVs, telemetry is typically referred to as the transmission of the vehicle status or sensor data. Within the context of the concepts tackled in this chapter, the authors also add the transmission of commands or queries by a control station or a peer node. This way, and for the purposes of this chapter, they define AUV telemetry as the underwater wireless data exchanges that support any AUV-based mission. The authors focus in this chapter on underwater wireless communications, covering from the propagation physics of the links up to the establishment of networks.

Chapter 10

The chapter 'Pressure-tolerant and pressure-neutral systems for AUV' by Gunnar Brink and Gerhard Körner introduces the state of the art of pressure-tolerant electronics and discusses the advantages and disadvantages of different approaches. In almost all industries, savings in raw materials, costs and energy play an increasingly important role in the manufacturing, usage and recycling of products. This means that the overall device can be made smaller and lighter. The savings in operating the optimized product are often partly offset by additional development, manufacturing and assembly costs. Consequently, lightweight construction is important in all industries, especially in vehicle construction, shipbuilding and aircraft construction. This is not different with subsea systems. A submersible vehicle, which is lighter and smaller than previous models for the same performance figures, is easier to launch and recover from the sea. A smaller cross section leads to less water resistance, reduces the energy consumption of the engine and increases the possible speed or reduces either the necessary battery capacity which must be carried along or increases the range of the vehicle. In deep-sea vehicles, pressure vessels offer a particularly worthwhile starting point for reducing vehicle weight and vehicle size. Pressure vessels are needed because on-board batteries, engines, optics and, above all, electronic components like computers have been developed for operation in air and must be protected from corrosion and ambient pressure in the sea. Traditionally, most electronic components, instruments and, in some cases, energy storage devices are housed in pressure vessels. They protect the components from seawater and high ambient pressure. As early as in the 1960s, the US Navy initiated the first tests with pressure-compensated electronic components. In the 1990s, technology development in this field resumed, particularly with the maturity of lithium cells for series production.

Part IV: Cooperation

Chapter 11

In the chapter 'Optimization of waterspace management and scheduling for multiple unmanned vehicles' by Matthew J. Bays and Thomas A. Wettergren, the authors describe an approach for accomplishing the high-level mission planning required for a heterogeneous team of autonomous vehicles performing surveys in multiple areas, such as required for mine countermeasure missions. The high-level mission scheduling and waterspace management requires sequencing the order and location of lower-level tasks to be completed by each vehicle in the heterogeneous team: unmanned surface vessels (USVs) and unmanned underwater vehicles (UUVs). In this context, the USVs serve as transport vehicles while the UUVs perform the actual surveys and execute any other local actions. The authors develop a solution to this complex sequencing operation by leveraging unique information processing, communication, refuelling and planning windows that form constraints within the system within a formal scheduling optimization framework. Mixed-integer linear programming is used as a solution method for the associated optimization problem. Using such a standard optimization approach has two advantages: optimality guarantees on the solution and utilisation of extensive commercial numerical solvers. In addition to developing the numerical optimization problem, the authors also include methods to account for relative uncertainty in the duration of planned tasks in such a manner to balance the risk of schedule slips or conflicts to the risk of creating an overly conservative and suboptimal schedule.

Chapter 12

In the chapter 'MONSUN: a swarm AUV for environmental monitoring and inspection' by Erik Maehle, Benjamin Meyer, Cedric Isokeit and Ulrich Behrje, the authors present the development of the small swarm-capable AUV MONSUN and its use for environmental monitoring and inspection tasks. A summarizing description of robot development in hardware and software is given, whereby efforts of the design process are explained in detail. After the focus on underwater communication techniques, swarm behaviours based on robust and scaleable localization principles are presented. Finally, the flexibility and functionality of the system are shown with the help of various experiments and field tests.

Chapter 13

In the chapter 'Reconfigurable, adaptable, multi-modality, mobile, wireless, energy-efficient, underwater sensor network of hover-capable AUVs' by Vladimir Djapic, a detailed description of the combination of cooperative AUVs for the purposes of detection of various underwater targets is given, including accurate timing and clock synchronization, underwater sensors and acoustic communication. The overall theme is the novel use of the aforementioned systems to create a coordinated network of accurately positioned and navigated AUVs that can be used for seismic surveys, surveillance and marine mammals detection applications. This network can cooperate with another group of coordinated AUVs in order to maximize coverage rate while reducing the detection time. The following technologies contribute to Autonomous Ocean Sampling Network made with heterogeneous AUVs: Small (micro), size, Weight and Power (SWaP) AUVs, accurate Position, Navigation and Timing (PNT) sensors, Wireless Communication Modems (acoustic and optical) and Spiral Beacon Technology.

Part V: Methodology

Chapter 14

In the chapter 'Task specification and behavior verification for UUV behavior design' by Signe A. Redfield, Elizabeth I. Leonard and Jamie Lennon, the authors introduce a new design method based on a tabular representation called Capability Analysis

Tables. The robotics community is deeply interested in both platform design and behaviour design, but there is a lack of tools to connect the two. The platform, the behaviour design and the environment work together to determine the robot's actions, but available tools visualize the design of the hardware and the design of the behaviours separately. There are no tools that allow us visualize the relationship between the platform and the behaviour. To address this gap, a new design method based on a tabular representation, called Capability Analysis Tables, is introduced. The Capability Analysis Table enables the designer to define the constraints on a behaviour design based on the platform it will be used on, and to define the constraints on the platform based on the behaviour design. It gives the customer and the designer an opportunity to more clearly specify the desired behaviour. Environmental factors are implicit in the platform interface definitions – sensory perception filters environmental inputs (coloured objects can only be seen by sensors that produce colour information) and actuators filter environmental outputs (wheeled robots typically do not provide depth control). Communications are handled explicitly as outputs (e.g. 'send message') and as either operator inputs (e.g. 'new task') or remotely gathered sensory data (e.g. 'other vehicle pose').

Chapter 15

In the chapter 'Advanced AUV fault management' by M.L. Seto and K. Svendsen, the authors discuss how fault management enables AUVs to provide persistent ocean observation – i.e. the ability to deploy for extended periods on complex missions in unstructured environments at great ranges from operators. It refers to an on-board response to an AUV fault that could be reactive or deliberative with the intent to recover from, or mitigate, the fault. Otherwise, the AUV is unavailable for the entire duration of the mission. On long deployments, there will inevitably be hardware failures and data corruption. Cascading failures could jeopardize a mission or the vehicle. The objective is to provide some ability for the AUV to manage the faults triggered by the AUV, the environment around it or its interaction with the environment. The advanced fault management within the on-board autonomy is an enabler to realize the potential for the complex roles expected of AUVs.

Chapter 16

In the chapter 'Designing, building and testing a fleet of novel AUV under the elevated requirements of an XPRIZE competition' by G. Brink, the author describes the application of agile management methods in large hardware projects using the example of the development of AUVs. The case study concerns the development, construction and testing of a fleet of five new AUVs with associated unmanned surface vehicles over a period of less than 2 years. First, however, the most important terms of classical project management and agile methods are compared. The *Agile Manifesto* and agile management methods including SCRUM and Kanban, in comparison with SCRUM alone, as well as eXtreme programming from software development, are presented. The possibilities of agile methods in the development of physically based systems such as automobiles are shown by the example of eXtreme manufacturing (i.e. Wikispeed). The experiences of the project itself, along with its participation in the Shell Ocean Discovery XPRIZE, are presented and discussed, deriving from it a development methodology.

Part VI: Application

Chapter 17

The chapter 'AUV navigation, guidance, and control for geoseismic data acquisition' by Lorenzo Pollini, Gianluca Antonelli, Filippo Arrichiello, Andrea Caiti, Giuseppe Casalino, Daniela De Palma, Giovanni Indiveri, Matteo Razzanelli and Enrico Simetti describes the main challenges, and the corresponding solutions, encountered during the development of the Guidance Navigation and Control Systems for the AUVs and the autonomous surface vehicles involved in the Widely scalable Mobile Underwater Sonar Technology (WiMUST) project. WiMUST is a project funded by the European Community within the H2020 Framework, which started in February 2015 and ended in January 2018. The ultimate goal of the project was to design and test a system of cooperating AUVs able to perform innovative geotechnical surveying operations. In particular, the WiMUST system is composed of a small fleet of AUVs carrying hydrophones to acquire sub-bottom profiling acoustic data. Contrary to the classical technology based on ship towed streamers, the WiMUST solution allowed a greater survey flexibility since the geometry, position and depth of the acoustic antenna could be changed during the mission: something that had not been achieved in practice before and represented a potential to significantly improve ocean surveying.

Chapter 18

In the chapter 'Adaptive sampling and energy-efficient navigation in time-varying flows' by Tahiya Salam, Dhanushka Kularatne, Eric Forgoston and M. Ani Hsieh, the authors present a strategy to enable a team of mobile robots to adaptively sample and track a dynamic spatio-temporal process. They propose a distributed strategy where robots collect sparse sensor measurements, create a reduced-order model of the spatio-temporal process and use this model to estimate field values for areas without sensor measurements of the dynamic process. The robots then use these estimates of the field, or inferences about the process, to adapt the model and reconfigure their sensing locations. This method is used to obtain an estimate for the underlying flow field and use that to plan optimal energy paths for robots to travel between sensing locations. The authors show that the errors due to the reduced-order modelling scheme are bounded, and they illustrate the application of the proposed solution in simulation and compare it to centralized and global approaches. The approach is then tested with physical marine robots sampling a spatially non-uniform time-varying process in a water tank.

Part I

Control

Chapter 2

Sensor-based motion control of autonomous underwater vehicles, part I: modeling and low-complexity state estimation

George C. Karras[1,2], Charalampos P. Bechlioulis[2], Panos Marantos[2], Shahab Heshmati-alamdari[3] and Kostas J. Kyriakopoulos[2]

2.1 Introduction

In view of the increasingly demanding and complex missions envisioned for underwater robots, especially in the presence of environmental constraints, navigation and guidance systems with enhanced robustness need to be developed. Therefore, an accurate dynamic model of the underwater vehicle is essential in order to develop efficient control schemes. However, the complexity of the dynamic model may vary depending of the vehicle configuration, actuation capabilities, operating speed and planes of symmetry [1]. In that sense, the identification problem of underwater vehicles continues to pose considerable challenges to control system designers. Although the generic form of approximate lumped-parameter finite-dimensional plant models for underwater vehicles is known [1], actual plant parameters (e.g., hydrodynamic added mass, buoyancy and drag parameters) have to be determined.

Thus far, four commonly used methods have been proposed to identify the dynamic model of underwater robotic vehicles. Captive model testing with a Planar Motion Mechanism (PMM) or other facilities is regarded as the most reliable one [2,3]. However, expensive infrastructure and very long and complex testings are required. On the other hand, estimation with empirical formulas is the most practical and convenient one [4], but provides reasonable results only for streamlined vehicles. Alternatively, numerical calculation based on computational fluid dynamics is a powerful modeling method [5], but its precision cannot be guaranteed a priori. Recently, System Identification (SI) approaches have become popular. Unlike the traditional naval architecture methods, this technique drastically reduces experimental time and

[1]Department of Computer Science and Telecommunications, University of Thessaly, Lamia, Greece
[2]School of Mechanical Engineering, Control Systems Lab, National Technical University of Athens, Athens, Greece
[3]Division of Decision and Control Systems, School of Electrical Engineering and Computer Science, KTH Royal Institute of Technology, Stockholm, Sweden

is preferable because it makes use solely of the on-board sensors' data as well as of the excitation signals. Additionally, the technique is cost-effective and the repeatability is high. Thus, it is very suitable for underwater robotic vehicles of variable configuration, in which the payload and shape may change according to the mission.

Although most of the offline techniques may provide sufficiently accurate results, they suffer from two major issues: (i) in case the configuration of the vehicle changes, the identification procedure should be repeated and (ii) the evaluation of the estimated model follows the experimental procedure, so poor identification results necessitate further experiments with different excitation signals. On the contrary, online identification is an on-the-fly process, where the estimation of the dynamic parameters occurs concurrently with the data acquisition process during the real-time operation of the system. Smallwood and Whitcomb in [6] as well as McFarland and Whitcomb in [7] describe an online adaptive technique with local stability proof for the identification of a finite-dimensional dynamical model of dynamically positioned vehicles. Van de Ven *et al.* [8] and Ishii *et al.* [9] discuss the use of neural networks with online adaptation laws for identifying the dynamic model of underwater robotic vehicles. Eng *et al.* [10] propose an online system identification procedure based on a state variable filter and a recursive least-square (SVF-RLS) estimator in order to determine the unknown dynamic parameters related to the yaw degrees of freedom (DoF) of an unmanned underwater vehicle. Finally, an online identification scheme based on a global derivative-free optimization algorithm was presented in [11].

Efficient methods for online system identification have also been built based on the well-known Kalman filter [12]. More specifically, a Kalman filter with a maximum likelihood estimation approach was presented in [13]. A description of the unscented Kalman filter's (UKF's) applicability for the state and parameter estimation of nonlinear systems, as well as an analytical justification of the superiority of the UKF over the extended Kalman filter (EKF) parameter estimation techniques, has been provided by der Merwe and Wan in [14] and VanDyke *et al.* [15]. In the same direction, Karras *et al.* [16] described a dual UKF algorithm for online state and parameter estimation of underwater vehicles. Regarding state estimation of underwater vehicles, interesting solutions based on the fusion of ultra-short baseline (USBL)/inertial measurement unit (IMU) and Doppler Velocity Log (DVL)/IMU appear in [17] and [18], respectively.

In this chapter, we describe the kinematic and dynamic equations of underwater robotic vehicles including standard assumptions and simplifications that may lead to low-complexity but accurate and operational models. Furthermore, we describe an RLS method for the online identification of the vehicle dynamic parameters. In addition, we present a state estimation technique which, according to the available sensors, can be adopted for either shallow or deep water operation. Keeping in mind low-complexity requirements, the proposed state estimation method is based on the complementary filter (CF) notion. The goal of this chapter is to provide the reader with the appropriate analytical methods, in order to derive a simple yet accurate vehicle model, design a state estimation algorithm that could be easily integrated into the embedded system framework of an underwater robotic vehicle and perform a fast online dynamic parameter identification.

2.2 Underwater vehicle kinematics and dynamics

2.2.1 General representation

Let us define a body-fixed frame $\mathcal{B} = \{e_x, e_y, e_z\}$ attached to the vehicle's center of gravity, as shown in Figure 2.1, and an inertial frame $\mathcal{I} = \{e_F, e_R, e_D\}$ located at a fix position $O_\mathcal{I}$ within the workspace of the vehicle. Following standard modeling techniques [1], the dynamic model of the vehicle in the body-fixed frame may be derived from the general Newton–Euler motion equations of a 6-DoF rigid body subject to external forces and torques in a fluid medium as follows:

$$\begin{aligned} \mathbf{M}\dot{\boldsymbol{\nu}} + \mathbf{C}(\boldsymbol{\nu})\boldsymbol{\nu} + \mathbf{D}(\boldsymbol{\nu})\boldsymbol{\nu} + \mathbf{g}(\boldsymbol{\eta}) &= \boldsymbol{\tau}_E + \boldsymbol{\tau} \\ \dot{\boldsymbol{\eta}} &= \mathbf{J}(\boldsymbol{\eta})\boldsymbol{\nu} \end{aligned} \tag{2.1}$$

where

- $\boldsymbol{\eta} \triangleq \left[\boldsymbol{\eta}_1^T\ \boldsymbol{\eta}_2^T\right]^T \in \mathbb{R}^6$ is the pose vector expressed in $\mathcal{I}$ that involves the position (i.e., $\boldsymbol{\eta}_1 \triangleq [x\ y\ z]^T$) and orientation (i.e., $\boldsymbol{\eta}_2 \triangleq [\phi\ \theta\ \psi]^T$) vectors;
- $\boldsymbol{\nu} \triangleq \left[\boldsymbol{\nu}_1^T\ \boldsymbol{\nu}_2^T\right]^T \in \mathbb{R}^6$ is the velocity vector expressed in $\mathcal{B}$ that involves the linear (i.e., $\boldsymbol{\nu}_1 \triangleq [u\ v\ w]^T$) and angular (i.e., $\boldsymbol{\nu}_2 \triangleq [p\ q\ r]^T$) velocity vectors;
- $\boldsymbol{\tau}_E \in \mathbb{R}^6$ is the total environmental force/torque vector expressed in $\mathcal{B}$ that is applied on the vehicle;
- $\boldsymbol{\tau} \triangleq [\tau_X,\ \tau_Y,\ \tau_Z,\ \tau_K,\ \tau_M,\ \tau_N]^T \in \mathbb{R}^6$ is the total propulsion vector (i.e., the body forces τ_X, τ_Y, τ_Z and torques τ_K, τ_M, τ_N generated by the actuators) applied on the vehicle and expressed in $\mathcal{B}$;
- $\mathbf{M} \triangleq \mathbf{M}_{RB} + \mathbf{M}_A$, where $\mathbf{M}_{RB} \in \mathbb{R}^{6\times 6}$ and $\mathbf{M}_A \in \mathbb{R}^{6\times 6}$ are the rigid body and added mass inertia matrices, respectively;

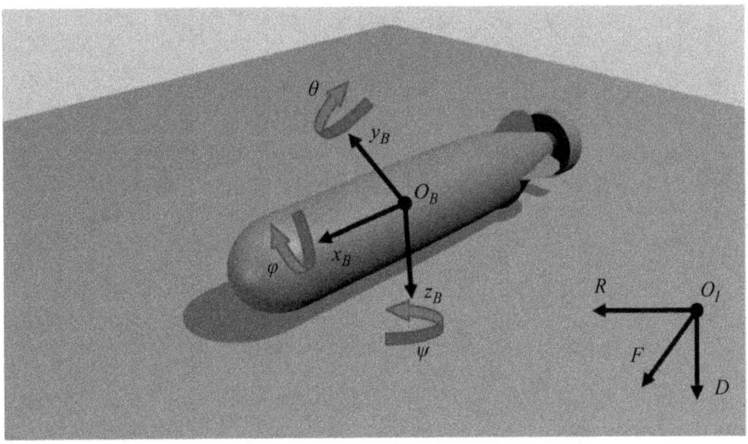

Figure 2.1 Underwater vehicle frames

- $\mathbf{C}(\boldsymbol{\nu}) \triangleq \mathbf{C}_{RB}(\boldsymbol{\nu}) + \mathbf{C}_A(\boldsymbol{\nu})$, where $\mathbf{C}_{RB}(\boldsymbol{\nu}) \in \mathbb{R}^{6\times6}$ and $\mathbf{C}_A(\boldsymbol{\nu}) \in \mathbb{R}^{6\times6}$ are the rigid body and added mass matrices that model the Coriolis and centrifugal effects, respectively;
- $\mathbf{D}(\boldsymbol{\nu}) \triangleq \mathbf{D}_{quad}(\boldsymbol{\nu}) + \mathbf{D}_{lin}$, where $\mathbf{D}_{quad}(\boldsymbol{\nu}) \in \mathbb{R}^{6\times6}$ and $\mathbf{D}_{lin} \in \mathbb{R}^{6\times6}$ denote the quadratic and linear drag matrices, respectively;
- $\boldsymbol{g}(\boldsymbol{\eta}) \in \mathbb{R}^6$ is the hydrostatic restoring force vector;
- $\mathbf{J}(\boldsymbol{\eta}) \triangleq \begin{bmatrix} \mathbf{J}_1(\boldsymbol{\eta}_2) & \mathbf{O}_3 \\ \mathbf{O}_3 & \mathbf{J}_2(\boldsymbol{\eta}_2) \end{bmatrix}$ is the Jacobian matrix transforming the velocities from the body-fixed frame $\mathcal{B}$ to the inertial frame $\mathcal{I}$, in which $\mathbf{J}_1(\boldsymbol{\eta}_2) \in SO(3)$ stands for the rotation matrix and $\mathbf{J}_2(\boldsymbol{\eta}_2) \in \mathbb{R}^{3\times3}$ denotes the lumped transformation matrix.

The expanded forms of the matrices appeared in (2.1), for a generic 6-DoF underwater vehicle, are given as follows [1]:

- $$\mathbf{M}_{RB} = \begin{bmatrix} m & 0 & 0 & 0 & mz_G & -my_G \\ 0 & m & 0 & -mz_G & 0 & mx_G \\ 0 & 0 & m & my_G & -mx_G & 0 \\ 0 & -mz_G & my_G & I_x & -I_{xy} & -I_{xz} \\ mz_G & 0 & -mx_G & -I_{yx} & I_y & -I_{yz} \\ -my_G & mx_G & 0 & -I_{zx} & -I_{zy} & I_z \end{bmatrix}$$
- $$\mathbf{M}_A = -\begin{bmatrix} X_{\dot{u}} & X_{\dot{v}} & X_{\dot{w}} & X_{\dot{p}} & X_{\dot{q}} & X_{\dot{r}} \\ Y_{\dot{u}} & Y_{\dot{v}} & Y_{\dot{w}} & Y_{\dot{p}} & Y_{\dot{q}} & Y_{\dot{r}} \\ Z_{\dot{u}} & Z_{\dot{v}} & Z_{\dot{w}} & Z_{\dot{p}} & Z_{\dot{q}} & Z_{\dot{r}} \\ K_{\dot{u}} & K_{\dot{v}} & K_{\dot{w}} & K_{\dot{p}} & K_{\dot{q}} & K_{\dot{r}} \\ M_{\dot{u}} & M_{\dot{v}} & M_{\dot{w}} & M_{\dot{p}} & M_{\dot{q}} & M_{\dot{r}} \\ N_{\dot{u}} & N_{\dot{v}} & N_{\dot{w}} & N_{\dot{p}} & N_{\dot{q}} & N_{\dot{r}} \end{bmatrix}$$
- $$C_{RB} = \begin{bmatrix} 0 & 0 & 0 & m(y_Gq + z_Gr) & -m(x_Gq - w) & -m(x_Gr + v) \\ 0 & 0 & 0 & -m(y_Gp + w) & m(z_Gr + x_Gp) & -m(y_Gr - u) \\ 0 & 0 & 0 & -m(z_Gp - v) & -m(z_Gq + u) & m(x_Gp + y_Gq) \\ -m(y_Gq + z_Gr) & m(y_Gp + w) & m(z_Gp - v) & 0 & -I_{yz}q - I_{xz}p + I_zr & I_{yz}r + I_{xy}p - I_yq \\ m(x_Gq - w) & -m(z_Gr + x_Gp) & m(z_Gq + u) & I_{yz}q + I_{xz}p - I_zr & 0 & -I_{xz}r - I_{xy}q + I_xp \\ m(x_Gr + v) & m(y_Gr - u) & -m(x_Gp + y_Gq) & -I_{yz}r - I_{xy}p + I_yq & I_{xz}r + I_{xy}q - I_xp & 0 \end{bmatrix}$$
- $$\mathbf{C}_A(\boldsymbol{\nu}) = \begin{bmatrix} 0 & 0 & 0 & 0 & -\alpha_3 & \alpha_2 \\ 0 & 0 & 0 & \alpha_3 & 0 & -\alpha_1 \\ 0 & 0 & 0 & -\alpha_2 & \alpha_1 & 0 \\ 0 & -\alpha_3 & \alpha_2 & 0 & -b_3 & b_2 \\ \alpha_3 & 0 & -\alpha_1 & b_3 & 0 & -b_1 \\ -\alpha_2 & \alpha_1 & 0 & -b_2 & b_1 & 0 \end{bmatrix}$$

where

$$\alpha_1 = X_{\dot{u}}u + X_{\dot{v}}v + X_{\dot{w}}w + X_{\dot{p}}p + X_{\dot{q}}q + X_{\dot{r}}r$$

$$\alpha_2 = X_{\dot{v}}u + Y_{\dot{v}}v + Y_{\dot{w}}w + Y_{\dot{p}}p + Y_{\dot{q}}q + Y_{\dot{r}}r$$

$$\alpha_3 = X_{\dot{w}}u + Y_{\dot{w}}v + Z_{\dot{w}}w + Z_{\dot{p}}p + Z_{\dot{q}}q + Z_{\dot{r}}r$$

$$b_1 = X_{\dot{p}}u + Y_{\dot{p}}v + Z_{\dot{p}}w + K_{\dot{p}}p + K_{\dot{q}}q + K_{\dot{r}}r$$

$$b_2 = X_{\dot{q}}u + Y_{\dot{q}}v + Z_{\dot{q}}w + K_{\dot{q}}p + M_{\dot{q}}q + M_{\dot{r}}r$$

$$b_2 = X_{\dot{r}}u + Y_{\dot{r}}v + Z_{\dot{r}}w + K_{\dot{r}}p + M_{\dot{r}}q + N_{\dot{r}}r$$

- $\mathbf{D}(\boldsymbol{v}) \triangleq \mathbf{D}_{lin} + \mathbf{D}_{quad}(\boldsymbol{v})$
  where

$$\mathbf{D}_{lin} = -diag\{X_u,\ Y_v,\ Z_w,\ K_p,\ M_q,\ N_r\}$$

$$\mathbf{D}_{quad} = -diag\{X_u,\ X_{u|u|},\ Y_v,\ Y_{v|v|},\ Z_w,\ Z_{w|w|},\ K_p,\ K_{p|p|},\ M_q,\ M_{q|q|},\ N_r,\ N_{r|r|}\}$$

- $$\boldsymbol{g}(\boldsymbol{\eta}) = -\begin{bmatrix} \mathbf{J}_1^T(\boldsymbol{\eta}_2)\left(m\boldsymbol{g}_{\mathcal{I}} + \boldsymbol{b}_{\mathcal{I}}\right) \\ \boldsymbol{r}_G \times \mathbf{J}_1^T(\boldsymbol{\eta}_2)m\boldsymbol{g}_{\mathcal{I}} + \boldsymbol{r}_B \times \mathbf{J}_1^T(\boldsymbol{\eta}_2)\boldsymbol{b}_{\mathcal{I}} \end{bmatrix}$$

- $$\mathbf{J}_1(\boldsymbol{\eta}_2) = \begin{bmatrix} c\psi c\theta & -s\psi c\varphi + c\psi s\theta s\varphi & s\psi s\varphi + c\psi c\varphi s\theta \\ s\psi c\theta & c\psi c\varphi + s\varphi s\theta s\psi & -c\psi s\varphi + s\theta s\psi c\varphi \\ -s\theta & c\theta s\varphi & c\theta c\varphi \end{bmatrix}$$

- $$\mathbf{J}_2(\boldsymbol{\eta}_2) = \begin{bmatrix} 1 & s\varphi t\theta & c\varphi t\theta \\ 0 & c\varphi & -s\varphi \\ 0 & s\varphi/c\theta & c\varphi/c\theta \end{bmatrix}$$

where

- m is the mass, $\boldsymbol{r}_G \triangleq [x_G, y_G, z_G]^T$ and $\boldsymbol{r}_B \triangleq [x_B, y_B, z_B]^T$ denotes the position of the gravity and buoyancy center with respect to the body frame $\mathcal{B}$.
- I_x, I_y, I_z, I_{xy}, I_{yx}, I_{xz}, I_{zx}, I_{yz} and I_{zy} denote the moments of inertia along the respective axes.
- $(x)_{\dot{y}}$, for $x \in \{X, Y, Z, K, M, N\}$ and $y \in \{u, v, w, p, q, r\}$ are the added mass terms according to [1]. Added mass should be understood as pressure-induced forces and moments due to a forced harmonic motion of the body which are proportional to the acceleration of the body. Consequently, the added mass forces and the acceleration will be 180 degrees out of phase to the forced harmonic motion [1]. More specifically, it stands: $(x)_{\dot{y}} \triangleq \frac{\partial(x)}{\partial\dot{y}}$.
- X_u, $X_{u|u|}$, Y_v, $Y_{v|v|}$, Z_w, $Z_{w|w|}$, K_p, $K_{p|p|}$, M_q, $M_{q|q|}$, N_r and $N_{r|r|}$ are the first- and second-order drag parameters according to [1].
- $\boldsymbol{g}_{\mathcal{I}}$ and $\boldsymbol{b}_{\mathcal{I}}$ are the gravitational acceleration and buoyancy vectors expressed in $\mathcal{I}$.
- $s\cdot = \sin(\cdot)$, $c\cdot = \cos(\cdot)$, $t\cdot = \tan(\cdot)$.

Remark 1. *Let us define a constant rotation matrix $\boldsymbol{R_{IN}}$ transforming any vector from the standard North-East-Down (NED) frame $\mathcal{N}$ to the inertial frame $\mathcal{I}$. Then, $\boldsymbol{g_{\mathcal{I}}} = \boldsymbol{R_{IN}} \boldsymbol{g_{\mathcal{N}}} = \boldsymbol{R_{IN}}[0, 0, g]^T$ and $\boldsymbol{b_{\mathcal{I}}} = \boldsymbol{R_{IN}}[0, 0, -B]^T$, where g is the gravitational acceleration and B the buoyancy force.*

Remark 2. *In most cases, the underwater vehicles have three planes of symmetry. In addition to that, the z-axis of the inertial frame $\mathcal{I}$ is selected to be aligned with the z-axis of the NED frame $\mathcal{N}$. Thus, the representation of matrices $\boldsymbol{M_A}$, $\boldsymbol{C_A}(\boldsymbol{\nu})$ and $\mathbf{g}(\boldsymbol{\eta})$ can be further simplified:*

- $\boldsymbol{M_A} = diag\{X_{\dot{u}}, Y_{\dot{v}}, Z_{\dot{w}}, K_{\dot{p}}, M_{\dot{q}}, N_{\dot{r}}\}$
- $\boldsymbol{C_A}(\boldsymbol{\nu}) = \begin{bmatrix} 0 & 0 & 0 & 0 & -Z_{\dot{w}}w & Y_{\dot{\upsilon}}\upsilon \\ 0 & 0 & 0 & Z_{\dot{w}}w & 0 & -X_{\dot{u}}u \\ 0 & 0 & 0 & -Y_{\dot{\upsilon}}\upsilon & X_{\dot{u}}u & 0 \\ 0 & -Z_{\dot{w}}w & Y_{\dot{\upsilon}}\upsilon & 0 & -N_{\dot{r}}r & M_{\dot{q}}q \\ Z_{\dot{w}}w & 0 & -X_{\dot{u}}u & N_{\dot{r}}r & 0 & -K_{\dot{p}}p \\ -Y_{\dot{\upsilon}}\upsilon & X_{\dot{u}}u & 0 & -M_{\dot{q}}q & K_{\dot{p}}p & 0 \end{bmatrix}$
- $\mathbf{g}(\boldsymbol{\eta}) = \begin{bmatrix} (mg - B)s\theta \\ -(mg - B)c\theta s\varphi \\ -(mg - B)c\theta c\varphi \\ -(y_G mg - y_B B)c\theta c\varphi + (z_G mg - z_B B)c\theta s\varphi \\ (z_G mg - z_B B)s\theta + (x_G mg - x_B B)c\theta c\varphi \\ -(x_G mg - x_B B)c\theta s\varphi - (y_G mg - y_B B)s\theta \end{bmatrix}$

Although (2.1) is highly nonlinear, by applying proper algebraic manipulations, the dynamic model of the vehicle in the body-fixed frame may be written in the following linear form with respect to the dynamic parameters form:

$$\boldsymbol{\tau} = \mathbf{Y}(\boldsymbol{\eta}, \boldsymbol{\nu}, \dot{\boldsymbol{\nu}})\boldsymbol{\pi} \tag{2.2}$$

where the elements of the vector $\boldsymbol{\pi}$ involve the unknown parameters of (i) the inertia matrix $\mathbf{M}$, (ii) the Coriolis and centrifugal matrix $\mathbf{C}(\boldsymbol{\nu})$, (iii) the drag matrix $\mathbf{D}(\boldsymbol{\nu})$ and (iv) the hydrostatic force vector $\mathbf{g}(\boldsymbol{\eta})$.

Remark 3. *The form appearing in (2.2) can be very practical for SI procedures, since linear representation encourages the employment of least squares approximation methods, which are of low complexity and may deliver accurate results given a proper set of input-output measurements.*

Next we describe the calculation of the total propulsion vector $\boldsymbol{\tau}$ for thruster-actuated underwater vehicles. Consider a 6-DoF actuated vehicle with $i = 1, \ldots, N$

thrusters. Each thruster produces the following vector of forces and torques with respect to the body-fixed frame $\mathcal{B}$:

$$ {}^{i}\boldsymbol{\tau} = \begin{bmatrix} {}^{i}T \\ {}^{i}Q \end{bmatrix} = \begin{bmatrix} {}^{i}T\,{}^{i}e \\ {}^{i}T({}^{i}L \times {}^{i}e) \end{bmatrix} = \begin{bmatrix} {}^{i}e_x \\ {}^{i}e_y \\ {}^{i}e_z \\ ({}^{i}L \times {}^{i}e)_x \\ ({}^{i}L \times {}^{i}e)_y \\ ({}^{i}L \times {}^{i}e)_z \end{bmatrix} {}^{i}T \tag{2.3} $$

where ${}^{i}e = [\,{}^{i}e_x \;\; {}^{i}e_y \;\; {}^{i}e_z\,]^T = \begin{bmatrix} \cos\psi_T^i \cos\theta_T^i \\ \sin\psi_T^i \cos\theta_T^i \\ -\sin\theta_T^i \end{bmatrix}$ is the orientation of the thruster i relative to the body-fixed frame $\mathcal{B}$ and ${}^{i}L = \left[\, l_i^x \;\; l_i^y \;\; l_i^z \,\right]^T$ determines the position of the point of attack for ${}^{i}T$ relatively to $\mathcal{B}$. Superposition of the individual contributions ${}^{i}\boldsymbol{\tau}$ leads to total vector of propulsion forces and torques:

$$ \boldsymbol{\tau} = \begin{bmatrix} X \\ Y \\ Z \\ K \\ M \\ N \end{bmatrix} = \sum_{i}^{N} {}^{i}\tau = \sum_{i}^{N} \begin{bmatrix} {}^{i}e \\ ({}^{i}L \times {}^{i}e) \end{bmatrix} {}^{i}T = \begin{bmatrix} {}^{1}e_x & & {}^{i}e_x & & {}^{N}e_x \\ {}^{1}e_y & & {}^{i}e_y & & {}^{N}e_y \\ {}^{1}e_z & & {}^{i}e_z & & {}^{N}e_z \\ ({}^{1}L \times {}^{1}e)_x & \cdots & ({}^{i}L \times {}^{i}e)_x & \cdots & ({}^{N}L \times {}^{N}e)_x \\ ({}^{1}L \times {}^{1}e)_y & & ({}^{i}L \times {}^{i}e)_y & & ({}^{N}L \times {}^{N}e)_y \\ ({}^{1}L \times {}^{1}e)_z & & ({}^{i}L \times {}^{i}e)_z & & ({}^{N}L \times {}^{N}e)_z \end{bmatrix} \begin{bmatrix} {}^{1}T \\ \cdot \\ \cdot \\ \cdot \\ {}^{i}T \\ \cdot \\ \cdot \\ \cdot \\ {}^{N}T \end{bmatrix} \tag{2.4} $$

2.2.2 Modeling and online parameter identification example: 4-DoF underwater vehicle

In this section, we provide a modeling and online parameter identification example of a 4-DoF underwater robotic vehicle, Seabotix LBV150 [19], as shown in Figure 2.2, actuated in Surge, Sway, Heave and Yaw via a four-thruster set configuration. We begin from the general 6-DoF representation, and according to particular assumptions and simplifications that apply to the specific vehicle, we derive the dynamic and kinematic equations describing its motion. Then, we reformulate the system into a linear form with respect to its dynamic parameters form and employ a conventional constrained RLS method [20]. For the vehicle under investigation, the following properties hold:

- the gravity center and the origin of $\mathcal{B}$ coincide; thus, $x_G = y_G = z_G = 0$.
- the buoyancy and gravity center coincide; thus, $x_B = y_B = z_B = 0$.
- the vehicle mass is symmetric in all axis; thus, the properties of $\mathbf{M}_A$ and $\mathbf{C}_A(\boldsymbol{\nu})$ as described in Remark 2 hold and $I_{xy} = I_{xz} = I_{yx} = I_{yz} = I_{zx} = I_{zy} = 0$.
- the vehicle is underactuated about Roll and Pitch axes. Thus, the fourth and fifth rows and columns of $\mathbf{M}_{RB}$, $\mathbf{M}_A$, $\mathbf{C}_{RB}$, $\mathbf{C}_A(\boldsymbol{\nu})$ and $\mathbf{D}(\boldsymbol{\nu})$, and the fourth and fifth rows of $\boldsymbol{g}(\boldsymbol{\eta})$ are omitted.
- the z-axis of $\mathcal{I}$ is aligned to the z-axis of $\mathcal{N}$. Thus, the properties of $\boldsymbol{g}(\boldsymbol{\eta})$ as described in Remark 2 hold.

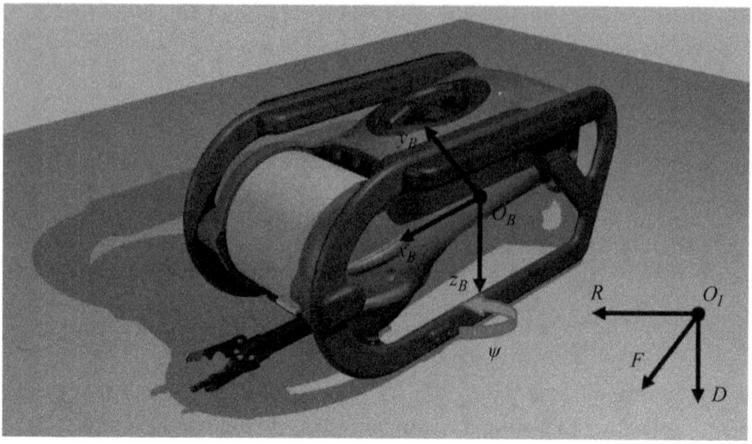

Figure 2.2 Seabotix LBV frames

Considering the above properties, the 4-DoF vehicle equations of motion are given in analytical form as follows:

$$
\begin{aligned}
&\dot{u}(m + X_{\dot{u}}) - vr(m + Y_{\dot{v}}) + qw(m + Z_{\dot{w}}) - uX_u - u|u|\, X_{u|u|} - \sin\theta(mg - B) = \tau_x \\
&\dot{v}(m + Y_{\dot{v}}) + ur(m + X_{\dot{u}}) - pw(m + Z_{\dot{w}}) - vY_v - v|v|\, Y_{v|v|} + \sin\varphi\cos\theta(mg - B) = \tau_y \\
&\dot{w}(m + Z_{\dot{w}}) - uq(m + X_{\dot{u}}) + uv(m + Y_{\dot{v}}) - wZ_w - w|w|\, Z_{w|w|} + \cos\varphi\cos\theta(mg - B) = \tau_z \\
&\dot{r}(I_z + N_{\dot{r}}) - uv(m + X_{\dot{u}}) + uv(m + Y_{\dot{v}}) - rN_r - r|r|\, N_{r|r|} = \tau_N
\end{aligned}
\tag{2.5}
$$

In terms of actuation, the Seabotix LBV150 is equipped with four thrusters, i.e., Port (p), Starboard (s), Vertical (v) and Lateral (l), which are effective in Surge (X), Sway (Y), Heave (Z) and Yaw (N) motion, as shown in Figure 2.3. Hence, the generalized force/torque input vector $\boldsymbol{\tau}$ is given by

$$
\boldsymbol{\tau} = \begin{bmatrix} X \\ Y \\ Z \\ K \\ M \\ N \end{bmatrix} = \begin{bmatrix} {}^{p}e_x & {}^{s}e_x & {}^{v}e_x & {}^{l}e_x \\ {}^{p}e_y & {}^{s}e_y & {}^{v}e_y & {}^{l}e_y \\ {}^{p}e_z & {}^{s}e_z & {}^{v}e_z & {}^{l}e_z \\ ({}^{p}L \times {}^{p}e)_x & ({}^{s}L \times {}^{s}e)_x & ({}^{v}L \times {}^{v}e)_x & \left({}^{l}L \times {}^{l}e\right)_x \\ ({}^{p}L \times {}^{p}e)_y & ({}^{s}L \times {}^{s}e)_y & ({}^{v}L \times {}^{v}e)_y & \left({}^{l}L \times {}^{l}e\right)_y \\ ({}^{p}L \times {}^{p}e)_z & ({}^{s}L \times {}^{s}e)_z & ({}^{v}L \times {}^{v}e)_z & \left({}^{l}L \times {}^{l}e\right)_z \end{bmatrix} \begin{bmatrix} \tau_p \\ \tau_s \\ \tau_v \\ \tau_l \end{bmatrix}
\tag{2.6}
$$

where $X = \tau_X$, $Y = \tau_Y$, $Z = \tau_Z$ and $N = \tau_N$.

By substituting in (2.6) the actual position and orientation of each thruster with respect to body frame $\mathcal{B}$, we derive the thruster allocation matrix $\boldsymbol{T}_{\mathbb{A}}$ for Seabotix LBV150:

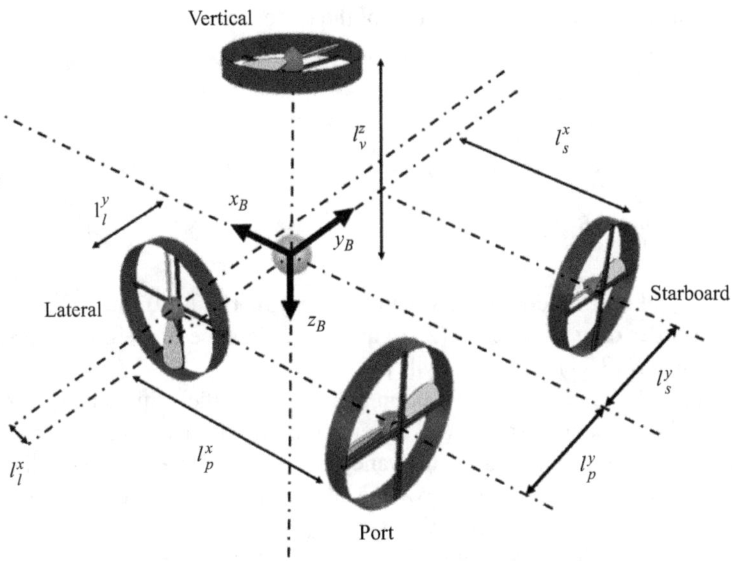

Figure 2.3 The location of the Seabotix LVB thrusters

- ${}^{p}L = \begin{bmatrix} -0.135 & -0.0475 & 0 \end{bmatrix}^T, {}^{s}L = \begin{bmatrix} -0.135 & 0.0475 & 0 \end{bmatrix}^T,$ ${}^{v}L = \begin{bmatrix} 0 & 0 & -0.05 \end{bmatrix}^T, {}^{L}L = \begin{bmatrix} 0.05 & 0.05 & 0 \end{bmatrix}^T$
- ${}^{p}e = \begin{bmatrix} 1 & 0 & 0 \end{bmatrix}^T, {}^{s}e = \begin{bmatrix} 1 & 0 & 0 \end{bmatrix}^T, {}^{l}e = \begin{bmatrix} 0 & 1 & 0 \end{bmatrix}^T, {}^{v}e = \begin{bmatrix} 0 & 0 & 1 \end{bmatrix}^T$

$$\boldsymbol{T}_{\mathbb{A}} = \begin{bmatrix} 1 & 1 & 0 & 0 \\ 0 & 0 & 1 & 0 \\ 0 & 0 & 0 & 1 \\ 0 & 0 & 0 & 0 \\ 0 & 0 & 0 & 0 \\ 0.0475 & -0.0475 & 0.05 & 0 \end{bmatrix} \tag{2.7}$$

By employing the appropriate algebraic manipulation, we can reformulate (2.5) to a linearly independent form with respect to the dynamic parameters form (2.2), as follows:

$$\overbrace{\begin{bmatrix} \tau_X \\ \tau_Y \\ \tau_Z \\ \tau_N \end{bmatrix}}^{\tau_{4d}} = \overbrace{\begin{bmatrix} \dot{\hat{u}} & -\hat{v}r & q\hat{w} & 0 & -\hat{u} & -\hat{u}|\hat{u}| & 0 & 0 & 0 & 0 & 0 & 0 & -\mathbf{s}_\theta \\ \hat{u}r & \dot{\hat{v}} & -p\hat{w} & 0 & 0 & 0 & -\hat{v} & -\hat{v}|\hat{v}| & 0 & 0 & 0 & 0 & \mathbf{s}_\phi \mathbf{c}_\theta \\ -\hat{u}q & \hat{v}p & \dot{\hat{w}} & 0 & 0 & 0 & 0 & 0 & -\hat{w} & -\hat{w}|\hat{w}| & 0 & 0 & \mathbf{c}_\phi \mathbf{c}_\theta \\ -\hat{u}\hat{v} & \hat{u}\hat{v} & 0 & \dot{r} & 0 & 0 & 0 & 0 & 0 & 0 & -r & -r|r| & 0 \end{bmatrix}}^{\mathbf{Y}_{4d}} \overbrace{\begin{bmatrix} \hat{\boldsymbol{\pi}}_{\mathbf{MC}} \\ \hat{\boldsymbol{\pi}}_{\mathbf{D}} \\ \hat{\boldsymbol{\pi}}_{g} \end{bmatrix}}^{\hat{\pi}} \tag{2.8}$$

where $\boldsymbol{\tau}_{4d}$ stands for the overall actuation of the system and $\hat{\boldsymbol{\pi}}$ represents the estimate of the unknown parameters $\boldsymbol{\pi} = [\boldsymbol{\pi}_{\mathbf{MC}}^T, \boldsymbol{\pi}_{\mathbf{D}}^T, \boldsymbol{\pi}_g^T]^T$ with

$$\boldsymbol{\pi}_{\mathbf{MC}} = [m + X_{\dot{u}},\ m + Y_{\dot{v}},\ m + Z_{\dot{w}} I_z + N_{\dot{r}}]^T \in \mathbb{R}^4 \tag{2.9}$$

$$\boldsymbol{\pi}_{\mathbf{D}} = \left[X_u,\ X_{|u|u},\ Y_v,\ Y_{|v|v},\ Z_w,\ Z_{|w|w},\ N_r,\ N_{|r|r}\right]^T \in \mathbb{R}^8 \tag{2.10}$$

$$\boldsymbol{\pi}_g = mg - B \tag{2.11}$$

denoting the unknown (added) masses and (added) inertia (2.9), the unknown first- and second-order drag parameters (2.10) and the difference of weight force mg and buoyancy force B (2.11), respectively [1].

Even though the equations of motion are nonlinear, the vehicle is input to state stable. Therefore, bounded inputs will never exhibit instability. This comes from the hydrodynamic terms and the neutral buoyancy properties which make the vehicle passive by construction. For that reason, we are allowed to employ open-loop sinusoidal and step inputs for the excitation of the system.

Following [20], the constraint RLS algorithm is formulated as

$$\mathbf{K}_k = \mathbf{P}_{k-1}\mathbf{Y}_{4d,k-1}^T\left(\mathbf{Y}_{4d,k-1}\mathbf{P}_{k-1}\mathbf{Y}_{4d,k-1}^T + \mathbf{R}_\tau\right)^{-1} \tag{2.12}$$

$$\mathbf{P}_k = \mathbf{P}_{k-1} - \mathbf{K}_k\mathbf{Y}_{4d,k-1}^T\mathbf{P}_{k-1} \tag{2.13}$$

$$\hat{\boldsymbol{\pi}}_k = \mathbf{Proj}\left\{\hat{\boldsymbol{\pi}}_{k-1} + \mathbf{K}_k\left(\boldsymbol{\tau}_{4d,k-1} - \mathbf{Y}_{4d,k-1}^T\hat{\boldsymbol{\pi}}_{k-1}\right)\right\} \tag{2.14}$$

where $\mathbf{P} \in \mathbb{R}^{13\times13}$ denotes the weighted sample covariance matrix, $\mathbf{R}_\tau \in \mathbb{R}^{4\times4}$ is a constant diagonal matrix of the noise of the system and the projection operator $\mathbf{Proj}\{\cdot\}$ is adopted to enforce $\hat{\boldsymbol{\pi}}_{\mathbf{MC}} > 0$, $\hat{\boldsymbol{\pi}}_{\mathbf{D}} < 0$ and $\hat{\boldsymbol{\pi}}_g < 0$.

It is well recognized that reliable, accurate and efficient SI requires appropriately designed experiments. When designing an identification experiment, it is essential to ensure that the excitation is sufficient to provide accurate and fast parameter estimation in the presence of measurement noise and actuator disturbances. In our approach, the excitation module is in charge of delivering the appropriate inputs (i.e., excitation trajectories) in the form of body forces and torques during the identification procedure. We propose a decision-making algorithm (see Figure 2.4) which employs four different excitation modes depending on the DoF that is currently under investigation, as well as an extra mode, where the vehicle is excited along all actuated DoFs simultaneously. Thus, we identify the following modes of excitation: (1) *modeYaw*, (2) *modeSurge*, (3) *modeSway*, (4) *modeHeave*, (5) *modeTotal*. In modes *1–4*, the vehicle is actuated only along the corresponding DoF, whereas in *modeTotal* all actuated DoFs are simultaneously excited. The transition from a mode of excitation to its subsequent is decided by monitoring the convergence of the dynamic parameters that are related to the individual DoF. More specifically, the convergence metric is evaluated by adding the appropriate diagonal elements of the covariance matrix $\mathbf{P}_k$ defined in (2.13), as follows:

$$
\begin{aligned}
P^k_{surge} &= P^k_{m+X_{\dot{u}}} + P^k_{X_u} + P^k_{X_{u|u|}} \\
P^k_{sway} &= P^k_{m+Y_{\dot{v}}} + P^k_{Y_v} + P^k_{Y_{v|v|}} \\
P^k_{heave} &= P^k_{m+Z_{\dot{w}}} + P^k_{Z_w} + P^k_{Z_{w|w|}} + P^k_{mg-B} \\
P^k_{yaw} &= P^k_{m+N_{\dot{r}}} + P^k_{N_r} + P^k_{N_{r|r|}} \\
P^k_{total} &= trace(\mathbf{P}_k)
\end{aligned}
\tag{2.15}
$$

where P^k_{surge}, P^k_{sway}, P^k_{heave} and P^k_{yaw} are the estimates of the covariances related to Surge, Sway, Heave and Yaw DoFs, respectively. The term P^k_{total} is the trace of the overall covariance matrix $\mathbf{P}_k$ and it is used for the final assessment of the identified dynamic model via concurrent excitation along all actuated DoFs. In order to proceed to the next excitation mode, the metric P^k_i, $i \in \{surge, sway, heave, yaw, total\}$, which is calculated during the online identification should become smaller than a default threshold value $P^\star_i$, $i \in \{surge, sway, heave, yaw, total\}$, as depicted in Figure 2.4.

Each of the aforementioned excitation modes employs an internal functionality which comprises three distinct submodules, namely *subMode0, subMode1* and *subMode2*. In particular, *subMode0* provides periodical inputs to the system with random amplitudes and frequencies in order to identify the parameters related to the (added) inertia, (added) mass and gravity/buoyancy terms of the vehicle. After a finite predefined time frame, *subMode1* is engaged, which in turn provides step inputs with random and relatively small amplitudes, in order to identify the first-order hydrodynamic coefficients. Finally, *subMode2* is used to provide inputs for the identification of the second-order hydrodynamic coefficients, via step inputs with random but larger

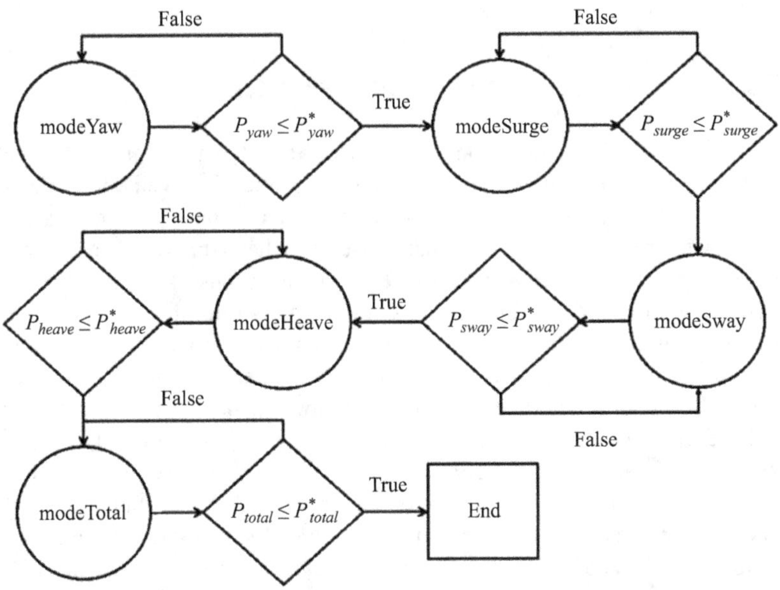

Figure 2.4 The decision-making algorithm of the proposed excitation module

Algorithm 1: The internal mode functionality

$SMcounter \leftarrow SMcounter + 1$ ▷ Increment submode counter (e.g. subMode0) by 1
if $SMcounter \geq SMcounterMax$ **then** ▷ Check if mode counter has exceeded the maximum number of iterations
  $submode \leftarrow (submode + 1)\%3$ ▷ If so progress to next mode (e.g. from subMode0 to subMode1)
  $SMcounter \leftarrow 0$ ▷ Set submode counter to zero
  $INTcounter \leftarrow INTcounterMax$
end if
if $submode = 0$ True **then** ▷ Check if subMode0 is enabled
  if $INTcounter \geq INTcounterMax$ **then**
    $f \leftarrow rand(0, f_{max})$ ▷ Calculate a random frequency between $(0, f_{max})$
    $a \leftarrow \tau_{max}$ ▷ Assign a maximum amplitude value
    $\tau \leftarrow a \cdot \cos(f \cdot t)$ ▷ Calculate sinusoidal input
    $INTcounter \leftarrow 0$
  end if
  $INTcounter \leftarrow INTcounter + 1$
end if
if $submode = 1$ True **then** ▷ Check if subMode1 is enabled
  if $INTcounter \geq INTcounterMax$ **then**
    $\tau \leftarrow rand(0, \tau_{max}^{low}) \cdot rand(sign)$ ▷ Calculate random step signal (low speeds) either positive or negative
    $INTcounter \leftarrow 0$
  end if
  $INTcounter \leftarrow INTcounter + 1$
end if
if $submode = 2$ True **then** ▷ Check if subMode2 is enabled
  if $INTcounter \geq INTcounterMax$ **then**
    $\tau \leftarrow rand(\tau_{max}^{low}, \tau_{max}^{high}) \cdot rand(sign)$ ▷ Calculate random step signal (high speeds) either positive or negative
    $INTcounter \leftarrow 0$
  end if
  $INTcounter \leftarrow INTcounter + 1$
end if

amplitudes. An algorithmic representation of the aforementioned internal functionality is described in Algorithm 1. The procedure iterates infinitely until the higher-level module reaches its performance goal $P_i^\star$, $i \in \{surge, sway, heave, yaw, total\}$, and proceeds to the subsequent mode of excitation as decided by the decision-making algorithm presented previously.

2.3 State estimation for underwater vehicles

The design of a state estimation module is highly affected by the available navigation sensors and the vehicle model (kinematic and/or dynamic). It is well known that GPS sensors cannot operate properly in the underwater environment. Hence, acoustic, inertial and vision-based sensors are usually fused together to provide an accurate state estimation of the underwater vehicle. However, acoustic sensors, such as USBL and DVL sensors, exhibit poor performance in shallow waters and constrained workspaces due to scattering effects. On the other hand, vision sensors may perform poorly in low-visibility conditions which often occur in deep water operations. Thus, in most cases different sensor suites are employed for shallow and deep water missions. In the rest of this chapter, we present state estimation techniques based on the CF notion for both shallow and deep water operations. In both cases, we assume that the vehicle is equipped with an attitude and heading reference system (AHRS), which successfully delivers three-dimensional (3D) orientation, angular velocities and accelerations with respect to the vehicle body frame. Additionally, we assume that a depth sensor is also installed on the underwater vehicle. The aforementioned suite is quite standard in almost all commercial and noncommercial underwater vehicles. The nature of the operating environment (i.e., shallow or deep waters) highly affects the choice of additional navigation sensors.

2.3.1 Introduction to CFs

CFs provide a solution of fusing two or more different sensors that measure the same quantity or its rate. The solution exploits the fact that the various sensors or models exhibit better behavior and performance in different bands of frequency. For example, for a simple integrator system (i.e., $\dot{x} = u$) and two sensors that measure x and u with measurement models defined as $z_x = x + n_x$ and $z_u = u + n_u + b(t)$, where $n_{x,u}$ are the measurements noise terms and b is a bias term that affects the z_u at low frequencies, the CF can be written in frequency domain as $\hat{x} = L(s)z_x + H(s)\frac{z_u}{s}$, where $L(s)$ is a low-pass filter and $H(s)$ a high-pass filter that follow the equation $L(s) + H(s) = 1$, which is the main property of the CFs. Additionally, the selection of the order of $L(s)$ and $H(s)$ determines the order of the filter, respectively. In the majority of use cases, first-order filters are used by selecting $L(s) = \frac{1}{1+T_c s}$ and $H(s) = \frac{T_c s}{1+T_c s}$, where $T_c = \frac{2\pi}{\omega_c}$ is the selected cutoff frequency for both filters. Using this estimator, the rate sensor z_u is first integrated to get the value of x. This could give unreliable results due to the bias term integration. For that reason, second-order filters are used by selecting $L(s) = \frac{2\zeta\omega_c s+\omega_c^2}{s^2+2\zeta\omega_c s+\omega_c^2}$ and $H(s) = \frac{s^2}{s^2+2\zeta\omega_c s+\omega_c^2}$, where ζ is the damping factor of the filter. In practice, the discrete form of the CFs is used. Thus, the first-order filter is written as

$$\hat{x}_k = \hat{x}_k^- + k\left(z_x^k - \hat{x}_k^-\right) \tag{2.16}$$

where $\hat{x}_k^- = \hat{x}_{k-1} + \Delta t z_u^{k-1}$ is the a priori estimation of $\hat{x}$ and $k = \frac{\Delta t}{\Delta t + T_c}$ is the selected gain, while the second-order filters are written as

$$\hat{x}_k = \hat{x}_k^- + k_p\left(z_x^k - \hat{x}_k^-\right), \ \hat{b}_k = \hat{b}_{k-1} - k_i\left(z_x^k - \hat{x}_k^-\right) \tag{2.17}$$

where the a priori estimation has been fixed by the bias term (i.e., $\hat{x}_k^- = \hat{x}_{k-1} + \Delta t z_u^{k-1} - \Delta t \hat{b}_{k-1}$) and $k_p = \frac{(2\zeta\omega_c + \Delta t\omega_c^2)\Delta t}{1+(2\zeta\omega_c+\Delta t\omega_c^2)\Delta t}$ and $k_i = \Delta t\omega_c^2\left(1 - k_p\right)$ are the selected gains. Simple block diagrams that accompany the aforementioned discussion are illustrated in Figures 2.5 and 2.6.

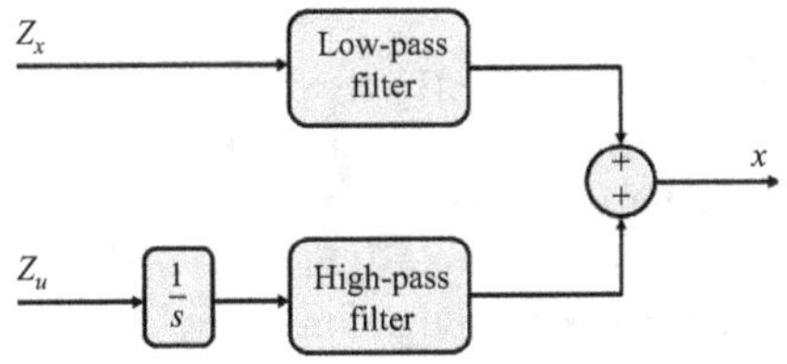

Figure 2.5 Simple first-order CF block diagram

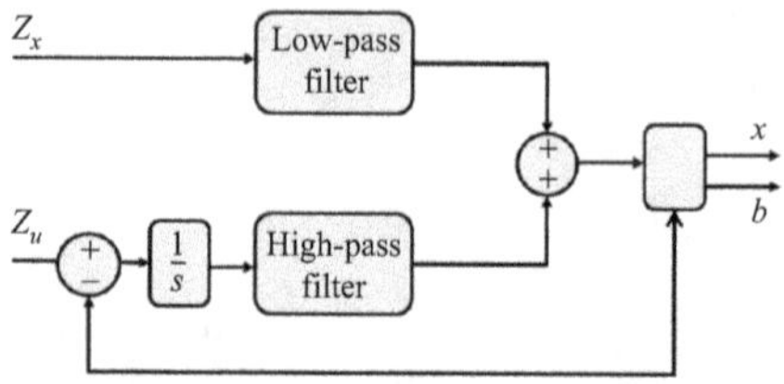

Figure 2.6 Simple second-order CF block diagram

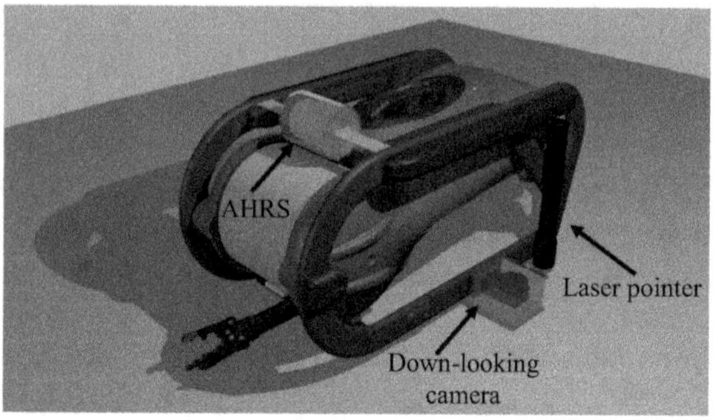

Figure 2.7 Sensor setup for shallow waters operation

2.3.2 CF design

2.3.2.1 Shallow waters operation

In shallow waters, acoustic sensors such as DVL and USBL may perform poorly. A fairly accurate and low-cost solution is the installation of a down-looking camera accompanied by a rigidly attached laser pointer which projects a dot inside the camera's field of view. The proposed setup is depicted in Figure 2.7. Assuming fair visibility and lighting conditions, the proposed laser vision system is able to provide 3D translational velocities with respect to the vehicle body frame $\mathcal{B}$, via computer vision optical flow techniques, as well as pose estimation of the vehicle in case specific visual landmarks are located within the operating workspace. Next, we present the design of a complete state estimation algorithm, based on the CF notion, assuming the vehicle is equipped with a down-looking camera, a laser pointer and an AHRS, in a landmark-based environment.

Following the concept of CFs as in [21], we propose a sensor fusion methodology that estimates the translational kinematics of the underwater vehicle (i.e., the position $\boldsymbol{\eta}_1$, the linear velocity $\boldsymbol{v}_1$ and the linear acceleration $\dot{\boldsymbol{v}}_1$). We should note that the AHRS is assumed to deliver accurate and reliable 3D angles and angular velocities, due to the on-board—usually EKF-based—navigation filter. Thus, AHRS measurements are considered as inputs to the system. The overall architecture of the proposed state estimation module is depicted in Figure 2.8, while a detailed analysis follows next.

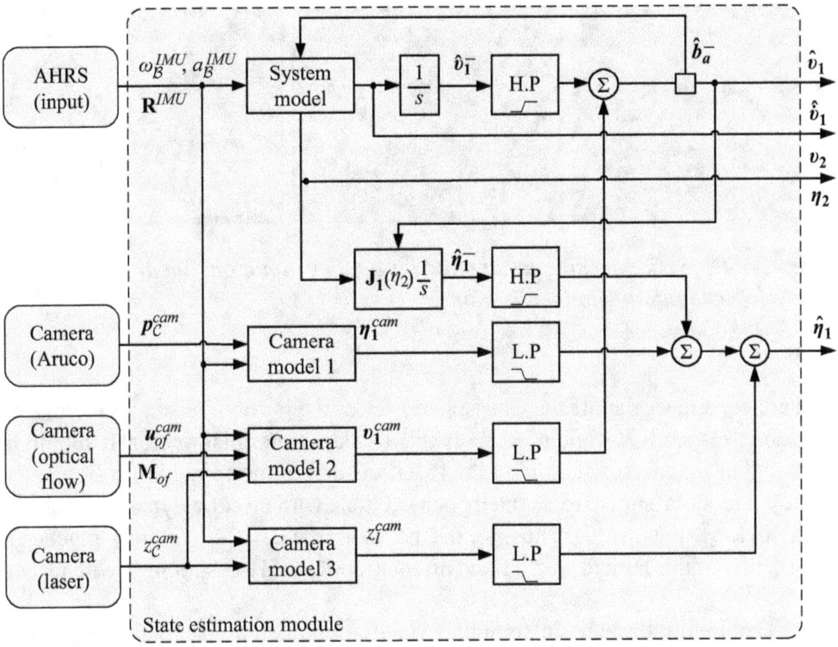

Figure 2.8 The proposed state estimation module for shallow waters

Owing to the high sampling frequency of the AHRS compared to the slow motion of the underwater vehicle, the translational kinematics can be written in discrete form as follows:

$$\boldsymbol{\eta}_1^{k+1} = \boldsymbol{\eta}_1^k + \Delta t \mathbf{J}_1\left(\boldsymbol{n}_2^k\right)\boldsymbol{v}_1^k \tag{2.18}$$

$$\boldsymbol{v}_1^{k+1} = \boldsymbol{v}_1^k + \Delta t \boldsymbol{a}_{\mathcal{B}}^k \tag{2.19}$$

where $\boldsymbol{a}_{\mathcal{B}}^k \in \mathbb{R}^3$ is the acceleration of the vehicle expressed in $\mathcal{B}$ and Δt denotes the time step of integration. The measurement model for each sensor is defined as follows:

- The AHRS of the vehicle consists of a 9-DoF IMU that delivers the angular velocity measurement $\boldsymbol{\omega}_{\mathcal{B}}^{IMU}$, the acceleration measurement $\boldsymbol{a}_{\mathcal{B}}^{IMU}$ and the orientation of the vehicle expressed as a rotation matrix $\mathbf{R}^{IMU} \in SO(3)$. Thus,

$$\boldsymbol{\omega}_{\mathcal{B}}^{IMU} = \boldsymbol{v}_2 + \boldsymbol{w}_{\omega} \tag{2.20}$$

$$\boldsymbol{a}_{\mathcal{B}}^{IMU} = \boldsymbol{a}_{\mathcal{B}} + \boldsymbol{b}_a + \mathbf{R}^{IMU}\boldsymbol{g}_N + \boldsymbol{w}_a \tag{2.21}$$

$$\mathbf{R}^{IMU} = \mathbf{R}_{IN}^T \mathbf{J}_1\left(\boldsymbol{\eta}_2\right) \tag{2.22}$$

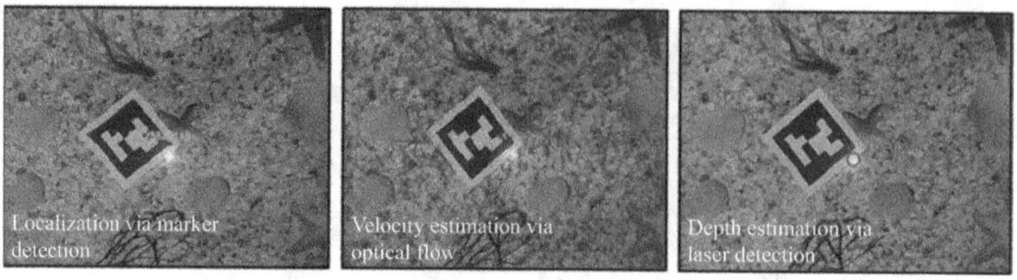

Figure 2.9 The vision techniques applied for depth estimation, linear velocity estimation and localization

where $\boldsymbol{w}_\omega$ and $\boldsymbol{w}_a$ denote the Gaussian noises of the gyroscope and accelerometer sensors, respectively, and $\boldsymbol{b}_a \in \mathbb{R}^3$ is the accelerometer bias vector. It should be noted that for sake of simplicity the IMU sensor's frame coincides to $\mathcal{B}$.

- The camera is assumed to be fixed on the vehicle with no extra dynamics included and looking downwards. Thus, we define the frame $\mathcal{C} = \{e_{cx}, e_{cy}, e_{cz}\}$ attached at the center of the camera with a fixed orientation $\mathbf{R}_\mathcal{C}$ and translation $\boldsymbol{l}_\mathcal{C}$ with respect to $\mathcal{B}$.

Employing three basic computer vision tools, i.e., a visual landmark-based localization algorithm, the optical flow technique and a laser detection system [21–23] (see Figure 2.9), we obtain the following essential state estimates: (a) the vehicle position vector expressed in $\mathcal{I}$, (b) the vehicle linear velocity expressed in $\mathcal{B}$ and (c) the range between the vehicle and the sea bottom, respectively. Hence, the camera measurement models can be defined as follows:

$$\boldsymbol{p}_\mathcal{C}^{cam} = \mathbf{R}_\mathcal{C}\left(\mathbf{J}_1^T(\eta_2)\eta_1 + \boldsymbol{l}_\mathcal{C}\right) + \boldsymbol{w}_{\boldsymbol{p}}^{cam} \tag{2.23}$$

$$\boldsymbol{v}_{of}^{cam} = \frac{1}{z_\mathcal{C}^{cam}}\mathbf{R}_\mathcal{C}(\boldsymbol{v}_1 + \boldsymbol{v}_2 \times \boldsymbol{l}_\mathcal{C}) + \mathbf{M}_{of}\mathbf{R}_\mathcal{C}\boldsymbol{v}_2 + \boldsymbol{w}_{of}^{cam} \tag{2.24}$$

$$z_\mathcal{C}^{cam} = -\frac{z + \boldsymbol{l}_{\mathcal{C},3}}{J_{3,3}} + n_z^{cam} \tag{2.25}$$

where $\boldsymbol{p}_\mathcal{C}^{cam}$ is the position measurement obtained from the landmark localization algorithm that denotes the position of the camera with respect to $\mathcal{I}$ expressed in $\mathcal{C}$, $\boldsymbol{v}_{of}^{cam}$ denotes the optical flow velocity with gyro compensation from the optical flow data $\mathbf{M}_{of}$ and $z_\mathcal{C}^{cam}$ is the range between the camera and the sea bottom measured by the laser detection technique. Moreover, $\boldsymbol{l}_{\mathcal{C},3}$ and $J_{3,3}$ denote the third element of $\boldsymbol{l}_\mathcal{C}$ and the third row-third column element of $\mathbf{J}_1(\eta_2)$, respectively. Finally, $\boldsymbol{w}_{\boldsymbol{p}}^{cam}$, $\boldsymbol{w}_{of}^{cam}$ and n_z^{cam} represent the Gaussian noises of the camera sensor for the position, optical flow and range measurements.

The proposed sensor fusion structure (see Figure 2.8) builds an estimated state vector $\hat{\boldsymbol{s}} = \left[\hat{\eta}_1^T\ \hat{\boldsymbol{v}}_1^T\ \hat{\boldsymbol{b}}_a^T\right]^T$ that involves the position $\hat{\eta}_1 = \left[\hat{x}\ \hat{y}\ \hat{z}\right]^T$ expressed in $\mathcal{I}$ as well as the linear velocity $\hat{\boldsymbol{v}}_1 = \left[\hat{u}\ \hat{v}\ \hat{w}\right]^T$ and the accelerometer bias $\hat{\boldsymbol{b}}_a = \left[\hat{b}_x\ \hat{b}_y\ \hat{b}_z\right]^T$

both expressed in $\mathcal{B}$. Recalling (2.18)–(2.25) and following [21], the proposed state estimation module is summarized as

$$\begin{aligned}
\hat{x}^{k+1} &= \hat{x}^{k,-} + k^{x}_{cam}\left(x^{cam,k} - \hat{x}^{k,-}\right) \\
\hat{y}^{k+1} &= \hat{y}^{k,-} + k^{y}_{cam}\left(y^{cam,k} - \hat{y}^{k,-}\right) \\
\hat{z}^{k+1} &= \hat{z}^{k,-} + k^{z}_{cam}\left(z^{cam,k} - \hat{z}^{k,-}\right) + k^{z}_{l}\left(z^{cam,k}_{l} - \hat{z}^{k,-}\right) \\
\hat{\boldsymbol{v}}^{k+1}_{1} &= \hat{\boldsymbol{v}}^{k,-}_{1} + \mathbf{K}^{\boldsymbol{v}_1}_{cam}\left(\boldsymbol{v}^{cam,k}_{1} - \hat{\boldsymbol{v}}^{k,-}_{1}\right) \\
\hat{\boldsymbol{b}}^{k+1}_{\boldsymbol{a}} &= \hat{\boldsymbol{b}}^{k,-}_{\boldsymbol{a}} - \mathbf{K}^{\boldsymbol{b_a}}_{cam}\left(\boldsymbol{v}^{cam,k}_{1} - \hat{\boldsymbol{v}}^{k,-}_{1}\right)
\end{aligned} \tag{2.26}$$

where

$$\begin{aligned}
\begin{bmatrix} x^{cam,k} \\ y^{cam,k} \\ z^{cam,k} \end{bmatrix} &= \boldsymbol{\eta}^{cam,k}_{1} = \mathbf{R}_{IN}\mathbf{R}^{IMU,k}\left(\mathbf{R}^{T}_{C}\boldsymbol{p}^{cam,k}_{C} - \boldsymbol{l}_{C}\right) \\
z^{cam,k}_{l} &= -(J_{3,3}z^{cam,k}_{C} - \boldsymbol{l}_{C,3}) \\
\boldsymbol{v}^{cam,k}_{1} &= z^{cam,k}_{C}\mathbf{R}^{T}_{C}\left(\boldsymbol{v}^{cam,k}_{of} - \mathbf{M}_{of}\mathbf{R}_{C}\boldsymbol{\omega}^{IMU}_{\mathcal{B}}\right) - \boldsymbol{\omega}^{IMU}_{\mathcal{B}} \times \boldsymbol{l}_{C}
\end{aligned} \tag{2.27}$$

and the a priori estimate $\hat{\boldsymbol{s}}^{k,-} = [\hat{\boldsymbol{\eta}}^{k,-^T}_{1}, \hat{\boldsymbol{v}}^{k,-^T}_{1}, \hat{\boldsymbol{b}}^{k,-^T}_{\boldsymbol{a}}]^T$ is calculated by

$$\begin{aligned}
\hat{\boldsymbol{\eta}}^{k,-}_{1} &= \hat{\boldsymbol{\eta}}^{k}_{1} + \Delta t\mathbf{R}_{IN}\mathbf{R}^{IMU,k}\hat{\boldsymbol{v}}^{k}_{1} \\
\hat{\boldsymbol{v}}^{k,-}_{1} &= \hat{\boldsymbol{v}}^{k,-}_{1} + \Delta t\left(\boldsymbol{a}^{IMU}_{\mathcal{B},k} - \hat{\boldsymbol{b}}^{k}_{\boldsymbol{a}} + \mathbf{R}^{IMU,k}\boldsymbol{g}_{N}\right) \\
\hat{\boldsymbol{b}}^{k,-}_{\boldsymbol{a}} &= \hat{\boldsymbol{b}}^{k}_{\boldsymbol{a}}
\end{aligned} \tag{2.28}$$

Finally, k^{x}_{cam}, k^{y}_{cam}, k^{z}_{cam} and k^{z}_{l} are positive gains that depend on the selected cutoff frequencies of the camera position and range measurements, and $\mathbf{K}^{\boldsymbol{v}_1}_{cam}$ and $\mathbf{K}^{\boldsymbol{b_a}}_{cam}$ are positive diagonal matrices that depend on the selected cutoff frequency and damping of the optical flow measurements. It should also be noted that based on standard gain tuning techniques for CFs [21], the overall stability of the estimation scheme is guaranteed under the following conditions:

$$\begin{aligned}
k^{i}_{cam} &< 1,\ i \in \{x, y\} \\
k^{z}_{cam} + k^{z}_{l} &< 1 \\
\mathbf{K}^{\boldsymbol{v}_1}_{cam} &< \mathbf{I}_3 \\
\mathbf{K}^{\boldsymbol{b_a}}_{cam} &< \Delta t\boldsymbol{\Omega}_c
\end{aligned} \tag{2.29}$$

where $\boldsymbol{\Omega}_c$ is a diagonal matrix that involves the selected cutoff frequencies of the optical flow measurements in all axes.

2.3.2.2 Deep water operation

In deep waters, vision sensors may present reduced performance due to low visibility and poor lighting conditions. In that case, the standard AHRS-depth sensor suite is

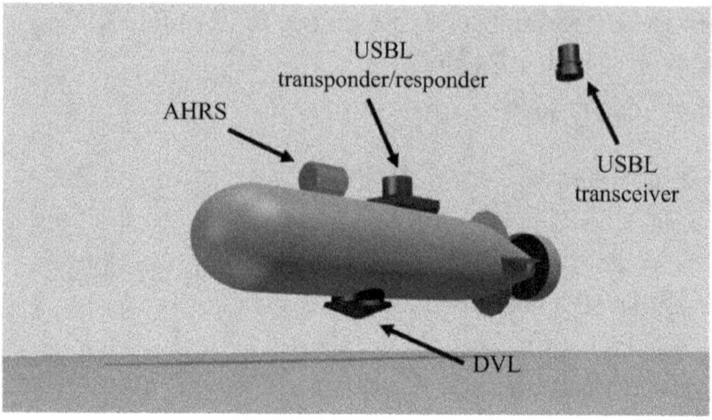

Figure 2.10 Sensor setup for deep water operation

enriched with additional acoustic navigation sensors, such as a DVL and/or a USBL. The DVL sensor delivers measurements of the vehicle linear velocity with respect to the seabed. The principal operation of the sensor is based on the transmission of acoustic pings through the water and measurement of the reflections on the seabed [24] via a multi-beam echo sounder. A Doppler shift appears in the reflections which can be used to calculate the velocity of the underwater vehicle. The output measurements are given with respect to the sensor-fixed coordinate system $\mathcal{D} = \{x_D, y_D, z_D\}$. Assuming that the distance between the DVL sensor and the body frame $\mathcal{B}$ is $\boldsymbol{l}_{\mathcal{D}}$ and $\mathbf{R}_{DB}$ is the rotation matrix that transforms any vector from the body frame $\mathcal{B}$ to the DVL frame $\mathcal{D}$, the following measurement equation for the linear velocity stands:

$$\boldsymbol{\nu}_{\mathcal{D}} = \mathbf{R}_{DB}\,(\boldsymbol{\nu}_1 + \boldsymbol{\nu}_2 \times \mathbf{l}_{\mathcal{D}}) + \boldsymbol{w}_{\nu_{\mathcal{D}}} \tag{2.30}$$

where $\boldsymbol{w}_{\nu_{\mathcal{D}}}$ is measurement noise due to a scale error produced by the uncertainty of the speed of sound within the water, which in turn is highly affected by the water salinity.

A USBL system consists of a transceiver, which is mounted on a fixed position (e.g. under a ship) and a transponder or responder which is rigidly attached on the underwater vehicle, as shown in Figure 2.10. The 3D position of the vehicle is calculated by the ranges and bearings measured by the transceiver. An acoustic pulse is transmitted by the transceiver and detected by the on-board transponder, which in turn replies with its own acoustic pulse. This latter is detected by the transceiver. The time duration from the initial transmission until the detected reply is converted into range. The bearings are measured by the transceiver, which contains an array of transducers (usually three or more).

For the sake of simplicity, we assume that the inertial frame $\mathcal{I}$ coincides with the transceiver's frame. The USBL delivers position measurements $\boldsymbol{p}_U$ of the responder with reference frame $\mathcal{U} = \{x_u, y_u, z_u\}$ with respect to $\mathcal{I}$. Assuming that the distance between the responder and the body frame $\mathcal{B}$ is $\boldsymbol{l}_{\mathcal{U}}$ and $\mathbf{R}_{UB}$ is the rotation matrix that

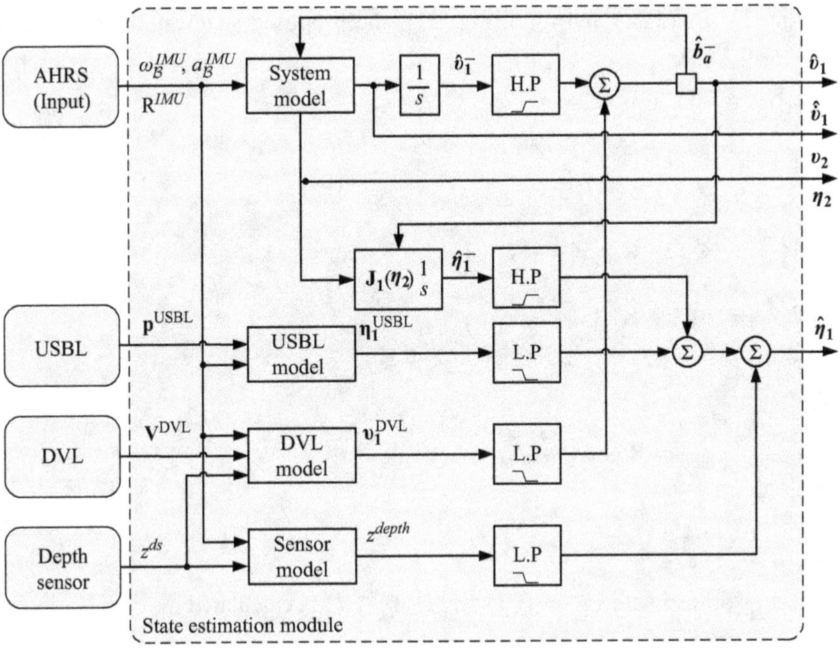

Figure 2.11 The proposed state estimation module for deep waters

transforms any vector from the body frame $\mathcal{B}$ to the responder frame $\mathcal{U}$, the following measurement equation for the position of the vehicle stands:

$$\boldsymbol{p}_{\mathcal{U}} = \mathbf{R}_{\mathcal{UB}}\left(\mathbf{J}_1^T(\boldsymbol{\eta}_2)\boldsymbol{\eta}_1 + \boldsymbol{l}_{\mathcal{U}}\right) + \boldsymbol{w}_{p_{\mathcal{U}}} \tag{2.31}$$

where $\boldsymbol{w}_{p_{\mathcal{U}}}$ denotes the sensor's measurement noise.

As for the depth sensor measurements, they are based on the difference between the current pressure and the atmospheric one. Assuming that the depth sensor frame coincides with the body frame $\mathcal{B}$, the relation between the depth sensor measurements and the vertical position of the vehicle is straightforward:

$$z_d = z + w_{z_d} \tag{2.32}$$

where z_d is the measured depth and w_{z_d} represents the Gaussian noise of the depth sensor for the vertical position measurements. The state model and the AHRS measurement model are exactly the same as in (2.18–2.19) and (2.20–2.22), respectively.

The proposed sensor fusion structure (see Figure 2.11) builds an estimated state vector $\hat{\boldsymbol{s}} = \left[\hat{\boldsymbol{\eta}}_1^T\ \hat{\boldsymbol{v}}_1^T\ \hat{\boldsymbol{b}}_{\boldsymbol{a}}^T\right]^T$ that involves the position $\hat{\boldsymbol{\eta}}_1 = \left[\hat{x}\ \hat{y}\ \hat{z}\right]^T$ expressed in $\mathcal{I}$, which in this case coincides with the transceiver's frame, as well as the linear velocity $\hat{\boldsymbol{v}}_1 = \left[\hat{u}\ \hat{v}\ \hat{w}\right]^T$ and the accelerometer bias $\hat{\boldsymbol{b}}_{\boldsymbol{a}} = \left[\hat{b}_x\ \hat{b}_y\ \hat{b}_z\right]^T$ both expressed in $\mathcal{B}$.

Recalling (2.18)–(2.22) and following [21], the proposed state estimation module is summarized as

$$\begin{aligned}
\hat{x}^{k+1} &= \hat{x}^{k,-} + k_U^x\left(x_u^k - \hat{x}^{k,-}\right) \\
\hat{y}^{k+1} &= \hat{y}^{k,-} + k_U^y\left(y_u^k - \hat{y}^{k,-}\right) \\
\hat{z}^{k+1} &= \hat{z}^{k,-} + k_U^z\left(z_u^k - \hat{z}^{k,-}\right) + k_d^z\left(z_d^k - \hat{z}^{k,-}\right) \\
\hat{\boldsymbol{v}}_1^{k+1} &= \hat{\boldsymbol{v}}_1^{k,-} + \mathbf{K}_D^{v_1}\left(\boldsymbol{v}_1^{D,k} - \hat{\boldsymbol{v}}_1^{k,-}\right) \\
\hat{\boldsymbol{b}}_a^{k+1} &= \hat{\boldsymbol{b}}_a^{k,-} - \mathbf{K}_D^{b_a}\left(\boldsymbol{v}_1^{D,k} - \hat{\boldsymbol{v}}_1^{k,-}\right)
\end{aligned} \tag{2.33}$$

where

$$\begin{aligned}
\begin{bmatrix} x_u^k \\ y_u^k \\ z_u^k \end{bmatrix} &= \mathbf{R}_{IN}\mathbf{R}^{IMU,k}\left(\mathbf{R}_{UB}^T\boldsymbol{p}_{\mathcal{U}}^k - \boldsymbol{l}_{\mathcal{U}}\right) \\
\boldsymbol{v}_1^{D,k} &= \mathbf{R}_{DB}^T\boldsymbol{v}_{\mathcal{D}}^k - \boldsymbol{v}_2^k \times \boldsymbol{l}_{\mathcal{D}}
\end{aligned} \tag{2.34}$$

and the a priori estimate $\hat{\boldsymbol{s}}^{k,-} = [\hat{\boldsymbol{\eta}}_1^{k,-T}, \hat{\boldsymbol{v}}_1^{k,-T}, \hat{\boldsymbol{b}}_a^{k,-T}]^T$ is calculated by

$$\begin{aligned}
\hat{\boldsymbol{\eta}}_1^{k,-} &= \hat{\boldsymbol{\eta}}_1^k + \Delta t\mathbf{R}_{IN}\mathbf{R}^{IMU,k}\hat{\boldsymbol{v}}_1^k \\
\hat{\boldsymbol{v}}_1^{k,-} &= \hat{\boldsymbol{v}}_1^{k,-} + \Delta t\left(\boldsymbol{a}_{\mathcal{B},k}{}^{IMU} - \hat{\boldsymbol{b}}_a^k + \mathbf{R}^{IMU,k}\boldsymbol{g}_N\right) \\
\hat{\boldsymbol{b}}_a^{k,-} &= \hat{\boldsymbol{b}}_a^k
\end{aligned} \tag{2.35}$$

Finally, k_U^x, k_U^y, k_U^z and k_d^z are positive gains that depend on the selected cutoff frequencies of the USBL and depth sensor measurements, and $\mathbf{K}_D^{v_1}$ and $\mathbf{K}_D^{b_a}$ are positive diagonal matrices that depend on the selected cutoff frequency and damping of the DVL measurements. It should also be noted that based on standard gain tuning techniques for CFs [21], the overall stability of the estimation scheme is guaranteed under the following conditions:

$$\begin{aligned}
k_U^i &< 1,\ i \in \{x, y\} \\
k_U^z + k_d^z &< 1 \\
\mathbf{K}_D^{v_1} &< \mathbf{I}_3 \\
\mathbf{K}_D^{b_a} &< \Delta t\boldsymbol{\Omega}_{\mathcal{D}}
\end{aligned} \tag{2.36}$$

where $\boldsymbol{\Omega}_{\mathcal{D}}$ is a diagonal matrix that involves the selected cutoff frequencies of the DVL measurements in all axes.

2.4 Results

In this section, we present experimental results related to the state estimation and SI of a small underwater vehicle operating in shallow waters. More specifically, the experiments were carried out inside the *NTUA, Control Systems Lab* test tank, with dimensions 5 m × 3 m × 1.5 m. The bottom of the tank is covered by a custom-made poster with various visual features and markers. The vehicle used in this work is a 4-DoF Seabotix LBV [19], actuated in Surge, Sway, Heave and Yaw via a four-thruster set configuration. The vehicle is equipped with a down-looking Sony PlayStation Eye camera, with 640 × 480 pixels at 30 frames per second (fps) enclosed in a waterproof housing. An underwater laser pointer projecting a green dot at the bottom of the test tank is rigidly attached on the vehicle with its axes aligned to the down-looking camera axis. The vehicle is also equipped with an *SBG IG* − 500*A* AHRS, delivering temperature-compensated 3D acceleration, angular velocity and orientation measurements at 100 Hz. Finally, the marker localization system is based on the *ArUco* library [22] and the software implementation of the proposed identification scheme as well as of the motion control schemes was conducted in C++ and Python under the Robot Operating System [25].

2.4.1 State estimation results

Notice that the 3D acceleration, angular velocity and orientation measurements are acquired by the on-board AHRS. Thus, the goal of the state estimation module is to deliver the 3D position and linear velocities of the vehicle. Based on CF technique, the selection of the appropriate cutoff frequencies is based on the reliability of both sensors, system model and the camera algorithms in certain bands of frequency. In our case, the accelerometer sensor suffers from bias below 4 Hz, while optical flow algorithm gives reliable measurements up to 5 Hz for x-y-axes and up to 3 Hz frequency for z-axis. Furthermore, due to unreliable data of optical flow in z-axis against x-y-axes, we select $\zeta_z = 0.5$ and $\zeta_{x,y} = 3$; thus, $\mathbf{K}^{v_1}_{cam} = diag\{0.3,\ 0.3,\ 0.06\}$ and $\mathbf{K}^{b_a}_{cam} = diag\{0.2,\ 0.2,\ 0.2\}$. Regarding the F and R axes position filters, the marker-based localization measurements (Aruco) were proven to exhibit unsatisfactory behavior over 6 Hz; thus, we selected $k^x_{cam} = 0.09$ and $k^y_{cam} = 0.09$. Finally, the depth measurements based on laser algorithm proved to be of good quality in a low bandwidth up to 2 Hz, while the same data captured by the Aruco algorithm were more reliable in higher frequencies (4 Hz); thus, we selected $k^z_{cam} = 0.05$ and $k^z_l = 0.003$. It should be noted here that the sensors and the appropriate camera algorithms run in different time frames. Thus, the aforementioned gains were selected accordingly.

Using the aforementioned gains, in Figure 2.12 the estimated horizontal position of the vehicle is depicted along with the position measurement provided by the marker localization algorithm. It is worth noting that even when there is no absolute localization measurement (no marker inside the camera field of view), the CF algorithm that is adopted in the state estimation module is able to provide valid estimations of the position vector η_1 without significant drift, improving in this way the robustness of the state estimation. The estimated longitudinal and lateral velocities are depicted in Figure 2.13 along with the corresponding velocity measurements obtained by the optical flow technique. The estimated depth and vertical velocity of the vehicle are depicted in Figure 2.14 along with the corresponding measurements acquired by the

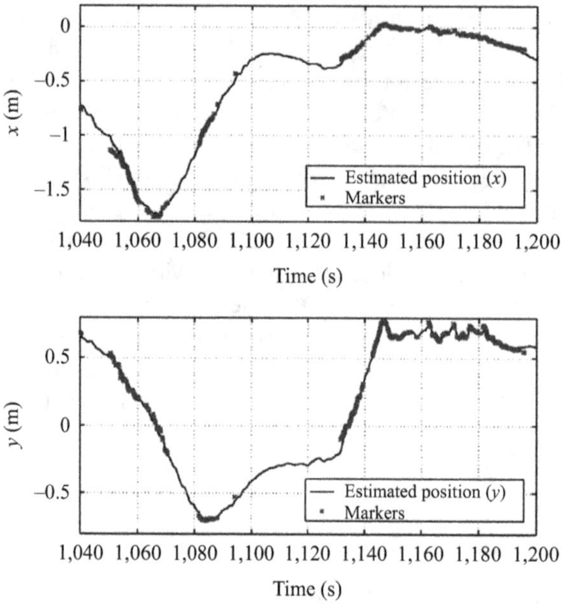

Figure 2.12 The estimated horizontal position of the vehicle

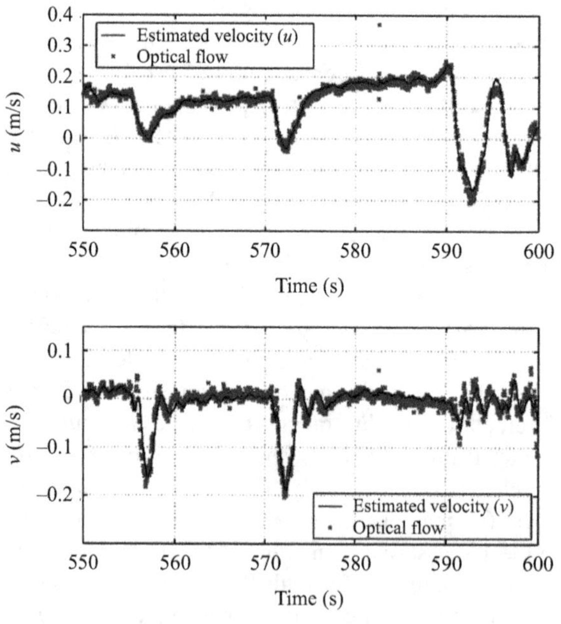

Figure 2.13 The estimated longitudinal and lateral velocities of the vehicle

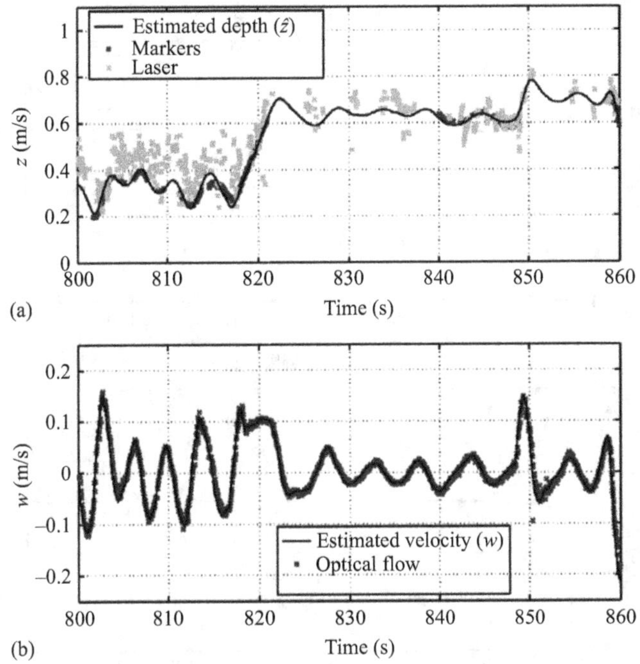

Figure 2.14 The estimated depth and vertical velocity of the vehicle

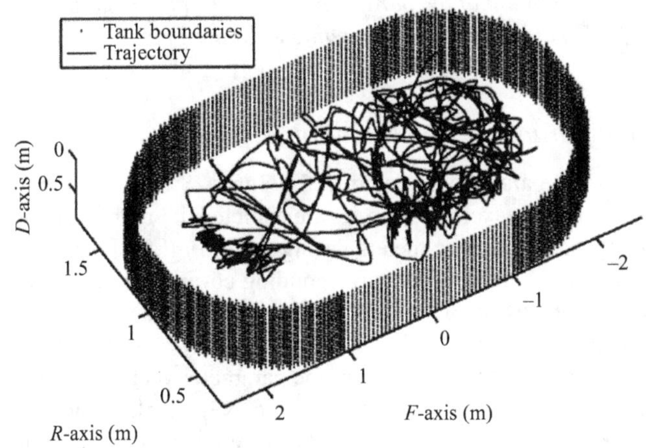

Figure 2.15 The trajectory of the vehicle during the identification procedure

laser-based technique, the marker localization framework and the optical flow technique. In the same figure, the estimated vertical velocity of the vehicle is provided along with the velocity measurements acquired by the optical flow technique. The trajectory of the vehicle during the identification procedure is shown in Figure 2.15.

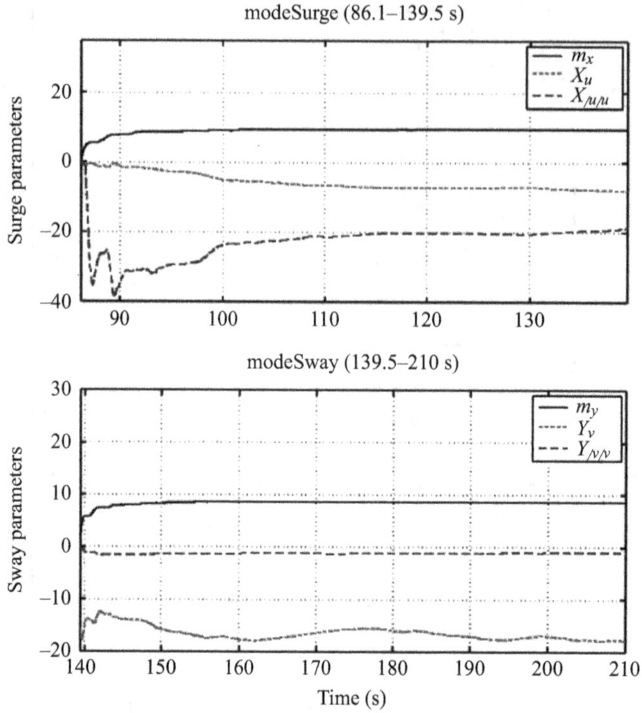

Figure 2.16 The evolution of the parameter estimates for the Surge and Sway DoFs, during the corresponding excitation modes

2.4.2 Online parameter identification results

The evolution of the parameters related to the Surge, Sway, Heave and Yaw DoFs during the corresponding excitation modes are depicted in Figures 2.16 and 2.17. The ultimate convergence of the parameters during *modeTotal* is shown in Figures 2.18 and 2.19. Moreover, the trace of the corresponding covariance matrices P_{surge}, P_{sway}, P_{heave} and P_{yaw} is depicted in Figure 2.20, from which we may observe that the trace for each DoF reduces gradually and converges ultimately to a small positive value. It is worth noting that no prior knowledge of the parameter vector was employed since the initial values for all parameters were set to zero. Nevertheless, the evolution of the parameter estimates shows that all parameters converged to a steady-state value after a relatively short period of time. The estimated parameter set is shown in Table 2.1.

As it can be observed from the experimental results, the total duration of the procedure for the demonstrated experiment was approximately 330 s. The duration varies, since certain factors affect the final result. More specifically, apart from the covariance matrix trace convergence, the characteristics of the input signals (frequencies, amplitudes) may affect the total duration of the identification procedure. In our case, the last were chosen in order to produce full iterations in every trajectory set.

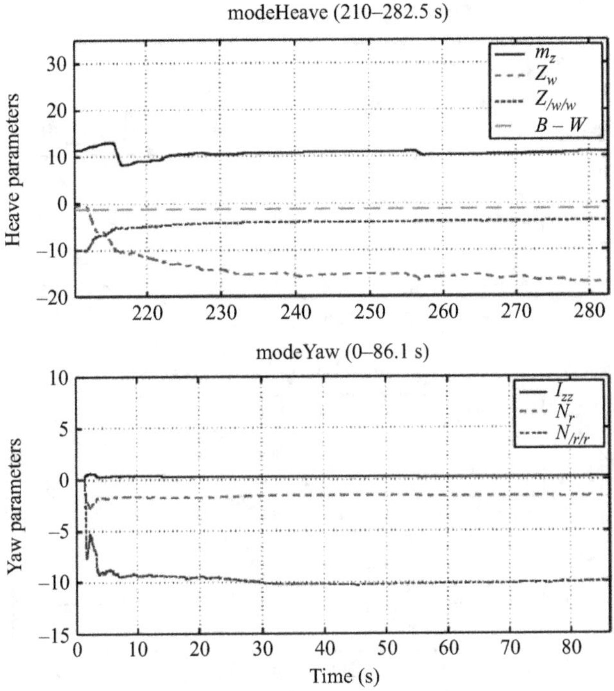

Figure 2.17 The evolution of the parameter estimates for the Heave and Yaw DoFs, during the corresponding excitation modes

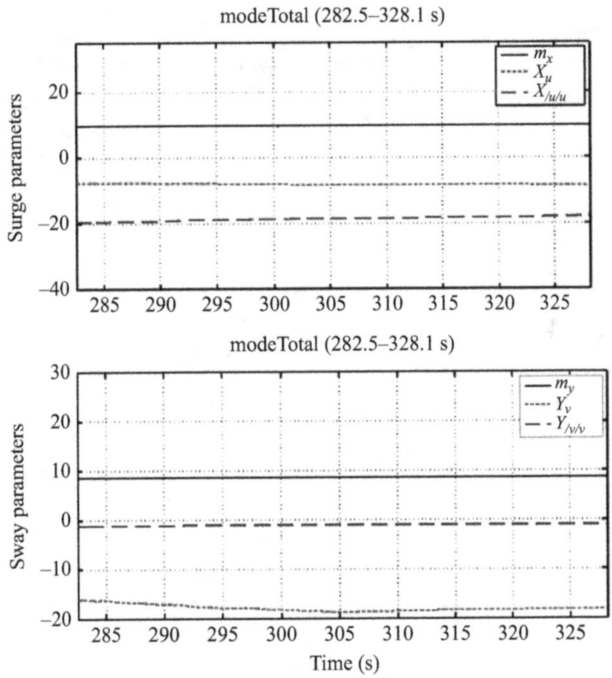

Figure 2.18 The evolution of the parameter estimates for the Surge and Sway DoFs, during modeTotal

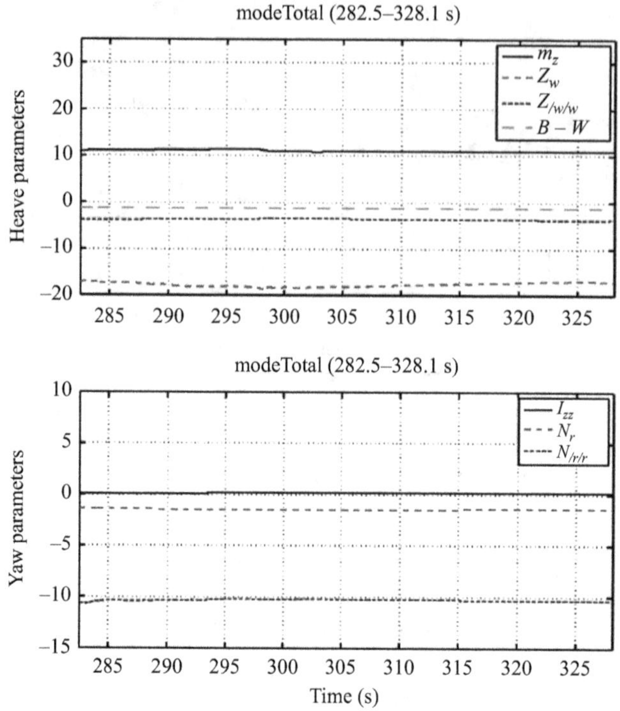

Figure 2.19 The evolution of the parameter estimates for the Heave and Yaw DoFs, during modeTotal

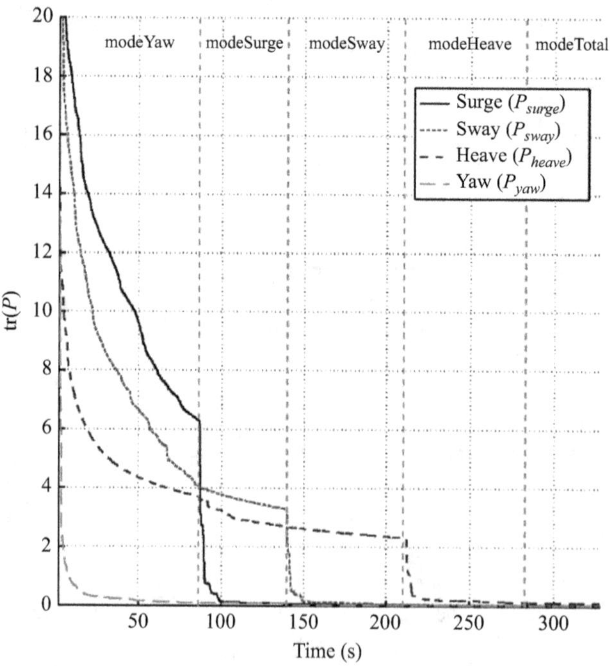

Figure 2.20 The evolution of the trace of the covariance matrices for the Surge, Sway, Heave and Yaw DoFs

Table 2.1 *The identified parameter set*

m_x	m_y	m_z	I_z
9.7532	8.6636	10.898	0.1589
X_u	Y_v	Z_w	N_r
−8.6040	−18.1106	−17.1828	−1.4146
$X_{\|u\|u}$	$Y_{\|v\|v}$	$Z_{\|w\|w}$	$N_{\|r\|r}$
−17.8534	−1.0594	−3.6482	−10.3483
		$W - B$	
		−1.1881	

The effect of the tether and the initial configuration may also slightly affect the final result. However, since the model parameters are not time varying and the excitation sequence is the same in every experiment, Yaw, Surge, Sway Heave, Total, the parameters converge to close values each time. Additionally, the random inputs exerted by the excitation module secure that the regression matrix Y is always full rank; thus, we do not anticipate significant deviations on the results of each identification procedure.

When a mode is completed, as indicated by the appropriate metric, the algorithm pauses the further update of the parameters related to that specific mode. But, when modeTotal is engaged, the estimation resumes for all the parameters, until the final convergence. As it can be seen by observing Figures 2.16–2.19, the final values estimated by modeTotal are not too far from the values calculated by the individual modes.

The terminating values for each mode are $P^\star_{yaw} = 0.08$, $P^\star_{surge} = 0.25$, $P^\star_{sway} = 0.25$, $P^\star_{heave} = 0.5$ and $P^\star_{total} = 0.3$. As it can be observed, $P^\star_{yaw}$ was assigned with the smallest value since the angle and angular velocity are directly obtained from the IMU which has very small measurement noise. On the contrary, the motion along Surge and Sway are estimated via the CF, hence the terminating values are higher. The same stands for the motion estimation along Heave axis, where additionally the measurements of the less reliable sensors (Depth Sensor and Laser) are employed, hence the terminating value is larger. The value of $P^\star_{total}$ was selected to be approximately 30% of the sum of the individual modes.

2.5 Summary

In this chapter, we described the kinematic and dynamic equations of underwater robotic vehicles considering standard assumptions and simplifications that may lead to low-complexity but accurate and operational models. Additionally, we presented an

RLS method for the online identification of the vehicle dynamic parameters. Moreover, we presented a state estimation technique which, depending on the available sensors, can be adopted for either shallow or deep water operation. The proposed state estimation method is based on the CF notion, hence low-complexity computational requirements are satisfied. The goal of this chapter was to provide the reader with the appropriate analytical methods, in order to derive a simple yet accurate vehicle model, design a state estimation algorithm that can be easily integrated into the embedded system framework of an underwater robotic vehicle and perform a fast online dynamic parameter identification.

References

[1] Fossen T. Guidance and Control of Ocean Vehicles. New York: Wiley; 1994.

[2] Bishop R, Parkinson AG. On the planar motion mechanism used in ship model testing. Phil Trans Roy Soc London Ser A Math Phys Sci. 1970;266(1171): 35–61.

[3] Avila JPJ, Adamowski JC. Experimental evaluation of the hydrodynamic coefficients of a ROV through Morison's equation. Ocean Engineering. 2011;38(17–18):2162–2170.

[4] Evans J, Nahon M. Dynamics modeling and performance evaluation of an autonomous underwater vehicle. Ocean Engineering. 2004;31(14–15):1835–1858.

[5] Sahin I, Crane JW, Watson KP. Application of a panel method to hydrodynamics of underwater vehicles. Ocean Engineering. 1997;24(6):501–512.

[6] Smallwood D, Whitcomb L. Adaptive identification of dynamically positioned underwater robotic vehicles. IEEE Transactions on Control Systems Technology. 2003;11(4):505–515.

[7] McFarland CJ, Whitcomb LL. Comparative experimental evaluation of a new adaptive identifier for underwater vehicles. In: Proceedings - IEEE International Conference on Robotics and Automation; 2013. pp. 4614–4620.

[8] van de Ven P, Johansen T, Sorensen A, *et al.* Neural network augmented identification of underwater vehicle models. Control Engineering Practice. 2007;15(6):715–725.

[9] Ishii K, Fujii T, Ura T. An on-line adaptation method in a neural network based control system for AUV's. IEEE Journal of Oceanic Engineering. 1995;20(3):221–228.

[10] Eng YH, Teo KM, Chitre M, *et al.* Online system identification of an autonomous underwater vehicle via in-field experiments. IEEE Journal of Oceanic Engineering. 2016;41(1):5–17.

[11] Karras GC, Bechlioulis CP, Leonetti M, *et al.* On-line identification of autonomous underwater vehicles through global derivative-free optimization. In: IEEE International Conference on Intelligent Robots and Systems; 2013. pp. 3859–3864.

[12] Julier S, Uhlmann J. A new extension of the Kalman filter to nonlinear systems. Proc. of the Int Symp Aerospace/Defense Sensing, Simul and Controls. 1997.

[13] Cutipa Luque JC, Donha DC. AUV identification and robust control. In: IFAC Proceedings Volumes (IFAC-PapersOnline). 2011;18:14735–14741.

[14] der Merwe RV, Wan E. The square-root unscented Kalman filter for state and parameter-estimation. IEEE International Conference on Acoustics, Speech, and Signal Processing. 2001;6:3461–3464.

[15] VanDyke M, Schwartz J, Hall C. Unscented Kalman filtering for spacecraft attitude state and parameter estimation. Advances in the Astronautical Sciences. 2004;217–228.

[16] Karras GC, Loizou SG, Kyriakopoulos KJ. Towards semi-autonomous operation of under-actuated underwater vehicles: sensor fusion, on-line identification and visual servo control. Autonomous Robots. 2011;31:67–86.

[17] Morgado M, Batista P, Oliveira P, *et al.* Position and velocity USBL/IMU sensor-based navigation filter. IFAC Proceedings Volumes. 2011;44(1):13642–13647. 18th IFAC World Congress. Available from: http://www.sciencedirect.com/science/article/pii/S1474667016458167.

[18] Skoglund MA, Gustafsson F, Jönsson K. Modeling and sensor fusion of a remotely operated underwater vehicle. In: 2012 15th International Conference on Information Fusion; 2012.

[19] Seabotix LBV150. Available from: http://www.teledynemarine.com/lbv150-4/?BrandID=19.

[20] Ioannou P, Sun J. Robust Adaptive Control. New Jersey, USA: Prentice Hall; 1996.

[21] Marantos P, Koveos Y, Kyriakopoulos KJ. UAV state estimation using adaptive complementary filters. Control Systems Technology, IEEE Transactions on. 2016;24(4):1214–1226.

[22] Garrido-Jurado S, noz Salinas RM, Madrid-Cuevas FJ, *et al.* Automatic generation and detection of highly reliable fiducial markers under occlusion. Pattern Recognition. 2014;47(6):2280–2292.

[23] Karras GC, Kyriakopoulos KJ, Karavas GK. Towards cooperation of underwater vehicles: a leader-follower scheme using vision-based implicit communications. In: Intelligent Robots and Systems (IROS), 2015 IEEE/RSJ International Conference on; 2015. pp. 4852–4857.

[24] Instruments R. Navigator ADCP/DVL Technical Manual.

[25] Robot Operating System. Available from: http://www.ros.org/.

Chapter 3

Sensor-based motion control of autonomous underwater vehicles, part II: robust motion control strategies

Charalampos P. Bechlioulis[1], Shahab Heshmati-alamdari[2], George C. Karras[1,3], Panos Marantos[1] and Kostas J. Kyriakopoulos[1]

3.1 Introduction

During the last decades, underwater robotic vehicles have been widely used in a variety of marine activities such as monitoring and inspection, surveillance of underwater facilities and oceanography. In most of these cases, the vehicles operate under the influence of strong external disturbances caused from ocean currents and waves. Moreover, the vehicle's dynamic parameters are likely to change because different types of sensors and tools are utilized, depending on the requirements of each mission, for the autonomous underwater vehicle (AUV) state estimation as well as the perception of the environment. Furthermore, the problem becomes much more intense especially when the vehicles are underactuated. Thus, the motion control scheme of an AUV should (i) be robust against external disturbances, (ii) comply with the kinematic constraints of the vehicle and (iii) up to a certain extent, be independent from the vehicle's dynamics. Additionally, the aforementioned applications require the underwater robot to operate under various constraints and increased level of autonomy, in terms of energy consumption and endurance. Thus, recent research directions point towards the development of motion control strategies that are able to handle complex missions under operational constraints with reduced energy requirements.

The motion control problem for AUVs has been an active research field for the past two decades and continues to pose considerable challenges to control designers especially when the vehicles exhibit large model uncertainty and are affected by environmental disturbances. A typical motion control problem is trajectory tracking which is concerned with the design of control laws that force a vehicle to reach and follow a reference trajectory. Classical approaches such as local linearization and

[1]School of Mechanical Engineering, Control Systems Lab, National Technical University of Athens, Athens, Greece
[2]Division of Decision and Control Systems, School of Electrical Engineering and Computer Science, KTH Royal Institute of Technology, Stockholm, Sweden
[3]Department of Computer Science and Telecommunications, University of Thessaly, Lamia, Greece

input-output decoupling have been used in the past to design tracking controllers for underactuated vehicles [1]. Nevertheless, the aforementioned methods yielded poor closed-loop performance and the results were local, around only certain selected operating points. Alternative approaches involved output feedback linearization combined with Proportional Integral Derivative (PID) control, linear quadratic optimal control and H_∞ control [2] which, however, was not always possible. Finally, based on a combined approach involving Lyapunov theory and the backstepping technique, various nonlinear model-based trajectory-tracking controllers have been reported during the last two decades [3–5]. However, these schemes demand a very accurate knowledge of the vehicle dynamic parameters which in most cases is quite difficult to obtain, and despite the recent progress in the tracking control for underwater vehicles, certain issues still remain open. Notice that even in case the actual vehicle model is considered accurately known, external disturbances affect severely the tracking performance, thus making the problem of guaranteeing prescribed performance difficult or impossible in certain situations.

In general, the motion control of underwater vehicles is a highly nonlinear problem, where multiple input and state constraints are imposed to the system. Investigating the corresponding literature, it can be concluded that in most research works the input (generalized body forces/torques or thrust) and the state (three-dimensional (3D) obstacles, velocities, etc.) constraints are not considered. Hence, nonlinear model predictive control (NMPC) [6] can be considered an ideal approach for complex underwater missions, as it is able to combine motion planning, obstacle avoidance and workspace restrictions, while handling efficiently input and state constraints. The model predictive control (MPC) has been already used in underwater motion planning. For instance, in order to optimize saw-tooth paths for an AUV, an MPC scheme with a least-squares cost function is presented in [7]. Interesting results, including estimated ocean wave profiles into an NMPC scheme, with an emphasis on real-time execution are presented in [8]. However, the effects of noise and disturbances are not theoretically considered, but instead are demonstrated through simulation testing. In addition, an MPC approach with reduced dynamic model is presented in [9], where simplified linear models for the vertical and horizontal directions were considered, in order to avoid computational complexity. In the aforementioned works, only the kinematic model of the vehicle is considered. Moreover, the validation of the proposed strategies was conducted via simulation tests. An early experimental result of a visually aided NMPC scheme for an underwater robotic system was presented in [10], where simple kinematic equations of the system was considered.

The first section of this chapter presents an NMPC strategy for underwater robotic vehicles operating under various constraints. The purpose of the controller is to guide the vehicle towards specific way-points. Various constraints such as obstacles, workspace boundaries and control input saturation as well as predefined upper bound of the vehicle velocity (requirements for several underwater tasks such as seabed inspection scenario and mosaicking) are considered during the control design. The proposed scheme incorporates the full dynamics of the vehicle in which the ocean currents are also involved. The controller is designed in order to find the optimal thrusts required for minimizing the way-point tracking error. Moreover, the control inputs calculated by the proposed approach are formulated in a way that the vehicle will exploit the ocean currents, when they are in favor of the way-point tracking mission, which results in reduced energy consumption by the thrusters.

In the second part of this chapter, novel position- and trajectory-tracking control schemes for AUVs are presented. The proposed controllers do not utilize the vehicle's dynamic model parameters and guarantee prescribed transient and steady-state performance despite the presence of external disturbances and kinematic constraints for the case of underactuated vehicles. Moreover, through the appropriate selection of certain performance functions, the proposed scheme can also guarantee the satisfaction of motion and performance constraints imposed by the desired task.

3.2 MPC for underwater vehicles

3.2.1 Preliminaries and problem formulation

3.2.1.1 Notation

In this work, the vectors are denoted with lower bold letters whereas the matrices by capital bold letters. We define as $\mathcal{B}(\boldsymbol{c}, r) = \{\boldsymbol{x} \in \mathbb{R}^3 : \|\boldsymbol{x} - \boldsymbol{c}\| \leq r\}$ the closed sphere with radius r and center $\boldsymbol{c}$. For a given set $A \subset \mathbb{R}^n$, we define as $\mathrm{cl}(A), \mathrm{int}(A)$ and $\partial S = \mathrm{cl}(A) \backslash \mathrm{int}(A)$ its closure, interior and boundary, respectively.

3.2.1.2 Mathematical modeling

The pose vector of the vehicle with respect to (w.r.t.) the inertial frame $\mathcal{I}$ is denoted by $\boldsymbol{\eta} = \left[\boldsymbol{\eta}_1^T\ \boldsymbol{\eta}_2^T\right]^\top \in \mathbb{R}^6$ including the position (i.e., $\boldsymbol{\eta}_1 = [x\ y\ z]^\top$) and orientation (i.e., $\boldsymbol{\eta}_2 = [\phi\ \theta\ \psi]^\top$) vectors. Moreover, $\boldsymbol{v} = \left[\boldsymbol{v}_1^T\ \boldsymbol{v}_2^T\right]^\top \in \mathbb{R}^6$ is the velocity vector of the vehicle expressed in fixed-body frame $\mathcal{V}$ and includes the linear (i.e., $\boldsymbol{v}_1 = [u\ v\ w]^\top$) and angular (i.e., $\boldsymbol{v}_2 = [p\ q\ r]^\top$) velocity vectors. In this section, we consider that the vehicle operates under the influence of bounded irrotational ocean currents w.r.t. the inertial frame $\mathcal{I}$. An estimation of the ocean currents can be achieved by employing the data obtained from Naval Coastal Ocean Model (NCOM) [11] and Regional Ocean Model Systems (ROMS) [12] or by employing a locally and appropriate estimator [13]. In this spirit, we denote the bounded irrotational ocean current velocities w.r.t. the inertial frame $\mathcal{I}$ by $\boldsymbol{v}_c^{\mathcal{I}} = [(\boldsymbol{v}_{c_1}^{\mathcal{I}})^\top, \boldsymbol{0}_{1\times 3}]^\top \in \mathbb{R}^6$ with $\boldsymbol{v}_{c_1}^{\mathcal{I}} = [u_c^{\mathcal{I}}, v_c^{\mathcal{I}}, w_c^{\mathcal{I}}]^\top$ to be the vector of linear velocity terms. Therefore, we can define the vehicle velocity vector relative to the water expressed in body frame $\mathcal{V}$ as

$$\boldsymbol{v}_r = \boldsymbol{v} - \boldsymbol{v}_c \tag{3.1}$$

Notice that the vector $\boldsymbol{v}_c = [u_c, v_c, w_c, \boldsymbol{0}_{1\times 3}]^\top$ indicates the expression of the ocean currents w.r.t. the body frame $\mathcal{V}$. Without loss of generality, assuming that the current velocity is slowly varying w.r.t. the inertial frame (e.g., $\frac{\partial \boldsymbol{v}_c^{\mathcal{I}}}{\partial t} \approxeq 0$), and the vehicle is operating at relative low speeds, the dynamic equations of the vehicle can be given as [14, Eqs. (3.110)–(3.116)]:

$$\dot{\boldsymbol{\eta}} = \boldsymbol{J}(\boldsymbol{\eta})\boldsymbol{v}_r + \boldsymbol{v}_c^{\mathcal{I}} \tag{3.2a}$$

$$M\dot{\boldsymbol{v}}_r + C(\boldsymbol{v}_r)\boldsymbol{v}_r + D(\boldsymbol{v}_r)\boldsymbol{v}_r + \boldsymbol{g}(\boldsymbol{\eta}) = \boldsymbol{\tau}_{\mathcal{V}} \tag{3.2b}$$

where $\boldsymbol{\tau}_{\mathcal{V}}$ is the total propulsion force/torque vector; $\boldsymbol{M}$ is the inertial matrix; $\boldsymbol{C}(\boldsymbol{v}_r)$ represents Coriolis and centrifugal terms; $\boldsymbol{D}(\boldsymbol{v}_r)$ models dissipative effects; $\boldsymbol{g}(\boldsymbol{\eta}) \in \mathbb{R}^6$ is the hydrostatic restoring force vector; and $\boldsymbol{J}(\boldsymbol{\eta})$ is the well-known Jacobian matrix transforming the velocities from the body-fixed ($\mathcal{V}$) to the inertial ($\mathcal{I}$) frame.

Remark 1. *The transformation from ocean current velocity defined in the inertial frame $\mathcal{I}$ (i.e., $\boldsymbol{v}_c^{\mathcal{I}}$) into body-fixed one (i.e., $\boldsymbol{v}_c$) is achieved using the transposed rotation matrix, i.e., $\boldsymbol{v}_c = \boldsymbol{J}^\top(\boldsymbol{\eta})\boldsymbol{v}_c^{\mathcal{I}}$ (see [14]).*

Remark 2. *In* (3.2), *the total propulsion force/torque vector ($\boldsymbol{\tau}_{\mathcal{V}}$) is computed using the thruster allocator matrix. Thus, for the vehicle used in the section's experimental results (i.e., a* 4*-DoF Seabotix LBV150), we can define a new thrust vector $\boldsymbol{\tau} = \left[\tau_{p_o}, \tau_s, \tau_{v_e}, \tau_l\right]^\top \in \mathbb{R}^4$ (notations p_o, s, v_e and l denote the Port (p_o), Starboard (s), Vertical (v_e) and Lateral (l) directions, respectively) and the appropriate thruster allocator matrix ($\boldsymbol{T}_A \in \mathbb{R}^{4\times4}$) such as*

$$\boldsymbol{\tau}_{\mathcal{V}}^{LBV} = \boldsymbol{T}_A\boldsymbol{\tau} \tag{3.3}$$

where $\boldsymbol{\tau}_{\mathcal{V}}^{LBV}\ [X,\ Y,\ Z,\ N]^\top \in \mathbb{R}^4$.

Remark 3. *In the vehicle used in this work (i.e., a* 4*-DoF Seabotix LBV150), the angles ϕ and θ and the angular velocities p and q are negligible, and we can consider them to be equal to zero. Thus, from now on, $\boldsymbol{\eta} = [x, y, z, \psi]$ and $\boldsymbol{v} = [u, v, w, r]$.*

3.2.1.3 Problem formulation

In this section, we formally address the problem under consideration:

Problem 1. *Given an underwater vehicle with dynamics as described in* (3.2), *design a robust feedback control law for the autonomous guidance towards a set of way-points $\boldsymbol{\eta}_i^d$, $i = \{1, \ldots, n\}$, while guaranteeing the following specifications:*

- *Avoid the workspace boundaries and a limited set of obstacles within.*
- *Respect operational limitations in the form of state (e.g., velocity bounds) and input (thrust saturation) constraints.*
- *The energy consumed from thrusters to be retained in a reduced level.*

3.2.2 Methodology

In this section, we present in detail the methodologies proposed in order to guide the vehicle towards a set of way-points $\boldsymbol{\eta}_i^d$, $i = \{1, \ldots, n\}$.

3.2.2.1 Geometry of workspace

Consider an underwater robotic vehicle which operates inside the workspace $\mathcal{W} \subset \mathbb{R}^3$ with boundaries $\partial\mathcal{W} = \{\boldsymbol{p} \in \mathbb{R}^3 : \boldsymbol{p} \in \mathrm{cl}(\mathcal{W})\backslash\mathrm{int}(\mathcal{W})\}$ and scattered obstacles located within. Without any loss of the generality, adopting spherical world representation

[16], the robot and the obstacles are all modeled by spheres. In this spirit, let $\mathcal{B}(\boldsymbol{\eta}_1, \bar{r})$ to be a closed sphere that covers all the vehicle volume including main body and additional equipments. Moreover, the $\mathcal{M}$ static obstacles within the workspace are defined as closed spheres described by $\pi_m = \mathcal{B}(\boldsymbol{p}_{\pi_m}, r_{\pi_m})$, $m \in \{1, \ldots, \mathcal{M}\}$, where $\boldsymbol{p}_{\pi_m} \in \mathbb{R}^3$ is the center and $r_{\pi_m} > 0$ the radius of the obstacle π_m. Additionally, based on the property of spherical world [16], for each pair of obstacles $m, m' \in \{1, \ldots, \mathcal{M}\}$ the following inequality holds:

$$||\pi_m - \pi_{m'}|| > 2\bar{r} + r_{\pi_m} + r_{\pi'_m} \tag{3.4}$$

which intuitively means that the obstacles m and m' are disjoint in such a way that the entire volume of the vehicle can pass through the free space between them. Therefore, we assume that there exists a feasible trajectory $\boldsymbol{\eta}(t)$ for the vehicle that connects the initial configuration $\boldsymbol{\eta}(t_0)$ with $\boldsymbol{\eta}^d$.

3.2.2.2 Dynamical system

Due to the aforementioned assumptions and following standard simplifications due to symmetries in the mass configuration [14], the dynamic equation (3.2) for the vehicle under consideration can be written in discrete-time form as

$$\boldsymbol{x}_{k+1} = f(\boldsymbol{x}_k, \boldsymbol{\tau}_k) \Rightarrow \boldsymbol{x}_{k+1} = \boldsymbol{x}_k + \mathcal{A}(\boldsymbol{x}_k)dt + \mathcal{C}(\boldsymbol{\tau}_k)dt \tag{3.5}$$

where

$$\mathcal{A}(\boldsymbol{x}_k) = \begin{bmatrix} u_{r_k} c\psi_k - v_{r_k} s\psi_k + u_c^{\mathcal{I}} \\ u_{r_k} s\psi_k + v_{r_k} c\psi_k + v_c^{\mathcal{I}} \\ w_{r_k} + w_c^{\mathcal{I}} \\ r_{r_k} \\ \frac{1}{m_{11}}(m_{22} v_{r_k} r_{r_k} + X_u u_{r_k} + X_{u|u|}|u_{r_k}|u_{r_k}) \\ \frac{1}{m_{22}}(-m_{11} u_{r_k} r_{r_k} + Y_v v_{r_k} + Y_{v|v|}|v_{r_k}|v_{r_k}) \\ \frac{1}{m_{33}}(Z_w w_{r_k} + Z_{w|w|}|w_{r_k}|w_{r_k}) \\ \frac{1}{m_{44}}((m_{11} - m_{22}) u_{r_k} v_{r_k} + N_r r_{r_k} + N_{r|r|}|r_{r_k}|r_{r_k}) \end{bmatrix}, \quad \mathcal{C}(\boldsymbol{\tau}_k) = \begin{bmatrix} \boldsymbol{0}_{4\times 1} \\ \boldsymbol{T}_A \boldsymbol{\tau}_k \end{bmatrix}$$

with $c(\cdot) = \cos(\cdot)$ and $s(\cdot) = \sin(\cdot)$, and $\boldsymbol{x}_k = [\boldsymbol{\eta}_k^\top, \boldsymbol{v}_{r_k}^\top]^\top = [x_k, y_k, z_k, \psi_k, u_{r_k}, v_{r_k}, w_{r_k}, r_{r_k}]^\top \in \mathbb{R}^8$ denotes the state vector at the time step k which includes the position and orientation of the vehicle w.r.t. the inertial frame $\mathcal{I}$ and the relative linear and angular velocities of the vehicle w.r.t. the water. In addition, $m_{ii}, i = 1, \ldots, 4$, are the mass terms including added mass, X_u, Y_v, Z_w, $N_r < 0$ are the linear drag terms, $X_{u|u|}$, $Y_{v|v|}$, $Z_{w|w|}$, $N_{r|r|} < 0$ are the quadratic drag terms, while dt denotes the sampling period. The control input of the system $\boldsymbol{\tau}_k = [\tau_{p_k}, \tau_{s_k}, \tau_{v_k}, \tau_{l_k}]^\top \in \mathbb{R}^4$ consists of the thrusters' forces.

3.2.2.3 Constraints

State constraints

We consider that the robot must avoid the obstacles and the workspace boundaries (test tank). Moreover, for the needs of several common underwater tasks (e.g., seabed inspection and mosaicking), the vehicle is required to move with relatively low speeds with upper bound denoted by the velocity vector $\boldsymbol{v}_p = [u_p\ v_p\ w_p\ r_p]^\top$. These requirements are captured by the state constraint set X of the system, given by

$$\boldsymbol{x}_k \in X \subset \mathbb{R}^8 \tag{3.6}$$

which is formed by the following constraints:

$$u_p + v_p - |u_r + v_r| \geq 0 \tag{3.7a}$$

$$w_p - |w_r| \geq 0 \tag{3.7b}$$

$$r_p - |r_r| \geq 0 \tag{3.7c}$$

$$\mathcal{B}(\boldsymbol{\eta}_1(t), \bar{r}) \cap \{\mathcal{B}(\boldsymbol{p}_{\pi_m}, r_{\pi_m}) \cup \partial\mathcal{W}\} = \emptyset, m \in \{1, \ldots, \mathcal{M}\} \tag{3.7d}$$

Input constraints
It is well known the forces generated by the thrusters. Thus, we define the control constraint set T as follows:

$$\boldsymbol{\tau}_k = [\tau_{po_k},\ \tau_{s_k},\ \tau_{ve_k},\ \tau_{l_k}]^\top \in T \subseteq \mathbb{R}^4 \tag{3.8}$$

These constraints are of the forms $|\tau_{po_k}| \leq \bar{\tau}_{po}$, $|\tau_{s_k}| \leq \bar{\tau}_s$, $|\tau_{ve_k}| \leq \bar{\tau}_{ve}$ and $|\tau_{l_k}| \leq \bar{\tau}_l$; therefore, we get $\|\backslash\boldsymbol{\tau}_k\| \leq \bar{T}$, where $\bar{T} = (\tau_{po}^2 + \tau_s^2 + \tau_{ve}^2 + \tau_l^2)^{\frac{1}{2}}$ and $\bar{\tau}_{po}, \bar{\tau}_s, \bar{\tau}_{ve}, \bar{\tau}_l \in \mathbb{R}_{\geq 0}$.

3.2.2.4 Control design

The control objective is to guide the vehicle towards regions around the desired way-points ${}^i\boldsymbol{x}^d \triangleq [({}^i\boldsymbol{\eta}^d)^\top, ({}^i\boldsymbol{v}_r^d)^\top]^\top = [{}^ix_d, {}^iy_d, {}^iz_d, {}^i\psi_d, {}^iu_d, {}^iv_d, {}^iw_d, {}^ir_d]^\top \in X$, $i = \{1, \ldots, n\}$, while respecting the state constraints (3.7a)–(3.7d) as well as the input constraints (3.8). A predictive controller is employed in order to achieve this task. In particular, at a given time instant k, the NMPC is assigned to solve an optimal control problem (OCP) w.r.t. a control sequence $\boldsymbol{\tau}_f(k) \triangleq [\boldsymbol{\tau}(k|k), \boldsymbol{\tau}(k+1|k), \ldots, \boldsymbol{\tau}(k+N-1|k)]$, for a prediction horizon N. The OCP of the NMPC is given as follows:

$$\min_{\boldsymbol{\tau}_f(k)} J_N(\boldsymbol{x}_k, \boldsymbol{\tau}_f(k)) = \min_{\boldsymbol{\tau}_f(k)} \sum_{j=0}^{N-1} F(\hat{\boldsymbol{x}}(k+j|k), \boldsymbol{\tau}(k+j|k)) + E(\hat{\boldsymbol{x}}(k+N|k)) \tag{3.9a}$$

subject to

$$\hat{\boldsymbol{x}}(k+j|k) \in X_j, \qquad \forall j = 1, \ldots, N-1 \tag{3.9b}$$

$$\boldsymbol{\tau}(k+j|k) \in T, \qquad \forall j = 0, \ldots, N-1 \tag{3.9c}$$

$$\hat{\boldsymbol{x}}(k+N|k) \in \mathcal{E}_f \tag{3.9d}$$

where $\mathcal{E}_f$ is the terminal set and F and E are the running and terminal cost functions, respectively. At time instant k, the solution of the OCP (3.9a)–(3.9d) is providing an optimal control sequence, denoted as

$$\boldsymbol{\tau}_f^*(k) = [\boldsymbol{\tau}(k|k), \boldsymbol{\tau}(k+1|k), \ldots, \boldsymbol{\tau}(k+N-1|k)] \tag{3.10}$$

where the first control vector (i.e., $\boldsymbol{\tau}(k|k)$) is applied to the system. Notice we use the double subscript notation for the predicted state of system (3.5) inside the OCP of the NMPC:

$$\hat{\boldsymbol{x}}(k+j|k) = f(\hat{\boldsymbol{x}}(k+j-1|k), \boldsymbol{\tau}(k+j-1|k)) \tag{3.11}$$

where the vector $\hat{\boldsymbol{x}}(k+j|k)$ denotes the predicted state of the system (3.5) at sampling time $k+j$ with $j \in \mathbb{Z}_{\geq 0}$. The predicted state is based on the measurement of the state $\boldsymbol{x}_k$ of the system at sampling time k (i.e., provided by the on-board navigation system), while applying a sequence of control inputs $[\boldsymbol{\tau}(k|k), \boldsymbol{\tau}(k+1|k), \ldots, \boldsymbol{\tau}(k+j-1|k)]$. It holds that $\hat{\boldsymbol{x}}(k|k) \equiv \boldsymbol{x}_k$. The cost function $F(\cdot)$, as well as the terminal cost $E(\cdot)$, are both of quadratic form, i.e., $F(\hat{\boldsymbol{x}}, \boldsymbol{\tau}) = \hat{\boldsymbol{x}}^\top Q\hat{\boldsymbol{x}} + \boldsymbol{\tau}^\top R\boldsymbol{\tau}$ and $E(\hat{\boldsymbol{x}}) = \hat{\boldsymbol{x}}^\top P\hat{\boldsymbol{x}}$, respectively, with P, Q and R being positive definite matrices. Particularly we define $Q = \text{diag}\{q_1, \ldots, q_8\}$, $R = \text{diag}\{r_1, \ldots, r_4\}$ and $P = \text{diag}\{p_1, \ldots, p_8\}$.

3.2.3 Experimental results

This subsection demonstrates the efficacy of the proposed motion control scheme via a set of real-time experiments employing a small underwater robotic vehicle.

3.2.3.1 Setup

The experiments were carried out inside the *NTUA, Control Systems Lab* test tank, with dimensions 5 m × 3 m × 1.5 m (Figure 3.1). The bottom of the tank is covered by a custom-made poster with various visual features and markers. Two cylindrical objects with known position and dimensions are placed inside the tank and considered as static obstacles. The vehicle used in this work is a 4-DoF Seabotix LBV, actuated in Surge, Sway, Heave and Yaw via a four-thruster set configuration. The vehicle is equipped with a down-looking Sony PlayStation Eye camera, with 640 × 480 pixels at 30 frames per second (fps) enclosed in a waterproof housing. An underwater laser pointer projecting a green dot at the bottom of the test tank is rigidly attached on the vehicle with its axes aligned to the down-looking camera axis. The vehicle is also equipped with an *SBG IG – 500A* AHRS, delivering temperature-compensated 3D

Figure 3.1 Experimental setup and problem formulation: the purpose of the controller is to guide the vehicle inside a constrained workspace

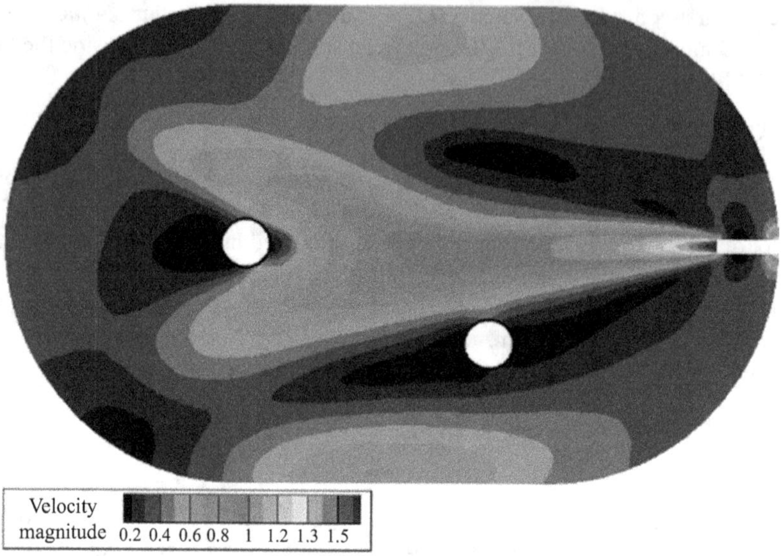

Figure 3.2 Distribution of the flow field inside the experimental water tank as computed by the CFD software presented in [15]

acceleration, angular velocity and orientation measurements at 100 Hz. The marker localization system is based on the *ArUco* library [17] and the overall state vector of the vehicle (3D position, orientation and velocity) is available via a sensor fusion and state estimation module based on the complementary filter notion presented in our previous results [18]. The vehicle's dynamic parameters have been identified via a proper identification scheme. The software implementation of the proposed motion control scheme was conducted in C++ and Python under the Robot Operating System [19]. The disturbances in the form of water currents were induced using a $BTD150$ thruster properly mounted inside the water tank. The generated flow field (i.e., assumed ocean current profile) was computed using a GPU-enabled computational fluid dynamics (CFD) software [15] developed in the Parallel CFD and Optimization Unit of the School of Mechanical Engineering of NTUA. The flow field distribution inside the water tank is depicted in Figure 3.2.

3.2.3.2 Results

In order to prove the efficacy of the proposed controller, two experiments are presented (Scenarios A and B). In both experiments, the objective is to follow a set of predefined way-points while simultaneously avoiding two static obstacles and respecting the workspace (test tank) boundaries. The location and geometry of the obstacles are considered known. More specific, the position of the obstacles w.r.t. the Inertial Frame $\mathcal{I}$ in x–y plane is given by $obs_1 = [-0.625, \ -0.625]$, $obs_2 = [0.9375, \ 0]$. The state constraints of (3.7a)–(3.7d) which must be satisfied during all the experiments are

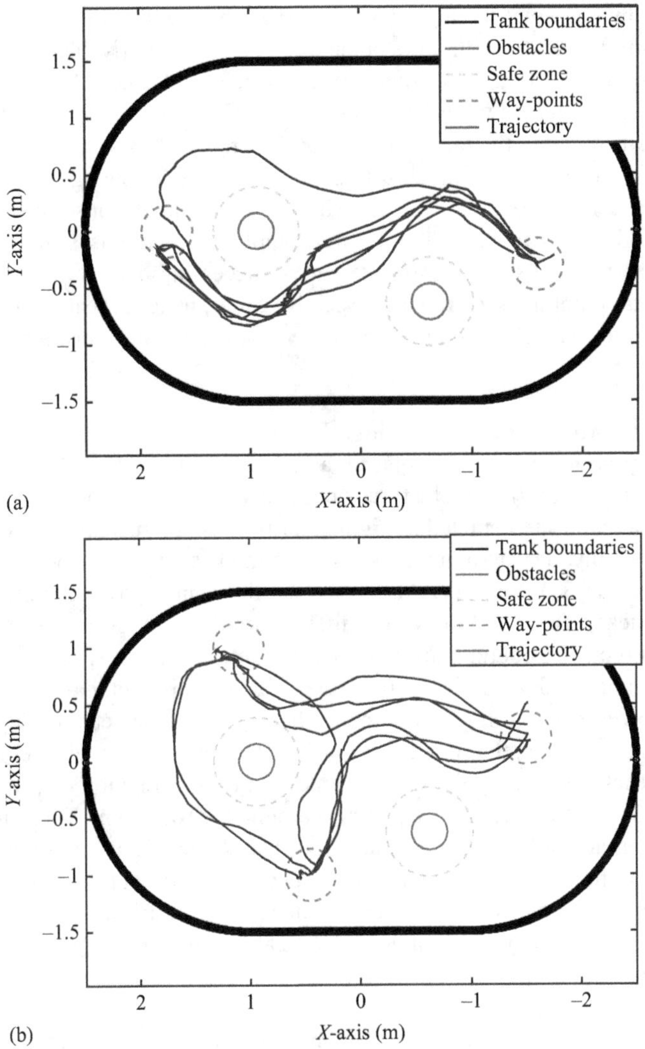

Figure 3.3 Vehicle trajectory in horizontal plane: (a) two-way-point tracking scenario and (b)three-way-point tracking scenario

analytically formulated as follows: (i) The obstacles are cylinders (see Figure 3.1) with radius $r_{\pi_i} = 0.16\,\text{m}$, $i = \{1, 2\}$, and are modeled together with the workspace boundaries according to the spherical world representations as consecutive spheres. (ii) The radius of the sphere $\mathcal{B}(\boldsymbol{\eta}_1, \bar{r})$ which covers all the vehicle volume (i.e., main body and additional equipment) is defined as $\bar{r} = 0.3\,\text{m}$. However, for the clarity of presentation, we depict it as a safe zone around the obstacles where the vehicle center

η_1 (denoted by blue line; see Figure 3.3(a) and 3.3(b)) should not violate it. (iii) The vertical position must be within the range of $0 < z < 1.2$ m. (iv) The vehicle's body velocity norm of (3.7a) $|u_r + v_r|$ (planar motion) must not exceed 0.5 m/s. (v) Heave velocity must be retained within the range of $-0.25 < w_r < 0.25$. (vi) Yaw velocity must be retained within the range of $-1 < r_r < 1$ rad/s.

Moreover, each of the four thrusters must obey the following input constraint: $-12 < \tau_i < 12\,\mathrm{N}$, $i = \{p_o,\ s,\ v_e,\ l\}$. The state and input constraints in figures 3.4–3.7 are depicted in dashed lines were applicable. At this point, we should mention that each mission is considered as successful only if the vehicle performs the way-point tracking three consecutive times, hence repeatability is proved. In all times, the vehicle is under the influence of the water currents depicted in Figure 3.2.

Scenario A: Two-way-point tracking. In this scenario, the vehicle must travel via two way-points which are placed at $\eta_1^d = [-1.6, -0.35, 0.45, 0]$ and $\eta_2^d = [1.75, 0, 0.3, \pi]$, respectively. The three consecutive trajectories of the vehicle along the horizontal plane are depicted in Figure 3.3(a). It can be seen that the vehicle performs successfully the way-point tracking while safely avoiding the obstacles and the test tank boundaries. We observe that in one case the vehicle traveled from the second way-point back to first one following a different trajectory. This can be explained by the fact that the MPC found a different optimal solution at the specific time frame, due to the unmodeled dynamics of the tether which significantly affects the vehicle motion. The vertical and angular motion of the vehicle are depicted in Figure 3.4(a), where it can be seen that the state constraints are always satisfied. The vehicle is considered to reach each way-point if it has entered a terminal region $\mathscr{E}_f$ (i.e., spherical region of 0.3 m and an offset of ± 0.15 rad) around the way-point. These regions are depicted in circles in Figures 3.3(a) and 3.4(a). In Figure 3.5(a), the body velocity norm in planar motion is depicted and the respective constraint is satisfied. The same stands for the heave and yaw velocities as shown in Figure 3.6(a). In Figure 3.7(a), the vehicle's thruster inputs are shown. As it can be seen, the input constraints are also satisfied.

Scenario B: Three-way-point tracking. This scenario is similar to the previous one except that the vehicle must travel along three way-points, which makes the mission more challenging considering the narrow workspace. The locations of the three way-points are given by $\eta_1^d = [-1.50\,\mathrm{m}, 0.30\,\mathrm{m}, 0.40\,\mathrm{m}, -\frac{\pi}{2}\,\mathrm{rad}]$, $\eta_2^d = [0.45, -1, 0.25, 0]$ and $\eta_3^d = [1.2, 1, 0.3, -\pi]$. The three consecutive trajectories of the vehicle along the horizontal plane are depicted in Figure 3.3(b). Although this scenario is more complicated, the vehicle again carries it out successfully. As it can be observed, the vehicle also in this mission follows different trajectories, for the same reasons explained in Scenario A. The vertical and angular motion of the vehicle are depicted in Figure 3.4(b), while in Figure 3.5(b) the body velocity norm in planar motion is shown. The heave and yaw velocities are presented in Figure 3.6(b), while in Figure 3.7(b) the thruster inputs are shown. As it can be observed, the vehicle achieved all desired way-points while simultaneously satisfied all respective state and input constraints.

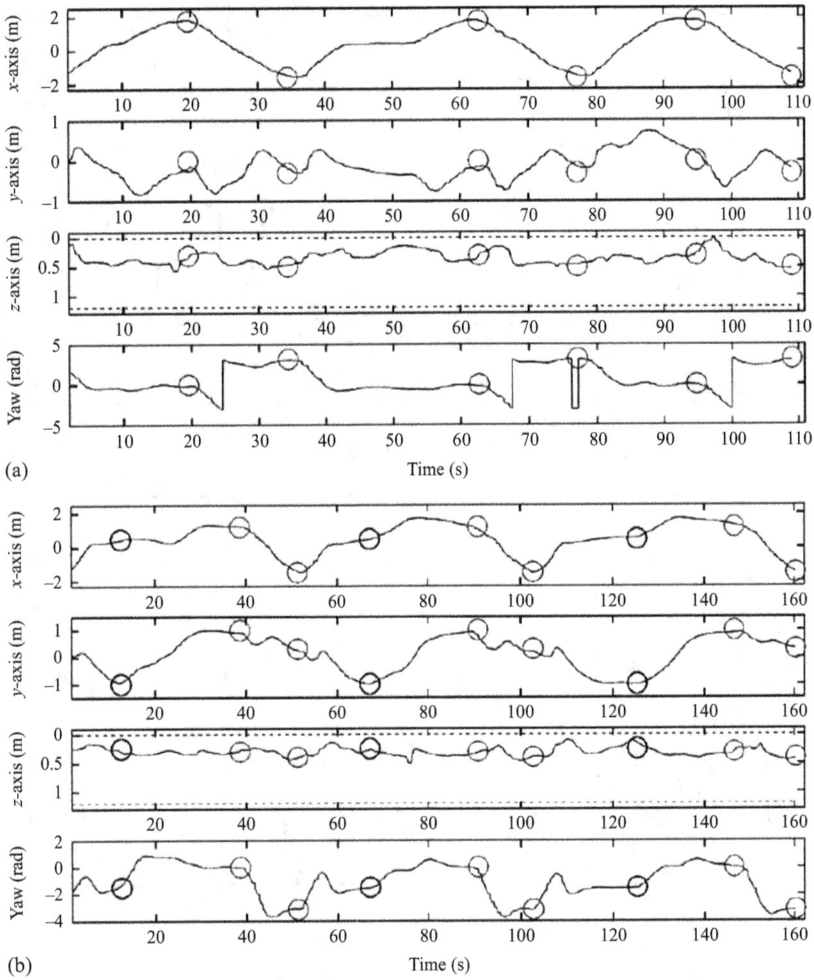

Figure 3.4 Vehicle vertical and angular motion: (a) two-way-point tracking scenario and (b) three-way-point tracking scenario

3.3 Model-free control for underwater vehicles

3.3.1 Fully actuated underwater vehicles

3.3.1.1 AUV kinematics and dynamics

In this section, we consider a simplified 3D dynamic model (in surge, sway, heave and yaw) of an AUV in accordance to the standard underwater vehicle modeling properties [1]. The roll and pitch degrees of freedom are neglected for the clarity of the presentation. However, the stability of both degrees of freedom can be secured owing to their input-to-state stability properties, without compromising the achieved results. In this respect, consider the vehicle modeled as a rigid body subject to external forces

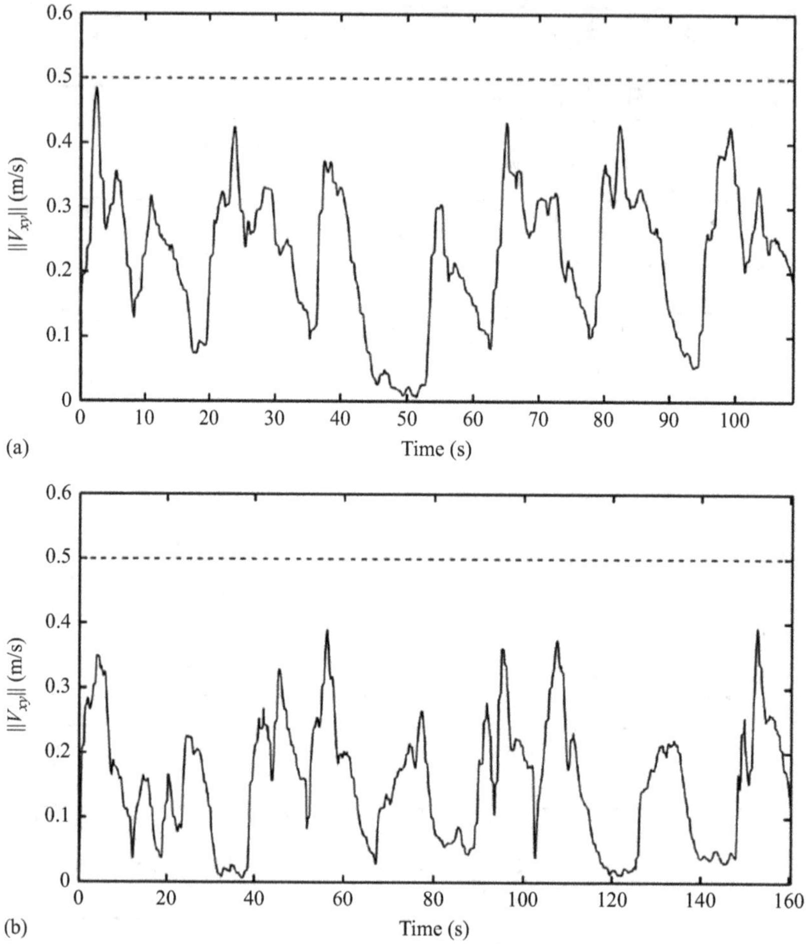

Figure 3.5 Vehicle body velocity norm $|u_r + v_r|$: (a) two-way-point tracking scenario and (b) three-way-point tracking scenario

and torques. Let $\{I\}$ be an inertial coordinate frame and $\{B\}$ a body-fixed coordinate frame, whose origin O_B is located at the center of mass of the vehicle. Furthermore, let (x, y, z) be the position of O_B in $\{I\}$ and ψ denote the yaw angle. Let (u, v, w) be the longitudinal (surge), transverse (sway) and vertical (heave) velocities of O_B w.r.t. $\{I\}$ expressed in $\{B\}$ and r be the vehicle's angular speed (yaw) around the vertical axis. Thus, the kinematic equations of motion for the considered vehicle can be written as

$$\dot{x} = u\cos\psi - v\sin\psi + \delta_x(t) \tag{3.12}$$

$$\dot{y} = u\sin\psi + v\cos\psi + \delta_y(t) \tag{3.13}$$

$$\dot{z} = w + \delta_z(t) \tag{3.14}$$

$$\dot{\psi} = r + \delta_\psi(t) \tag{3.15}$$

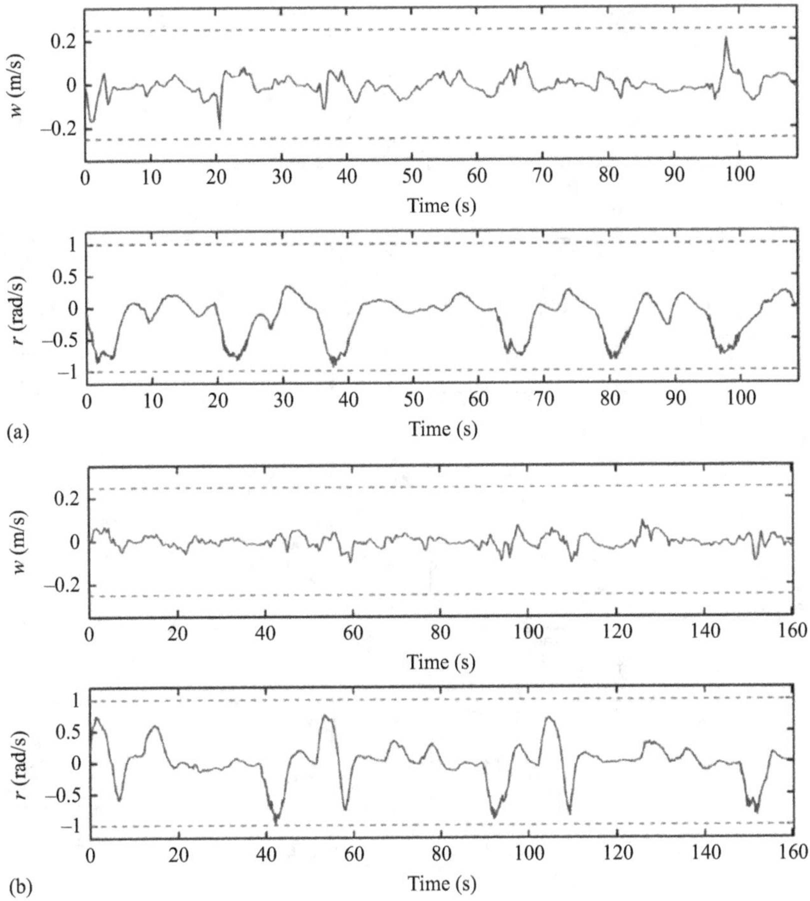

Figure 3.6 Vehicle heave and yaw velocities: (a) two-way-point tracking scenario and (b) three-way-point tracking scenario

where $\delta_x(t)$, $\delta_y(t)$, $\delta_z(t)$ and $\delta_\psi(t)$ denote bounded ocean currents. Neglecting the motion in roll and pitch, the simplified equations for the surge, sway, heave and yaw can be written as

$$m_u\dot{u} = m_v vr + X_u u + X_{|u|u}|u|u + X + \delta_u(t) \tag{3.16}$$

$$m_v\dot{v} = -m_u ur + Y_v v + Y_{|v|v}|v|v + Y + \delta_v(t) \tag{3.17}$$

$$m_w\dot{w} = Z_w w + Z_{|w|w}|w|w + (W - B) + Z + \delta_w(t) \tag{3.18}$$

$$m_r\dot{r} = (m_u - m_v)uv + N_r r + N_{|r|r}|r|r + N + \delta_r(t) \tag{3.19}$$

where m_u, m_v, m_w and m_r denote the vehicle's mass, moment of inertia, added mass and moment of inertia terms, respectively; X_u, $X_{|u|u}$, Y_v, $Y_{|v|v}$, Z_w, $Z_{|w|w}$, N_r and $N_{|r|r}$ are negative hydrodynamic damping coefficients of first and second order; W and B

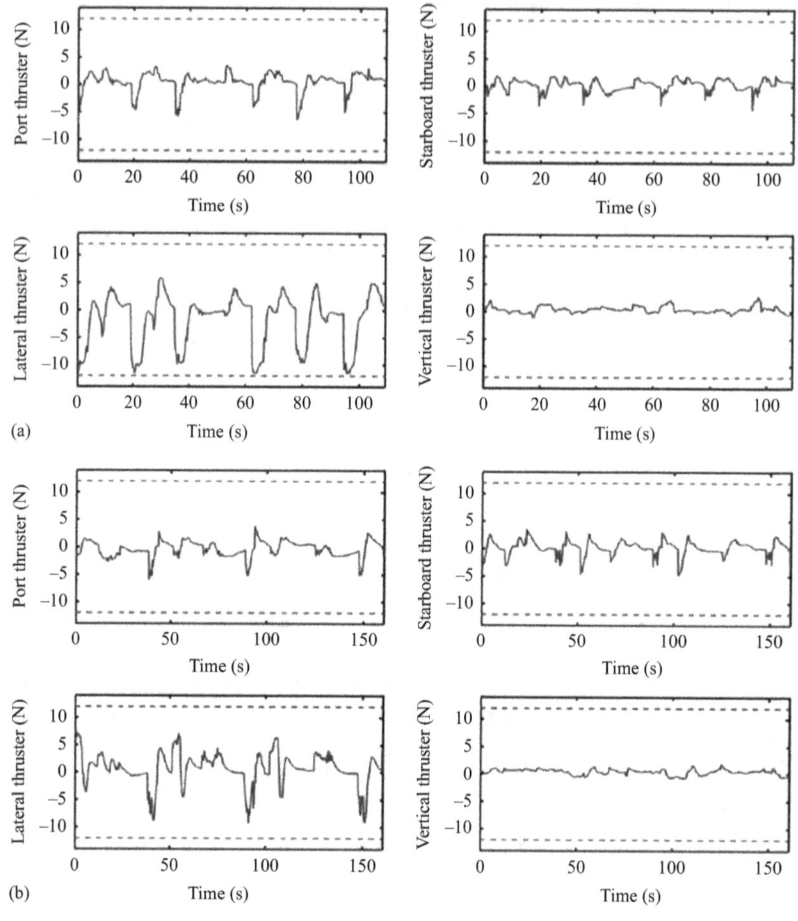

Figure 3.7 Thruster commands: (a) two-way-point tracking scenario and (b) three-way-point tracking scenario

are the vehicle weight and buoyancy, respectively; $\delta_u(t)$, $\delta_v(t)$, $\delta_w(t)$ and $\delta_r(t)$ denote bounded exogenous forces and torques acting on surge, sway, heave and around yaw owing to ocean waves; and X, Y, Z and N denote the control input forces and torque, respectively, that are applied by the thrusters in order to produce the desired motion of the body-fixed frame.

3.3.1.2 Control scheme

Let $x_d(t)$, $y_d(t)$, $z_d(t)$ and $\psi_d(t)$ denote a desired smooth and bounded trajectory as well as yaw orientation, respectively. The objective is to design a controller, without incorporating any information regarding the vehicle model, such that (a) it tracks the desired trajectory with the desired orientation via bounded closed-loop signals and

(b) it achieves prescribed transient and steady-state performance, despite the presence of exogenous disturbances representing ocean currents and waves.

Let us define the position errors:

$$e_x = x - x_d,\ e_y = y - y_d,\ e_z = z - z_d \tag{3.20}$$

as well as the orientation error:

$$e_\psi = \psi - \psi_d \tag{3.21}$$

Given the desired trajectories $x_d(t)$, $y_d(t)$ and $z_d(t)$ and orientation $\psi_d(t)$ as well as the position/orientation errors (3.20) and (3.21), we proceed as follows:

Kinematic controller

Select the exponentially decaying position/orientation performance functions $\rho_x(t)$, $\rho_y(t)$, $\rho_z(t)$ and $\rho_\psi(t)$ that (i) satisfy

a. $\lvert e_x(0)\rvert < \rho_x(0)$	$0 < \rho_x(t)$	$0 < \lim_{t\to\infty} \rho_x(t)$
b. $\lvert e_y(0)\rvert < \rho_y(0)$	$0 < \rho_y(t)$	$0 < \lim_{t\to\infty} \rho_y(t)$
c. $\lvert e_z(0)\rvert < \rho_z(0)$	$0 < \rho_z(t)$	$0 < \lim_{t\to\infty} \rho_z(t)$
d. $\lvert e_\psi(0)\rvert < \rho_\psi(0)$	$0 < \rho_\psi(t)$	$0 < \lim_{t\to\infty} \rho_\psi(t)$

and (ii) incorporate the desired performance specifications regarding the steady-state error and the speed of convergence, and design the desired velocities:

$$\begin{bmatrix} u_d \\ v_d \\ w_d \end{bmatrix} = \begin{bmatrix} \cos\psi & -\sin\psi & 0 \\ \sin\psi & \cos\psi & 0 \\ 0 & 0 & 1 \end{bmatrix}^{-1} \begin{bmatrix} -k_x \frac{e_x}{\rho_x(t)} \\ -k_y \frac{e_y}{\rho_y(t)} \\ -k_z \frac{e_z}{\rho_z(t)} \end{bmatrix} \tag{3.22}$$

$$r_d = -k_\psi \frac{e_\psi}{\rho_\psi(t)} \tag{3.23}$$

with positive control gains k_x, k_y, k_z and k_ψ.

Dynamic controller

Select exponentially decreasing velocity performance functions $\rho_u(t)$, $\rho_v(t)$, $\rho_w(n)$, $\rho_r(t)$ that satisfy

a. $\lvert u(0) - u_d(0)\rvert < \rho_u(0)$	$0 < \rho_u(t)$	$0 < \lim_{t\to\infty} \rho_u(t)$
b. $\lvert v(0) - v_d(0)\rvert < \rho_v(0)$	$0 < \rho_v(t)$	$0 < \lim_{t\to\infty} \rho_v(t)$
c. $\lvert w(0) - w_d(0)\rvert < \rho_w(0)$	$0 < \rho_w(t)$	$0 < \lim_{t\to\infty} \rho_w(t)$
d. $\lvert r(0) - r_d(0)\rvert < \rho_r(0)$	$0 < \rho_r(t)$	$0 < \lim_{t\to\infty} \rho_r(t)$

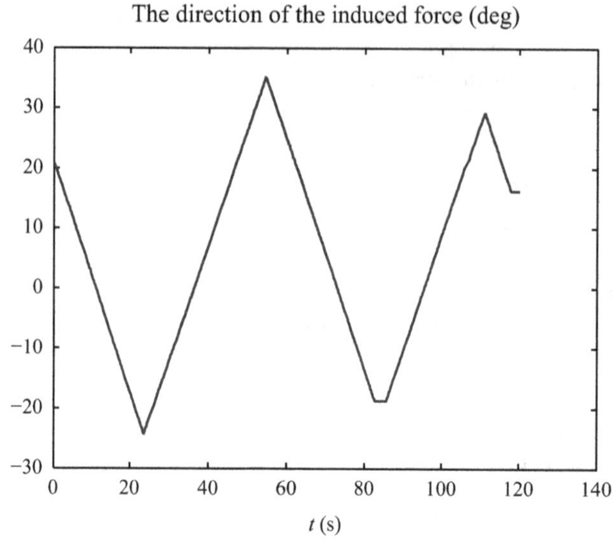

Figure 3.8 Direction of the external disturbance induced by the water jet during the navigation and stabilization towards a specific configuration

and design the external forces in the surge, sway and heave as well as the external torque around yaw as

$$X = -k_u \frac{u - u_d}{\rho_u(t)},\ Y = -k_v \frac{u - u_d}{\rho_u(t)},\ Z = -k_w \frac{w - w_d}{\rho_w(t)},\ N = -k_r \frac{r - r_d}{\rho_r(n)} \tag{3.24}$$

with positive control gains k_u, k_v, k_w and k_r.

3.3.1.3 Experiments

To illustrate the performance of the proposed control scheme, two experimental procedures were carried out: (a) navigation and stabilization to a specific configuration and (b) meandrus-like trajectory tracking. The experiments took place inside a water tank using the Girona500 AUV. A panel consisting of valves and handles located inside the pool was used as the visual target during the experiments, and a navigation module was responsible for providing the state feedback to the closed-loop system in both experiments. In order to create strong external disturbances to the system, a water jet mechanism was installed on the AUV 4-DoF robotic manipulator. The water jet was installed on the second rotational DoF of the manipulator which was moving constantly in both experiments in order to create time-varying and multidirectional external disturbances, as shown in Figures 3.8 and 3.9.

Navigation and stabilization towards a specific configuration

In this experiment, the vehicle starts from an arbitrary initial configuration in order to navigate and stabilize towards a specific configuration (in front of a control panel).

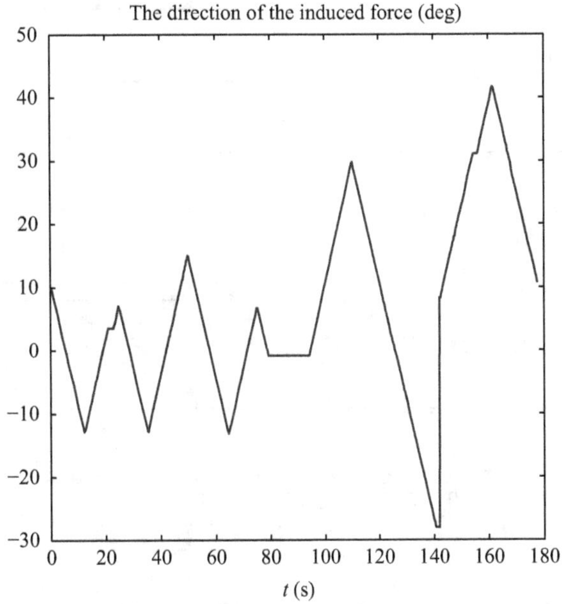

Figure 3.9 Direction of the external disturbance induced by the water jet during the meandrus-like trajectory tracking

During the motion, the vehicle is constantly under the effect of time-varying and multidirectional external disturbances caused by the water jet mechanism. The required transient and steady-state specifications (i.e., maximum steady-state position errors 0.05 m, maximum steady-state orientation error 5° and exponential convergence $e^{-0.1t}$) are described by the following performance functions: $\rho_x(t) = (4.5 - 0.05)e^{-0.1t} + 0.05$, $\rho_y(t) = (3.0 - 0.05)e^{-0.1t} + 0.05$, $\rho_z(t) = (1 - 0.05)e^{-0.1t} + 0.05$, $\rho_\psi(t) = (85 - 5)e^{-0.1t} + 5$, $\rho_u(t) = (1 - 0.1)e^{-0.1t} + 0.1$, $\rho_v(t) = (1 - 0.1)e^{-0.1t} + 0.1$, $\rho_w(t) = (1 - 0.1)e^{-0.1t} + 0.1$ and $\rho_r(t) = (1 - 0.1)e^{-0.1t} + 0.1$. Finally, the control gains were chosen as follows: $k_x = 1$, $k_y = 0.5$, $k_z = 0.5$, $k_\psi = 0.5$, $k_u = 15$, $k_v = 7.5$, $k_w = 15$ and $k_r = 7.5$.

The tracking error evolution is depicted in Figure 3.10. The dashed lines indicate the desired performance bounds and the solid lines indicate the evolution of the tracking errors $e_x(t)$, $e_y(n)$, $e_z(t)$ and $e_\psi(t)$, respectively. All the errors have met the transient and steady-state specifications imposed by the previously selected performance functions. The 3D trajectory of the vehicle is shown in Figure 3.11. As it can be seen from the results, the control objective has been achieved under the influence of external disturbances.

Meandrus-like trajectory tracking

In this experiment, the vehicle starts from an arbitrary initial configuration in order to perform a meandrus-like trajectory tracking under the effect of time-varying and

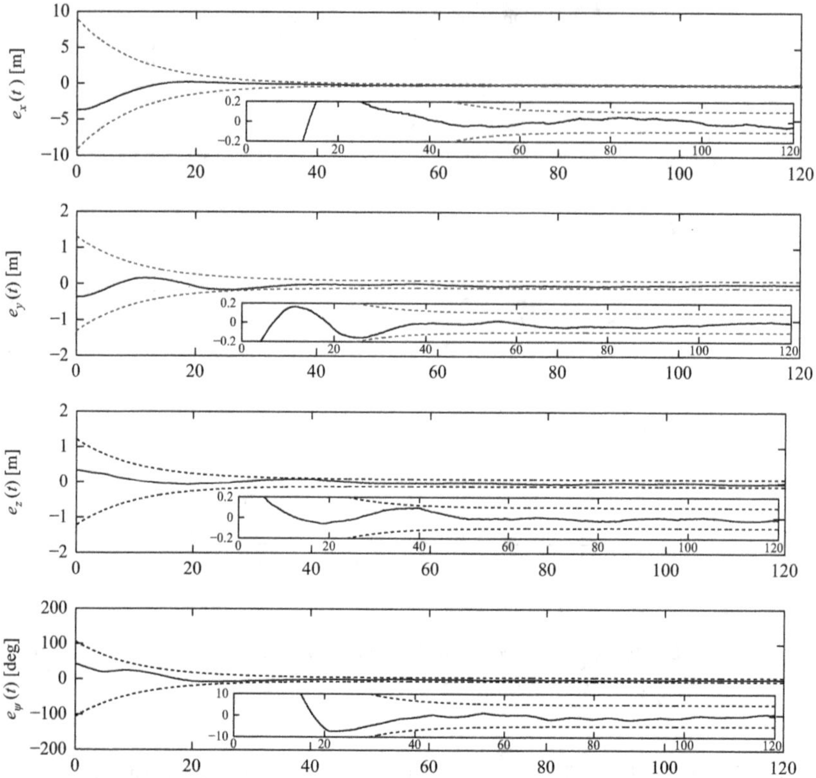

Figure 3.10 Tracking error evolution during navigation and stabilization towards a specific configuration. The dashed lines indicate the desired performance bounds. The solid lines indicate the evolution of $e_x(t)$, $e_y(t)$, $e_z(t)$ and $e_\psi(t)$

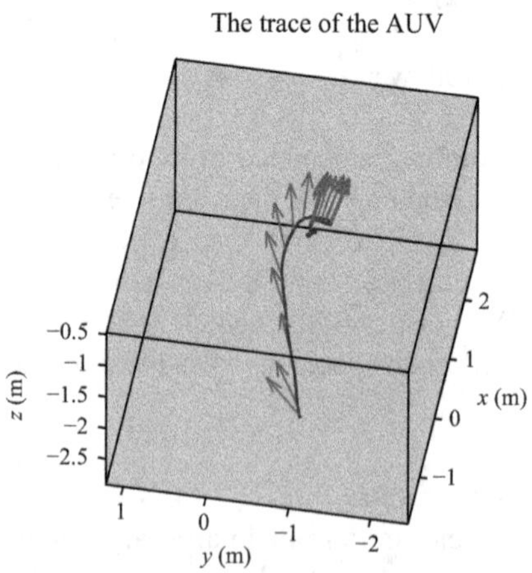

Figure 3.11 Three-dimensional trajectory of the AUV while performing navigation and stabilization towards a specific configuration

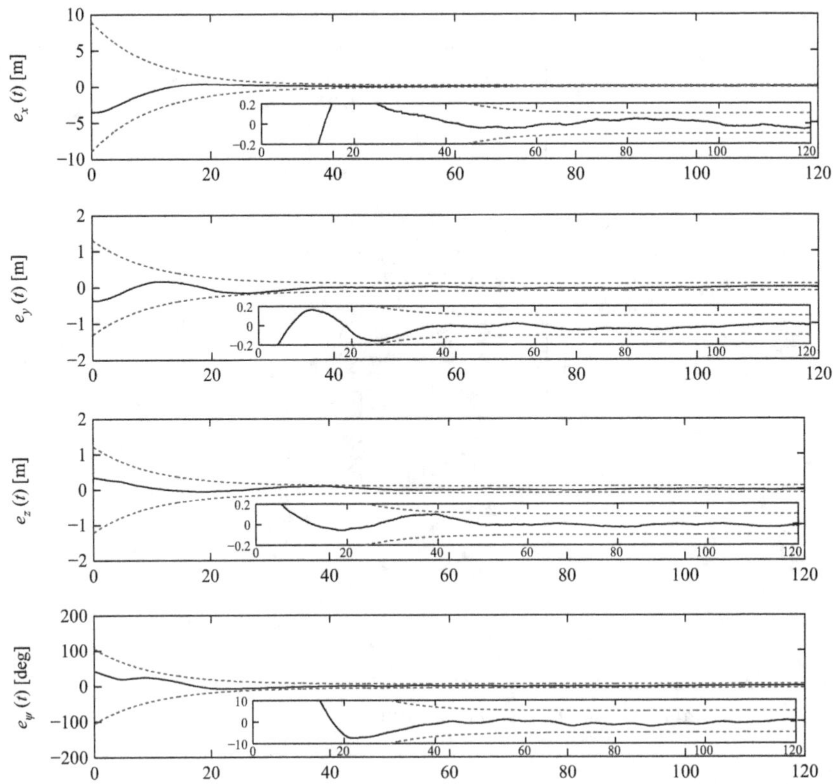

Figure 3.12 Tracking error evolution during meandrus-like trajectory. The dashed lines indicate the desired performance bounds. The solid lines indicate the evolution of $e_x(t)$, $e_y(t)$, $e_z(t)$ and $e_\psi(t)$

multidirectional external disturbances. The required transient and steady-state specifications are the same with the previous experiment. Therefore, we employed the same performance functions and control gains in the control scheme.

The tracking error evolution during the steady state is depicted in Figure 3.12. The dashed lines indicate the desired performance bounds and the solid lines indicate the evolution of the tracking errors $e_x(t)$, $e_y(t)$, $e_z(t)$ and $e_\psi(n)$, respectively. The 3D trajectory of the vehicle is shown in Figure 3.13. The gray line depicts the desired trajectory, while the bold one depicts the actual performed by the AUV. As it is demonstrated by the experiments and predicted from the theoretical analysis, the control objective has been achieved in the presence of external disturbances.

3.3.2 Underactuated underwater vehicles: Case I

A simplified 3D dynamic model of an underwater robotic vehicle is presented in this section and a rigorous formulation of the tracking problem of a desired trajectory is

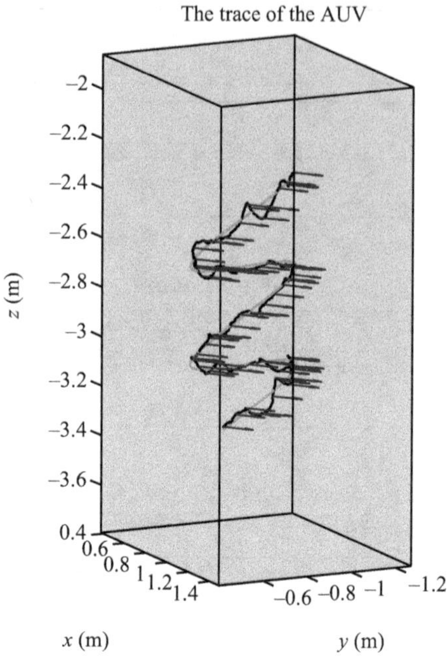

Figure 3.13 Three-dimensional trajectory of the AUV while performing the meandrus-like trajectory tracking

introduced. The vehicle has two identical rear thrusters, mounted symmetrically w.r.t. its longitudinal axis as well as a thruster mounted along the vertical axis. The common and the differential action modes of the rear thrusters result in a force X along the vehicle's longitudinal axis and a torque N about its vertical axis, respectively, whereas the action of the vertical thruster results in a force Z along the corresponding axis. The aforementioned forces X, Z and torque N define the input control variables of the dynamic system. In this respect, the vehicle is underactuated since it lacks a lateral thruster. Finally, the roll and pitch degrees of freedom are neglected for the clarity of the presentation. However, both degrees of freedom are stable and do not affect the analysis.

3.3.2.1 Vehicle kinematics and dynamics

Consider an underactuated underwater vehicle modeled as a rigid body subject to external forces and torques. Let $\{I\}$ be an inertial coordinate frame and $\{B\}$ a body-fixed coordinate frame, whose origin O_B is located at the center of mass of the vehicle. Furthermore, let (x, y, z) be the position of O_B in $\{I\}$ and ψ denote the yaw angle. Let (u, v, w) be the longitudinal (surge), transverse (sway) and vertical (heave) velocities of O_B w.r.t. $\{I\}$ expressed in $\{B\}$ and r be the vehicle's angular speed (yaw) around the vertical axis. Hence, the kinematic equations of motion for the considered vehicle can be written as

$$\dot{x} = u\cos\psi - v\sin\psi + \delta_x(t) \tag{3.25}$$

$$\dot{y} = u\sin\psi + v\cos\psi + \delta_y(t) \tag{3.26}$$

$$\dot{z} = w + \delta_z(t) \tag{3.27}$$

$$\dot{\psi} = r \tag{3.28}$$

where $\delta_x(t)$, $\delta_y(t)$ and $\delta_z(t)$ denote bounded ocean currents. Neglecting the motion in roll and pitch, the simplified equations for the surge, sway, heave and yaw can be written as

$$m_u\dot{u} = m_v vr + X_u u + X_{|u|u}|u|u + X + \delta_u(t) \tag{3.29}$$

$$m_v\dot{v} = -m_u ur + Y_v v + Y_{|v|v}|v|v + \delta_v(t) \tag{3.30}$$

$$m_w\dot{w} = Z_w w + Z_{|w|w}|w|w + Z + \delta_w(t) \tag{3.31}$$

$$m_r\dot{r} = (m_u - m_v)uv + N_r r + N_{|r|r}|r|r + N + \delta_r(t) \tag{3.32}$$

where m_u, m_v, m_w and m_r denote the vehicle's mass, moment of inertia, added mass and moment of inertia terms, X_u, $X_{|u|u}$, Y_v, $Y_{|v|v}$, Z_w, $Z_{|w|w}$ and N_r, $N_{|r|r}$ are negative hydrodynamic damping coefficients of first and second order, $\delta_u(t)$, $\delta_v(t)$, $\delta_w(t)$ and $\delta_r(t)$ denote bounded exogenous forces and torques acting on surge, sway, heave and yaw owing to ocean waves and X, Z and N denote the control input forces and torque, respectively, that are applied by the thrusters in order to produce the desired motion of the body-fixed frame. Finally, notice that there is no lateral thruster to control the sway motion. Hence, (3.30) is uncontrolled making thus (3.25)–(3.32) an underactuated dynamical system. It is important to stress that the dynamics of the sway velocity v must be explicitly taken into account since the presence of the sway velocity in the kinematics is not negligible contrary to the case of wheeled mobile robots.

3.3.2.2 Error coordinates

Let $x_d(t)$, $y_d(t)$ and $z_d(t)$ denote a given smooth bounded desired trajectory with bounded time derivatives. The objective of this work is to design a controller without incorporating any information regarding the vehicle model such that it tracks the desired trajectory with bounded closed-loop signals and prescribed performance (concerning the steady-state error and the speed of convergence) despite the presence of exogenous disturbances representing ocean currents and waves. Let us now define the position errors

$$e_x = x - x_d(t),\ e_y = y - y_d(t),\ e_z = z - z_d(t) \tag{3.33}$$

the projected error on the horizontal plane

$$e_d = \sqrt{e_x^2 + e_y^2} \tag{3.34}$$

as well as the projected on the horizontal plane orientation error

$$e_o = -\frac{e_x}{e_d}\sin(\psi) + \frac{e_y}{e_d}\cos(\psi) = \sin(\psi_e) \tag{3.35}$$

where ψ_e is the angle measured from the longitudinal axis of the vehicle defined by the 2D vector $[\cos(\psi),\ \sin(\psi)]^T$ to the vector $\left[-\frac{e_x}{e_d},\ -\frac{e_y}{e_d}\right]^T$. Equation (3.35) can be easily verified if we consider the outer product of the unity vectors $\left[-\frac{e_x}{e_d},\ -\frac{e_y}{e_d}\right]^T$ and $[\cos(\psi),\ \sin(\psi)]^T$ which equals to the sinus of the angle the aforementioned vectors define. Hence, the tracking problem stated above is solved if the projected error on the horizontal plane e_d, the vertical error e_z and the orientation error e_o reduce to zero (i.e., the vehicle is heading to the desired trajectory). However, notice that the orientation error e_o is well defined only for nonzero values of e_d, since for $e_d = 0$ the angle ψ_e is unidentified. Therefore, in this work, we will design a controller that also guarantees $e_d(t) > \underline{\rho}_d > 0, \forall t \geq 0$, with $\underline{\rho}_d$ an arbitrarily small positive design constant, to avoid the aforementioned singularity issue.

Differentiating (3.34) and (3.35) to derive the kinematics of the underwater vehicle in the error coordinates, we obtain

$$\begin{aligned}\dot{e}_d &= -u\cos(\psi_e) - v\sin(\psi_e) - (\delta_x(t) - \dot{x}_d(t))\cos(\psi + \psi_e)\\ &\quad -(\delta_y(t) - \dot{y}_d(t))\sin(\psi + \psi_e) && (3.36)\\ \dot{e}_z &= w + \delta_z(t) - \dot{z}_d(t) && (3.37)\\ \dot{e}_o &= -r\cos(\psi_e) + u\frac{e_o\cos(\psi_e)}{e_d} - v\frac{1-e_o^2}{e_d} + (\delta_x(t) - \dot{x}_d(t))\frac{\sin(\psi) + e_o\cos(\psi + \psi_e)}{e_d}\\ &\quad - \left(\delta_y(t) - \dot{y}_d(t)\right)\frac{\cos(\psi) - e_o\sin(\psi + \psi_e)}{e_d} && (3.38)\end{aligned}$$

where the following equalities have been utilized:

$$\begin{aligned}e_x &= -e_d\cos(\psi + \psi_e)\\ e_y &= -e_d\sin(\psi + \psi_e)\end{aligned}$$

Finally, to solve the tracking problem stated above, we assume that the initial projected error on the horizontal plane $e_d(0)$ and orientation heading $\psi_e(0)$ satisfy (a) $e_d(0) > \underline{\rho}_d$ and (b) $|\psi_e(0)| < \frac{\pi}{2}$.

3.3.2.3 Control scheme

Given a smooth bounded desired trajectory $x_d(t)$, $y_d(t)$ and $z_d(t)$ with bounded time derivatives and any initial system configuration satisfying Assumption 1 for an arbitrarily small positive design constant $\underline{\rho}_d$, we proceed as follows:

Kinematic controller

Select position/orientation performance functions $\rho_d(t)$, $\rho_z(t)$ and $\rho_o(t)$ that (i) satisfy

a. $e_d(0) < \rho_d(0)$	$\underline{\rho}_d < \rho_d(t)$	$\underline{\rho}_d < \lim_{t\to\infty} \rho_d(t)$
b. $\lvert e_z(0)\rvert < \rho_z(0)$	$0 < \rho_z(t)$	$0 < \lim_{t\to\infty} \rho_z(t)$
c. $\lvert e_o(0)\rvert < \rho_o(0) < 1$	$0 < \rho_o(t)$	$0 < \lim_{t\to\infty} \rho_o(t)$

and (ii) incorporate the desired performance specifications regarding the steady-state error and the speed of convergence, and design the desired velocities

$$u_d = k_d \ln\left(\frac{1+\frac{e_d - \frac{\rho_d(t)+\underline{\rho}_d}{2}}{\frac{\rho_d(t)-\underline{\rho}_d}{2}}}{1-\frac{e_d - \frac{\rho_d(t)+\underline{\rho}_d}{2}}{\frac{\rho_d(t)-\underline{\rho}_d}{2}}}\right), \quad w_d = -k_z \ln\left(\frac{1+\frac{e_z}{\rho_z(t)}}{1-\frac{e_z}{\rho_z(t)}}\right), \quad r_d = k_o \ln\left(\frac{1+\frac{e_o}{\rho_o(t)}}{1-\frac{e_o}{\rho_o(t)}}\right) \tag{3.39}$$

with positive control gains k_d, k_z and k_o.

Dynamic controller

Select velocity performance functions $\rho_u(t)$, $\rho_w(t)$ and $\rho_r(t)$ that satisfy

a. $\lvert u(0) - u_d(0)\rvert < \rho_u(0)$	$0 < \rho_u(t)$	$0 < \lim_{t\to\infty} \rho_u(t)$
b. $\lvert w(0) - w_d(0)\rvert < \rho_w(0)$	$0 < \rho_w(t)$	$0 < \lim_{t\to\infty} \rho_w(t)$
c. $\lvert r(0) - r_d(0)\rvert < \rho_r(0)$	$0 < \rho_r(t)$	$0 < \lim_{t\to\infty} \rho_r(t)$

and design the external forces in the surge and heave as well as the external torque in yaw as

$$X = -k_u \ln\left(\frac{1+\frac{u-u_d}{\rho_u(t)}}{1-\frac{u-u_d}{\rho_u(t)}}\right), \quad Z = -k_w \ln\left(\frac{1+\frac{w-w_d}{\rho_w(t)}}{1-\frac{w-w_d}{\rho_w(t)}}\right),$$

$$N = -k_r \ln\left(\frac{1+\frac{r-r_d}{\rho_r(t)}}{1-\frac{r-r_d}{\rho_r(t)}}\right) \tag{3.40}$$

with positive control gains k_u, k_w and k_r.

3.3.2.4 Experiments

To illustrate the tracking performance and the robustness of the proposed scheme, an experimental procedure was carried out. The experiments took place inside a water tank using a small underwater vehicle.

System components

The ROV used (VideoRay PRO, VideoRay LLC) is equipped with three thrusters, affecting only the surge, heave and yaw motion. Hence, the vehicle is underactuated along the sway axis. The angles ϕ and θ and angular velocities p and q are negligible. The ROV is symmetric about x–z plane and close to symmetric about y–z plane.

Therefore, we can safely assume that motions in heave, roll and pitch are decoupled [1]. Although the vehicle is not symmetric about x–y plane, heave motion can be considered decoupled from surge and sway because the vehicle is operating at relative low speeds, where coupling effects are considered to be negligible. Thus, the ROV used for the experimental procedure meets the required properties. The vehicle is also equipped with a control unit and a CCD camera. The control unit is connected with a personal computer (PC) through a serial communication interface (RS-232). A position sensor (Polhemus, Isotrak) was used as pose feedback sensor for the motion control scheme. The Isotrak tracking system uses electromagnetic fields to determine the position and orientation of a remote tracker. The Isotrak interfaces to the host computer via RS-232 serial communication at 30 Hz. The Isotrak system consists of a transmitter and a receiver. The transmitter is placed at a fixed position outside the tank and the receiver (tracker) is placed on the vehicle. The system software is implemented in C++. The operating system is Linux.

Results

The ROV starts at rest from initial configuration $x(0) = 0.15$ m, $y(0) = 0.00$ m, $z(0) = 0.25$ m, $\psi(0) = -120^\circ$ and is requested to track the trajectory $x_d(t) = -0.4\sin(0.1\pi t)$, $y_d(n) = -0.4\cos(0.1\pi t)$, $z_d(t) = 0.3 + 0.015t$, with maximum steady-state errors $e_{d_{\max}} = 0.10$, $e_{z_{\max}} = 0.05$, $e_{o_{\max}} = 0.05$ and minimum convergence rate as obtained by the exponential $e^{-0.25t}$. Notice that the aforementioned initial configuration satisfies Assumption 1 for $\underline{\rho}_d = 0.025$. Hence, we selected the performance functions $\rho_d(t) = (0.52 - 0.1)e^{-0.25t} + 0.1$, $\rho_z(t) = (0.7 - 0.05)e^{-0.25t} + 0.05$, $\rho_o(t) = (0.7 - 0.05)e^{-0.25t} + 0.05$, $\rho_u(t) = (2 - 0.1)e^{-0.25t} + 0.1$, $\rho_w(t) = (2 - 0.1)e^{-0.25t} + 0.1$, $\rho_r(t) = (2 - 0.1)e^{-0.25t} + 0.1$ and the control gains $k_d = 0.3$, $k_z = 1$, $k_o = 1$, $k_u = 7.5$, $k_w = 5.0$, $k_r = 0.25$ such that (3.40) produce reasonable control effort.

As it was predicted by the theoretical analysis and is actually depicted in Figure 3.15(a), tracking with prescribed performance is successfully achieved with the errors evolving strictly within the predefined performance bounds, despite the lack of knowledge regarding the ROV dynamic model parameters. The trace of the ROV along with the desired trajectory are illustrated in Figure 3.14. Additionally, as it is pictured in Figure 3.15(b) the required control effort was satisfactorily smooth and did not impose any saturation to the thrusters.

Unfortunately, with the existing setup we were not able to create external disturbances in the form of waves and currents. However, the tether attached to the vehicle was creating nontrivial external disturbances in the form of random forces and torques while the vehicle was constantly changing configuration during the experiment. As it can be seen from Figure 3.15(a) and the overall performance of the controller, these external disturbances were rejected successfully proving thus the efficiency and robustness of the proposed control scheme.

3.3.3 Underactuated underwater vehicles: Case II

A 6-DoF dynamic model of a torpedo-like AUV is presented in this section and a rigorous formulation of the problem of tracking a desired trajectory is introduced.

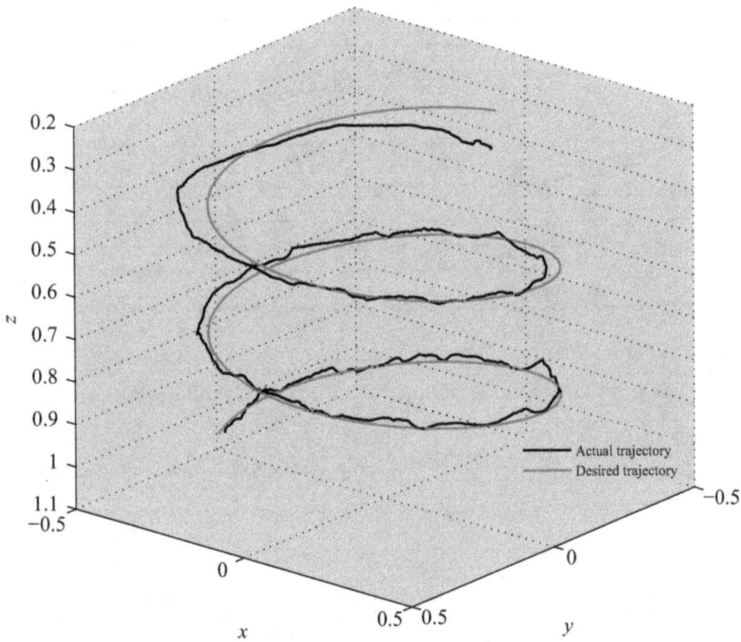

Figure 3.14 The trace along with the desired trajectory

The vehicle considered in this work is actuated by a force X along its longitudinal axis (surge), by a torque M about its transverse axis (pitch) as well as by a torque N about its vertical axis (yaw). The aforementioned force X and torques M, N define the input control variables of the dynamic system. In this respect, the vehicle is underactuated since only three of its six DoFs are actuated (sway, heave and roll are unactuated).

3.3.3.1 AUV kinematics and dynamics

Consider a neutrally buoyant AUV modeled as a rigid body subject to external forces and torques. Let $\{I\}$ be an inertial coordinate frame and $\{B\}$ a body-fixed coordinate frame with orthonormal axes $\bar{n} = [n_x, n_y, n_z]^T$, $\bar{o} = [o_x, o_y, o_z]^T$ and $\bar{t} = [t_x, t_y, t_z]^T$ relatively to $\{I\}$, whose origin O_B is located at the center of mass of the vehicle that coexists with the center of buoyancy. Further let $\boldsymbol{p} = [x, y, z]^T$ be the position of O_B in $\{I\}$ and $R = [\bar{n}, \bar{o}, \bar{t}]$ a rotation matrix that describes the orientation of the vehicle. Let $\boldsymbol{v} = [u, v, w]^T$ be the linear velocity (u, v and w are the longitudinal-surge, transverse-sway and vertical-heave velocities, respectively) and $\boldsymbol{w} = [p, q, r]^T$ be the angular velocity (p, q and r are the angular velocities around the longitudinal-roll, the transverse-pitch and vertical-yaw axis, respectively) of O_B w.r.t. to $\{I\}$ expressed in $\{B\}$.

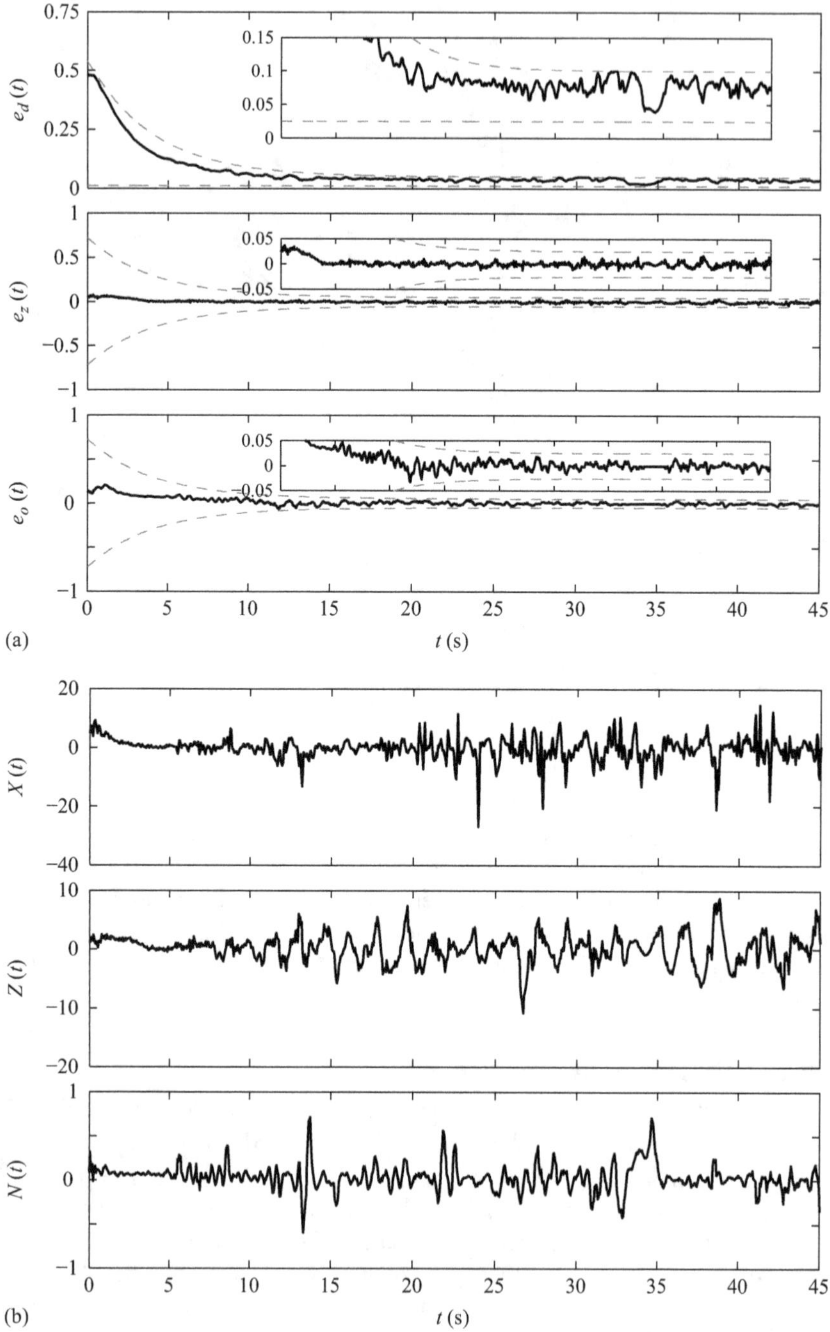

Figure 3.15 Experimental results. (a) The gray dashed lines indicate the desired performance bounds. The black solid lines indicate the evolution of $e_d(t)$, $e_z(t)$ and $e_o(t)$. (b) The required control effort along surge (N), heave (N) and about yaw (Nm) during the experiment

Hence, the kinematic equations of motion for the considered AUV can be written as

$$\dot{\boldsymbol{p}} = R\boldsymbol{v} + \delta_{\boldsymbol{v}}(t) \tag{3.41}$$

$$\dot{R} = RS(\boldsymbol{w}) \tag{3.42}$$

where $\delta_{\boldsymbol{v}}(t) = \left[\delta_x(n), \delta_y(t), \delta_z(t)\right]^T$ denotes bounded ocean currents and $S(\boldsymbol{w}) = \begin{bmatrix} 0 & -r & q \\ r & 0 & -p \\ -q & p & 0 \end{bmatrix}$.

The dynamic equations of motion for the underactuated AUV considered in this work can be written as

$$m_u\dot{u} = m_v vr - m_w wq + X_u u + X + \delta_u(t) \tag{3.43}$$

$$m_v\dot{v} = m_w wp - m_u ur + Y_v v + \delta_v(t) \tag{3.44}$$

$$m_w\dot{w} = m_u uq - m_v vp + Z_w w + \delta_w(t) \tag{3.45}$$

$$m_p\dot{p} = m_{vw} vw + m_{qr} qr + K_p p + \delta_p(t) \tag{3.46}$$

$$m_q\dot{q} = m_{wu} wu + m_{rp} rp + M_q q + M + \delta_q(t) \tag{3.47}$$

$$m_r\dot{r} = m_{uv} uv + m_{pq} qq + N_r r + N + \delta_r(t) \tag{3.48}$$

where m_u, m_v, m_w, m_p, m_q and m_r denote the AUV's mass/moment of inertia and added mass/moment of inertia terms with $m_{vw} = m_v - m_w$, $m_{wu} = m_w - m_u$, $m_{uv} = m_u - m_v$, $m_{qr} = m_q - m_r$, $m_{rp} = m_r - m_p$, $m_{pq} = m_p - m_q$ and X_u, Y_v, Z_w, K_p, M_q and N_r are negative hydrodynamic damping coefficients. Moreover, $\delta_u(t)$, $\delta_v(n)$, $\delta_w(t)$, $\delta_p(t)$, $\delta_q(t)$ and $\delta_r(t)$ denote bounded exogenous forces and torques acting on surge, sway, heave, roll, pitch and yaw owing to ocean waves and X, M and N denote the control input force and torques, respectively, that are applied in order to produce the desired motion of the body-fixed frame. Furthermore, notice that there is no control input in the sway, heave and roll equation of motion. Hence, (3.44)–(3.46) are uncontrolled making thus (3.41)–(3.48) an underactuated dynamical system. Finally, it should be noted that contrary to the majority of the works in the current AUV literature, where the Euler angle's parameterization is mainly employed owing to its physical meaning, our approach, which is based on the rotation matrix parameterization, does not suffer from geometric singularities when the pitch angle approaches ±90°.

3.3.3.2 Error coordinates

Let $x_d(t)$, $y_d(t)$ and $z_d(n)$ denote a smooth, bounded desired trajectory with bounded time derivatives. The objective of this work is to design a robust controller, without incorporating any information regarding the AUV model, such that it tracks the desired trajectory with bounded closed-loop signals and prescribed performance (concerning the steady-state error and the speed of convergence) despite the presence of exogenous disturbances representing ocean currents and waves. Let us now define the position error:

$$e_d = \sqrt{e_x^2 + e_y^2 + e_z^2} \tag{3.49}$$

where

$$e_x = x_d(t) - x,\ e_y = y_d(t) - y,\ e_z = z_d(t) - z \tag{3.50}$$

as well as the orientation errors:

$$e_o = \frac{e_x}{e_d}o_x + \frac{e_y}{e_d}o_y + \frac{e_z}{e_d}o_z = \cos(\theta_o) \tag{3.51}$$

$$e_t = \frac{e_x}{e_d}t_x + \frac{e_y}{e_d}t_y + \frac{e_z}{e_d}t_z = \cos(\theta_t) \tag{3.52}$$

where θ_o and θ_t are the angles measured from the vector $\bar{e}_d = \left[\frac{e_x}{e_d}, \frac{e_y}{e_d}, \frac{e_z}{e_d}\right]^T$ to the transverse $\bar{o}$ and vertical $\bar{t}$ axis of the AUV, respectively. Equations (3.51) and (3.52) can be easily verified if we consider the inner product of the unite vector $\bar{e}_d$ with the unite vectors of the transverse $\bar{o}$ and vertical $\bar{t}$ axis, respectively, which equals to the cosine of the angle defined by the aforementioned vectors. Hence, the tracking problem stated above is solved if the position error e_d and the orientation errors e_o and e_t reduce to zero (i.e., the AUV is heading to the desired trajectory since in such case the unite vector $\bar{e}_d$ tends to be normal to both the transverse $\bar{o}$ axis and vertical $\bar{t}$ axis of the AUV and consequently aligned to its longitudinal $\bar{n}$ axis). However, notice that the orientation errors e_o and e_t are well defined only for nonzero values of e_d, since for $e_d = 0$ the angles θ_o and θ_t are unidentified. Thus, in this work, we will design a controller that further guarantees that $e_d(t) > \underline{\rho}_d > 0, \forall t \geq 0$, for an arbitrarily small positive design constant $\underline{\rho}_d$, to avoid the aforementioned singularity issue.

Differentiating (3.49)–(3.52) and employing (3.41) and (3.42) to obtain the kinematics of the AUV in the error coordinates, we arrive at

$$\dot{e}_d = -u\cos(\theta_n) - v\cos(\theta_o) - w\cos(\theta_t) + \frac{e_x(\dot{x}_d - \delta_x)}{e_d} + \frac{e_y(\dot{y}_d - \delta_y)}{e_d} + \frac{e_z(\dot{z}_d 2 - \delta_z)}{e_d} \tag{3.53}$$

$$\begin{aligned}\dot{e}_o = {} & -r\cos(\theta_n) + p\cos(\theta_t) + u\frac{\cos(\theta_n)\cos(\theta_o)}{e_d} + (\dot{x}_d - \delta_x)\frac{e_d o_x + e_x\cos(\theta_o)}{e_d^2} - v\frac{\sin^2(\theta_o)}{e_d} \\ & + (\dot{y}_d - \delta_y)\frac{e_d o_y + e_y\cos(\theta_o)}{e_d^2} + w\frac{\cos(\theta_t)\cos(\theta_o)}{e_d} + (\dot{z}_d - \delta_z)\frac{e_d o_z + e_z\cos(\theta_o)}{e_d^2}\end{aligned} \tag{3.54}$$

$$\begin{aligned}\dot{e}_t = {} & q\cos(\theta_n) - p\cos(\theta_o) + u\frac{\cos(\theta_n)\cos(\theta_t)}{e_d} + (\dot{x}_d - \delta_x)\frac{e_d t_x + e_x\cos(\theta_t)}{e_d^2} + v\frac{\cos(\theta_o)\cos(\theta_t)}{e_d} \\ & + (\dot{y}_d - \delta_y)\frac{e_d t_y + e_y\cos(\theta_t)}{e_d^2} - w\frac{\sin^2(\theta_t)}{e_d} + (\dot{z}_d - \delta_z)\frac{e_d t_z + e_z\cos(\theta_t)}{e_d^2}\end{aligned} \tag{3.55}$$

where the following equality has been utilized:

$$\frac{e_x}{e_d}n_x + \frac{e_x}{e_d}n_y + \frac{e_z}{e_d}n_z = \cos(\theta_n)$$

with θ_n denoting the angle measured from the vector $\bar{e}_d$ to the longitudinal $\bar{n}$ axis of the AUV. Finally, to solve the tracking problem stated above, we assume that the initial position error $e_d(0)$ and orientation heading $\theta_n(0)$ satisfy (a) $e_d(0) > \underline{\rho}_d$ and (b) $|\theta_n(0)| < \frac{\pi}{2}$.

3.3.3.3 Control scheme

Given a smooth, bounded desired trajectory $x_d(t)$, $y_d(t)$ and $z_d(t)$ with bounded time derivatives and any initial system configuration satisfying Assumption 1 for an arbitrarily small positive design constant $\underline{\rho}_d$, we proceed as follows:

Kinematic controller

Select position/orientation performance functions $\rho_d(n)$, $\rho_o(t)$ and $\rho_t(n)$ that (i) satisfy

a. $e_d(0) < \rho_d(0)$	$\underline{\rho}_d < \rho_d(t)$	$\underline{\rho}_d < \lim_{t\to\infty} \rho_d(t)$
b. $\lvert e_o(0)\rvert < \rho_o(0) < 1$	$0 < \rho_o(t)$	$0 < \lim_{t\to\infty} \rho_o(t)$
c. $\lvert e_t(0)\rvert < \rho_t(0) < 1$	$0 < \rho_t(t)$	$0 < \lim_{t\to\infty} \rho_t(t)$
d. $\rho_o^2(t) + \rho_t^2(t) \leq \bar{\rho} < 1, \forall t \geq 0$		

for a positive constant $\bar{\rho} < 1$ and (ii) incorporate the desired performance specifications regarding the steady-state error and the speed of convergence. Subsequently, design the desired velocities

$$u_d = k_d \ln\left(\frac{1+\frac{e_d-\frac{\rho_d(t)+\underline{\rho}_d}{2}}{\frac{\rho_d(t)-\underline{\rho}_d}{2}}}{1-\frac{e_d-\frac{\rho_d(t)+\underline{\rho}_d}{2}}{\frac{\rho_d(t)-\underline{\rho}_d}{2}}}\right), q_d = -k_t \ln\left(\frac{1+\frac{e_t}{\rho_t(t)}}{1-\frac{e_t}{\rho_t(t)}}\right), r_d = k_o \ln\left(\frac{1+\frac{e_o}{\rho_o(t)}}{1-\frac{e_o}{\rho_o(t)}}\right) \quad (3.56)$$

with positive control gains k_d, k_o and k_t.

Dynamic controller

Select velocity performance functions $\rho_u(t)$, $\rho_q(t)$, $\rho_r(t)$ that satisfy

a. $\lvert u(0) - u_d(0)\rvert < \rho_u(0)$	$0 < \rho_u(t)$	$0 < \lim_{t\to\infty} \rho_u(t)$
b. $\lvert q(0) - q_d(0)\rvert < \rho_q(0)$	$0 < \rho_q(t)$	$0 < \lim_{t\to\infty} \rho_q(t)$
c. $\lvert r(0) - r_d(0)\rvert < \rho_r(0)$	$0 < \rho_r(t)$	$0 < \lim_{t\to\infty} \rho_r(t)$

and design the external force in the surge as well as the external torques in pitch and yaw as

$$X = -k_u \ln\left(\frac{1+\frac{u-u_d}{\rho_u(t)}}{1-\frac{u-u_d}{\rho_u(t)}}\right), M = -k_q \ln\left(\frac{1+\frac{q-q_d}{\rho_q(t)}}{1-\frac{q-q_d}{\rho_q(t)}}\right), N = -k_r \ln\left(\frac{1+\frac{r-r_d}{\rho_r(t)}}{1-\frac{r-r_d}{\rho_r(t)}}\right) \quad (3.57)$$

with positive control gains k_u, k_q and k_r.

3.3.3.4 Simulation results

To illustrate the tracking performance and the robustness of the proposed scheme against external disturbances representing ocean currents and waves, two simulation studies were carried out on a 6-DOF torpedo-like AUV actuated in surge, pitch and yaw. The vehicle dynamic model used in the simulation can be found in [20]. The AUV starts at rest from initial configuration $x(0) = -65$ m, $y(0) = -15$ m, $z(0) = 5$ m, $R(0) = [0\ 1\ 0; 1\ 0\ 0; 0\ 0\ 1]$ and is requested to track the trajectory $x_d(t) = -50\cos(0.02\pi t)$ m, $y_d(t) = 0.15\pi t$ m, $z_d(t) = 50\sin(0.02\pi t)$ m, with maximum steady-state errors $e_{d\max} = 0.1$, $e_{o\max} = 0.01$ and $e_{t\max} = 0.01$ and minimum convergence rate as obtained by the exponentials $e^{-0.1t}$, e^{-t} and e^{-t}. Notice that the aforementioned initial configuration satisfies Assumption 1 for $\underline{\rho}_d = 0.05$. Hence, following. Subsection 3.3.3, we select the performance functions $\rho_d(t) = (40 - 0.1)e^{-0.1t} + 0.1$, $\rho_o(t) = (0.45 - 0.01)e^{-t} + 0.01$, $\rho_t(t) = (0.45 - 0.01)e^{-t} + 0.01$, $\rho_u(t) = (2 - 0.1)e^{-t} + 0.1$, $\rho_q(t) = (2 - 0.1)e^{-t} + 0.1$ and $\rho_r(t) = (2 - 0.1)e^{-t} + 0.1$ and the control gains $k_d = 5$, $k_o = 1$, $k_t = 1$, $k_u = 500$, $k_q = 100$ and $k_r = 100$ such that (3.57) yield reasonable control effort.

In the first study, we considered a disturbance-free case. On the contrary, the second study captured the case where the ocean current is $d_x(t) = 0$, $d_y(t) = -2.5$, $d_z(t) = 0$ and the wave terms include three sinusoids with amplitude and frequency equally distributed in $[0.1, 1]$ and $[0.01, 0.1]$, respectively. The trajectory tracking is shown in Figures 3.16(a) and 3.17(a). The output error evolution along with the performance bounds are pictured in Figures 3.16(b) and 3.17(b). As it was predicted, tracking with prescribed performance is achieved despite the presence of external disturbances and the lack of knowledge of the AUV model parameters.

3.4 Summary

In this chapter, we presented model-based and model-free motion control strategies for underwater robotic vehicles. Regarding model-based control, we presented an NMPC strategy for underwater robotic vehicles operating under various constraints. The purpose of the controller is to guide the vehicle towards specific way-points while simultaneously satisfying various state and inputs constraints. The proposed scheme incorporates the full dynamics of the vehicle in which the ocean currents are also involved. The controller is designed in order to find the optimal thrusts required for minimizing the way-point tracking error. Furthermore, the control inputs calculated by the proposed approach are formulated in a way that the vehicle will exploit the ocean currents, when they are in favor of the way-point tracking mission, which results in reduced energy consumption by the thrusters. Regarding model-free control strategies, we presented position- and trajectory-tracking control schemes for AUVs. The proposed controllers do not utilize the vehicle's dynamic model parameters and guarantee prescribed transient and steady-state performance despite the presence of external disturbances and kinematic constraints. Moreover, through the appropriate selection of certain performance functions, the proposed control strategies can also guarantee the satisfaction of motion and performance constraints imposed by the desired task.

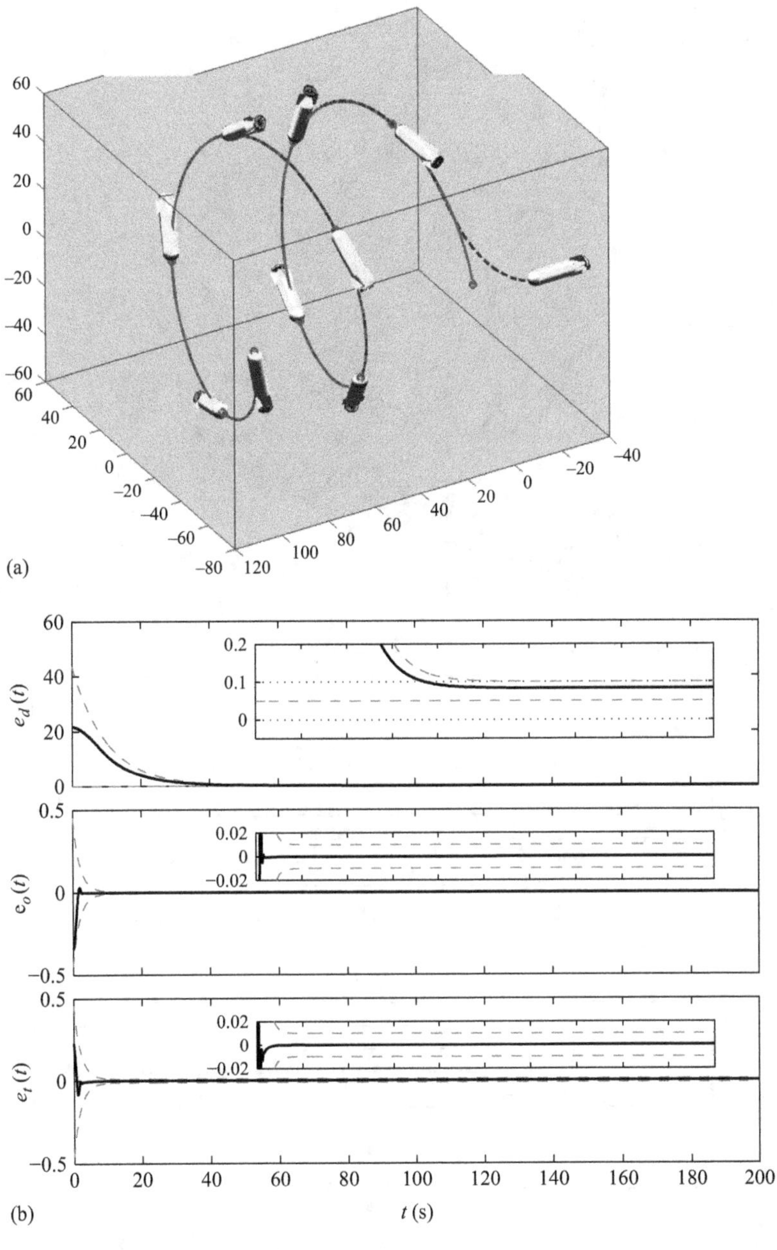

Figure 3.16 Disturbance-free case. (a) Trajectory tracking in 3D space. The solid line indicates the desired trajectory and the dashed line indicates the AUV trajectory. (b) Tracking error evolution. The gray dashed lines indicate the desired performance bounds. The black solid lines indicate the evolution of $e_d(t)$, $e_o(t)$ and $e_t(t)$. The subplots give details at the steady state

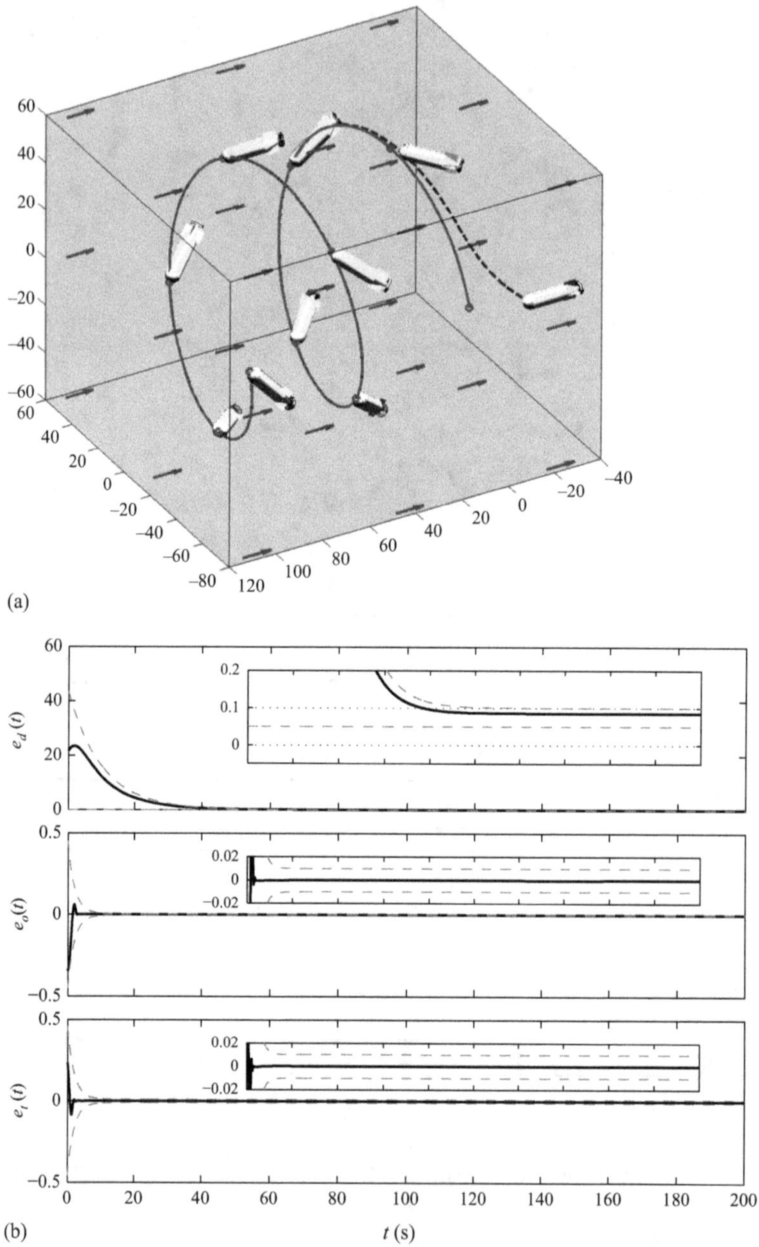

Figure 3.17 *Case of ocean currents and waves. (a) Trajectory tracking in 3D space. The solid line indicates the desired trajectory and the dashed line indicates the AUV trajectory. (b) Tracking error evolution. The gray dashed lines indicate the desired performance bounds. The black solid lines indicate the evolution of $e_d(t)$, $e_o(t)$ and $e_t(t)$. The subplots give details at the steady state*

References

[1] Fossen TI. Guidance and Control of Ocean Vehicles. Chichester, UK: Wiley; 1994.

[2] Lawton JRT, Beard RW, Young BJ. A decentralized approach to formation maneuvers. IEEE Transactions on Robotics and Automation. 2003;19(6): 933–941.

[3] Repoulias F, Papadopoulos E. Planar trajectory planning and tracking control design for underactuated AUVs. Ocean Engineering. 2007;34(11–12): 1650–1667.

[4] Aguiar AP, Hespanha JP. Trajectory-tracking and path-following of underactuated autonomous vehicles with parametric modeling uncertainty. IEEE Transactions on Automatic Control. 2007;52(8):1362–1379.

[5] Santhakumar M, Asokan T. Investigations on the hybrid tracking control of an underactuated autonomous underwater robot. Advanced Robotics. 2010;24(11):1529–1556.

[6] Allgöwer F, Findeisen R, Nagy ZK. Nonlinear model predictive control: From theory to application. The Chinese Institute of Chemical Engineers. 2004;35(3):299–315.

[7] Medagoda L, Williams SB. Model predictive control of an autonomous underwater vehicle in an in situ estimated water current profile. Program Book – OCEANS 2012 MTS/IEEE Yeosu: The Living Ocean and Coast – Diversity of Resources and Sustainable Activities. 2012.

[8] Fernandez DC, Hollinger GA. Model predictive control for underwater robots in ocean waves. IEEE Robotics and Automation Letters. 2017;2(1):88–95.

[9] Jagtap P, Raut P, Kumar P, *et al.* Control of autonomous underwater vehicle using reduced order model predictive control in three-dimensional space. IFAC-PapersOnLine. 2016;49(1):772–777.

[10] Heshmati-Alamdari S, Eqtami A, Karras GC, *et al.* A self-triggered visual servoing model predictive control scheme for under-actuated underwater robotic vehicles. Proceedings – IEEE International Conference on Robotics and Automation. 2014; pp. 3826–3831.

[11] National HF RADAR network—surface currents. Available from: http://cordc.ucsd.edu/projects/mapping/maps/.

[12] Smith RN, Chao Y, Li PP, *et al.* Planning and implementing trajectories for autonomous underwater vehicles to track evolving ocean processes based on predictions from a regional ocean model. International Journal of Robotics Research. 2010;29(12):1475–1497.

[13] Aguiar AP, Pascoal AM. Dynamic positioning and way-point tracking of underactuated AUVs in the presence of ocean currents. International Journal of Control. 2007;80(7):1092–1108.

[14] Fossen TI. Guidance and Control of Ocean Vehicles. New York: Wiley; 1994.

[15] Asouti VG, Trompoukis XS, Kampolis IC, *et al.* Unsteady CFD computations using vertex-centered finite volumes for unstructured grids on graphics

processing units. International Journal for Numerical Methods in Fluids. 2011;67(2):232–246.

[16] Koditschek DE, Rimon E. Robot navigation functions on manifolds with boundary. Advances in Applied Mathematics. 1990;11(4):412–442.

[17] Garrido-Jurado S, noz Salinas RM, Madrid-Cuevas FJ, *et al.* Automatic generation and detection of highly reliable fiducial markers under occlusion. Pattern Recognition. 2014;47(6):2280–2292.

[18] Marantos P, Koveos Y, Kyriakopoulos KJ. UAV State estimation using adaptive complementary filters. IEEE Transactions on Control Systems Technology. 2016;24(4):1214–1226.

[19] Quigley M, Gerkey B, Conley K, *et al.* ROS: An open-source Robot Operating System. In: Proc. of the IEEE Intl. Conf. on Robotics and Automation (ICRA) Workshop on Open Source Robotics. Kobe, Japan; 2009.

[20] Pettersen KY, Egeland O. Time-varying exponential stabilization of the position and attitude of an underactuated autonomous underwater vehicle. IEEE Transactions on Automatic Control. 1999;44(1):112–115.

Chapter 4

Adaptive–robust control of autonomous underwater vehicle with unknown system dynamics

Spandan Roy[1]*, Sayan Basu Roy*[2] *and Indra Narayan Kar*[3]

This chapter proposes a new adaptive–robust control (ARC) architecture for autonomous underwater vehicle (AUV) in the face of uncertain as well as completely unknown system dynamics parameters. The conventional ARC mechanisms either presume that the overall uncertainties or its time derivative is norm bounded by a constant or require structural knowledge of the system. However, for an AUV, the upper bound of the uncertainty has explicit presence of the systems states. Thus, assigning constant upper bound to the uncertainty (or its time derivative) is restrictive if one considers a system-like AUV. Moreover, due to the highly non-linear and coupled dynamics, true structural knowledge as well as parametric values of the AUV system are not always available. Considering such prevailing issues, adaptive switching gain-based robust control (ASRC) is conceptualized here for AUV, which avoids any presumption of constant upper bound on the overall uncertainties. The proposed control law does not utilize any prior knowledge of the parameters of the system dynamics, while negotiating the uncertainties regardless of the fact whether the uncertainties appear as linear in parameters (LIP) or non-linear in parameters (NLIP). Besides, the proposed adaptive law for switching gain alleviates the over- and underestimation issues commonly observed in the literature of ARC. Extensive simulation studies also validate the effectiveness of the proposed ASRC law in uncertain scenarios compared to the conventional adaptive sliding mode design.

4.1 Background and motivation

With the scientific and technological advancements, AUVs have found greater applications ranging from seafloor surveying, ocean cable inspection and maintenance to petroleum drilling, lost asset location and underwater rescue. The modelling aspect

[1]Robotics Research Center, International Institute of Information Technology, Hyderabad (IIIT-H), India
[2]Department of Electronics and Communication Engineering, Indraprastha Institute of Information Technology, Delhi (IIIT-D), India
[3]Department of Electrical Engineering, Indian Institute of Technology Delhi, New Delhi, India

of an AUV becomes tedious due the highly coupled and non-linear structure of its dynamics. Further, uncertainty in the hydrodynamic parameters adds up to the challenge of devising an efficient tracking controller for AUV. In real-life circumstances, AUVs are associated with impregnated uncertainties which arise from payload changes affecting the inertia, centre of mass and centre of buoyancy. These factors often result in inexact cancellation of non-linearities if exact system model-based (e.g. feedback linearization) control law is employed. Apart from the above-mentioned uncertainties, external disturbances such as sea current necessitates development of an advanced control framework for AUV systems.

The tracking control problem of AUVs under uncertain situation is commonly dealt with ARC. An adaptive controller is reported in [1] under the assumption that the vehicle model holds the linearity property inside the operating range space. The authors in [2] used recursive least-square methodology for online update of uncertain parameters adaptively. Multivariable self-tuning autopilots are proposed in [3] to deal with model uncertainties. In [4,5], adaptive controllers are developed for MIMO systems with application towards AUV. To avoid singularity, [6] proposed adaptive controller to tackle unknown parameters and used quaternions in the representation of attitude error. The adaptive controller reported by [7] tackles the effects due to the uncertainty in hydrodynamic parameters. Adaptive control, compared to robust control, does not require any prior bound of the uncertainties and thus advantageous. However, online computation of unknown parameters for complex uncertain systems like AUV becomes computationally intensive.

Considering the limitation of adaptive control, many robust control techniques are developed. Sliding mode control (SMC) is one such robust control which has been exploited extensively. The SMC devised by [8] provides robustness against the perturbations due to uncertain hydrodynamics terms. However, cross-coupling terms were ignored in this work. The SMC proposed by [9] considers the sliding surface as a function of error in states instead of error in output. Separate autopilots were designed for various subsystems under the assumption that the subsystems are non-interacting or having loose interaction. Adaptive SMC is reported in [10,11] by linearizing the system dynamics about the operating point. A discrete time quasi-SMC is reported in [12] with enhanced controller performance in lieu of larger sampling time. The authors in [13] conducted a performance comparison between SMC and LMI methodologies on a linearized model of AUV. A linear parametric varying control technique is developed by [14] based on LMI method. A robust model predictive control based on linearized system dynamics is reported in [15]. However, for a complex and highly non-linear system like AUV, controller development-based system linearization does not incorporate the true effect of unmodelled dynamics.

A hybrid controller with SMC and PID control is considered by [16]. SMC is used in reaching or transient phase and PID controller is used for the steady-state part. However, the true essence of SMC is realized in sliding phase rather than in reaching phase, and PID control is not itself robust against the uncertainties. The main disadvantage of a classical SMC lies in its high switching gain and consequently higher control input. The effectiveness of robust outer loop control (ROC) method was studied in detail in [17] for trajectory tracking control of AUV-150 under various time varying parametric uncertainties. The authors [18,19] exploited time delay control (TDC) for tracking problem of AUV where knowledge of all uncertain system parameters are not required a priori for the controller. It is shown by [20,21] that under bounded dynamic uncertainties, TDC fails to provide robustness as it is unable to cancel the approximation error and robust controller provides better tracking accuracy than TDC through its switching control law. An SMC-based controller is reported by [22] to negotiate the approximation error of TDC. However, it requires a predefined bound of the approximation to evaluate the switching gain which is again not always possible to obtain in practical scenario. For any conventional robust controller like SMC and ROC, an estimation of uncertainty bound is required a priori for the design purpose of the robust control method which in turn needs complete knowledge of system model of the complex AUV system. Further, to increase the working paradigm of the controller, often higher bound estimations are made, which leads to higher switching gains. This increment in switching gain necessitates broadening of the error bandwidth to avoid chattering and as a fall out, tracking accuracy gets compromised.

To reduce dependency on the system modelling, researchers have applied various artificial intelligence methods [23–25], but at the cost of computation burden. Therefore, a lot of recent focus is given on adaptive–robust controllers [26–35] (though none of these are verified for an AUV system). These types of controllers are not computationally expensive (as well as in terms of memory requirement) due to their primary nature of robust controller framework; further, to remove any prior requirement of the bound on the uncertainties, adaptive laws are employed to evaluate the switching gain online. Nevertheless, the control laws [26,27,31–34] still require knowledge of system dynamics parameters to build the control law [26,27] or maintain system stability [30–34]. Moreover, the control laws [26–29,36] presume that the uncertainties or its time derivative is upper bounded by a constant as well as suffer from the over- and underestimation problems of switching gain [35]. However, such assumption is restrictive for AUV systems since the upper bound of uncertainty (or its time derivative) has explicit dependency on the system states.

4.1.1 Contributions

Considering the pertaining gaps in the literature, this chapter proposes a new ASRC law for tracking control problem of AUV. The proposed framework *does not require any prior knowledge of the system parameters*. Further, unlike [26–29], ASRC does not presume the system uncertainty or its time derivative to be upper bounded by a constant; rather, it exploits the LIP structure of the upper bound of the uncertainty to devise the structure of the switching gain. *It is to be noted here that the uncertainty itself can be LIP or NLIP; however, the upper bound of the uncertainty always satisfies an LIP structure for AUV.* Therefore, the designed controller is insensitive towards the nature or characteristics (LIP or NLIP) of the uncertainty. Besides, the adaptive law of ASRC prevents the switching gain from becoming a monotonically increasing function of time as well as alleviates the over- and underestimation problems of switching gain. The efficacy of the proposed controller is extensively studied in simulation using the parameters of the AUV used in [17].

4.1.2 Organization and notations

The remainder of the chapter is organized as follows: Section 4.2 outlines the brief AUV dynamics and the control problem. The proposed ASRC framework is detailed in Section 4.3, while the corresponding stability analysis is provided in Section 4.4. This is followed by the simulation results of ASRC in Section 4.5. Section 4.6 presents the concluding remarks with a future direction.

The following notations are used in this chapter: $||\cdot||$ and $\lambda_{\min}(\cdot)$ represent the Euclidean norm and minimum eigenvalue of $(\cdot)$, respectively, and I denotes the identity matrix of appropriate dimension.

4.2 The AUV dynamics and problem formulation

4.2.1 Brief dynamics of AUV

In general, a six-degrees-of-freedom equation of motion of an AUV are formulated by employing Kirchhoff's equations and this can be written in the body coordinate frame as follows [37]:

$$\bar{\mathbf{M}}\dot{\nu} + \bar{\mathbf{C}}(\nu)\nu + \bar{\mathbf{g}}(\mathbf{q}) + \bar{\mathbf{D}}(\nu)\nu = \bar{\tau} \tag{4.1}$$

where $\bar{\mathbf{M}} \in \mathbb{R}^{6\times 6}$ and $\bar{\mathbf{C}} \in \mathbb{R}^{6\times 6}$ represent the mass matrix and Coriolis matrix, respectively; $\bar{\mathbf{D}} \in \mathbb{R}^{6\times 6}$ denotes the damping matrix; $\bar{\mathbf{g}} \in \mathbb{R}^6$ is the gravity and buoyancy force vector; $\bar{\tau} \in \mathbb{R}^{6\times 6}$ represents generalized force; linear positions (x, y, z) and attitude angles are expressed by $\mathbf{q} = [x, y, z, \varphi, \omega, \psi]^T$ in inertial frame and $\nu = [u, v, w, p, \kappa, r]^T$ is the vector representation of the linear and angular velocities in body coordinate frame. Here, φ, ω and ψ denote the roll, pitch and yaw/heading of the AUV, respectively.

Further, the kinematic equation representing the transformation between the body and inertial frame through the linear and angular velocities is expressed as

$$\dot{\mathbf{q}} = \mathbf{J}(\mathbf{q})\nu \tag{4.2}$$

where $\mathbf{J}(\mathbf{q}) = diag\{\mathbf{J}_1(\mathbf{q}), \mathbf{J}_2(\mathbf{q})\}$ represents the Jacobian matrix. The detailed structures of these matrices and vectors can be found in [37].

Using (4.1) and (4.2) and considering the bounded external disturbance $d_s(t)$, the AUV system dynamics can be represented in the inertial frame as

$$\mathbf{M}(\mathbf{q})\ddot{\mathbf{q}} + \mathbf{C}(\mathbf{q}, \nu)\dot{\mathbf{q}} + \mathbf{g}(\mathbf{q}) + \mathbf{D}(\mathbf{q}, \nu)\dot{\mathbf{q}} + d_s = \tau \tag{4.3}$$

where

$$\begin{aligned} \mathbf{M} &= \mathbf{J}^{-T}\bar{\mathbf{M}}\mathbf{J}^{-1}, \mathbf{C} = \mathbf{J}^{-T}[\bar{\mathbf{C}} - \bar{\mathbf{M}}\mathbf{J}^{-1}\dot{\mathbf{J}}]\mathbf{J}^{-1} \\ \mathbf{D} &= \mathbf{J}^{-T}\bar{\mathbf{D}}\mathbf{J}^{-1}, \mathbf{g} = \mathbf{J}^{-T}\bar{\mathbf{g}}, \tau = \mathbf{J}^{-T}\bar{\tau} \end{aligned}$$

Here $d_s(t)$ can be considered to be uncertain sea current. Based on the dynamics structure of AUV as given in [37] (i.e. considering slow moving vehicle) and using (4.1) and (4.2), it can be verified that the system (4.3) possesses the following properties:

Property 1: The matrix $(\dot{\mathbf{M}} - 2\mathbf{C})$ is skew symmetric.

Property 2: $\exists g_b, f_b, \bar{d} \in \mathbb{R}^+$ such that $||\mathbf{g}(\mathbf{q})|| \leq g_b$, $||\mathbf{D}(\mathbf{q}, \nu)|| \leq f_b||\dot{\mathbf{q}}||$ and $||d_s(t)|| \leq \bar{d}$.

Property 3: The matrix $\mathbf{M}(\mathbf{q})$ is uniformly positive definite and there exist two positive constants μ_1, μ_2 such that

$$0 < \mu_1\mathbf{I} \leq \mathbf{M}(\mathbf{q}) \leq \mu_2\mathbf{I} \tag{4.4}$$

Property 4: $\exists C_b \in \mathbb{R}^+$ such that $||\mathbf{C}(\mathbf{q}, \nu)|| \leq C_b||\dot{\mathbf{q}}||$.

The terms $C_b, f_b, g_b, \bar{d}, \mu_1$ and μ_2 are considered to be unknown.

4.2.2 Problem formulation

Let $\mathbf{q}^d(t)$ is the desired trajectory to be followed and it is defined such that $\dot{\mathbf{q}}^d, \ddot{\mathbf{q}}^d \in \mathscr{L}_\infty$. Let $\mathbf{e}(t) \triangleq \mathbf{q}(t) - \mathbf{q}^d(t)$ be the tracking error and $\mathbf{e}_f$ be the filtered tracking error:

$$\mathbf{e}_f \triangleq \dot{\mathbf{e}} + \Omega\mathbf{e} \Rightarrow \mathbf{e}_f = \Gamma\xi \tag{4.5}$$

where $\Gamma \triangleq [\Omega\,\mathbf{I}]$, $\xi \triangleq [\mathbf{e}^T\ \dot{\mathbf{e}}^T]^T$ and $\Omega \in \mathbb{R}^{n\times n}$ is a positive definite matrix. Multiplying the time derivative of (4.5) by $\mathbf{M}$ and using (4.3) yields

$$\begin{aligned}\mathbf{M}\dot{\mathbf{e}}_f &= \mathbf{M}(\ddot{\mathbf{q}} - \ddot{\mathbf{q}}^d + \Omega\dot{\mathbf{e}})\\ &= \tau - \mathbf{C}(\mathbf{q}, \nu)\mathbf{e}_f + \sigma\end{aligned} \tag{4.6}$$

where $\sigma \triangleq -(\mathbf{C}(\mathbf{q}, \nu)\dot{\mathbf{q}} + \mathbf{g}(\mathbf{q}) + \mathbf{D}(\mathbf{q}, \nu)\dot{\mathbf{q}} + d_s + \mathbf{M}\ddot{\mathbf{q}}^d - \mathbf{M}\Omega\dot{\mathbf{e}} - \mathbf{C}(\mathbf{q}, \nu)\mathbf{e}_f)$ represents the overall uncertainty. Further, $\xi = [\mathbf{e}^T\ \dot{\mathbf{e}}^T]^T$ implies $||\xi|| \geq ||\mathbf{e}||, ||\xi|| \geq ||\dot{\mathbf{e}}||$.

Lemma 1. *The term $||\sigma||$ can be upper bounded with an LIP structure as follows:*

$$||\sigma|| \leq \theta_0^* + \theta_1^*||\xi|| + \theta_2^*||\xi||^2 \triangleq \mathbf{Y}(\xi)^T\Theta^* \tag{4.7}$$

where $\theta_0^* \triangleq C_b||\dot{\mathbf{q}}^d||^2 + g_b + f_b||\dot{\mathbf{q}}^d||^2 + \bar{d} + \mu_2||\ddot{\mathbf{q}}^d||$

$$\theta_1^* \triangleq 2f_b||\dot{\mathbf{q}}^d|| + \mu_2||\Omega|| + C_b||\dot{\mathbf{q}}^d||(1 + ||\Omega||)$$

$$\theta_2^* \triangleq C_b||\Omega|| + f_b,\ \mathbf{Y}(\xi) = [1||\xi||||\xi||^2]^T, \Theta^* = [\theta_0^*\ \theta_1^*\ \theta_2^*]^T \tag{4.8}$$

Proof. Relation (4.5) and the system Property 4 yield

$$\begin{aligned}||\mathbf{C}\mathbf{e}_f - \mathbf{C}\dot{\mathbf{q}}|| &= ||\mathbf{C}(\dot{\mathbf{e}} + \Omega\mathbf{e}) - \mathbf{C}\dot{\mathbf{q}}|| = ||-\mathbf{C}\dot{\mathbf{q}}^d + \mathbf{C}\Omega\mathbf{e}||\\ &\leq C_b||\dot{\mathbf{q}}||||\dot{\mathbf{q}}^d|| + C_b||\dot{\mathbf{q}}||||\Omega||||\mathbf{e}||\\ &\leq C_b||\dot{\mathbf{e}} + \dot{\mathbf{q}}^d||||\dot{\mathbf{q}}^d|| + C_b||\dot{\mathbf{e}} + \dot{\mathbf{q}}^d||||\Omega||||\xi||\\ &\leq C_b\{||\xi||||\dot{\mathbf{q}}^d|| + ||\dot{\mathbf{q}}^d||^2 + ||\xi||^2||\Omega|| + ||\dot{\mathbf{q}}^d||||\Omega||||\xi||\}\end{aligned} \tag{4.9}$$

Further, the system Properties 2 and 3 provide the following:

$$\begin{aligned}||\mathbf{g}+\mathbf{D}+d_s+\mathbf{M}\ddot{\mathbf{q}}^d-\mathbf{M}\Omega\dot{\mathbf{e}}|| &\leq g_b+f_b||\dot{\mathbf{q}}||||\dot{\mathbf{q}}||+\bar{d}+\mu_2||\ddot{\mathbf{q}}^d||+\mu_2||\Omega||||\dot{\mathbf{e}}|| \\ &\leq g_b+f_b(||\dot{\mathbf{e}}+\dot{\mathbf{q}}^d||||\dot{\mathbf{e}}+\dot{\mathbf{q}}^d||)+\bar{d}+\mu_2||\ddot{\mathbf{q}}^d||+\mu_2||\Omega||||\xi|| \\ &\leq g_b+f_b(||\xi||^2+||\dot{\mathbf{q}}^d||^2+2||\xi||||\dot{\mathbf{q}}^d||)+\bar{d}+\mu_2||\ddot{\mathbf{q}}^d||+\mu_2||\Omega||||\xi||\end{aligned} \tag{4.10}$$

Since $\dot{\mathbf{q}}^d, \ddot{\mathbf{q}}^d \in \mathscr{L}_\infty$, it can be verified using (4.9)–(4.10) that $\exists\theta_i^* \in \mathbb{R}^+$ $i=0,1,2$, such that the upper bound of $||\sigma||$ holds the LIP structure (4.7). The values of θ_i^* are considered to be unknown; only their existences are ensured. □

The control laws developed in [26,27] and [28,29] assume that σ and $\dot{\sigma}$ are upper bounded by constants, respectively. It can be easily inferred from the structure of $||\sigma||$ from (4.7) that considerations of such constant bound on the uncertainties a priori, whether known or unknown, constraint the states beforehand. Moreover, the switching gain in [26,27] suffers from overestimation–underestimation problem.

4.3 Controller design

This section details the proposed ASRC framework with the major aims being the following:

- To track the desired trajectory $\mathbf{q}^d$ while compensating the uncertainty σ without any knowledge of the system dynamics parameters.
- To alleviate the overestimation–underestimation problems of switching gain.

The control input of the proposed ASRC is designed as

$$\tau = -\mathbf{e}-\mathbf{G}\mathbf{e}_f-\Delta\tau,\ \Delta\tau=\eta\hat{\rho}||\mathbf{e}_f||^{\varepsilon}\frac{\mathbf{e}_f}{||\mathbf{e}_f||} \tag{4.11}$$

$$\hat{\rho} = \hat{\theta}_0+\hat{\theta}_1||\xi||+\hat{\theta}_2||\xi||^2+\gamma \triangleq \mathbf{Y}(\xi)^T\hat{\Theta}+\gamma \tag{4.12}$$

where $\Delta\tau$ provides robustness against σ through $\hat{\rho}$; $\mathbf{G}$ is a positive definite user-defined matrix; $\hat{\Theta}=[\hat{\theta}_0\ \hat{\theta}_1\ \hat{\theta}_2]^T$ is the estimate of Θ; and γ is an auxiliary gain. The importance of γ will be explained later; $\eta \geq 1$ is a scalar and $0<\varepsilon<1$ is small scalar which helps to avoid chattering. The gains $\gamma, \hat{\theta}_i$, $i=0,1,2$, are evaluated using the following adaptive laws:

(i) for $||\mathbf{e}_f|| \geq \varpi$,

$$\dot{\hat{\theta}}_i = \begin{cases} \alpha_i||\xi||^i||\mathbf{e}_f|| \text{ if } \left(\{\mathbf{e}^T\dot{\mathbf{e}}>0\}\vee\{\bigcup_{i=0}^2\hat{\theta}_i\leq 0\}\vee\{\gamma\leq\beta\}\right) \\ -\alpha_i||\xi||^i||\mathbf{e}_f|| \text{ otherwise}\end{cases} \tag{4.13}$$

$$\dot{\gamma} = \begin{cases} \alpha_3||\mathbf{e}_f|| \text{ if } \left(\{\mathbf{e}^T\dot{\mathbf{e}}>0\}\vee\{\bigcup_{i=0}^2\hat{\theta}_i\leq 0\}\cup\{\gamma\leq\beta\}\right) \\ -\varsigma\alpha_3||\xi||^4 \text{ otherwise}\end{cases} \tag{4.14}$$

(ii) for $||\mathbf{e}_f|| < \varpi$,

$$\dot{\hat{\theta}}_i = 0, \dot{\gamma} = 0 \tag{4.15}$$

$$\text{with} \quad \hat{\theta}_i(t_0) > 0, i = 0, 1, 2, \ \gamma(t_0) > \beta, \text{ and} \tag{4.16}$$

$$\varpi \triangleq \left(\frac{1}{\eta}\right)^{\frac{1}{\varepsilon}} \tag{4.17}$$

Here t_0 is the initial time and $\beta, \varsigma, \alpha_0, \alpha_1, \alpha_2$ and $\alpha_3 \in \mathbb{R}^+$ are user-defined scalars. Substituting (4.11) into (4.6), the closed-loop system is formed as

$$\mathbf{M}\dot{\mathbf{e}}_f = -\mathbf{e} - \mathbf{G}\mathbf{e}_f - \Delta\tau - \mathbf{C}\mathbf{e}_f + \sigma \tag{4.18}$$

Remark 1. *It can be observed from (4.13)–(4.14) that, when $||\mathbf{e}_f|| \geq \varpi$, the gains $\hat{\theta}_i, \gamma$ increase if error trajectories move away from $||\mathbf{e}|| = 0$ (governed by $\mathbf{e}^T\dot{\mathbf{e}} > 0$) and decrease if error trajectories do not move away from $||\mathbf{e}|| = 0$ (governed by the 'otherwise' condition in (4.13)–(4.14) which implies $(\{\mathbf{e}^T\dot{\mathbf{e}} \leq 0\} \wedge \{\bigcup_{i=0}^{2} \hat{\theta}_i > 0\} \wedge \{\gamma > \beta\})$. Therefore, the proposed law stops the switching gain from becoming a monotonically increasing function of time and consequently alleviates the overestimation problem.*

Remark 2. *The condition $||\mathbf{e}_f|| < \varpi$ implies that the tracking error remains bounded inside the ball $B_\varpi \triangleq \{||\Gamma\xi|| < \varpi\}$ by virtue of the relation $\mathbf{e}_f = \Gamma\xi$. Hence, the gains are kept unchanged for the case $||\mathbf{e}_f|| < \varpi$ considering the switching gains to be sufficient enough to keep the error within B_ϖ. One may select small ε to improve tracking accuracy (this reduces B_ϖ) as long as such selection does not invoke chattering.*

Remark 3. *The gains are initiated as $\hat{\theta}_i(t_0) > 0$, $\gamma(t_0) > \beta$. Further, when $||\mathbf{e}_f|| \geq \varpi$, the gains are forced to increase by the adaptive laws (4.13)–(4.14) if either of them attempt to go below their designated lower bounds (governed by $\{\bigcup_{i=0}^{2} \hat{\theta}_i \leq 0\} \vee \{\gamma \leq \beta\}$). This implies that $\gamma(t) \nless \beta$, $\hat{\theta}_i(t) \nless 0 \ \forall i = 0, 1, 2$, when $||\mathbf{e}_f|| \geq \varpi$. Again, the gains do not change when $||\mathbf{e}_f|| < \varpi$. Therefore, combining the above conditions yields*

$$\hat{\theta}_i(t) \geq 0 \ \forall i = 0, 1, 2 \text{ and } \gamma(t) \geq \beta \ \ \forall t \geq t_0 \tag{4.19}$$

The condition (4.19) is later exploited in stability analysis.

Remark 4. *The crucial point to notice from (4.11)–(4.17) is that the overall control architecture of the proposed ASRC is built without any knowledge of the system parameters.*

The following theorem guarantees the alleviation of the overestimation problem of switching gain by ensuring that $\hat{\theta}_i, \gamma$ decrease within a finite time.

Theorem 1. *Let $t = t_{in}$ be any time instant when gains start increasing. Then there exist finite times $t_1, t_2, t_3, \delta t$ such that the gains $\hat{\theta}_0, \hat{\theta}_1, \hat{\theta}_2, \gamma$ decrease for $t \geq t_{in} + T$, where $T \leq \bar{t} + \delta t$, $\bar{t} = max\{t_1, t_2, t_3\}$. These times are obtained as*

$$t_1 \leq \frac{\theta_0^*}{(\alpha_0 + \alpha_3)\varpi}, t_2 \leq \frac{\theta_1^*||\Gamma||}{\alpha_1 \varpi^2}, t_3 \leq \frac{\theta_2^*||\Gamma||^2}{\alpha_2 \varpi^3} \tag{4.20}$$

$$\delta t \leq \frac{1}{\rho} ln \left(\frac{2V(\bar{t} + t_{in})}{||\mathbf{e}(\bar{t} + t_{in})||^2} \right) \tag{4.21}$$

where

$$V = \frac{1}{2}\mathbf{e}_f^T \mathbf{M} \mathbf{e}_f + \frac{1}{2}\mathbf{e}^T \mathbf{e}, \rho \triangleq \frac{\min\{\lambda_{\min}(\mathbf{G}), \lambda_{\min}(\Omega)\}}{\max\{\mu_2, 1\}} \tag{4.22}$$

Proof. The objective of this theorem is to find when the gains start to decrease. The adaptive laws (4.13)–(4.15) suggest that the gains increase only when $||\mathbf{e}_f|| \geq \varpi$; hence, it is sufficient to only explore the condition $||\mathbf{e}_f|| \geq \varpi$. Further, it is to be noted that t_{in} can represent any time when the gains begin to increase and its usage is solely restricted for analysis. Moreover, using $\mathbf{e}_f = \Gamma\xi$ from (4.5), one has

$$\varpi \leq ||\mathbf{e}_f|| \leq ||\Gamma|| ||\xi|| \Rightarrow ||\xi|| \geq (\varpi/||\Gamma||) \tag{4.23}$$

So, the first laws of (4.13) and (4.14) and the condition (4.23) yield

$$\dot{\hat{\theta}}_0 \geq \alpha_0 \varpi, \dot{\hat{\theta}}_1 \geq (\alpha_1 \varpi^2)/||\Gamma||, \dot{\hat{\theta}}_2 \geq (\alpha_2 \varpi^3)/||\Gamma||^2, \dot{\gamma} \geq \alpha_3 \varpi \tag{4.24}$$

Consider the Lyapunov function V in (4.22). Using (4.18) and the relation $\mathbf{e}^T \dot{\mathbf{e}} = \mathbf{e}^T(\mathbf{e}_f - \Omega\mathbf{e})$ (from (4.5)), the time derivative of V yields

$$\begin{aligned} \dot{V} &= \mathbf{e}_f^T \mathbf{M} \dot{\mathbf{e}}_f + \frac{1}{2}\mathbf{e}_f^T \dot{\mathbf{M}} \mathbf{e}_f + \mathbf{e}^T \dot{\mathbf{e}} \\ &= \mathbf{e}_f^T(-\mathbf{e} - \mathbf{G}\mathbf{e}_f - \Delta\tau + \sigma) + \frac{1}{2}\mathbf{e}_f^T(\dot{\mathbf{M}} - 2\mathbf{C})\mathbf{e}_f + \mathbf{e}^T(\mathbf{e}_f - \Omega\mathbf{e}) \end{aligned} \tag{4.25}$$

One can verify that $\eta \mathbf{e}_f^T ||\mathbf{e}_f||^{\varepsilon} \frac{\mathbf{e}_f}{||\mathbf{e}_f||} \geq ||\mathbf{e}_f||$ is satisfied when $||\mathbf{e}_f|| \geq \varpi$. Further, substituting (4.11) into (4.25) and using Property 1 (implying $\mathbf{e}_f^T(\dot{\mathbf{M}} - 2\mathbf{C})\mathbf{e}_f = 0$), $\dot{V}$ is simplified as

$$\begin{aligned} \dot{V} &= -\mathbf{e}_f^T \mathbf{G} \mathbf{e}_f - \mathbf{e}^T \Omega \mathbf{e} + \mathbf{e}_f^T \left(-\eta\hat{\rho}||\mathbf{e}_f||^{\varepsilon} \frac{\mathbf{e}_f}{||\mathbf{e}_f||} + \sigma \right) \\ &\leq -\mathbf{e}_f^T \mathbf{G} \mathbf{e}_f - \mathbf{e}^T \Omega \mathbf{e} - (\mathbf{Y}(\xi)^T \hat{\Theta} + \gamma)||\mathbf{e}_f|| + \mathbf{Y}(\xi)^T \Theta^* ||\mathbf{e}_f|| \\ &\leq -\lambda_{\min}(\mathbf{G})||\mathbf{e}_f||^2 - \lambda_{\min}(\Omega)||\mathbf{e}||^2 - \{(\hat{\theta}_0 + \gamma - \theta_0^*) + (\hat{\theta}_1 - \theta_1^*)||\xi|| \\ &\quad + (\hat{\theta}_2 - \theta_2^*)||\xi||^2\}||\mathbf{e}_f|| \end{aligned} \tag{4.26}$$

Thus, the sufficient condition to achieve $\dot{V} < 0$ would be

$$\hat{\theta}_0 + \gamma \geq \theta_0^*, \ \hat{\theta}_1 \geq \theta_1^* \text{ and } \hat{\theta}_2 \geq \theta_2^* \tag{4.27}$$

Let the system (4.3) does not possess finite time escape (holds for an Euler–Lagrange system like AUV in practice [38]). Then integrating both sides of the inequalities in (4.24) and using those results in (4.27) lead to the expressions of t_1, t_2, t_3 in (4.20). So, for $t \geq t_{in} + \bar{t}$,

$$\begin{aligned} \dot{V} &\leq -\lambda_{\min}(\mathbf{G})||\mathbf{e}_f||^2 - \lambda_{\min}(\Omega)||\mathbf{e}||^2 \\ &\leq -\rho_m(||\mathbf{e}_f||^2 + ||\mathbf{e}||^2) \end{aligned} \tag{4.28}$$

where $\rho_m \triangleq \min\{\lambda_{\min}(\mathbf{G}), \lambda_{\min}(\Omega)\}$. Further, the definition of V yields

$$V \leq \rho_M(||\mathbf{e}_f||^2 + ||\mathbf{e}||^2) \tag{4.29}$$

where $\rho_M \triangleq \max\{\mu_2, 1\}$. Substituting (4.29) into (4.28) and using the comparison lemma [39] yields

$$\dot{V} \leq -\rho V \Rightarrow V(t) \leq V(t_{in} + \bar{t})e^{-\rho(t-\bar{t})} \ \forall t \geq t_{in} + \bar{t} \tag{4.30}$$

where $\rho \triangleq \rho_m/\rho_M$. Here $\hat{\theta}_i > 0, \gamma > \beta$ as gains were increasing. Therefore, to satisfy the '*otherwise*' condition (i.e. $\{e^T\dot{e} \leq 0\} \wedge \{\bigcup_{i=0}^{2} \hat{\theta}_i > 0\} \wedge \{\gamma > \beta\}$), $\mathbf{e}^T\dot{\mathbf{e}} \leq 0$ (implying $||\mathbf{e}(t)||$ does not increase) needs to be ensured. The upper bound of e can be found from the definition of V as

$$V(t) \geq \frac{1}{2}||\mathbf{e}(t)||^2 \Rightarrow ||\mathbf{e}(t)|| \leq \sqrt{2V(t)} \ \forall t \geq t_0 \tag{4.31}$$

Let $||\mathbf{e}(t_{in} + \bar{t})|| \triangleq \psi$, which implies $V(t_{in} + \bar{t}) \geq \psi^2/2$ from (4.31). As $V(t)$ decreases exponentially $\forall t \geq t_{in} + \bar{t}$ following (4.30), there exists a finite time $\delta t \triangleq t - (t_{in} + \bar{t})$ such that $V(t_{in} + \bar{t} + \delta t) = \psi^2/2$ implying $||\mathbf{e}(t_{in} + \bar{t} + \delta t)|| \leq \psi$. So, $\{\mathbf{e}^T\dot{\mathbf{e}} \leq 0\} \wedge \{\bigcup_{i=0}^{2} \hat{\theta}_i > 0\} \wedge \{\gamma > \beta\}$ would occur at $t \geq t_{in} + T$ where $T \leq \bar{t} + \delta t$ and $\hat{\theta}_i, \gamma$ would start decreasing. The time δt is found from (4.30):

$$\begin{aligned} &\psi^2 \leq 2V(t_{in} + \bar{t})e^{-\rho\delta t}, \quad \forall t \geq t_{in} + \bar{t} \\ &\Rightarrow \delta t \leq \frac{1}{\rho}\ln\left(\frac{2V(t_{in} + \bar{t})}{\psi^2}\right) \end{aligned} \tag{4.32}$$

□

Remark 5. *The increment and decrement of $\hat{\theta}_i$, γ can occur several times depending on the error incurred by the system. However, time interval Δt between two successive decrement will always satisfy $\Delta t \leq \bar{t} + \delta t$. Moreover, high values of $\alpha_0, \alpha_1, \alpha_2$ and α_3 help to reduce $\bar{t}$ and achieve faster adaptation.*

4.4 Stability analysis of ASRC

Before stating the main stability result of the closed-loop system (4.18), an important preliminary definition is provided in the following:

Definition 1 (Globally uniformly ultimately bounded (UUB) stability [39]). *System (4.3) is globally UUB if there exists a convex and compact set Υ such that for every initial condition $(\mathbf{q}(t_0), \dot{\mathbf{q}}(t_0))$, there exists a finite time $T(\mathbf{q}(t_0), \dot{\mathbf{q}}(t_0))$ such that $(\mathbf{q}(t), \dot{\mathbf{q}}(t)) \in \Upsilon$ for all $t \geq T(\mathbf{q}(t_0), \dot{\mathbf{q}}(t_0))$.*

Exploring the structures of the adaptive laws (4.13)–(4.15), three possible scenarios are identified $\forall i = 0, 1, 2$:

Case (1): $||\mathbf{e}_f|| \geq \varpi$ and $\dot{\hat{\theta}}_i, \dot{\gamma}$ increase;

Case (2): $||\mathbf{e}_f|| \geq \varpi$ and $\dot{\hat{\theta}}_i, \dot{\gamma}$ decrease;

Case (3): $\dot{\hat{\theta}}_i = 0, \dot{\gamma} = 0$ when $||\mathbf{e}_f|| < \varpi$.

Theorem 2. *The closed-loop system (4.18) with control input (4.11)–(4.15) guarantees* $\mathbf{e}(t)$, $\mathbf{e}_f(t)$, $\tilde{\theta}_i(t)$ *and* $\gamma(t)$ *to be UUB where* $\tilde{\theta}_i \triangleq (\hat{\theta}_i - \theta_i^*)$, $i = 0, 1, 2$.

Proof. The stability of the closed loop is analysed for all the three cases mentioned earlier utilizing the common Lyapunov function considered as

$$V_1 = V + \sum_{i=0}^{2} \frac{1}{2\alpha_i}\tilde{\theta}_i^2 + \frac{1}{2\alpha_3}\gamma^2 \tag{4.33}$$

Case (1): $||\mathbf{e}_f|| \geq \varpi$ and $\dot{\hat{\theta}}_i, \dot{\gamma}$ increase $\forall i = 0, 1, 2$.

Note that $\sum_{i=0}^{2} \frac{1}{\alpha_i}\tilde{\theta}_i\dot{\hat{\theta}}_i = \mathbf{Y}(\xi)^T(\hat{\Theta} - \Theta^*)||\mathbf{e}_f||$. Then using (4.13) and (4.14) and following the procedure in (4.26), one obtains

$$\begin{aligned}
\dot{V}_1 &\leq -\mathbf{e}_f^T\mathbf{G}\mathbf{e}_f - \mathbf{e}^T\Omega\mathbf{e} + \mathbf{e}_f^T\left(-\eta\hat{\rho}||\mathbf{e}_f||^{\varepsilon}\frac{\mathbf{e}_f}{||\mathbf{e}_f||} + \sigma\right) \\
&\quad + \mathbf{Y}(\xi)^T(\hat{\Theta} - \Theta^*)||\mathbf{e}_f|| + \gamma||\mathbf{e}_f|| \\
&\leq -\mathbf{e}_f^T\mathbf{G}\mathbf{e}_f - \mathbf{e}^T\Omega\mathbf{e} - (\mathbf{Y}(\xi)^T\hat{\Theta} + \gamma)||\mathbf{e}_f|| + \mathbf{Y}(\xi)^T\Theta^*||\mathbf{e}_f|| \\
&\quad + \mathbf{Y}(\xi)^T(\hat{\Theta} - \Theta^*)||\mathbf{e}_f|| + \gamma||\mathbf{e}_f|| \\
&\leq -\lambda_{\min}(\mathbf{G})||\mathbf{e}_f||^2 - \lambda_{\min}(\Omega)||\mathbf{e}||^2 \leq 0
\end{aligned} \tag{4.34}$$

From (4.34), it can be inferred that $V_1(t) \in \mathcal{L}_\infty$. Thus, the definition of V_1 yields $\gamma(t)$, $\tilde{\theta}_i(t)$, $\mathbf{e}(t)$, $\mathbf{e}_f(t) \in \mathcal{L}_\infty \Rightarrow \xi(t), \hat{\theta}_i \in \mathcal{L}_\infty$. Thus, for this case, the closed-loop system remains stable.

The gains $\gamma, \hat{\theta}_i$, $i = 0, 1, 2$, remain bounded for Case (1), decrease for Case (2) and remain unchanged for Case (3). Hence, $\exists\, \bar{\gamma}, \bar{\theta}_i \in \mathbb{R}^+$ such that

$$\hat{\theta}_0(t) \leq \bar{\theta}_0, \hat{\theta}_1(t) \leq \bar{\theta}_1, \hat{\theta}_2(t) \leq \bar{\theta}_2, \gamma(t) \leq \bar{\gamma}\ \forall t \geq t_0 \tag{4.35}$$

Case (2): $||\mathbf{e}_f|| \geq \varpi$ and $\dot{\hat{\theta}}_i, \dot{\gamma}$ decrease $\forall i = 0, 1, 2$.

Using $\gamma \geq \beta$ (from (4.19)) and $||\mathbf{e}_f|| \leq ||\Gamma||||\xi||$ yields

$$\begin{aligned}
\dot{V}_1 &\leq -\mathbf{e}_f^T\mathbf{G}\mathbf{e}_f - \mathbf{e}^T\Omega\mathbf{e} + \mathbf{e}_f^T\left(-\eta\hat{\rho}||\mathbf{e}_f||^{\varepsilon}\frac{\mathbf{e}_f}{||\mathbf{e}_f||} + \sigma\right) \\
&\quad - \mathbf{Y}(\xi)^T(\hat{\Theta} - \Theta^*)||\mathbf{e}_f|| - \gamma\varsigma||\xi||^4
\end{aligned}$$

$$\leq -\mathbf{e}_f^T\mathbf{G}\mathbf{e}_f - \mathbf{e}^T\Omega\mathbf{e} - (\mathbf{Y}(\xi)^T\hat{\Theta} + \gamma)||\mathbf{e}_f|| + \mathbf{Y}(\xi)^T\Theta^*||\mathbf{e}_f||$$
$$-\varsigma\beta||\xi||^4 - \mathbf{Y}(\xi)^T(\hat{\Theta} - \Theta^*)||\mathbf{e}_f||$$
$$\leq -\varsigma\beta||\xi||^4 + 2||\Gamma||\{\theta_0^* + \theta_1^*||\xi|| + \theta_2^*||\xi||^2\}||\xi||$$
$$-\lambda_{\min}(\mathbf{G})||\mathbf{e}_f||^2 - \lambda_{\min}(\Omega)||\mathbf{e}||^2 \tag{4.36}$$

Since $0 \leq \hat{\theta}_i(t) \leq \bar{\theta}_i$, $\beta \leq \gamma \leq \bar{\gamma}$ (from (4.19) and (4.35)), the definition of V_1 in (4.33) yields

$$V_1 \leq \rho_M(||\mathbf{e}_f||^2 + ||\mathbf{e}||^2) + \zeta \tag{4.37}$$

where $\zeta \triangleq \sum_{i=0}^{2}\frac{1}{\alpha i}(\theta_i^{*2} + \bar{\theta}_i^2) + \frac{1}{\alpha_3}\bar{\gamma}^2$. Thus using (4.37),

$$-\lambda_{\min}(\mathbf{G})||\mathbf{e}_f||^2 - \lambda_{\min}(\Omega)||\mathbf{e}||^2 \leq -\rho V_1 + \rho\zeta \tag{4.38}$$

Substitution of (4.38) into (4.36) yields

$$\dot{V}_1 \leq -\rho V_1 + f_p(||\xi||) \tag{4.39}$$

where $f_p(||\xi||) = -\varsigma\beta||\xi||^4 + 2||\Gamma||\{\theta_0^*||\xi|| + \theta_1^*||\xi||^2 + \theta_2^*||\xi||^3\} + \rho\zeta$. Applying Descartes' rule of sign change [40], one can verify that $f_p(||\xi||)$ has maximum one positive real root. Further, it is to be noticed that $f_p(||\xi|| = 0) = \rho\zeta \in \mathbb{R}^+$ and $f_p(||\xi||) \to -\infty$ as $||\xi|| \to \infty$. Hence, according to Bolzano's Intermediate Value Theorem [41], $f_p(||\xi||)$ will have at least one positive real root. Consequently, combination of the Intermediate Value Theorem and Descartes' rule of sign change reveals that $f_p(||\xi||)$ has exactly one positive real root. Therefore, the nature of roots of $f_p(||\xi||)$ will be either (i) one positive real root and three negative real roots or (ii) one positive real root, one negative real root and a pair of complex conjugate roots.

Let $\iota \in \mathbb{R}^+$ be the positive real root of $f_p(||\xi||)$. Figure 4.1 depicts the nature of $f_p(||\xi||)$ depending on the various combination of roots [42]. It is to be noted that the actual graph and values of the roots of the polynomial $f_p(||\xi||)$ would depend on the values of the coefficients of $f_p(||\xi||)$. However, $\theta_0^*, \theta_1^*, \theta_2^*$ and ζ are unknown here. Nevertheless, to study the stability of the system, it is sufficient to analyse the nature of $f_p(||\xi||)$ (i.e. the instances when $f_p(||\xi||) > 0$ or $f_p(||\xi||) \leq 0$) rather than determining the values of the roots. The nature of any polynomial can be understood from the occurrence of its real roots [42]. Moreover, the leading coefficient of $f_p(||\xi||)$ (the coefficient of the highest degree term $||\xi||^4$) is negative (as $\varsigma, \beta \in \mathbb{R}^+$). As a matter of fact, one can notice from Figure 4.1 that $f_p(||\xi||) \leq 0$ when $||\xi|| \geq \iota$. Hence, the overall system would be UUB for this case. From the point of view of controller design, it is important to reduce ι such that better tracking accuracy can be achieved and this can be obtained by increasing ς.

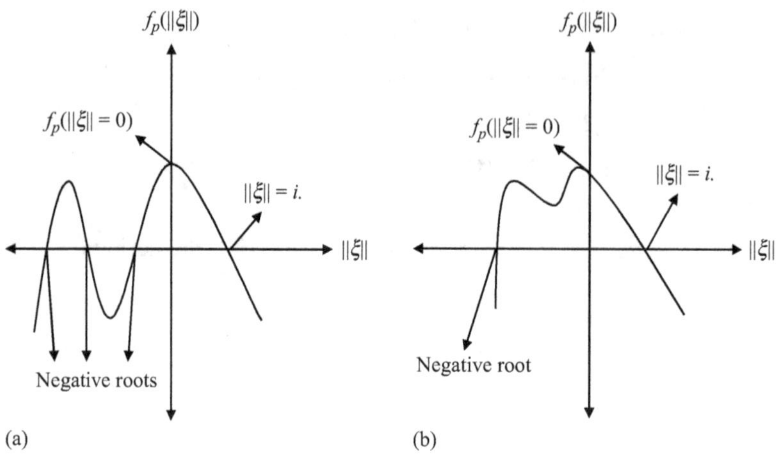

Figure 4.1 $f_p(||\xi||)$ *with (a) one positive real root and three negative real roots and (b) one positive real root, one negative real root and a pair of complex conjugate roots*

Case 3: $\dot{\hat{\theta}}_i = 0, \dot{\gamma} = 0, \forall i = 0, 1, 2$ when $||\mathbf{e}_f|| < \varpi$.
Similar to Case (1), $\dot{V}_1$ can be simplified for this case as

$$\begin{aligned}\dot{V}_1 &\le -\mathbf{e}_f^T\mathbf{G}\mathbf{e}_f - \mathbf{e}^T\Omega\mathbf{e} + \mathbf{e}_f^T\left(-\eta\hat{\rho}||\mathbf{e}_f||^\varepsilon\frac{\mathbf{e}_f}{||\mathbf{e}_f||} + \sigma\right)\\ &\le -\mathbf{e}_f^T\mathbf{G}\mathbf{e}_f - \mathbf{e}^T\Omega\mathbf{e} - (\mathbf{Y}(\xi)^T\hat{\Theta} + \gamma)(||\mathbf{e}_f||^2/\varpi) + \mathbf{Y}(\xi)^T\Theta^*||\mathbf{e}_f||\\ &\le -\lambda_{\min}(\mathbf{G})||\mathbf{e}_f||^2 - \lambda_{\min}(\Omega)||\mathbf{e}||^2 + \mathbf{Y}(\xi)^T\Theta^*||\mathbf{e}_f||\end{aligned} \tag{4.40}$$

For $||\mathbf{e}_f|| < \varpi$, the system remains bounded inside the ball $B_\varpi \triangleq \{||\Gamma\xi|| < \varpi\}$ as $\mathbf{e}_f = \Gamma\xi$. This implies that $\mathbf{Y}(||\xi||) \in \mathscr{L}_\infty$. Hence, for $||\mathbf{e}_f|| < \varpi$, $\exists\vartheta \in \mathbb{R}^+$ such that

$$||\mathbf{Y}(\xi)^T\Theta^*||||\mathbf{e}_f|| \le \varpi\vartheta \tag{4.41}$$

Let us define a scalar z as $0 < z < \lambda_{\min}(\mathbf{G})$. Then using (4.37) and (4.41), (4.40) is modified as

$$\begin{aligned}\dot{V}_1 &\le -\lambda_{\min}(\Omega)||\mathbf{e}||^2 - \{\lambda_{\min}(\mathbf{G}) - z\}||\mathbf{e}_f||^2 - z||\mathbf{e}_f||^2 + \varpi\vartheta\\ &\le -\rho_1 V_1 - z||\mathbf{e}_f||^2 + \rho_1\zeta + \varpi\vartheta\end{aligned} \tag{4.42}$$

where $\rho_1 \triangleq \{\min\{(\lambda_{\min}(\mathbf{G}) - z), \lambda_{\min}(\Omega)\}\}/\rho_M$. Hence, the system would be UUB when

$$||\mathbf{e}_f|| = ||\Gamma\xi|| \ge \sqrt{(\rho_1\zeta + \varpi\vartheta)/z} \tag{4.43}$$

Since the closed-loop system remains UUB for both the cases $||\mathbf{e}_f|| \geq \varpi$ and $||\mathbf{e}_f|| < \varpi$ using the common Lyapunov function (4.33), the overall closed-loop system also remains UUB [43]. □

Remark 6. *It is to be noted that, for the stability of the system, the condition (4.19) is necessary. Moreover, reduction in ι results in improved controller accuracy and this can be achieved with high values of ς. However, one needs to be cautious that the condition $\gamma \leq \beta$ may get excited with too high value of ς which may eventually lead to unwanted increment in all the gains $\gamma, \hat{\theta}_i, i = 0, 1, 2$. Further, the scalar terms $z, \vartheta, \psi, \mu_2, \zeta, \bar{\theta}_i$ and $\bar{\gamma}$ are solely utilized for the analytical purpose.*

Remark 7. *Theorems 1 and 2 help to realize the importance of the auxiliary gain γ. The expression (4.20) reveals that the presence of α_3 (contributed by $\dot{\gamma} > 0$) reduces t_1, which results in faster adaptation. Moreover, the leading negative term in $f_p(||\xi||)$ with degree four, i.e. $-\varsigma\beta||\xi||^4$ (contributed by $\dot{\gamma} < 0$) guarantees system stability for Case (2) by making $f_p(||\xi||) \leq 0$ for $||\xi|| \geq \iota$. This also rationalizes the selection of positive lower bound β for γ while the other gains $\hat{\theta}_i$, $i = 0, 1, 2$ are lower bounded by zero.*

Tuning of the parameter ϖ: It is beneficial to reduce ϖ so that $\hat{\theta}_i$'s and γ can be updated over a long range. This can be accomplished by simply increasing η, as can be noted from (4.17). Moreover, it is noteworthy that increment of η does not necessarily demand increased control input. This argument can be explained as follows: the control input $\Delta\tau$ can be written in the following way from (4.12) and (4.13) and (4.14) for $||\mathbf{e}_f|| \geq \varpi$:

$$\Delta\tau = \left(\sum_{i=0}^{2}\int_{t_{in}}^{t} \eta\alpha_i||\xi||^i||\mathbf{e}_f||\mathrm{d}t + \eta\alpha_3\int_{t_{in}}^{t}||\mathbf{e}_f||\mathrm{d}t\right)\left(||\mathbf{e}_f||^{\varepsilon}\frac{\mathbf{e}_f}{||\mathbf{e}_f||}\right), \quad \dot{\hat{\theta}}_i, \dot{\gamma} > 0$$

$$\Delta\tau = \left(\sum_{i=0}^{2}\int_{t_{in}}^{t} (-\eta\alpha_i||\xi||^i||\mathbf{e}_f||)\mathrm{d}t - \eta\varsigma\alpha_3\int_{t_{in}}^{t}||\xi||^4\mathrm{d}t\right)\left(||\mathbf{e}_f||^{\varepsilon}\frac{\mathbf{e}_f}{||\mathbf{e}_f||}\right), \quad \dot{\hat{\theta}}_i, \dot{\gamma} > 0$$

It is obvious that $\hat{\rho}$ would be non-negative by virtue of the ASRC law (4.13)–(4.14). The important aspect to observe here is that η effectively governs the rate of change of $\hat{\theta}_i$'s and γ along with α_i's. As a matter of fact, a designer can always split a desired value among η and α_i. For example, if a designer wants the gains to increase at a rate of 10 for $\hat{\theta}_0$, then it can be taken as $\eta = 10$ and $\alpha_0 = 1$ giving a combined value of $\eta\alpha_0 = 10$. The benefit of taking higher value of η is that, one can reduce ϖ for a fixed ε and can increase the adaptation range for the gains. However, such mathematical arrangement increases the theoretical upper bounds of t_1, t_2 and t_3 of Theorem 1; nevertheless, since these time values only signify upper bounds, the true reaching values remain the same.

4.5 Simulation results

The proposed ASRC framework is numerically evaluated using the parameters of AUV-150 [17,20]. Since one of the core applications of AUV is underwater survey, a lawn mower path [17] is selected for the AUV to follow. The desired trajectory for this path is given as follows:

$$x^d = 0.2a_1t + a_1\sin(0.2t),\ y^d = 15\cos(0.1t) \tag{4.44}$$

where a_1 and b_1 denote the lawn mower path's width and height, respectively. In simulated scenario, these parameters are selected as $a_1 = 3$ and $b_1 = 15$. In practical circumstances, sea current, wave, wind, etc. are inevitable phenomena and these adversely impact a controller's performance. Amongst them, sea wave and wind are always a factor while AUV is on the sea surface during launching operation and these have pervasive effect on the initial position and orientation of the AUV. Nevertheless, the effect of sea wave and wind can be neglected while AUV is submerged. Hence, for simplicity, the effects of sea wave and wind are neglected during simulation considering the AUV to be at a certain depth a priori before initiating the surveying motion. However, initial error in the positions are considered to emulate a more practical scenario during the simulation. Accordingly, the desired trajectory for the other degrees-of-freedom are selected as

$$z^d = 70,\ \phi^d = 0,\ \omega^d = 0,\ \psi^d = \tan^{-1}(y^d/x^d) \tag{4.45}$$

Here, for simplicity, the desired depth has been considered to be 70, while it can be taken as any other constant values. Considering the sea current velocity as 0.03 m/s in all three linear positions in the body coordinate frame, the effect of the sea current on the linear velocity measurement in the earth fixed frame is evaluated in the following manner [37]:

$$\{\dot{x},\dot{y},\dot{z}\}^T_{\text{actual}} = \{\dot{x},\dot{y},\dot{z}\}^T_{\text{measured}} - \mathbf{J}_1\{0.03, 0.03, 0.03\}^T \tag{4.46}$$

where the expression for the Jacobian $\mathbf{J}_1$ can be found in [37]. To substantiate the performance of the proposed controller, two different scenarios are constructed for the simulation purpose as given in the following:

Scenario 1: In this scenario, the AUV is directed to follow the lawn mower path (4.44) under the influence of the sea current (4.46). Moreover, this scenario studies the capability of the proposed ASRC in the alleviation of the over- and underestimation problems of switching gains.

Scenario 2: One of the important aspects that has been solved in this chapter is to devise an ARC law for AUV which has a state dependent structure in uncertainty. This particular issue is studied in this scenario, which is an extension of the Scenario 1, where efficacy of the ASRC framework is studied in comparison to one of the most widely used ARC framework-adaptive SMCs [26].

For both the scenarios, the controller parameters for ASRC are selected as $\mathbf{G} = 5\mathbf{I}$, $\eta = 2.5$, $\epsilon = 0.5$, $\alpha_0 = \alpha_1 = \alpha_2 = 10$, $\alpha_3 = 1$, $\varsigma = 10$ and $\beta = 0.0001$. The initial condition for the gains are selected as $\hat{\theta}_0(t_0) = \hat{\theta}_1(t_0) = \hat{\theta}_2(t_0) = \gamma(t_0) = 300$.

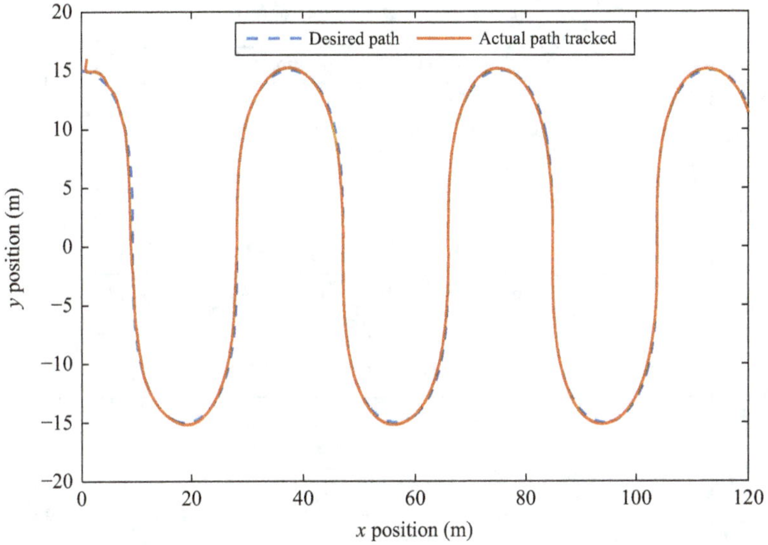

Figure 4.2 Lawn mower path tracked by AUV while employing the proposed ASRC

Moreover, the initial position error is considered as 1 m for each of the linear positions, i.e. x, y and z. Also, initial individual orientations in roll, pitch and heading are considered to be at 0 degree. Such initial conditions are kept same for both the scenarios.

4.5.1 Results and discussions for Scenario 1

The lawn mower path tracked by the AUV using the ASRC algorithm is shown in Figure 4.2. Further, to clearly comprehend the tracking performance, the tracking error in linear positions and attitude angles are shown in Figures 4.3 and 4.4, respectively. In Figure 4.3, the path error is defined as the Euclidean distance between the tracking error of linear positions. The control input requirement for the all the degrees-of-freedom are depicted in Figure 4.5.

One of the important attributes of an ARC is the evaluations of its switching gains. Accordingly, it would be crucial to observe how the overall switching gain $\hat{\rho}$ as well as its component gains $\hat{\theta}_0, \hat{\theta}_1, \hat{\theta}_2$ and γ vary. In this regard, the evaluations of $\hat{\theta}_0, \hat{\theta}_1, \hat{\theta}_2$ and γ are shown in Figure 4.6. According to (4.13) and (4.14), the sign of $\mathbf{e}^T\dot{\mathbf{e}}$ is an important determining factor for the evaluation of these gains when $||\mathbf{e}_f|| \geq \varpi$. In simplified terms, the gains $\hat{\theta}_i, \gamma$ should follow the pattern of $||\mathbf{e}||$ as long as $||\mathbf{e}_f|| \geq \varpi$ and they do not attempt to go below their respective lower bounds. To illustrate this fact, a magnified version of Figure 4.6 is shown in Figure 4.7.

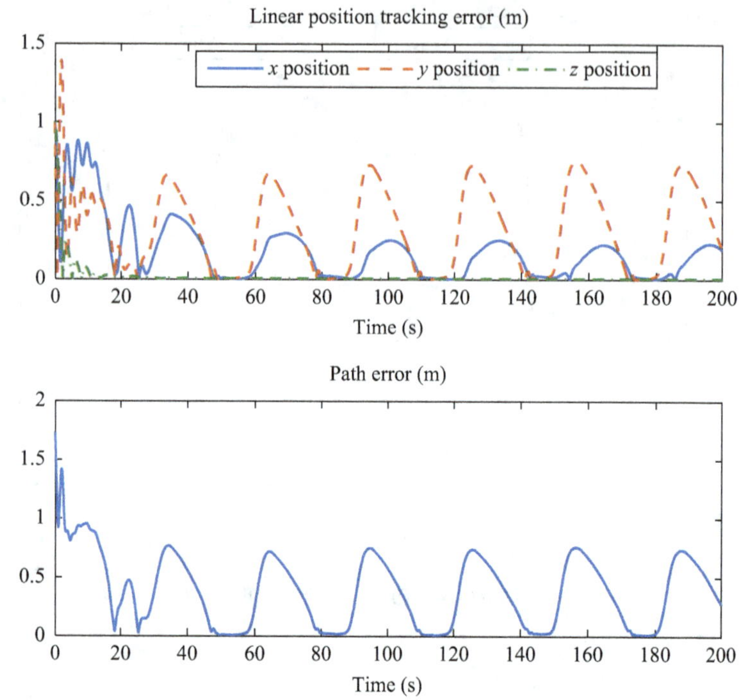

Figure 4.3 Tracking performance of AUV for liner positions

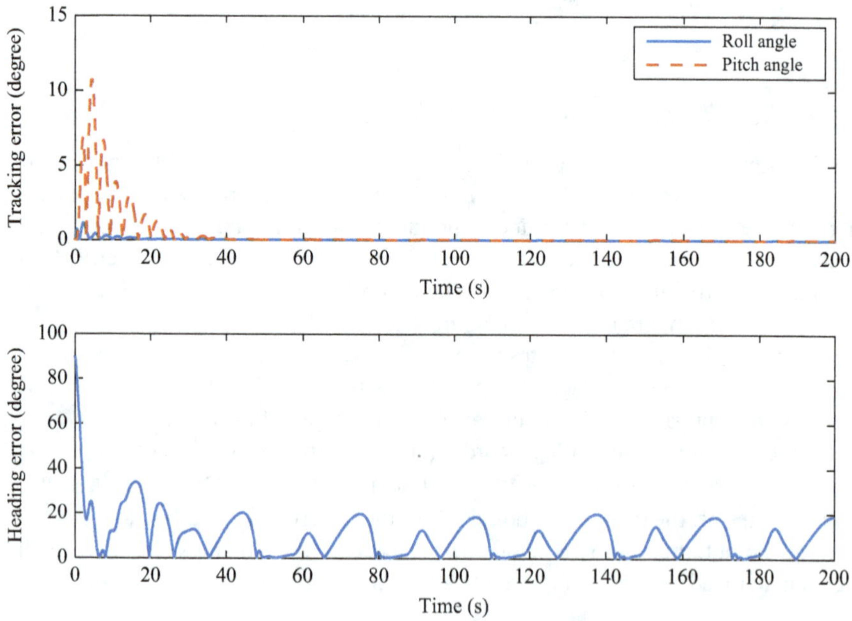

Figure 4.4 Tracking performance of AUV for attitude angles

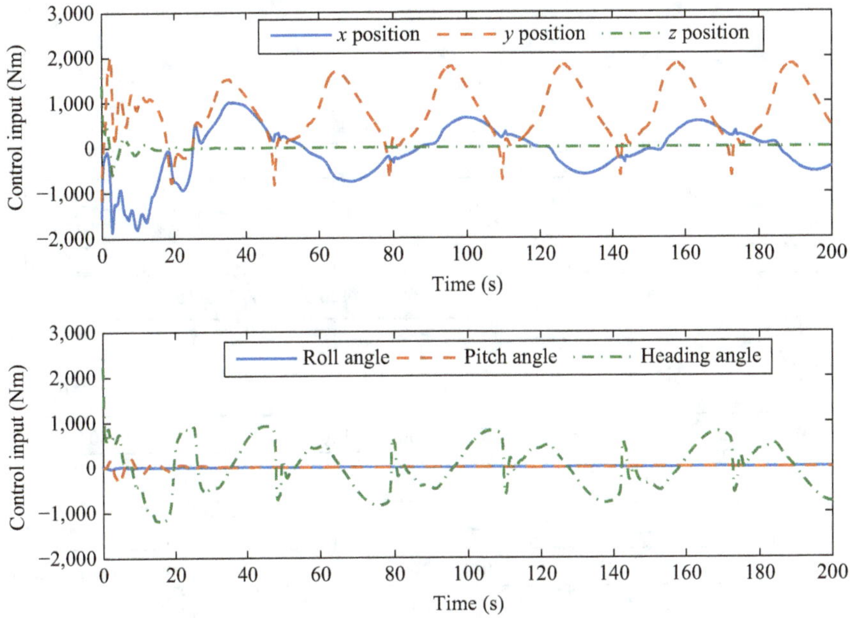

Figure 4.5 Control input requirements

It can be noticed from Figure 4.7 that the gains $\hat{\theta}_i, \gamma$ follow the variation of $||\mathbf{e}||$; that is, they increase and decrease whenever $||\mathbf{e}||$ increases or does not increase. Further, the gains do not change $||\mathbf{e}_f|| < \varpi$ during the time duration $t = 48$ s to $t = 58$ s (approx.). Similar phenomena can be noticed for other time durations also. This helps the ASRC to alleviate the over- and underestimation problems. Moreover, one can notice that during the initial time duration of $t = 1$ s to $t = 5$ s (approx.), the gains increased even when $||\mathbf{e}||$ was decreasing. This happened according to (4.13) and (4.14), where the gain γ, while decreasing, were attempting to breach its lower bound β. This situation can be avoided or limited by properly tuning the parameters α_i's and β. Nevertheless, during such situations, overestimation of gains cannot be avoided even though it sustains for a short-time duration. Hence, it is claimed that the proposed ASRC can *alleviate* rather than *eliminate* the over- and underestimation problems of switching gains.

It can be observed from Figure 4.6 that γ has an increasing nature. Nevertheless, it is the overall gain $\hat{\rho}$ which is of prime importance and $\hat{\rho}$ is a function of the gains $\hat{\theta}_i, \gamma$ as well as the tracking error variable ξ. The evaluation of $\hat{\rho}$ is given in Figure 4.8, which asserts the boundedness of $\hat{\rho}$.

4.5.2 Results and discussions for Scenario 2

The objective of this scenario is to validate the efficacy of the proposed ARC framework where state-dependent uncertainty is present in the system. For this purpose,

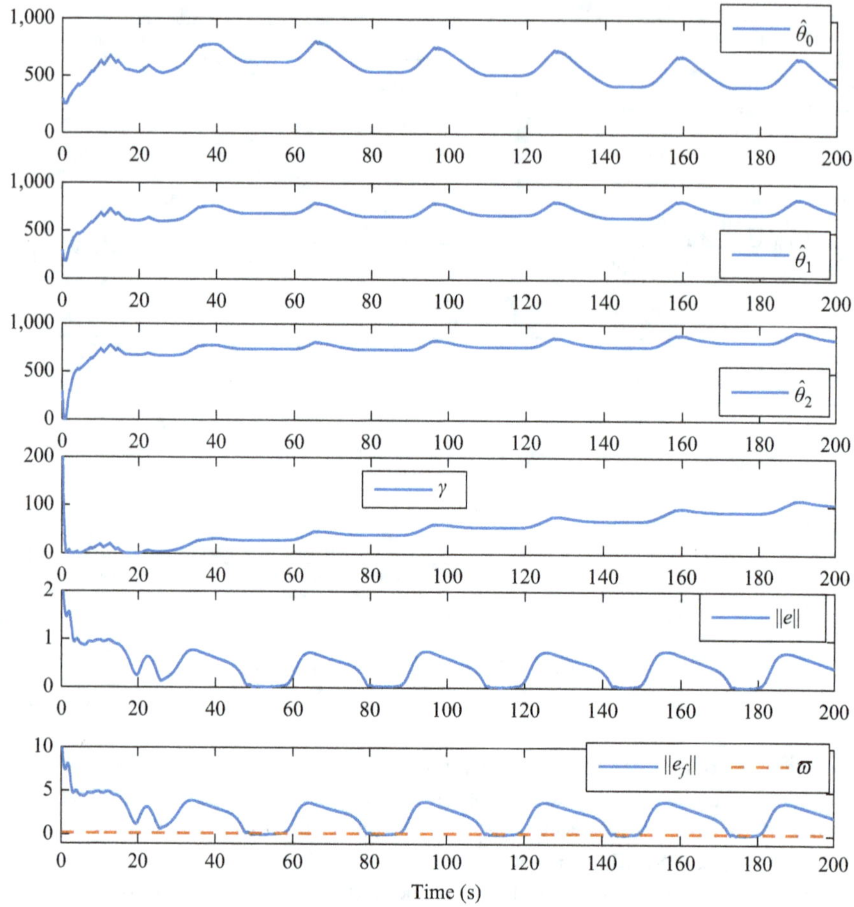

Figure 4.6 Evaluation of switching gains

the proposed ASRC is compared here with the adaptive sliding mode controller (ASMC) [26]. It is to be noted here that, to implement ASMC, the nominal knowledge of the system dynamics parameters are required. However, the proposed ASRC law does not require such prerequisite. Moreover, the adaptive law of ASMC suffers from the over- and underestimation problems [34]. Nevertheless, to have a fair comparison as much as possible, the ASMC is implemented by only replacing the adaptive law of ASRC (4.13)–(4.15) with that of [26] while keeping the same control structure (4.11).

However, ASMC [26] assumes that the overall uncertainty is upper bounded by a constant. This assumption implies, for AUV system (4.1), the parameters $\theta_1^* = \theta_2^* = 0$. Consequently, the overall switching gain becomes $\hat{\rho} = \hat{\theta}_0$ and the adaptive law is only used to evaluate $\hat{\theta}_0$. It can be comprehended from the structure of uncertainty

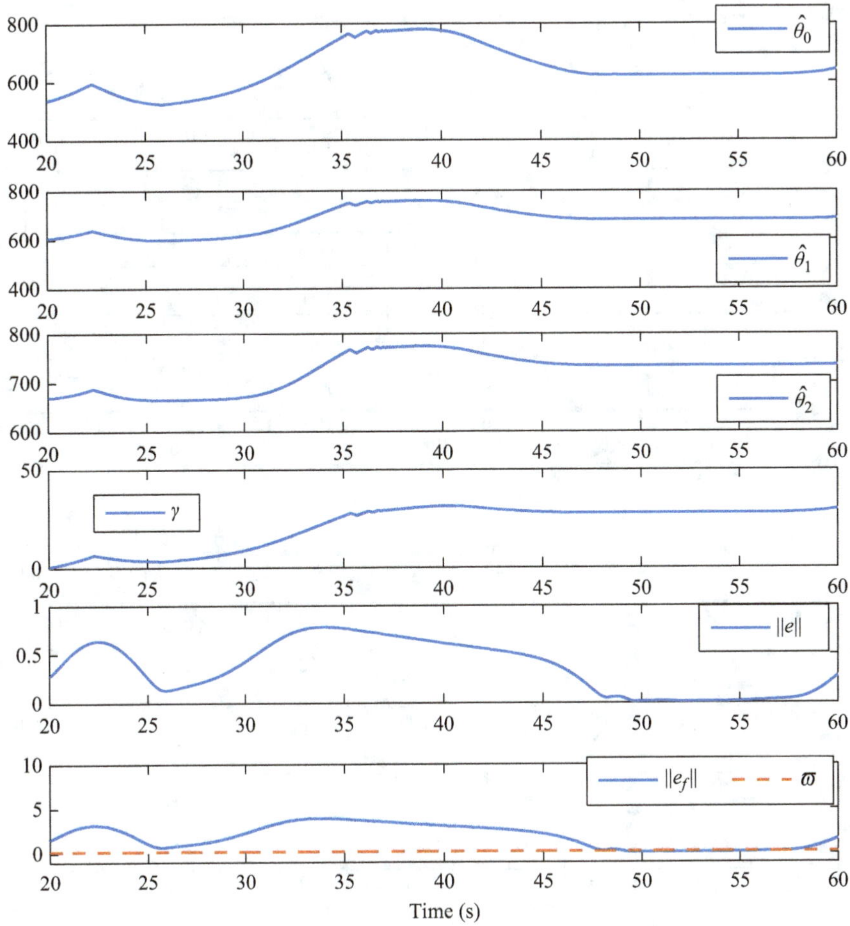

Figure 4.7 Evaluation of switching gains (magnified)

(4.7) that such assumption is restrictive (implying the states are bounded a priori) for an AUV system. To clearly infer the pervasive effects of such assumption on the performance of the controller, the tracking performance of ASRC and that of ASMC is illustrated in Figure 4.9. The simulation scenario is kept similar for both the controllers as in Scenario 1 with same choice of initial gain conditions. It can be noticed from Figure 4.9 that the tracking performance of the proposed ASRC is certainly superior to that of ASMC under the similar condition. As mentioned earlier, the primary reason for ASMC not providing similar control performance is down to its conservative assumption which, in effect, results in improper switching gain and consequently controller gain cannot negotiate state-dependent uncertainty properly.

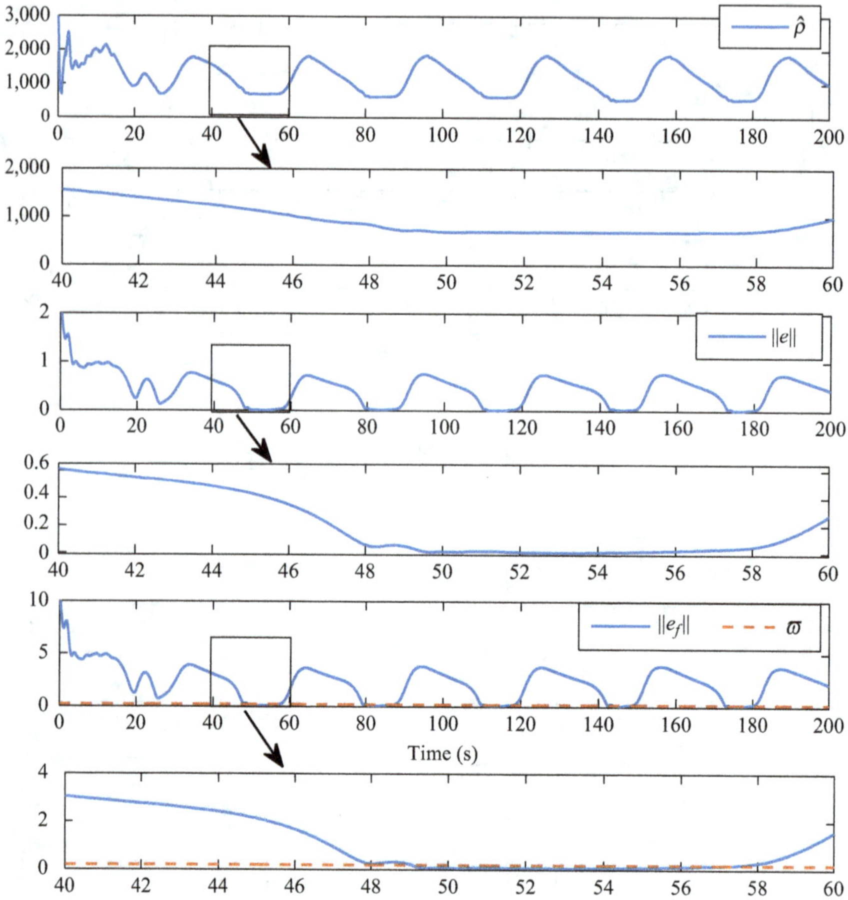

Figure 4.8 Evaluation of the overall gain $\hat{\rho}$

4.6 Conclusion and future direction

This chapter introduces a new class of ARC tracking control framework for AUV without any knowledge of the system dynamics parameters. Conventional ARC strategies either require complete/partial knowledge of the systems dynamics parameters or presume the overall uncertainty (or its time derivative) is upper bounded by a constant. However, such presumption is restrictive for systems like AUV which has explicit presence of states in its structure of system uncertainty. Besides, many of the existing ARC-based designs suffer from the over- and underestimation problems of switching gains. In such regards, the proposed ASRC law not only avoids restrictive presumptions of system uncertainty being upper bounded by a constant, it simultaneously alleviates the over- and underestimation problems of switching gain. Further, the

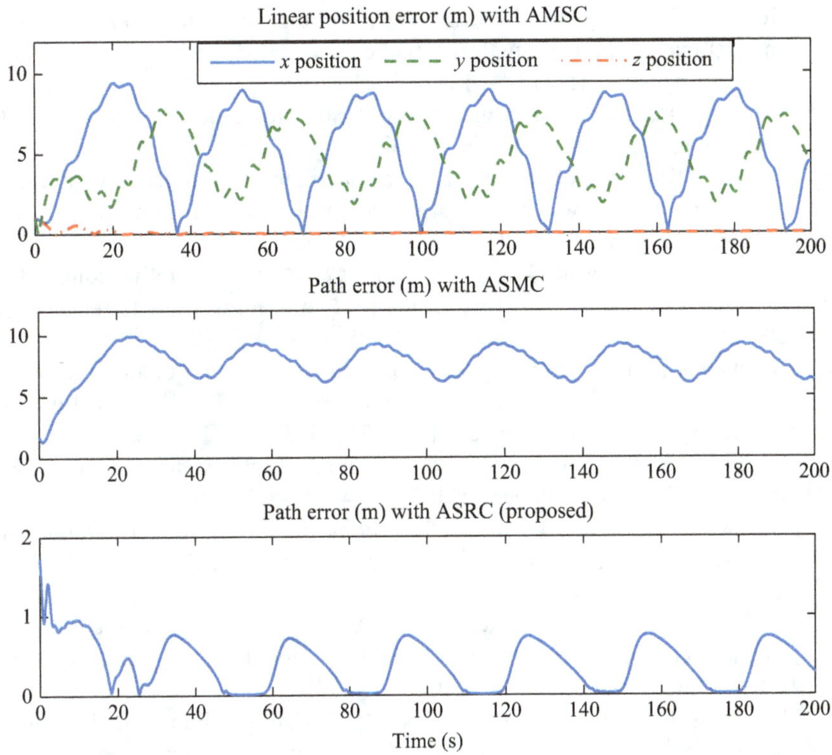

Figure 4.9 Tracking performance comparison between the ASMC

effectiveness of the ASRC law is verified in simulation in comparison with the existing ASMC law under the scenario of uncertain system dynamics. An important future work would be to formulate an ARC for AUV considering input/output delay since data acquisition and localization in underwater scenario is always challenging and a controller needs to take care of such circumstances.

References

[1] Cristi R, Healey A. Adaptive identification and control of an autonomous underwater vehicle. In: Unmanned Untethered Submersible Technology, 1989. Proceedings of the Sixth International Symposium on. IEEE; 1989. pp. 563–572.

[2] Fossen TI, Sagatun SI. Adaptive control of nonlinear underwater robotic systems. In: Robotics and Automation, 1991. Proceedings, 1991 IEEE International Conference on. IEEE; 1991. pp. 1687–1694.

[3] Goheen KR, Jefferys ER. Multivariable self-tuning autopilots for autonomous and remotely operated underwater vehicles. IEEE Journal of Oceanic Engineering. 1990;15(3):144–151.

[4] Yuh J. An adaptive and learning control system for underwater robots. IFAC Proceedings Volumes. 1996;29(1):145–150.

[5] Choi SK, Yuh J. Experimental study on a learning control system with bound estimation for underwater robots. In: Underwater Robots. Minneapolis, USA: Springer; 1996. pp. 113–120.

[6] Antonelli G, Caccavale F, Chiaverini S, *et al.* A novel adaptive control law for underwater vehicles. IEEE Transactions on Control Systems Technology. 2003;11(2):221–232.

[7] Antonelli G, Chiaverini S, Sarkar N, *et al.* Adaptive control of an autonomous underwater vehicle: experimental results on ODIN. IEEE Transactions on Control Systems Technology. 2001;9(5):756–765.

[8] Yoerger D, Slotine J. Robust trajectory control of underwater vehicles. IEEE Journal of Oceanic Engineering. 1985;10(4):462–470.

[9] Healey AJ, Lienard D. Multivariable sliding mode control for autonomous diving and steering of unmanned underwater vehicles. IEEE Journal of Oceanic Engineering. 1993;18(3):327–339.

[10] Cristi R, Papoulias FA, Healey AJ. Adaptive sliding mode control of autonomous underwater vehicles in the dive plane. IEEE Journal of Oceanic Engineering. 1990;15(3):152–160.

[11] Papoulias F, Cristi R, Marco D, *et al.* Modeling, sliding mode control design, and visual simulation of AUV dive plane dynamic response. In: Unmanned Untethered Submersible Technology, 1989. Proceedings of the Sixth International Symposium on. IEEE; 1989. pp. 536–547.

[12] Lee PM, Hong SW, Lim YK, *et al.* Discrete-time quasi-sliding mode control of an autonomous underwater vehicle. IEEE Journal of Oceanic Engineering. 1999;24(3):388–395.

[13] Innocenti M, Campa G. Robust control of underwater vehicles: sliding mode vs. LMI synthesis. In: American Control Conference, 1999. Proceedings of the 1999. Vol. 5. IEEE; 1999. pp. 3422–3426.

[14] Sutarto H, Budiyono A. Development of linear parameter varying control system for autonomous underwater vehicle. Indian Journal of Marine Sciences. 2011;40(2):275–286.

[15] Akçakaya H, Gören Sümer L. Robust control of variable speed autonomous underwater vehicle. Advanced Robotics. 2014;28(9):601–611.

[16] Santhakumar M, Asokan T. Investigations on the hybrid tracking control of an underactuated autonomous underwater robot. Advanced Robotics. 2010;24(11):1529–1556.

[17] Roy S, Shome S, Nandy S, *et al.* Trajectory following control of AUV: a robust approach. Journal of The Institution of Engineers (India): Series C. 2013;94(3):253–265.

[18] Kumar RP, Dasgupta A, Kumar C. Robust trajectory control of underwater vehicles using time delay control law. Ocean Engineering. 2007;34(5–6): 842–849.

[19] Kumar R, Kumar C, Sen D, *et al.* Discrete time-delay control of an autonomous underwater vehicle: theory and experimental results. Ocean Engineering. 2009;36(1):74–81.

[20] Roy S, Nandy S, Shome SN, *et al.* Robust position control of an autonomous underwater vehicle: a comparative study. In: Automation Science and Engineering (CASE), 2013 IEEE International Conference on. IEEE; 2013. pp. 1002–1007.

[21] Roy S, Nandy S, Ray R, *et al.* Robust diving and composite path tracking control of an autonomous underwater vehicle. International Journal of Offshore and Polar Engineering. 2015;25(04):305–313.

[22] Kim J, Joe H, Yu Sc, *et al.* Time-delay controller design for position control of autonomous underwater vehicle under disturbances. IEEE Transactions on Industrial Electronics. 2016;63(2):1052–1061.

[23] Cui R, Yang C, Li Y, *et al.* Adaptive neural network control of AUVs with control input nonlinearities using reinforcement learning. IEEE Transactions on Systems, Man, and Cybernetics: Systems. 2017;47(6):1019–1029.

[24] Wu H, Song S, You K, *et al.* Depth control of model-free AUVs via reinforcement learning. IEEE Transactions on Systems, Man, and Cybernetics: Systems. 2018.

[25] Chin CS, Lin WP. Robust genetic algorithm and fuzzy inference mechanism embedded in a sliding-mode controller for an uncertain underwater robot. IEEE/ASME Transactions on Mechatronics. 2018;23(2):655–666.

[26] Plestan F, Shtessel Y, Bregeault V, *et al.* New methodologies for adaptive sliding mode control. International Journal of Control. 2010;83(9):1907–1919.

[27] Plestan F, Shtessel Y, Bregeault V, *et al.* Sliding mode control with gain adaptation? Application to an electropneumatic actuator. Control Engineering Practice. 2013;21(5):679–688.

[28] Utkin VI, Poznyak AS. Adaptive sliding mode control with application to super-twist algorithm: equivalent control method. Automatica. 2013;49(1): 39–47.

[29] Moreno JA, Negrete DY, Torres-Gonzâlez V, *et al.* Adaptive continuous twisting algorithm. International Journal of Control. 2016;89(9):1798–1806.

[30] Roy S, Kar IN. Adaptive-robust control of uncertain Euler-Lagrange systems with past data: a time-delayed approach. In: Robotics and Automation (ICRA), 2016 IEEE International Conference on. IEEE; 2016. pp. 5715–5720.

[31] Roy S, Kar IN, Lee J, *et al.* Adaptive-robust time-delay control for a class of uncertain Euler–Lagrange Systems. IEEE Transactions on Industrial Electronics. 2017;64(9):7109–7119.

[32] Roy S, Kar IN. Adaptive robust tracking control of a class of nonlinear systems with input delay. Nonlinear Dynamics. 2016;85(2):1127–1139.

[33] Roy S, Kar IN, Lee J, *et al.* Adaptive-robust control of a class of EL systems with parametric variations using artificially delayed input and position feedback. IEEE Transactions on Control Systems Technology. 2019;27(2):603–615.

[34] Roy S, Kar IN. Adaptive sliding mode control of a class of nonlinear systems with artificial delay. Journal of the Franklin Institute. 2017;354(18): 8156–8179.

[35] Roy S, Roy SB, Kar IN. Adaptive–robust control of Euler–Lagrange systems with linearly parametrizable uncertainty bound. IEEE Transactions on Control Systems Technology. 2018;26(5):1842–1850.
[36] Patre B, Londhe P, Waghmare L, *et al.* Disturbance estimator-based non-singular fast fuzzy terminal sliding mode control of an autonomous underwater vehicle. Ocean Engineering. 2018;159:372–387.
[37] Fossen TI. Guidance and Control of Ocean Vehicles. New York: John Wiley & Sons Inc; 1994.
[38] Grimm WM. Robot non-linearity bounds evaluation techniques for robust control. International Journal of Adaptive Control. 1990;4(6):501–522.
[39] Khalil HK. Nonlinear Systems, 3rd edn. Upper Saddle River, New Jersey, Prentice Hall. 2002;9(4.2).
[40] Anderson B, Jackson J, Sitharam M. Descartes' rule of signs revisited. The American Mathematical Monthly. 1998;105(5):447–451.
[41] Russ S. A translation of Bolzano's paper on the intermediate value theorem. Historia Mathematica. 1980;7(2):156–185.
[42] Leung KT, Mok IAC, Suen SN. Polynomials and Functions. Hong Kong: Hong Kong University Press; 1992.
[43] Liberzon D. Switching in Systems and Control. Boston, USA: Springer Science & Business Media; 2012.

Part II

Navigation

Chapter 5

AUV navigation: challenges and solutions

Kjetil Bergh Ånonsen[1], Kenneth Gade[1] and Ove Kent Hagen[1]

5.1 Introduction

The term *navigation* is commonly used to describe several different but related concepts. Some use the term for the process of determining the position (in some reference frame) and the orientation of the vehicle only, whereas others also include the problem of how to get from a point *A* to a point *B* in the navigation concept. In this chapter, we focus solely on the former meaning of the term. In other words, we define navigation to mean estimation of the six degrees of freedom of a rigid body (any vehicle or device), i.e., the position and orientation of the body. The position can be expressed in various reference frames (local or global), and the orientation is defined as the spatial orientation of the body frame relative to a local reference tangential to the Earth ellipsoid. This is more precisely defined in Section 5.2. It is often customary to include the estimation of linear and angular velocities in the navigation problem. The uncertainties of the estimates are often also part of the navigation output, and the navigation can be performed in real time, or in post-processing.

Throughout the chapter, we will sometimes need to refer to the positioning part of the navigation problem only, and for this we use the term *positioning*. This is to be understood as the passive process of determining a position, not the more active process of physically placing an object in a certain position. The term *localisation* is used, e.g., in conjunction with the simultaneous localisation and mapping (SLAM) problem, and it is in this chapter to be understood as the same as positioning.

5.1.1 Why is underwater navigation challenging?

In air, land and sea surface applications, the advent of GNSS (Global Navigation Satellite System) has, in many applications, solved the positioning part of the navigation problem. In underwater applications, however, this is not the case, as GNSS signals are blocked by water and cannot be used as long as the vehicle GNSS antenna is submerged. Most autonomous underwater vehicles (AUVs) today, and all high-end AUVs, rely on inertial navigation – i.e., they use measurements from inertial sensors

[1]Norwegian Defence Research Establishment (FFI), Department of Defence Systems, Kjeller, Norway

like gyroscopes and accelerometers, in combination with aiding sensors to counter the inherent drift in the pure inertial navigation estimates. This chapter focuses on inertial navigation-based systems only. The fundamentals of inertial navigation and common aiding sensors and techniques are described in Sections 5.3 and 5.5, respectively.

As stated above, the navigation problem deals with estimating the six degrees of freedom for the body or vehicle. However, not all six degrees of freedom in an inertial navigation system (INS) are equally challenging to estimate. For INSs that are not in free fall, the gravity vector typically dominates the accelerometer measurements, and thus roll and pitch are often estimated to a sufficient accuracy. The horizontal orientation (the heading angle), however, is more complicated to estimate accurately, as discussed in Section 5.4.

When it comes to position estimation, the vertical position, i.e., the depth of the AUV, can be accurately measured by a pressure sensor and is thus usually rather straightforward to estimate [1] up to the level of a possibly unknown tide. The horizontal position of the body or vehicle can, however, in addition to the heading angle, represent a challenge for the navigation system. It is natural to distinguish between the cases when GNSS is available and when it is not. When GNSS is not available, which is the case for a submerged AUV, the navigation system is dependent on other aiding sensors to limit the inherent drift in the position estimate from the INS. Since some of the methods that can be used for heading estimation also involve GNSS, it is natural to divide the navigation problem into four different cases based on gyro quality and GNSS availability, as shown in Figure 5.1. For AUVs, at least when we talk about sophisticated commercial AUVs, the inertial sensors are usually of high quality, such that the heading can be estimated with sufficient accuracy from gyrocompassing (see Section 5.4.2.1). The main challenge for AUV navigation systems is therefore the horizontal position. This corresponds to Category A2 in Figure 5.1.

Green/italic = Often satisfactory Red/underlined = Challenging		GNSS (or similar) available Yes	No
Gyros with sufficient accuracy for gyro-compassing	Yes	Category A1: *Heading* *Horizontal position* Typical cases: Large/expensive vehicles (not submerged), e.g., airplanes, ships and helicopters	Category A2: *Heading* Horizontal position Typical cases: Underwater navigation of large/expensive vehicles, e.g., submarines and AUVs
	No	Category B1: Heading Horizontal position Typical cases: Light/small/cheap applications in air, land or at sea, e.g., UAVs, boats, robots, cameras and personnel	Category B2: Heading Horizontal position Typical cases: GNSS denied light/small/cheap applications, e.g., indoor navigation, applications under GNSS jamming and low-cost underwater navigation

Figure 5.1 All navigation systems can be classified into four different categories, based on gyro quality and GNSS availability. Taken from [2]

Because of this, most aiding techniques developed for AUVs either deal with limiting the position drift, e.g., with the help of velocity sensors like Doppler velocity logs (DVLs), or with obtaining submerged position fixes via acoustic positioning systems, using terrain information, etc. Different aiding techniques for AUVs are thoroughly discussed in Section 5.5.

5.1.2 Autonomous navigation

Precise and accurate navigation is a key prerequisite for true autonomous AUV operations, and the term *navigation autonomy* is sometimes used for the ability of the navigation system to operate autonomously [3]. Even though AUVs by definition are supposed to be autonomous, many AUV systems today rely heavily on assistance from surface vessels or infrastructure on the seafloor. It is therefore beneficial to define different levels of navigation autonomy, depending on the nature of the aiding methods used.

5.1.2.1 Supervised navigation

Supervised navigation is still a widely used operation mode when high position accuracy is needed, e.g., in commercial offshore surveying. When running in supervised mode, a surface vessel is following the AUV, providing it with acoustic aiding using a combination of the position of the surface vessel, which is based on GNSS, and acoustic positioning of the AUV relative to the surface vessel, using ultra-short baseline (USBL) positioning [4]. As long as the surface vessel has GNSS coverage, this results in a drift-free AUV position. However, following the AUV with a surface vessel is expensive and binds up the surface vessel, which could otherwise have been used for other operations. Also, in many scenarios, e.g., in covert military operations or when doing mine counter measures in an area where mines have been deployed, following the AUV is impossible or dangerous. An attractive alternative that has gained popularity recently is to replace the surface ship by an unmanned surface vehicle (USV). This solves some of the issues, but it complicates the overall USV-AUV operation and does not facilitate covertness.

5.1.2.2 Semi-autonomous navigation

When running in semi-autonomous mode, the vehicle is not dependent on aiding from a surface ship but uses pre-deployed infrastructure like underwater transponders to bound the navigation drift (see Figure 5.2). This might be done, e.g., using long baseline (LBL) navigation [5,6] or underwater transponder positioning (UTP) [7], as described in Section 5.5.3. After the necessary infrastructure has been deployed on the seafloor, the AUV can be left to conduct its mission completely autonomously, and this solution is therefore useful in scenarios where high navigation accuracy is needed, while the costs related to the infrastructure deployment are acceptable, e.g., in offshore underwater construction sites or in pipeline inspection scenarios.

5.1.2.3 Fully autonomous navigation

To be able to conduct fully autonomous operations with high submerged navigation accuracy, the AUV cannot depend on aiding from a surface vessel or pre-deployed

Figure 5.2 A Kongsberg Maritime HUGIN AUV utilising underwater transponders, an example of semi-autonomous navigation

infrastructure. To be fully navigation autonomous, the vehicle must use alternative submerged aiding methods. In Section 5.5, some alternatives are discussed, including terrain- and feature-based navigation. Modern AUVs usually carry advanced payload sensors, e.g., sonars and optical cameras, capable of recognising terrain and features that can be used for navigation purposes.

It is worth mentioning that surfacing for GNSS fixes, which is frequently used in some scenarios, e.g., in non-covert operations in shallow water or by underwater gliders, falls into the semi-autonomous navigation category in our definition, since external infrastructure is used. It is therefore possible for a vehicle to operate fully autonomously without being fully navigation autonomous.

5.1.3 Chapter outline

This chapter is structured as follows. First, in Section 5.2, a precise notation system for navigation purposes is outlined, before we in Section 5.3 discuss the basic principles of a general INS which are important for understanding the applicability and limitations of such systems for AUVs. Then Section 5.4 discusses some important properties of a basic INS in an AUV. In Section 5.5, we discuss various sensors and methods that can be used for aiding the INS, including geophysical navigation and SLAM, before making some concluding remarks in Section 5.6. We herein focus on single vehicle navigation only, but some of the methods discussed, especially those involving acoustic ranging, can be adapted to multi-vehicle scenarios by putting transponders on each vehicle and develop a communication protocol to enable the vehicles to exchange navigation information. An overview of work done on cooperative AUV navigation is given in [8]. A tractable alternative in many scenarios is to use a USV as the master node in the collaborative network [9,10].

5.2 Notation

A navigation system must keep track of several rotating coordinate frames, and to get correct answers and full understanding, a stringent and precise notation system is of utmost importance. A unified notation system for navigation is presented in Chapter 2 of [11]. A short, simplified summary of this system, with only the symbols relevant to this chapter, is included in the following.

A *coordinate frame* is defined as a combination of a point (origin), representing position, and a set of basis vectors, representing orientation. Thus, a coordinate frame has six degrees of freedom and can be used to represent the position and orientation of a rigid body.

A general vector is here represented in two different ways:

- $\vec{x}$ (lower case symbol with arrow): coordinate free/geometrical vector, i.e., not decomposed in any coordinate frame.
- $\boldsymbol{x}^A$ (bold lower case symbol with right superscript): vector decomposed/represented in a specific frame A, i.e., a column vector with three scalars, suited for computer implementation.

The relation between the two variants is illustrated in Figure 5.3. Let $\vec{b}_{A,1}$, $\vec{b}_{A,2}$ and $\vec{b}_{A,3}$ be an arbitrary set of basis vectors spanning A, such that $\vec{x}$ can be written as a linear combination of the basis vectors,

$$\vec{x} = x_1\vec{b}_{A,1} + x_2\vec{b}_{A,2} + x_3\vec{b}_{A,3} \tag{5.1}$$

which means that $\boldsymbol{x}^A = [x_1\ x_2\ x_3]^T$, where $[\cdots]^T$ denotes the matrix/vector transpose.

Basic quantities in navigation, such as position, (angular) velocity, etc., relate one coordinate frame to another. To make a quantity unique, the two frames in question are given as right subscripts, as shown in Table 5.1. The vectors in Table 5.1 are written in coordinate-free form since that is the most general form, but before implementation on a computer, they must be decomposed in a selected coordinate frame, e.g., the position vector $\vec{p}_{AB}$ decomposed in frame C is $\boldsymbol{p}^C_{AB}$.

Note that in most examples in Table 5.1 only the position or the orientation of the frame is relevant and the context should make it clear which of the two properties is relevant. For instance, for the quantity $\boldsymbol{p}^C_{AB}$ only the orientation of C is relevant, since

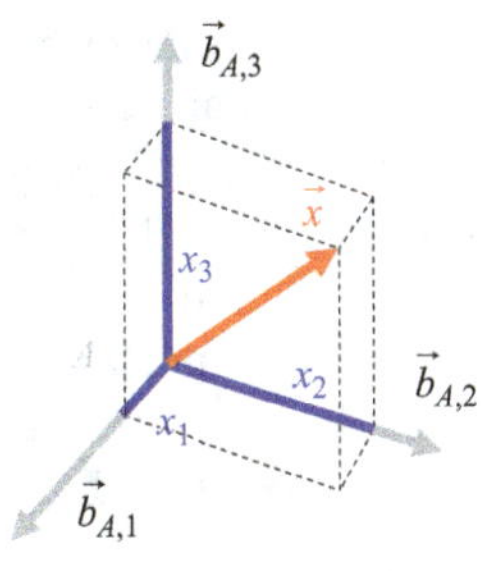

Figure 5.3 Decomposition of a geometrical vector in a coordinate frame

Table 5.1 Summary of the notation used in this chapter

Quantity	Symbol	Description
Position	$\vec{p}_{AB}$	A vector whose length and direction is such that it extends from the origin of frame *A* to the origin of frame *B*, i.e., the position of *B* relative to *A*
Velocity	$\vec{v}_{\underline{A}B}$	The velocity of the origin of frame *B*, relative to frame *A*. The underline indicates that both the position and orientation of *A* is relevant (whereas only the position of *B* matters). Thus, it emphasises that the order of the frames cannot be switched to get the negative vector (which they can for position and angular velocity)
Acceleration	$\vec{a}_{\underline{A}B}$	The acceleration of the origin of frame *B* relative to frame *A*. The underline has the same meaning as for velocity
Rotation matrix	$\boldsymbol{R}_{AB}$	A 3×3 direction cosine matrix describing the orientation of frame *B* relative to frame *A*
Angular velocity	$\vec{\omega}_{AB}$	The angular velocity of frame *B* relative to frame *A*
Horizontal position	$\boldsymbol{n}^E$	*n*-vector used to represent horizontal position without singularities (replaces latitude and longitude). For more details, see Section 5.3.2 and [12,13]

it denotes the frame of decomposition, where only the directions of the basis vectors matter. On the other hand, only the positions of *A* and *B* matter. If both the position and the orientation of the frame are relevant, the frame is underlined to emphasise that fact (see Section 2.3.2 in [11] for more details).

The four most central coordinate frames used in navigation are listed in Table 5.2 and shown in Figure 5.4.

5.3 INS fundamentals

5.3.1 Inertial sensors

In inertial navigation, the navigation is done by means of inertial sensors, i.e., accelerometers and gyros. The principle of an accelerometer is shown in Figure 5.5, where a proof mass is attached to springs, and the measured deflection is proportional to the acceleration. Practical accelerometers are obviously more complex, but the basic principle is the same. According to Einstein's equivalence principle [15], acceleration and gravitation cannot be separated (at one point in space), and hence the gravitation must be compensated for. The total measurement is called specific force and is a measurement of the following vector:

$$\vec{f}_{IB} = \vec{a}_{\underline{I}B} - \vec{g}_B = \vec{a}_{\underline{I}B} - \frac{\vec{F}_{B,\,\text{gravitation}}}{m} \tag{5.2}$$

Table 5.2 The central coordinate frames used in navigation

Symbol	Description
E	***Name***: Earth ***Position***: The origin coincides with the Earth's centre (geometrical centre of ellipsoid model) ***Orientation***: The x-axis is along the Earth's rotation axis, pointing north (the yz plane coincides with the equatorial plane), the y-axis points towards longitude $\pm 90°$ east ***Comments***: The frame is Earth-fixed (rotates and moves with the Earth). The choice of axis directions ensures that at zero latitude and longitude, N (described below) has the same orientation as E. If roll, pitch and yaw are zero, also B (described below) has this orientation. Note that these properties are not valid for another very common choice of the axis directions, here denoted e (lower case), which has z pointing north and x pointing to latitude = longitude = 0
B	***Name***: Body (typically of a vehicle) ***Position***: The origin is in the vehicle's reference point ***Orientation***: The x-axis points forwards, the y-axis to the right (starboard) and the z-axis in the vehicle's down direction ***Comments***: The frame is fixed to the vehicle
N	***Name***: North-East-Down (local level) ***Position***: The origin is directly beneath or above the vehicle (B), at the Earth's surface (surface of ellipsoid model) ***Orientation***: The x-axis points towards north, the y-axis towards east (both are horizontal) and the z-axis down ***Comments***: When moving relative to the Earth, the frame rotates about its z-axis to allow the x-axis to always point towards north. When getting close to a pole, this rotation rate increases, being infinite at the pole. The poles are thus singularities and the direction of the x- and y-axes are not defined here. Hence, this coordinate frame is not suitable for general calculations
L	***Name***: Local level, wander azimuth ***Position***: The origin is directly beneath or above the vehicle (B), at the Earth's surface (surface of ellipsoid model) ***Orientation***: The z-axis points down. Initially, the x-axis normally points towards north (if defined), and the y-axis points towards east, but as the vehicle moves they are not rotating about the z-axis (their angular velocity relative to the Earth has zero component along the z-axis) ***Comments***: The L frame is equal to the N frame except for the rotation about the z-axis, which is always zero for this frame (relative to E). Hence, at a given time, the only difference between the frames is an angle between the x-axis of L and the north direction; this angle is called the *wander azimuth* angle. The L frame is well suited for general calculations, as it is non-singular. See also [14]

where $\vec{g}_B$ is the gravitational acceleration at point B, $\vec{F}_{B,\text{gravitation}}$ is the force resulting from the gravitational attraction on the proof mass and m is the mass of the proof mass. In practice, most accelerometers measure specific force along a single axis, and hence three accelerometers are needed to measure the full 3D vector. Assuming the accelerometers are rigidly attached to our vehicle B (strapdown inertial sensors),

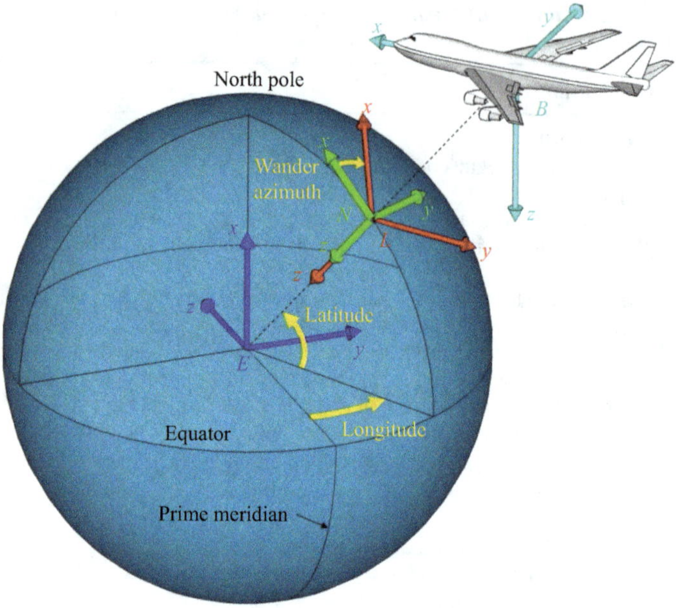

Figure 5.4 Coordinate frames used

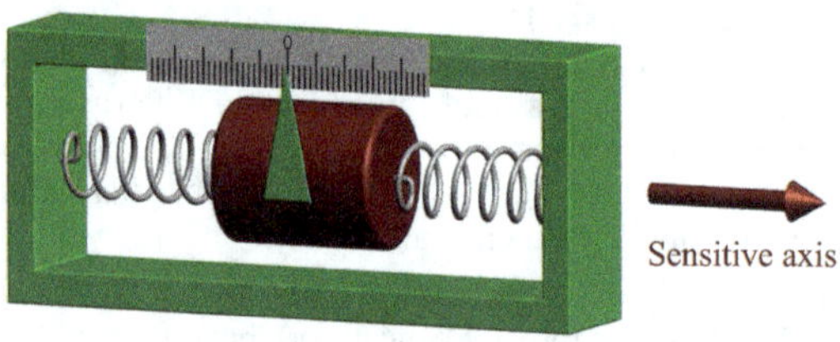

Figure 5.5 Principle of a simple accelerometer

we typically get an accelerometer measurement decomposed in the vehicle B system, as $\boldsymbol{f}_{IB}^{B}$.

The second important inertial sensor is the gyroscope, and several measurement principles can be used. The three most common principles are the following:

- **Spinning mass**: A spinning mass will resist any change in its angular momentum vector relative to the inertial space. Isolating the spinning mass from vehicle angular movements, e.g., by means of gimbals and then measure and output the gimbal positions, is the basic idea of a mechanical gyro.
- **Sagnac effect**: The inertial characteristics of light can also be utilised, by letting two beams of light travel in a loop in opposite directions. If the loop rotates clockwise relative to inertial space, the clockwise beam must travel a longer distance before finishing the loop. The opposite is true for the counter-clockwise

beam. Combining the two rays in a detector, an interference pattern can be formed, which depends on the angular velocity relative to inertial space. There are two main ways to implement the loop:

  - *Ring laser gyro* (RLG): The loop is implemented by three or four mirrors.
  - *Fibre optic gyro* (FOG): The loop is implemented by means of optical fibres.

- **Coriolis effect**: Assume a mass that is vibrating in the radial direction of a rotating system. Due to the Coriolis force working perpendicular to the original vibrating direction, a new vibration will take place in the tangential direction. The amplitude of this new vibration is a function of the angular velocity relative to inertial space:
  - *MEMS gyros* (microelectromechanical systems), 'tuning fork' and 'wine-glass' gyros utilise this principle. Coriolis-based gyros are typically cheaper, lighter and less accurate than mechanical gyros, RLGs or FOGs.

Regardless of the measurement principle used, the gyro measures the angular velocity of the (sensor) body frame relative to the inertial frame, $\vec{\omega}_{IB}$. As a strapdown gyro typically has one sensitive axis, three gyros are needed in order to measure the full angular velocity vector $\boldsymbol{\omega}_{IB}^{B}$.

In practice, three gyros and three accelerometers are typically assembled to form an inertial measurement unit (IMU). Most IMUs today are strapdown IMUs, in which the inertial sensors are rigidly attached to the unit (no mechanical movement). In a gimballed IMU, the gyros and accelerometers are isolated from vehicle angular movements by means of gimbals. IMUs are sometimes categorised based on the quality of the inertial sensors [16,17].

5.3.2 Inertial navigation

An IMU (giving $\boldsymbol{f}_{IB}^{B}$ and $\boldsymbol{\omega}_{IB}^{B}$) is sufficient to navigate relative to inertial space (assuming no gravitation is present), if initial values of velocity, position and orientation are given. Integrating the sensed acceleration gives the velocity, while a second integration gives the position. To integrate in the correct direction, the orientation is needed. This is obtained by integrating the sensed angular velocity.

A similar integration can be done near the Earth, with E as the reference system, but since E is not an inertial system and gravitation is present, some compensations must be made. For the gyros, the gyro measurement must be compensated with the angular rate of the Earth ($\boldsymbol{\omega}_{IE}^{B}$) to obtain the angular rate of the vehicle relative to the Earth:

$$\boldsymbol{\omega}_{EB}^{B} = \boldsymbol{\omega}_{IB}^{B} - \boldsymbol{\omega}_{IE}^{B} \tag{5.3}$$

For the accelerometers, the necessary compensations include gravitation, centrifugal force (due to the rotating Earth) and Coriolis force (due to movement in a rotating reference frame).* The equations that integrate the gyro and accelerometer measurements to velocity, position and orientation are called the *navigation equations*, as shown in Figure 5.6. The combination of an IMU and navigation equations is called an INS.

*The two latter are fictitious forces, seeming to act in a non-inertial system.

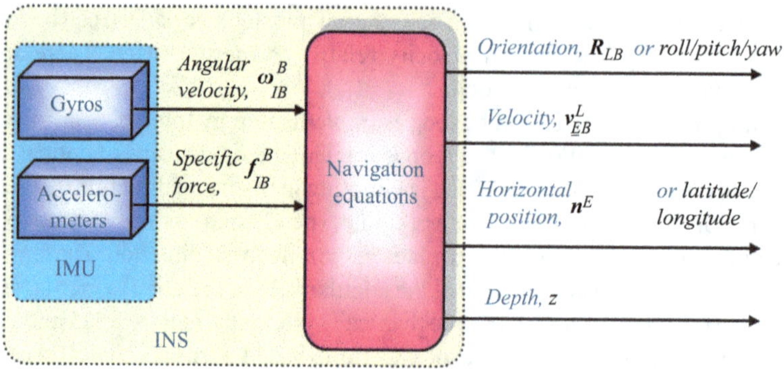

Figure 5.6 Inertial navigation system

As shown in Figure 5.6, horizontal and vertical positions (depth/height) are separated, which is often very useful in practice. Latitude and longitude can be used to represent horizontal position, but these are singular and can give other problems in calculations. An attractive alternative is to use *n*-vector (n^E) instead, a non-singular representation of horizontal position [12,13].

The measurement errors in gyros and accelerometers cause unlimited drift, in velocity, position and orientation, as measurements are integrated in the navigation equations. To reduce or stop this drift, other measurements (or information/knowledge) are needed, and we get an aided INS (AINS). High-end AUVs are typically equipped with so-called *navigation grade* or *aviation* [16] grade IMUs, which have a position drift on the order of one nautical mile per hour when the INS is run in *free inertial* mode, i.e., without any aiding sensors. These IMUs are typically equipped with RLG or FOG gyros (see Section 5.3.1) capable of gyrocompassing.

5.3.3 AINS – core AUV aiding sensors

Sensors well suited as aiding sensors for an AUV include GNSS receivers (used when the AUV is at the surface), DVLs and pressure sensors, and these can be seen as core AUV aiding sensors. The aiding sensors provide redundancy since they measure some of the integrated quantities (velocity, position or orientation) from the navigation equations. We have, e.g., position both from the INS and from the GNSS receiver. It is important to note that the IMU and GNSS have complementary characteristics:

- The IMU has a high measurement rate (typically more than 100 Hz), and hence the INS position accurately includes small and rapid movements of the vehicle (but it is drifting).
- For the GNSS, the properties are opposite: there is no drift, but typically it has relatively low rate and a significant noise in each measurement, hence small and rapid movements are not measured.

Combining the two sources gives a solution that has the best from both sources (i.e., stable, with high frequency and improved accuracy).

To blend the redundant information, an estimation algorithm is needed. The Kalman filter is the optimal estimator (i.e., minimum variance of the estimation errors), given certain assumptions that are usually quite good for an AINS. To give correct weight at the different sources of information, the Kalman filter needs models of the sensor errors, i.e., white noise, biases, scale factor errors, etc.

A general Kalman filter (see, e.g., [18,19]) is a recursive algorithm that combines two different kinds of information:

- The **sensor measurements** (e.g., the measurements from the gyros, GNSS and DVL) and
- a **mathematical model** of the system (e.g., models of sensor errors and a model of how IMU errors propagate through the navigation equations).

Let $\boldsymbol{x}_k$ denote the state vector to be estimated at time t_k, and $\boldsymbol{z}_k$ denote the measurement vector at the same time instant. This state estimation problem can be formulated mathematically by the following set of equations:

$$\boldsymbol{x}_{k+1} = \boldsymbol{f}(\boldsymbol{x}_k, \boldsymbol{u}_k) + \boldsymbol{v}_k \tag{5.4}$$

$$\boldsymbol{z}_k = \boldsymbol{h}(\boldsymbol{x}_k) + \boldsymbol{w}_k \tag{5.5}$$

where $\boldsymbol{f}$ is a known dynamical model of the system which depends on the state vector and a known deterministic input vector $\boldsymbol{u}_k$. The model also has an additive stochastic process noise vector denoted by $\boldsymbol{v}_k$. The known measurement model $\boldsymbol{h}$ also depends on the state vector with an additive stochastic noise vector $\boldsymbol{w}_k$.

In addition to the sensor measurement $\boldsymbol{z}_k$ and the models defined by the functions $\boldsymbol{f}$ and $\boldsymbol{h}$, the *accuracy* of both the sensor measurements and the mathematical model must be specified, corresponding to the variances of the noises $\boldsymbol{v}_k$ and $\boldsymbol{w}_k$. The Kalman filter algorithm can now be described by the following:

1. At time t_0, the Kalman filter is provided with an *initial estimate*, including its uncertainty (covariance matrix).
2. Based on the mathematical model and the initial estimate, a new estimate valid at time t_1 is *predicted*. The uncertainty of the *predicted estimate* is calculated based on the initial uncertainty, and the accuracy of the model (*process noise*).
3. Measurements valid at time t_1 give new information about the states. Based on the accuracy of the measurements (*measurement noise*) and the uncertainty in the predicted estimate, the two sources of information are weighted and a new *updated estimate* valid at t_1 is calculated. The uncertainty of this estimate is also calculated.
4. At t_2, a new estimate is predicted as in Step 2, but now based on the updated estimate from t_1.

The prediction and the following measurement update are repeated each time a new measurement arrives.

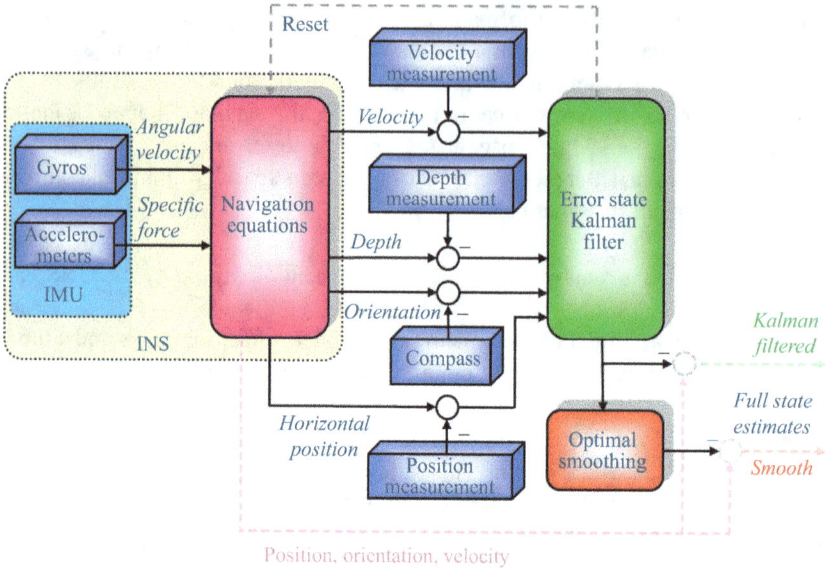

Figure 5.7 Error state Kalman filter-based AINS structure

When using a Kalman filter for inertial navigation, the error state structure is usually preferred†, as in Figure 5.7, which shows a possible structure of a Kalman filter-based navigation system. We see that the measurements from the aiding sensors are subtracted from the corresponding quantities from the navigation equations, and the differences are used as measurements in the error state Kalman filter. Estimates of the full states are achieved by subtracting the estimated errors from the navigation equations. Since the navigation equations have unlimited drift, they are reset regularly to be kept close to the true values, to keep the non-linearities in the models small. As the system is linearised around the most recent state estimate, this is an example of an extended Kalman filter (EKF).

In some cases, the non-linearities in the problem are too large for a linearised estimator like the EKF to handle properly. This is for example the case for terrain aided navigation, which is further described in Section 5.5.5.1. To handle these strong non-linearities, several approaches exist. One solution is to replace the Kalman filter in the INS with a non-linear estimator, like a particle filter (PF) [20] or unscented Kalman filter [21]. However, as non-linear estimators are often less robust and more computationally demanding than the Kalman filter, another possibility is to treat the non-linear measurement in a separate non-linear estimation algorithm, and to integrate it in the main Kalman filter-based INS in a loosely coupled fashion [22]. This possibility is further discussed in Section 5.5.5.1.

It should be noted that not only sensor measurements can be used to aid an INS, but knowledge about the vehicle's position, velocity or orientation can be used as well.

† Discussions of the advantages of error state versus full state are outside the scope of this chapter.

An example is an AUV without DVL in a period where no position measurements are available. In such a case, the horizontal velocity and position quickly drift off (the drift rate is determined by the IMU quality and the initial accuracy) and reaches speeds that are not feasible for the AUV. A practical way to include this knowledge is to make a synthetic velocity measurement based on a vehicle model that possibly uses inputs from propeller speed and rudder deflection [23].

5.3.3.1 Smoothing

The recursive Kalman filter is the optimal (given certain assumptions) estimator in real time, but if an estimate is needed for a specified time t_1 and measurements logged *after* t_1 are also available, a better estimator exists. In general, suppose we have measurements of the system in the interval $t \in [t_0, T]$, and let $Z_T = \{z_0, z_1, \ldots, z_T\}$ be the collection of all available measurements. Then the state estimate $\hat{\boldsymbol{x}}(t|Z_T)$ is called a *filter estimate* if $t = T$, a *prediction estimate* if $t > T$ and a *smoothed estimate* if $t < T$, and the corresponding estimators are called *filters*, *predictors* and *smoothers* [24].

For an AUV, we often have a recorded data set for a given period, t_0 to T, and the best estimator within this interval utilises the entire set of measurements, i.e., we use a (fixed interval) smoother (at $t = T$ the performance is equal to a Kalman filter, but for all other values of t, the performance is normally better). Figure 5.8 shows typical standard deviations of a Kalman filter and a smoother for a fixed interval. The figure is from NavLab, a simulation and post-processing tool developed by FFI [25]. The

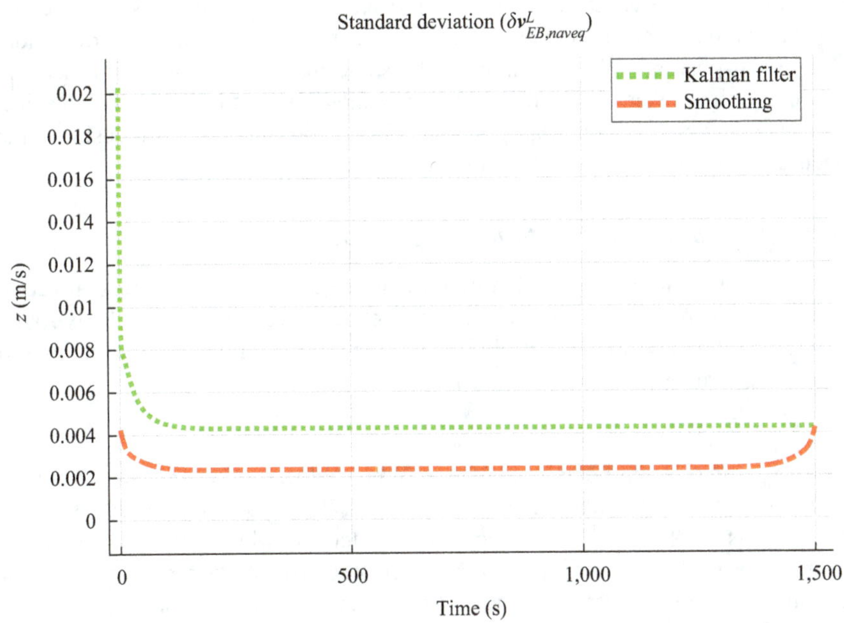

Figure 5.8 Example of the effect of smoothing on estimated standard deviation. Figure created by NavLab [25]

figure shows vertical velocity and we see that the initial uncertainty for the Kalman filter is 0.02 m/s. When the pressure sensor and DVL are used, the standard deviation converges to about 0.0043 m/s. In the last time step, the smoothing has no future measurements to utilise, and hence the standard deviation is equal to the standard deviation of the Kalman filter. In the middle section of the interval, the smoothing can utilise both past and future measurements, and hence it has a lower standard deviation. In the first part of the interval, the amount of measurements from the past is reduced, leading to an increased uncertainty for the smoothing.

Compared to a real-time filter, smoothing gives the following:

- improved accuracy (number of relevant measurements doubled),
- improved robustness,
- improved integrity,
- estimate in accordance with process model:
  - no measurements are unexpected and hence no sudden jumps occur (see Section 3.2 in [25] for details).

5.4 Properties of an AINS for an AUV

A navigation system gives estimates of velocity, position and orientation, and the accuracies of these estimates depend on several factors such as the accuracy of the sensors and vehicle manoeuvres. It can be difficult to understand how manoeuvres and accuracy in different sensors translate into, e.g., heading accuracy. In this section, we study the main factors that determine the accuracy of the estimated orientation, velocity and position. We also explain why some manoeuvres, such as a heading change, give improvements in some of the estimates. For an arbitrary vehicle/device, there are many possible relations and manoeuvres, but this section focuses on the effects that are most important for an AUV.

5.4.1 Estimation of orientation in general

To estimate orientation in general, a vector is needed; let us denote a general vector $\vec{x}$. The direction of the vector must be known both relative to the vehicle body frame B, i.e., $\boldsymbol{x}^B$, and relative to the Earth E, i.e., $\boldsymbol{x}^E$. The variants of the vector decomposed in the two coordinate frames are related through

$$\boldsymbol{x}^E = \boldsymbol{R}_{EB}\boldsymbol{x}^B \tag{5.6}$$

where $\boldsymbol{R}_{EB}$ represents the vehicle orientation. When both $\boldsymbol{x}^E$ and $\boldsymbol{x}^B$ are known (and not zero vectors), two of the three degrees of freedom in $\boldsymbol{R}_{EB}$ can be calculated. Orientation about an axis parallel with $\vec{x}$ is not possible to calculate, while orientation about the two axes perpendicular to $\vec{x}$ can both be calculated. To estimate the last degree of freedom, a new vector that is not parallel with $\vec{x}$ is needed. This is a special case of what is often referred to as Wahba's problem [26].

Assume that we want to determine the accuracy of the estimated orientation about an axis $\vec{y}$. This accuracy depends on the accuracies of $\boldsymbol{x}^E$ and $\boldsymbol{x}^B$ and the length of the component of $\vec{x}$ that is perpendicular to $\vec{y}$, denoted $\vec{x}_{\perp\vec{y}}$. When $\overrightarrow{\delta x_E}$ and $\overrightarrow{\delta x_B}$ denote the errors in the determination of $\vec{x}$ in E and B, only the components of the error vectors that are normal to both $\vec{x}$ and $\vec{y}$ contribute to the orientation error (when assuming small angular errors). If we let the operator $(\cdot)_{\text{contr}}$ return the length of the contributing component, we can write the standard deviation of these contributing errors as $\sigma((\overrightarrow{\delta x_E})_{\text{contr}})$ and $\sigma((\overrightarrow{\delta x_B})_{\text{contr}})$, respectively. The standard deviation of the resulting orientation error, $\delta\alpha$, is then given as follows (uncorrelated errors assumed, first order approximation):

$$\sigma(\delta\alpha) \approx \frac{\sqrt{\sigma((\overrightarrow{\delta x_E})_{\text{contr}})^2 + \sigma((\overrightarrow{\delta x_B})_{\text{contr}})^2}}{|\vec{x}_{\perp\vec{y}}|} \tag{5.7}$$

where $\delta\alpha$ denotes the orientation angle error in question.

5.4.2 Heading accuracy

Heading is defined as orientation about the vertical axis, and hence it corresponds to $\vec{y}$ being vertical. Since $\vec{y}$ is vertical, the gravity vector cannot be used to find heading, and thus this is often a challenging degree of freedom to estimate with sufficient accuracy, as was shown in Figure 5.1. In general, there are seven possible ways to estimate heading for any vehicle or device [2]. The seven methods are listed in Figure 5.9, together with some properties or examples of each method.

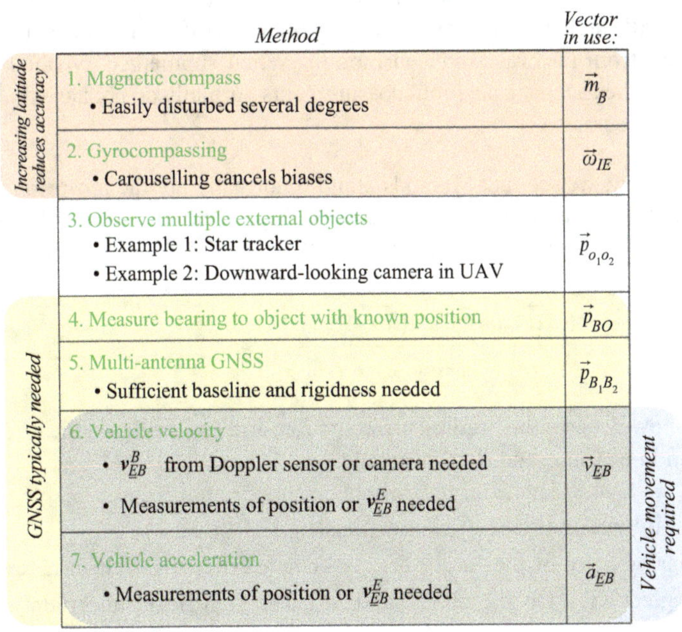

Figure 5.9 The seven fundamental ways to find the heading angle. Figure from [2]

All of the seven methods in Figure 5.9 are clearly not feasible for an AUV, so a closer look at each method is needed:

- Method 1, magnetic compass: A magnetic compass is in most cases too inaccurate and unreliable for a high-end AUV, e.g., due to local magnetic disturbances (both from own vehicle and ferromagnetic material in the seafloor), but it is frequently used in low-cost AUVs. See also Section 5.5.1.
- Method 2, gyrocompassing: This is the most important method to find heading for high-end AUVs. Expensive gyros are required, and the accuracy is reduced at high latitudes.
- Method 3, observing multiple external objects: This method is not feasible for an AUV.
- Method 4, measure bearing to an object with known position: In special cases, it might be possible for an AUV to use this method (e.g., towards an underwater transponder or when the AUV is at the surface), but it is clearly not an important method for AUVs.
- Method 5, multi-antenna GNSS: This method may be used before launch and possibly when the AUV is at the surface.
- Method 6, vehicle velocity vector: When an AUV has a DVL with bottom track and is followed by a surface vehicle with USBL, Method 6 can give relatively high heading accuracy (depending on the velocity and the accuracy of the DVL and USBL).
- Method 7, vehicle acceleration vector: This method can be used before launch and when the AUV is at the surface, if sufficient acceleration is present and GNSS with high accuracy and rate is available. Sufficient acceleration can be achieved from the wave-induced motion. When submerged, Method 7 is not feasible, since acoustic position measurements normally do not have the required rate and accuracy.

In summary, we see that Methods 2 and 6 normally are the most important for a submerged AUV. To find the accuracies of these methods, we use Equation (5) from [2] (which is a special case of (5.7)):

$$\sigma(\delta\psi) \approx \frac{\sqrt{\sigma((\overrightarrow{\delta x_E})_{\text{contr}})^2 + \sigma((\overrightarrow{\delta x_B})_{\text{contr}})^2}}{|\vec{x}_{\text{horizontal}}|} \tag{5.8}$$

where $\delta\psi$ now denotes the heading error and $\overrightarrow{\delta x_E}$ and $\overrightarrow{\delta x_B}$ are again the errors in the determination of $\vec{x}$ in E and B, respectively.

5.4.2.1 The accuracy of gyrocompassing (Method 2)

The angular velocity of the Earth, $\vec{\omega}_{IE}$, is known extremely well in E, and hence we can neglect $\overrightarrow{\delta x_E}$. On the other hand, $\boldsymbol{\omega}_{IE}^{B}$ has significant uncertainty and the magnitude of $\sigma((\overrightarrow{\delta x_B})_{\text{contr}})$ determines the accuracy of the gyrocompassing (together with $|\vec{x}_{\text{horizontal}}|$, which is simply given by $|\vec{\omega}_{IE}|\cos(\text{latitude})$). In practice, the main contributor to $\sigma((\overrightarrow{\delta x_B})_{\text{contr}})$ is often the (non-observable) gyro bias, $\delta\omega_{\text{gyrobias}}$, and we get the following typical accuracy for gyrocompassing:

$$\sigma(\delta\psi) \approx \frac{\sigma(\delta\omega_{\text{gyrobias}})}{|\vec{\omega}_{IE}|\cos(\text{latitude})} \tag{5.9}$$

There are also cases in which significant parts of the gyro biases are observable, which is typically the case when the vehicle is turning, and better accuracy than given by (5.9) can be achieved. There are two important effects improving the heading accuracy when turning:

1. Assume a stationary vehicle at zero latitude, with roll, pitch and yaw angles equal to 0. The y gyro then points towards east, and its bias is cause to a heading error according to (5.9). The x gyro measures (almost) the entire $\vec{\omega}_{IE}$, and since the length of this vector is well known, the gyro x bias is observable. If the vehicle now changes heading to 90°, the x gyro will now point towards east and hence be the one that determines the heading accuracy $(\overrightarrow{\delta x_B})_{\text{contr}}$. Since its bias is already estimated, the heading is now found much more accurately.
2. Assume the same situation as above, but that no gyro biases have yet been observed, while the vehicle is now constantly rotating about the z-axis with $\boldsymbol{\omega}_{EB}^{B} = [0\ 0\ 5]^{T}{}^{\circ}/\text{s}$ (i.e., the heading is constantly changing). The x and y gyros take turns to be the one that points east/west and if no z gyro was present, the heading error would alternate between the errors given by the $\pm x$ and $\pm y$ gyro biases. A simple example is that the y gyro bias is $+0.1$°/h pointing east initially and after 36 s it is pointing west. The rate of the heading error change during a 180° turn is then $\frac{0.2^{\circ}/h/(15.04^{\circ}/\text{h})}{36\,\text{s}} = 76^{\circ}/\text{h}$. If we now also have a z gyro with bias 0.1°/h, we see that the rate of change is clearly measurable, and hence the bias is observable. In practice, the estimation takes place many times during the 180° turn, and a linear combination of the two biases contributes for all angles, making both biases observable.

When an AUV turns, both the aforementioned effects contribute to reducing the heading error. Figure 5.10 shows the standard deviation of the heading estimate for 0.1°/ h gyros from a NavLab simulation of an AUV with a DVL. The initial heading uncertainty is 5°, and we see that it falls to about 0.4° (which is in accordance with (5.9)). At time $t = 500$ s, the vehicle starts turning with a rate $\boldsymbol{\omega}_{EB}^{B} = [0\ 0\ 5]^{T}{}^{\circ}/\ \text{s}$, and we see that the uncertainty drops further.

In Figure 5.11, the estimation of the gyro biases is shown, and we see that the initial uncertainty is 0.1°/ h (since the bias is initially unknown). As expected, the x gyro bias is estimated in the interval before the rotation, since it points along $\vec{\omega}_{IE}$, but the y and z biases are not observable (as described above). When the vehicle starts rotating, the y bias becomes observable, while the z bias is never observable in this scenario.

5.4.2.2 The accuracy of the velocity vector method (Method 6)

From (5.8), we see that the accuracy of the velocity vector method depends on the AUV speed, the DVL accuracy $((\overrightarrow{\delta x_B})_{\text{contr}})$ and the rate and accuracy of the position

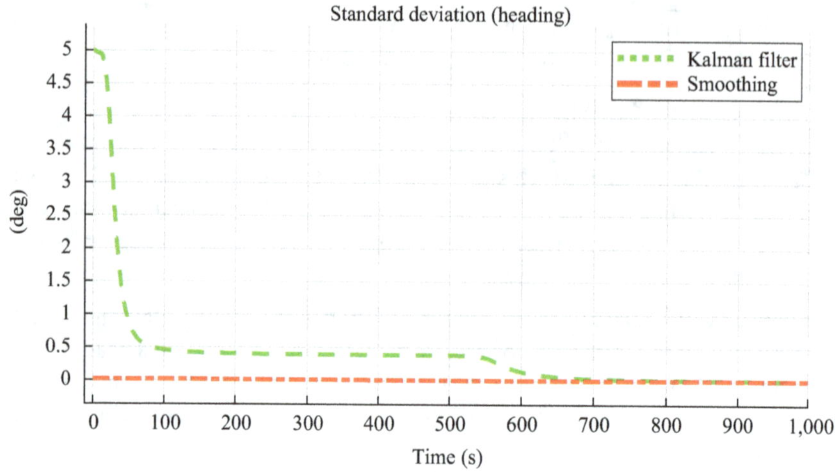

Figure 5.10 The standard deviation in heading when gyrocompassing and turning. Figure created by NavLab [25]

measurements ($(\overrightarrow{\delta x_E})_{\text{contr}}$). An example value from the HUGIN I AUV [27] is a DVL error of 0.015 m/s when the vehicle speed is 2.1 m/s, which contributes to a 0.41° heading error for one single DVL measurement. Some of the DVL error can be estimated, but $(\overrightarrow{\delta x_E})_{\text{contr}}$ also contributes, and the observed total heading error of Method 6 is about 0.5° in [27].

5.4.3 *Accuracy in roll/pitch*

The roll, pitch and yaw/heading angles are defined by rotations about new axes (intrinsic) in the order *z*-*y*-*x*. The three angles (called Euler angles or Tait–Bryan angles) are defined by the following procedure of successive rotations:

> Given coordinate frames N and B (from Table 5.2), consider a temporary frame T that initially coincides with N. In order to make T align with B, we first rotate T a yaw (or heading) angle about its *z*-axis (common axis for both N and T). Second, T is rotated a pitch angle about the new *y*-axis of T. Finally, T is rotated a roll angle about its *newest* *x*-axis (which is the same as the *x*-axis of B). The final orientation of T now coincides with the orientation of B. The signs of the angles are given by the directions of the axes and the right-hand rule.

Note that the N coordinate frame is singular at the poles, and an alternative L frame (such as wander azimuth) can be used instead of N, to avoid singularity in heading/yaw (as discussed in Section 5.2).

From (5.7), we saw the need for a vector to estimate orientation, and when estimating roll, the vector must have a component perpendicular to the vehicle body

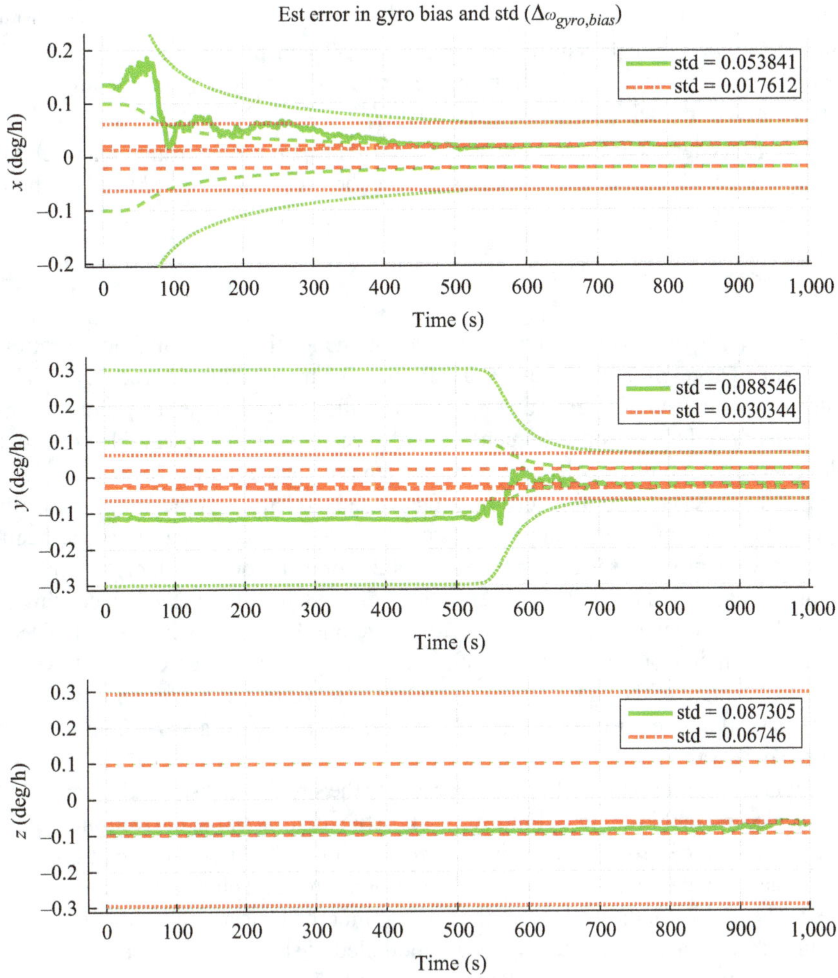

Figure 5.11 Gyro bias estimation. Dashed lines: $\pm 1\sigma$. Dotted lines: $\pm 3\sigma$. Solid/dashed-dotted lines: the actual true estimation error in this simulation. Green, solid lines are from the Kalman filter and red, dashed-dotted lines from the smoothing algorithm. Figure created by NavLab [25]

x-axis. For pitch estimation, the vector must have a component perpendicular to the horizontal y-axis (according to the definition of pitch). As long as the vehicle is not at a pitch of $\pm 90°$, we see that the gravity vector fulfils the requirements to estimate roll and pitch, and this vector is indeed the most important in practice when estimating these angles.

When discussing orientation estimation for an AUV, we typically treat heading separately, since the gravity vector cannot be used for estimating heading, and we treat roll and pitch together since they typically have similar accuracy and behaviour. From (5.7), we can calculate the accuracy of roll and pitch. The $\sigma((\overrightarrow{\delta x_E})_{\text{contr}})$ is normally small, since the direction of the gravitation is well known relative to E (a gravitational model, e.g., EGM96 [28] or EGM2008 [29] can be used when high accuracy is needed). The main contributor is $\sigma((\overrightarrow{\delta x_B})_{\text{contr}})$, and for a stationary vehicle this is given by the accelerometer biases. Hence the accuracy of roll (ϕ) and pitch (θ) is typically given by

$$\sigma(\delta\phi) = \sigma(\delta\theta) \approx \frac{\sigma(\delta f_{\text{accbias}})}{g} \tag{5.10}$$

where $\delta f_{\text{accbias}}$ is the accelerometer bias and g is the local gravitational acceleration. With 1 mg of accelerometer accuracy, a roll/pitch accuracy of 0.057° is achieved. This is also a typical accuracy for a moving vehicle with position or velocity aiding since horizontal drift due to wrong compensation of g vector is easily observable with such aiding sensors (and the accelerometer biases remain the main error source).

Note, however, that the turning of a vehicle (change of heading) typically improves the roll/pitch accuracy. The effect has a similar explanation as the heading improvement we described in Section 5.4.2. For a stationary vehicle at zero roll and pitch, biases in the x and y accelerometers give pitch and roll errors according to (5.10). Assume now that the vehicle rotates 90° about the z-axis over a period of 18 s. The accelerometer biases are moved, and hence also the roll/pitch errors. However, the rate of change of the roll/pitch errors (relative to I) due to the rotation is normally much faster than the errors in the x and y gyros, and hence the accelerometer biases become observable.

Figure 5.12 shows the roll/pitch accuracy for the simulation used in Section 5.4.2. The accelerometer biases are 0.1 mg, and before the rotation starts, the roll/pitch accuracies are about $5.7 \cdot 10^{-3}$ degrees, in accordance with (5.7). During the rotation, the x and y accelerometer biases are observable, and hence roll and pitch are accurately estimated. The reason for the backwards increasing smoothed uncertainty in the stationary period is that the accelerometer biases are modelled as slowly varying (and hence the biases estimated after time $t = 500$ s are less valid when getting closer to $t = 0$ s).

In Section 5.4.2, we discussed how the velocity vector (Method 6) was used for heading estimation. For an AUV with forward speed, this vector also contributes to the estimation of pitch, and this estimate may become rather accurate since position in the z direction is well known due to the pressure sensor. However, the accuracy of the DVL relative to the speed typically results in a significantly lower angular accuracy than what is achieved from (5.10); thus, this effect is normally not contributing significantly to improve the pitch accuracy.

5.4.4 Accuracy in velocity

If position measurements are available, their rate and accuracy greatly affect the resulting velocity accuracy. In the following, we discuss cases without any position updates, and then the presence of a DVL with bottom track is normally the most important factor for the velocity accuracy.

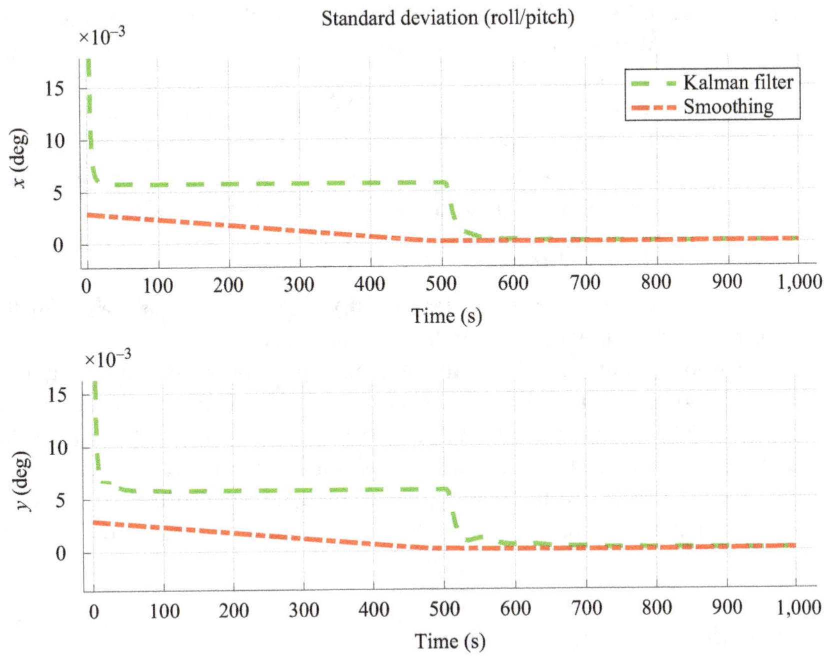

Figure 5.12 The standard deviation in roll/pitch from g vector, and when turning. Figure created by NavLab [25]

5.4.4.1 DVL with bottom track available

For AUVs equipped with a DVL that has bottom track, the accuracy of the velocity vector $\boldsymbol{v}_{\underline{EB}}^{B}$ is mainly determined by the DVL accuracy. However, body-fixed DVL biases may become observable during a turn. The reason is that the biases in the directions that are normal to the angular velocity vector (typically the x and y biases) correspond to a centripetal acceleration. Such a centripetal acceleration is measurable using the accelerometers, and hence the DVL biases become observable. In theory, this is an important effect, but for this effect to work in practice, the DVL biases must stay the same during turns as they are when the AUV travels in a straight line. From our practical experience with DVLs, the bias is typically different during turns, and hence this effect has in practice limited importance for the velocity accuracy.

The velocity that is integrated to position is $\boldsymbol{v}_{\underline{EB}}^{E}$, and thus it is errors in this quantity that are important for the position drift. The velocity $\boldsymbol{v}_{\underline{EB}}^{B}$ relates to $\boldsymbol{v}_{\underline{EB}}^{E}$ through the rotation matrix $\boldsymbol{R}_{EB}$, and as discussed in Section 5.4.2 the main uncertainty in $\boldsymbol{R}_{EB}$ results from the heading uncertainty, and for an AUV with DVL bottom track the heading error is consequently an important contributor to the cross track positional drift.

5.4.4.2 No DVL with bottom track available

If the DVL has no bottom track, but returns a velocity relative to the water (water track), the main error source is the unknown sea current. For AUVs without DVL, a vehicle model can be used together with information about propeller speed and rudder deflections. With a good vehicle model, the main error source is also here the unknown sea current [23].

Note that if some position measurements are available (during a limited period), the sea current can be estimated. Assuming the sea current is relatively constant, an improved velocity accuracy is in this case achieved for a long period after the position measurements were present (as long as the sea current estimate is valid).

5.4.5 Accuracy in position

The accuracy of the vertical position (the depth) is given by the accuracy of the pressure sensor and the pressure-to-depth conversion used [30].

If no position aiding is available, the horizontal position drifts at a speed given by the uncertainty in $\boldsymbol{v}^E_{EB}$. In [31], a conservative straight-line drift estimate of 0.08 per cent of travelled distance is given, for the Kongsberg HUGIN/MUNIN vehicles equipped with 300 kHz DVLs, though substantially better performance is often observed. For cases with position aiding, the accuracy is mainly determined by the type(s) of position aiding available. Possible types of position aiding are discussed in Section 5.5.

5.5 Additional aiding techniques

5.5.1 Magnetometer

The magnetometer measures the 3D magnetic field vector at the sensor location. The measurement is the sum of the local Earth magnetic field, the magnetic field of the AUV and any local magnetic anomalies in the area surrounding the AUV. The magnetic field of the AUV itself is usually body-fixed, and if assumed stationary, it can be found through a calibration procedure by rotating the AUV in a stationary homogeneous external magnetic field. If perfectly calibrated, the magnetic measurements lie on a sphere centred at the origin. If not calibrated, the measurements lie on an ellipsoid, whose centroid is given by the hard iron magnetic biases of the AUV, while the principle axis of the ellipsoid is given by the soft iron magnetic biases [32]. The calibration is then found by estimating an ellipsoid from the calibration measurements, and calibration is done by applying the transformation from the ellipsoid to the centred sphere, to the *in-situ* magnetometer measurements. If we assume a calibrated magnetometer and no local magnetic anomalies, we can describe the measurement model in (5.5) by [14]

$$\boldsymbol{h}(\boldsymbol{x}, t) = \boldsymbol{R}^T_{NB}\boldsymbol{m}^N_{EB}(t) \tag{5.11}$$

Here $\boldsymbol{m}^N_{EB}(t)$ denotes a model of the Earth magnetic field, at the AUV's position, decomposed in the local-level system N. Such models are available, e.g., from the World Magnetic Model [33]. Since the magnetic field $\boldsymbol{m}^N_{EB}$ has its largest component in the horizontal plane (except close to the magnetic poles, where the horizontal component is zero), the measurement described by (5.11) primarily gives information about the heading angle, in accordance with (5.7).

5.5.2 Surface vessel to AUV aiding

In situations in which it is possible to aid the AUV from a surface ship, several principles exist for acoustic aiding. In short-baseline (SBL) positioning, two transceivers are placed at either end of the ship. By comparing the sound signals received by each of the transceivers from a transponder on the AUV, the range and bearing between the ship and the AUV can be computed. In USBL, an array of transceivers is used and time of flight and phase-differencing of the received signal are applied, in order to obtain accurate range and bearing estimates. After combining the range and bearing estimates from the USBL or SBL system with the GNSS position of the surface ship, if necessary enhanced by using differential GNSS (DGNSS) or real-time kinematic augmentation, an accurate position estimate can be obtained, and this estimate can be sent to the AUV via acoustic communications, and integrated in the AUV INS [4]. Due to its higher performance, USBL systems are usually used in scenarios in which high AUV navigation accuracy is needed, e.g., in offshore surveying.

5.5.3 Transponder positioning

Transponder positioning utilises a set of transponders placed on the seafloor. Usually the AUV polls the transponders, and based on the time delay before the response signals are received, the AUV is able to compute the range to each transponder. This is known as a two-way travel time set-up [8]. If the transponders and the AUV have synchronised clocks, one-way travel time ranging is possible [34].

The main disadvantage of using transponders for position aiding is the need for costly deployment and positioning of the transponders prior to the mission. In some situations, this process can be simplified using SLAM-based methods to localise the transponders while simultaneously using them for navigation, a concept which is called *concurrent deployment and navigation* in [4] and further discussed in Section 5.5.6. Instead of transponders on the seafloor, it is also possible to use transponder buoys equipped with GNSS receivers on the sea surface or in sea ice [31,35].

5.5.3.1 LBL positioning

The traditional way of us using transponder positioning underwater is triangulation in an LBL network [36]. Theoretically, three transponders are needed to obtain a unique position fix, but usually four or more transponders are used, in order to ensure redundancy and favourable measurement geometry. The frequency used is a trade-off between desired range and accuracy, ranging from 12 kHz in full ocean depth scenarios to 300 kHz or higher when sub-centimetre accuracy is needed [37].

5.5.3.2 Underwater transponder positioning

In traditional LBL aiding, a position update is computed using the LBL measurements only, before using this as a position update in the INS. This is known as *loose coupling*. If, however, the ranges to the individual transponders are integrated directly in INS estimator, using so-called *tight coupling*, the information in the measurements is utilised more efficiently. This means that, by combining the information of the INS and transponder measurement, it is possible to obtain acceptable navigation accuracy using a single transponder. This aiding method is known as UTP [4,7] or *synthetic LBL* navigation [38]. For operations in a large area, UTP significantly reduces the number of transponders needed. In [7,39], real-time position accuracy of around 5 m was demonstrated on a HUGIN 1000 vehicle during an 8-h operation, using four transponders distributed along the vehicle trajectory in such a manner that a maximum of one transponder was within range at any given time. In-ice mounted UTP transponders were used in [31], to meet IHO S-44 Special Order [40] mapping requirements in a low-logistic commercial under-ice operation. Single transponder UTP navigation is also discussed in a SLAM setting in Section 5.5.6.1.

5.5.4 Delta-position aiding

A large class of aiding techniques comprises the methods that compute a difference in position from one time instant to another. The interval between the two time instants may be small, as in synthetic aperture sonar (SAS) micro navigation [41,42], in which consecutive sonar measurements over the same seafloor patch are used to estimate the vehicle displacement over the interval, or large, as when a previously seen feature on the seafloor is revisited and used to improve the position estimate, sometimes known as *macro delta-position aiding* [3]. The latter situation is also common in the SLAM problem, which is discussed in Section 5.5.6. Delta-position estimates can also be obtained, e.g., from comparing SAS data between repeated passes of a scene [43], from sonar-based change detection [44] or from optical camera mosaics.

Delta-position aiding is useful also in situations when the global position of the observed feature or bathymetry is uncertain, or even unknown. One example is when revisiting an area which has been surveyed previously. Even though the global navigation accuracy of the previous survey is low, delta-position aiding can be utilised to obtain accurate navigation relative to the observed feature, which can be useful to relocate previously detected objects in the area.

5.5.5 Geophysical navigation

In many situations, the sensors on board an AUV are able to recognise properties of the physical environment around the vehicle in a way that makes it possible to obtain a position fix when the data are compared to a previous reference database. If the database contains globally positioned data which are not expected to change over time, a global position fix can be obtained, provided the observed data are sufficiently unique within the database. On the other hand, if the database contains locally positioned data, a relative position update can be established. This is the principle used in SLAM, where data from the same mission are used as reference data to obtain a relative update via the so-called *loop closing* procedure. The SLAM problem is discussed in Section 5.5.6, whereas this section discusses the case in which a previous reference database exists. In both cases, the accuracy of the obtained position fix is bounded by the position accuracy of the reference database. The most common form of geophysical navigation is terrain-based navigation, but it is also possible to use other geophysical properties, like gravity and magnetic field.

5.5.5.1 Terrain navigation

The idea behind terrain navigation is to use a set of terrain measurements to build up a terrain profile, and then find a position estimate by comparing this profile to a prior map database. For this to work, a certain degree of terrain variations must be present. We here focus on so-called bathymetric terrain navigation, where the map database is a digital terrain model (DTM). The compared terrain profiles thus consist of a set of sea depths, as opposed to so-called *feature-based* navigation methods which operate on terrain features or other features on the seafloor, e.g., rocks and man-made objects.

Terrain navigation has been used for decades in aircraft and missile navigation [45–47], and several approaches exist. We here concentrate on the so-called search area methods (see [48] for a classification of methods) which define a grid around the current, possibly uncertain position estimate, and try to find the position within the search grid that gives the best match between the DTM and the terrain measurements. The simplest way to do this is to use a modification of the original Terrain Contour Matching method for cruise missiles [45] which simply finds the position that minimises the absolute difference between the measurements and the DTM along the measured terrain profile. However, in modern AUV navigation, more sophisticated methods are usually applied, like Bayesian estimators. Because of non-linearities in the comparison of the DTM and the measurement, non-linear Bayesian estimators, like point mass filters (PMFs) and PFs, are attractive estimators [49,50].

The Bayesian terrain navigation algorithms use a state-space model to model the temporal development of the system, as well as the terrain measurements. If we let the state vector $\boldsymbol{x}$ simply include the position of the vehicle decomposed in the Earth frame E, i.e., $\boldsymbol{x} = \boldsymbol{p}_{EB}^{E}$, the state-space model can be formulated as

$$\boldsymbol{x}_{k+1} = \boldsymbol{f}(\boldsymbol{x}_k, \boldsymbol{u}_k, \boldsymbol{v}_k) \tag{5.12}$$

$$z_k = \boldsymbol{h}(\boldsymbol{x}_k) + \boldsymbol{w}_k \tag{5.13}$$

where the process equation (5.12) describes the position of the vehicle at time step $k+1$ based on the previous position $\boldsymbol{x}_k$, an input $\boldsymbol{u}_k$ and the process noise $\boldsymbol{v}_k$. The function $\boldsymbol{f}$ may be derived, e.g., from the INS measurements or from a vehicle model. The measurement equation (5.13) describes the measured depths z_k from a terrain measuring sensor. The dimension of the measurement vector is determined by the type of sensor used. Multibeam echo sounders (MBEs) typically give several hundred measurement points from each ping. The function $\boldsymbol{h}(\cdot)$ gives the DTM depths at the sensor footprint, based on the horizontal position and depth of the vehicle, and it is often highly non-linear. The measurement noise $\boldsymbol{w}_k$ models the combined errors due to sensor measurement noise, DTM errors and vehicle depth uncertainty. Based on the statistics of the noise terms and the functions $\boldsymbol{f}$ and $\boldsymbol{h}$, the prediction kernel probability density function (PDF) $p(\boldsymbol{x}_{k+1}|\boldsymbol{x}_k)$ and the measurement likelihood PDF $p(z_k|\boldsymbol{x}_k)$ can be derived. If we assume that the initial state PDF $p(\boldsymbol{x}_0)$ is known and let $Z_k = \{z_1, z_2, \ldots, z_k\}$ denote the set of all measurements up to time step k, the posterior

PDF $p(\boldsymbol{x}_k|Z_k)$ can be computed recursively using the time update and measurement update equations

$$p(\boldsymbol{x}_{k+1}|Z_k) = \int_{\mathbb{R}^{n_x}} p(\boldsymbol{x}_{k+1}|\boldsymbol{x}_k)p(\boldsymbol{x}_k|Z_k)\, d\boldsymbol{x}_k \tag{5.14}$$

$$p(\boldsymbol{x}_{k+1}|Z_{k+1}) = \frac{p(z_{k+1}|\boldsymbol{x}_{k+1})p(\boldsymbol{x}_{k+1}|Z_k)}{\int_{\mathbb{R}^n_x} p(z_{k+1}|\boldsymbol{x}_{k+1})p(\boldsymbol{x}_{k+1}|Z_k)\, d\boldsymbol{x}_{k+1}} \tag{5.15}$$

Equations (5.14) and (5.15) constitute the optimal Bayes filter equations. Note that no assumptions have been made about the distributions of the noise functions, which may be of arbitrary form. As the integrals in the optimal Bayes filter in general cannot be solved analytically, approximative methods must be used in practice. The PMF uses numerical quadrature on a grid to approximate the integrals [49,51], whereas PFs use sequential Monte Carlo sampling [20,52]. An important difference between the approximate Bayes filter and linear estimators like the Kalman filter is that they try to approximate the complete posterior PDF, which in many cases is multi-modal. In the linear, Gaussian case, the Bayes filter reduces to the Kalman filter.

In the above description of the Bayesian terrain navigation problem, the state was chosen to include the vehicle 3D position only. It is also possible to include more states, e.g., states modelling possible depth biases or coloured process noise, but due to the computational demands of the PMF and PF algorithms, in practice one is often restricted to three to four states. It is possible to use Rao–Blackwellisation in these algorithms, separating the non-linear and linear states, which are treated separately by combining non-linear and linear (Kalman filter-based) estimators [53–55].

Integration of the terrain navigation method with the main navigation system of the AUVs can be done in several ways. Tight integration with the main INS Kalman filter can be done, following the lines of [54]. Due to robustness reasons, it is often common to use loose coupling of the INS and the terrain navigation system, using the converged terrain navigation fixes as regular position updates to the INS. The terrain navigation system then takes the role of a position measurement in Figure 5.7. This is the solution used in the Kongsberg HUGIN AUVs, which include real-time terrain navigation as an optional add-on [56]. As the main INS Kalman filter is only able to handle measurements with Gaussian noise, a common approach is to run the approximate Bayes filter algorithm until it converges to a distribution resembling a Gaussian, and use the mean and covariance of this distribution as a position update in the INS. Typically, sanity and integrity checks of the converged terrain navigation estimate are performed before using the update in the INS, to minimise the possibility of using so-called *false fixes*, i.e., erroneous positions with high estimated accuracy.

Terrain navigation for AUVs has matured significantly over the last decade, and can now be considered a standard aiding method. Several publications with real-time terrain navigation results have been published, e.g., [57–59]. Figure 5.13 shows results from the first full-scale demonstration of real-time terrain navigation with the HUGIN AUVs, conducted in the Barents Sea in May 2009. Terrain navigation was used as the only position aiding source during a 50-km-long straight-line transit,

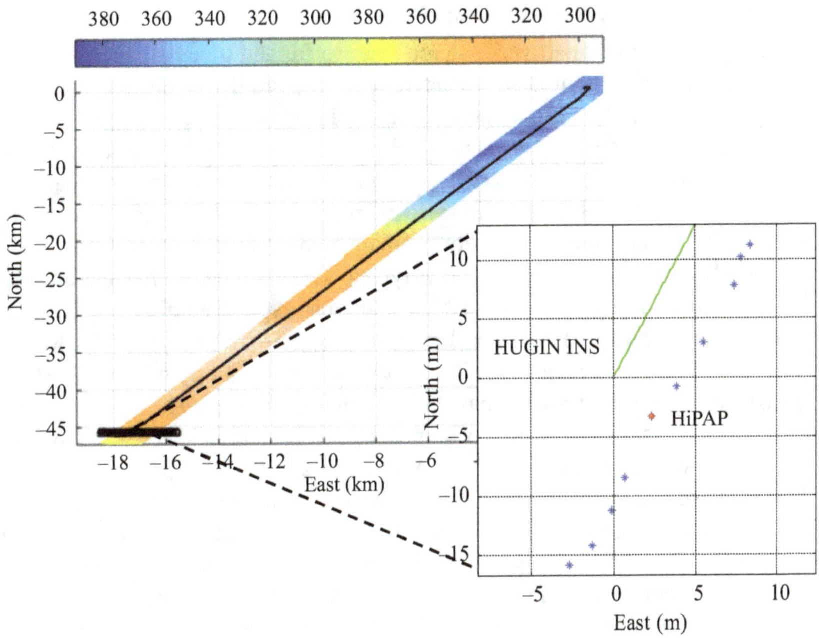

Figure 5.13 Results from real-time terrain navigation demonstration with HUGIN AUV from the Barents Sea in May 2009. Terrain navigation was used as the only position aiding method in the DVL-AINS. After a 50-km straight transit line, the position error was approximately 5 m, when compared to a reference solution based using GNSS-USBL updates from a surface ship

using a Kongsberg EM2000 MBE as terrain measuring sensor. The transit corridor was mapped prior to the mission, with a Kongsberg EM710 MBE from a surface vessel. The terrain in the area was relatively flat, but the seafloor was dominated by shallow iceberg scouring, which proved to be ideal for terrain navigation. While the real-time navigation in the AUV was run fully autonomously, the surface vessel followed the AUV, collecting GNSS-USBL measurements in order to make a high-quality reference solution. After almost 8 h of fully autonomous transit along a straight line, the position error was observed to be within 5 m. According to [3], a typical drift of a DVL AINS (using a 300 kHz DVL) is 0.1% of travelled distance, corresponding to 50 m in this scenario.

Traditional terrain navigation requires a prior DTM to work, which in many scenarios limits its applicability. However, it is possible to make the DTM early in the mission and later to return to the mapped area in order to obtain a terrain navigation update, in a SLAM-like fashion. These kinds of operations extend the area of applicability of terrain navigation significantly. Some examples of this from the HUGIN vehicles are outlined in [60–62].

5.5.5.2 Gravity-based navigation

The Earth's gravitational field varies across the world, and given an accurate gravitation model, this can be utilised for navigation in a similar manner as in terrain navigation. To use gravity measurements for navigation purposes, a database of gravity anomalies (the difference between the observed gravity and the gravity from a theoretical gravity model) is needed. Typical gravity anomalies are in the range of 0–100 mGals (1 mGal $= 1.0 \cdot 10^{-5}$ m/s^2). This requires a high accuracy both in the gravimeter used for building the gravity anomaly map, as well as for the sensor used on board the AUV for measuring gravity. A typical accelerometer in navigation grade IMUs, which are common in high-end AUVs, often has specified biases on the order of μg. As an example, the biases in accelerometers in the Honeywell HG9900 IMU are specified to be less than 25 μg $= 2.4510^{-5}$ m/s^2, which is slightly more than what is needed to be used for gravity-based navigation. Consequently, the AUV would have to carry a dedicated gravimeter, which is only an option for larger AUVs. Development of a gravimeter specially designed for AUVs was presented in [63].

Gravity-based navigation differs from terrain navigation in several aspects. One obvious difference is that it uses a passive sensor which might be favourable in some scenarios. The measured gravity also changes with the depth of the vehicle, which has to be compensated for. The high demands both for accurate measurements and accurate and high-resolution anomaly maps are probably the main reasons why gravity-based navigation is less frequently used for AUVs. However, some experimental results have been published, e.g., in [64–66].

5.5.5.3 Geomagnetic field-based navigation

The Earth's geomagnetic field is another geophysical quantity that can be utilised for navigation purposes. Like the gravitational field, the geomagnetic field varies across the Earth. The measured field is again dependent on the depth of the vehicle. In addition, geomagnetic measurements are vulnerable to electromagnetic disturbances from the surroundings, e.g., metal or electric currents. The largest source to such disturbances is the vehicle itself, through its magnetic signature. If the magnetic signature is known and assumed to be fairly constant, it is possible to compensate for this effect. Another source of variations in the geomagnetic field is the solar activity, which induces temporal variations in the geomagnetic field [67].

Like gravity-based navigation, geomagnetic navigation is not widely used in AUV applications, though some results have been published. In [68], a combination of terrain-based and geomagnetic navigation was applied, using a Rao–Blackwellised PF on simulated data. A similar combination of synthetic magnetic and bathymetric data for long-range AUV applications was discussed in [69]. The first work presenting experimental results of geomagnetic navigation in sea trials is [67], where a USV was used as a stand-in for an AUV, and the magnetometer was installed in a tow-fish behind the vehicle, in order to minimise the effect the vehicle self-noise. Using a magnetic field map with 1-m horizontal resolution in a small 300 m × 200 m lake, a bounded navigation accuracy better than 5 m was demonstrated over a 2-h period.

5.5.6 Classification of SLAM methods for AUVs

The requirement of *a-priori* detailed knowledge of the mission area cannot always be met. Still, it is possible to use sensor data to map the mission area, while at the same time using the constructed map to essentially constrain drift in the navigation error. Algorithms based on this principle are usually called SLAM algorithms [70]. SLAM is not a singular aiding technique but rather an estimation framework for solving these problems. In the traditional description, one considered a kidnapped robot waking up in an unknown place, exploring and mapping its surrounding, while using the same map for navigation. The robot has no need for a global reference for this exploration. In our case, we consider SLAM within a global reference, where the estimator is part of the INS, with which any globally referenced sensor measurements can be fused. The main mechanism behind SLAM is shown in Figure 5.14. When the AUV observes a landmark $\boldsymbol{m}_1$ for the first time through its sensor, the landmark is added to the estimation state vector, and the landmark's position is highly correlated with the AUV's current position. Now, the AUV's position error increases in time, when it sees landmark $\boldsymbol{m}_2$ and $\boldsymbol{m}_3$, while the position error of the landmarks stays the same. This changes when the AUV returns back to observe the first landmark $\boldsymbol{m}_1$, in what is called a *loop closure*. The loop-closure observation is used by the SLAM algorithm to reduce the AUV's position error down to the level of $\boldsymbol{m}_1$. Since the errors in $\boldsymbol{m}_2$ and $\boldsymbol{m}_3$ are both correlated with the AUV's error and thereby $\boldsymbol{m}_1$, the loop closure also reduces the position errors of these landmarks [70]. An overview of different SLAM algorithms is given in, e.g., [71], but here we concentrate on EKF-SLAM with known data association. We formulate the problem in a way to include different types of maps, in order to structure and classify the different SLAM-like techniques relevant for AUV navigation.

Let $\boldsymbol{m}(t) = \{m_i(t)\}_{i=1}^n$ denote a finite set of states that represents the map, and we assume that we know the temporal differential equation of the map states, and

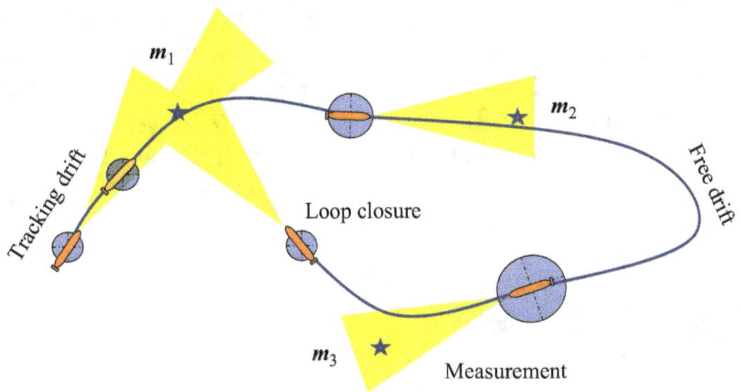

Figure 5.14 The AUV can use sensor data to track landmarks both to lower position drift, but also to regain original position accuracy after periods of drift through loop closure

augment the discretisation of the time development model for these states to the estimation problem in (5.4). The map is observed through a sensor measurement described by a measurement model replacing the one in (5.5)

$$z = \boldsymbol{h}(\boldsymbol{x}, \boldsymbol{m}, t) + \boldsymbol{w} \tag{5.16}$$

where $\boldsymbol{w}$ is the sensor noise and $\boldsymbol{h}$ is a known measurement function. We assume that the map states can be initialised through a local inversion of $\boldsymbol{h}$ given the current INS solution $\boldsymbol{x}$, and a sensor measurement, or that we have some initial estimate. We will now look at some concrete examples of using this framework.

5.5.6.1 Underwater transponder positioning SLAM

If the AUV carries and deploys a single transponder on the seafloor at a point T_p in the mission area, then the techniques in Section 5.5.3 can be formulated within this framework [4,72]. The map states are now given by the position of the transponder in an Earth-fixed frame

$$\boldsymbol{m} = \boldsymbol{p}^E_{ET_p} \tag{5.17}$$

There is no time dependency, so $\dot{\boldsymbol{m}} = 0$, and the measurement equation is given by the range between the transponder and the AUV

$$\boldsymbol{h}(\boldsymbol{x}, \boldsymbol{m}, t) = \left\| \boldsymbol{p}^E_{EB} - \boldsymbol{p}^E_{ET_p} \right\| \tag{5.18}$$

5.5.6.2 Sonar feature-based SLAM

AUVs equipped with imaging sonars can process sonar images of the seafloor and detect features that are recognisable on later images from the same area, e.g., like automatic target recognition used to detect and classify sea mines [73]. If the features are associated, this observation can be used to aid the INS. The positions of the targets T_i form the map $\boldsymbol{m} = \{\boldsymbol{p}^E_{ET_i}\}$, and the targets are stationary on the seafloor, i.e., $\dot{\boldsymbol{m}} = 0$. The measurement model depends on the type of sonar used: forward-looking, side-scan and synthetic apertures, possibly interferometric. If we assume that we can measure the range and two spherical polar angles $z_i = (r_i, \theta_i, \phi_i)^T$ to the target (i.e., we have an interferometric sonar), then the measurement function becomes

$$\boldsymbol{h}_i(\boldsymbol{x}, \boldsymbol{m}, t) = \begin{bmatrix} \left\| \boldsymbol{p}^E_{EB} - \boldsymbol{p}^E_{ET_i} \right\| \\ \tan^{-1}\left(\dfrac{\left\| \boldsymbol{p}^E_{EB,h} - \boldsymbol{p}^E_{ET_i,h} \right\|}{\left\| \boldsymbol{p}^E_{EB} - \boldsymbol{p}^E_{ET_i} \right\|} \right) \\ \tan^{-1}\left(\dfrac{\left\| \boldsymbol{p}^E_{EB,z} - \boldsymbol{p}^E_{ET_i,z} \right\|}{\left\| \boldsymbol{p}^E_{EB} - \boldsymbol{p}^E_{ET_i} \right\|} \right) \end{bmatrix} \tag{5.19}$$

where subscript h indicates that the horizontal component of the position vector is used, and subscript z denotes the vertical component.

5.5.6.3 Bathymetric SLAM

We can use *in-situ* bathymetric measurements from the AUV to generate a DTM that we can also use for terrain navigation within the same mission. Assume that the depth of any point on the seafloor can be estimated by a global interpolant

$$h(\boldsymbol{x}, \boldsymbol{m}) \approx \sum_{i=0}^{n} m_i B_i\left(\boldsymbol{p}_{EB}^{E}\right) \tag{5.20}$$

where the coefficients define our map states $\boldsymbol{m} = \{m_i\}$, and B_i is a kernel or basis function for the interpolant. Since there is no time dependency in the map states, $\dot{\boldsymbol{m}} = 0$. The global interpolant is part of our measurement function; see Section 5.5.5.

Large, high-resolution maps require a large number of map states, and combined with non-linearities of low-cost high-drift systems, this is challenging. In [74], an algorithm called BPSLAM (Bathymetric distributed Particle SLAM) is proposed, based on a Rao–Blackwellised PF combined with clever handling of map states. For low-drift systems, the map state reduction can be done using submaps, introduced in [75], and used as reference lines for autonomous mapping in [62]. One of the key results of bathymetric SLAM is automatically consistent maps; see the example in Figure 5.15.

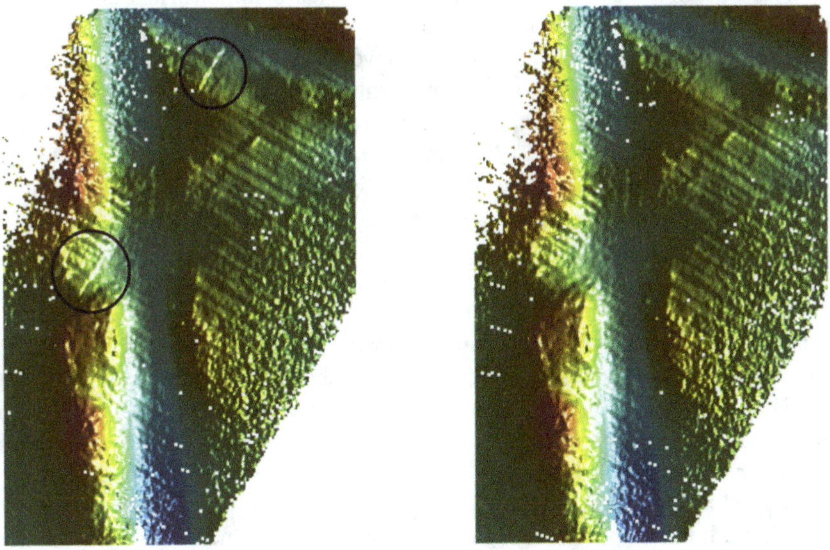

Figure 5.15 Notice the marked errors in the bathymetry measured from two adjacent lines (left) in a mission of the HUGIN AUV, corrected by relative terrain navigation (right), from [61]. The bathymetry has been regridded using the updated navigation solution, such that locations of the white areas with missing data are changed

5.5.6.4 Sea current SLAM

The map could also be made from the sea current vector field $V^E_{\underline{E}W}(p,t)$, using a global interpolant as for bathymetric SLAM. There are examples of work utilising spatial variations in the sea current field for localisation; see, e.g., [76], given initial input from global circulation models. Others have built a vertical sea current profile, during descent initialised by GNSS-aiding on the surface, and during ascent initialised by DVL-aiding with bottom track; see, e.g., [77]. But it is most common to model the sea current only in the vicinity of the AUV, which gives the map $m(t) = v^L_{\underline{E}W_B} = R^T_{EL} V^E_{\underline{E}W}(p^E_{EB})$. Since the sea current field is typically unknown, the time evolution of the map is usually approximated by, e.g., a Gauss–Markov process allowing for some time variations due to tides and other ocean processes, and any spatial variation experienced by the AUV travelling through the sea current field [39]. The water-referenced velocity measurement of the DVL can then be written as [39]

$$h(x, m) = R^T_{LB} \left(v^L_{\underline{E}W_B} - v^L_{\underline{E}B} \right) \tag{5.21}$$

5.5.6.5 Camera feature-based SLAM

The detection and association of features in mono camera images bear similarities to the processing of high-resolution sonar images. One major difference is, however, that we lack information on range to the feature, and we only have accurate directional measurements. The bearing measurements are then given by the spherical angles in (5.19). Because of the lacking information on range, it is common in land and air applications to re-parametrise the problem using, e.g., the inverse depth parametrisation described in [78]. This enables an EKF to converge even with completely unknown initial range. This is because the range efficiently becomes observable when fusing camera measurements with inertial measurements in a tightly integrated filter for an accelerating platform [79]. For an AUV, the situation is different. The range of optical sensors is severely limited under water, i.e., the initial range is more precisely known and could be initialised through a best guess given, e.g., the current altitude of the AUV above the seafloor.

5.5.6.6 Stochastic cloning

In all of the previous examples, the map states were explicitly added to the estimation state vector. For high-resolution sensors, the number of landmarks can grow fast. There is a structure-less way to formulate the SLAM problem that recognises that it is enough to estimate our own state at the observation time. This is a technique called *stochastic cloning* [80]. When the AUV makes the first measurement of a landmark, the navigation state vector is cloned and augmented to the current state vector (see Figure 5.16). It is now possible to process any general delta measurement in time for the cloned states. This also fits in the framework. The map in this case is the cloned historical states

$$m(t) = \begin{bmatrix} x(t_1) \\ x(t_2) \\ \vdots \\ x(t_n) \end{bmatrix}, \quad t_1 \leq \cdots \leq t_n \leq t \tag{5.22}$$

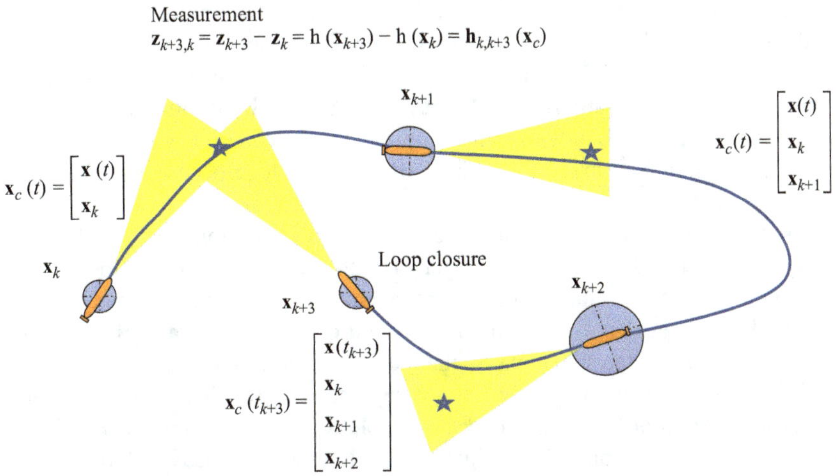

Figure 5.16 *The estimator states are augmented each time a new landmark is observed. When a loop-closer occurs, the measurement can now be used since the state vector contains a copy of the state at first observation*

The time evolution of the map is essentially that history does not change, $\dot{\boldsymbol{m}} = \mathbf{0}$. The measurement is the difference between the measurement z_k at loop closure and the initial measurement of the landmark z_0, and the measurement model is given by

$$\boldsymbol{h}_{0,k}(\boldsymbol{x}, \boldsymbol{m}) = \boldsymbol{h}(\boldsymbol{x}_k, \boldsymbol{m}_k) - \boldsymbol{h}(\boldsymbol{x}_0, \boldsymbol{m}_0) \tag{5.23}$$

In a real application, the state vector will still grow too large, but given the nature of the sensor measurements, only part of the state vector needs to cloned. In many cases, it is sufficient to clone only position, and maybe also the rotation states. The special case of delta-position measurements was covered earlier in Section 5.5.4.

5.6 Conclusions

This chapter has given an overview of single vehicle AUV navigation, which in all advanced vehicles is based on inertial navigation. We have seen how an INS is able to estimate the AUV's position, velocity and orientation, and how its accuracy is dependent on manoeuvres and the quality of the sensors. With basic aiding sensors such as pressure sensors and DVLs, one is able to significantly reduce the inherent drift in the INS, but to facilitate extensive submerged, fully autonomous navigation, additional aiding methods are needed. One possibility is to rely on transponder navigation of some kind (LBL or UTP), but this requires deployment and positioning of the transponders to be used. For the AUV to be completely independent of infrastructure or aiding from other vehicles (manned or unmanned), the only remaining option is to use observations of the underwater environment for navigation purposes, e.g., to use payload sensors (sonars, optical cameras, etc.) to obtain accurate delta-position estimates or to use some kind of geophysical navigation. Bathymetric terrain referenced navigation is by far the most mature geophysical navigation for AUVs today. In areas where no prior DTM exist, terrain- or feature-based techniques can be used as the basis for SLAM algorithms. However, as SLAM by definition is local, the global accuracy obtainable through the loop-closing procedure is bounded above by the global navigation accuracy at the first bathymetry or feature observation involved in the loop closure.

To achieve fully autonomous operations with high navigation accuracy, the autonomy system of the AUV must work together with the navigation systems to utilise all possible navigation aiding methods whenever needed, based on the constraints of the operation. For example, in a non-covert operation in shallow water, the autonomy system could make a decision to pause the mission execution in order to surface for a GNSS fix if the navigation accuracy is unacceptably low. In other scenarios, altering the mission plan in order to visit an area suitable for terrain navigation could be an option. Future research on AUV navigation should therefore focus both on developing new and better aiding methods, as well as optimally using the existing methods autonomously.

References

[1] Hagen OK, Jalving B. Vertical position estimation for underwater vehicles. Sea Technology. 2008;49(12):51–54.

[2] Gade K. The seven ways to find heading. The Journal of Navigation. 2016;69(05):955–970.

[3] Hagen PE, Hegrenæs Ø, Jalving B, Midtgaard Ø, Wiig M, Hagen OK. Making AUVs truly autonomous. In: Inzartsev AV, editor. Underwater Vehicles. In-Tech Education and Publishing: Vienna, Austria; 2009. pp. 129–152.

[4] Jalving B, Gade K, Hagen OK, Vestgård K. A toolbox of aiding techniques for the HUGIN AUV integrated inertial navigation system. In: Proceedings of the IEEE OCEANS 2003. San Diego, CA, USA; 2003. pp. 1146–1153.

[5] Kussat NH, Chadwell CD, Zimmerman R. Absolute positioning of an autonomous underwater vehicle using GPS and acoustic measurements. IEEE Journal of Oceanic Engineering. 2005;30(1):153–164.

[6] Jakuba MV, Roman CN, Singh H, *et al.* Long-baseline acoustic navigation for under-ice autonomous underwater vehicle operations. Journal of Field Robotics. 2008;25(11–12):861–869.

[7] Hegrenæs Ø, Gade K, Hagen OK, Hagen PE. Underwater transponder positioning and navigation. In: Proceedings of the MTS/IEEE Oceans Conference. Biloxi, MS; 2009. pp. 1–7.

[8] Paull L, Saeedi S, Seto M, Li H. AUV navigation and localization: A review. IEEE Journal of Oceanic Engineering. 2014;39(1):131–149.

[9] Fallon MF. Cooperative AUV navigation using a single maneuvering surface craft. The International Journal of Robotics Research. 2010;29(12):1461–1474.

[10] Vasilijević A, Nađ Đ, Mandić F, Miskovic N, Vukic Z. Coordinated navigation of surface and underwater marine robotic vehicles for ocean sampling and environmental monitoring. IEEE/ASME Transactions on Mechatronics. 2017;22(3):1174–1184.

[11] Gade K. Inertial Navigation – Theory and Applications [Dr. Philos. Thesis]. Norwegian University of Science and Technology, Department of Engineering Cybernetics. Trondheim, Norway; 2018.

[12] Gade K. A non-singular horizontal position representation. The Journal of Navigation. 2010;63(3):395–417. Available from: http://www.navlab.net/nvector.

[13] Gade K. The n-vector page; 2018. Accessed 15 Aug, 2018. Web page. Available from: http://www.navlab.net/nvector/.

[14] Titterton DH, Weston JL. Strapdown Inertial Navigation Technology. London, UK: Institution of Engineering and Technology; 2004.

[15] Einstein A. Relativity: The Special and General Theory. Henry Holt: New York; 1920.

[16] Groves PD. Principles of GNSS, Inertial, and Multisensor Integrated Navigation Systems. London, UK/Boston, MA: Artech House; 2013.

[17] Passaro V, Cuccovillo A, Vaiani L, De Carlo M, Campanella CE. Gyroscope technology and applications: A review in the industrial perspective. Sensors. 2017;17(10):2284.

[18] Gelb A, editor. Applied Optimal Estimation. The MIT Press: Cambridge, MA; 1974.

[19] Brown RG, Hwang PY. Introduction to Random Signals and Applied Kalman Filtering. Vol. 1. Wiley, New York; 1983.

[20] Andrieu C, Doucet A, Punskaya E. Sequential Monte Carlo methods for optimal filtering. In: Doucet A, de Freitas N, Gordon N, editors. Sequential Monte Carlo Methods in Practice. Springer-Verlag, New York; 2001. pp. 79–95.

[21] Julier SJ, Uhlmann JK. Unscented filtering and nonlinear estimation. In: Proceedings of the IEEE. Vol. 92; 2004. pp. 401–422.

[22] Hagen OK, Ånonsen KB, Mandt M. The HUGIN real-time terrain navigation system. In: Proceedings of the MTS/IEEE OCEANS 2010. Seattle, WA, USA; 2010. pp. 1–7.

[23] Hegrenæs Ø, Hallingstad O, Gade K. Towards model-aided navigation of underwater vehicles. Modeling, Identification and Control. 2007;28(4):113–123.

[24] Bar-Shalom Y, Li XR, Kirubarajan T. Estimation with Applications to Tracking and Navigation. John Wiley & Sons: New York; 2001.

[25] Gade K. NavLab, a Generic Simulation and post-processing tool for navigation. Modeling, Identification and Control. 2005;26(3):135.

[26] Wahba G. A least squares estimate of satellite attitude. SIAM Review. 1965;7(3):p. 409.

[27] Gade K, Jalving B. An aided navigation post processing filter for detailed seabed mapping. Modeling, Identification and Control. 1999;20(3): 165–176.

[28] Lemoine FG, Kenyon SC, Factor JK, *et al.* The development of the joint NASA GSFC and the National Imagery and Mapping Agency (NIMA) geopotential model EGM96. NASA Technical Report. 1998 July; 1998-206861.

[29] Pavlis NK, Holmes SA, Kenyon SC, Factor JK. The development and evaluation of the Earth Gravitational Model 2008 (EGM2008). Journal of Geophysical Research: Solid Earth. 2012;117(B4).

[30] Willumsen AB, Hagen OK, Boge PN. Filtering depth measurements in underwater vehicles for improved seabed imaging. In: OCEANS 2007-Europe. IEEE; 2007. pp. 1–6.

[31] Hegrenæs Ø, Wallace C, Børhaug E. Autonomous under-ice surveying using the MUNIN AUV and single-transponder navigation. In: OCEANS–Anchorage, 2017. IEEE; 2017. pp. 1–10.

[32] Vasconcelos JF, Elkaim G, Silvestre C, Oliveira P, Cardeira B. Geometric approach to strapdown magnetometer calibration in sensor frame. IEEE Transactions on Aerospace and Electronic Systems. 2011;47(2):1293–1306.

[33] Chulliat A, Macmillan S, Alken P, *et al.* The US/UK world magnetic model for 2015–2020, Technical Report, National Geophysical Data Center, NOAA. doi: 10.7289/V5TB14V7; 2015.

[34] Crosbie BP, Anderson M, Wolbrecht E, Canning J, Edwards DB. Synchronous navigation of AUVs using WHOI micro-modem 13-bit communications. In: OCEANS 2010. IEEE; 2010. pp. 1–5.

[35] Kaminski C, Crees T, Ferguson J, *et al.* 12 days under ice – An historic AUV deployment in the Canadian High Arctic. In: Autonomous Underwater Vehicles (AUV), 2010 IEEE/OES. IEEE; 2010. pp. 1–11.

[36] Milne PH. Underwater Acoustic Positioning Systems. TXGulf Publishing Company: Houston; 1983.

[37] Kinsey JC, Eustice RM, Whitcomb LL. A survey of underwater vehicle navigation: Recent advances and new challenges. In: IFAC Conference of Manoeuvering and Control of Marine Craft. Vol. 88; 2006. pp. 1–12.

[38] Larsen MB. Synthetic long baseline navigation of underwater vehicles. In: OCEANS 2000 MTS/IEEE Conference and Exhibition. Vol. 3. Providence, RI; 2000. pp. 2043–2050.

[39] Hegrenaes Ø, Hallingstad O. Model-aided INS with sea current estimation for robust underwater navigation. IEEE Journal of Oceanic Engineering. 2011;36(2):316–337.

[40] International Hydrographic Organization. IHO Standards for Hydrographic Surveys, 5th edition. Special publication No. 44. Monaco: International Hydrographic Organization; 2008.

[41] Bellettini A, Pinto MA. Theoretical accuracy of synthetic aperture sonar micronavigation using a displaced phase-center antenna. IEEE Journal of Oceanic Engineering. 2002;27(4):780–789.

[42] Hansen RE, Sæbø TO, Gade K, Chapman S. Signal processing for AUV-based interferometric synthetic aperture sonar. In: OCEANS 2003. Proceedings. Vol. 5. IEEE; 2003. pp. 2438–2444.

[43] Sæbø TO, Hansen R, Callow HJ, Synnes SA. Coregistration of synthetic aperture sonar image from repeated passes. In: Proceedings of Underwater Acoustic Measurements. Kos, Greece; 2011.

[44] Hansen R, Sæbø TO, Lorentzen OJ, Midtgaard Ø. Change detection in topographic structures using interferometric synthetic aperture sonar. In: Proceedings of the 2nd Underwater Acoustic Conference; 2014.

[45] Golden JP. Terrain contour matching (TERCOM): A cruise missile guidance aid. In: Wiener TF, editor. Image Processing for Missile Guidance. Vol. 238. San Diego, CA: The Society of Photo-Optical Instrumentation Engineers; 1980. pp. 10–18.

[46] Boozer D, Fellerhoff J. Terrain-aided navigation test results in the AFTI/F-16 aircraft. Journal of The Institute of Navigation. 1988;35(2):161–175.

[47] Hollowell J. Heli/SITAN: A terrain referenced navigation algorithm for helicopters. In: Proceedings of the IEEE Position Location and Navigation Symposium (PLANS). Las Vegas, NV; 1990. pp. 616–625.

[48] Ånonsen KB. Advances in Terrain Aided Navigation for Underwater Vehicles [PhD thesis]. Norwegian University of Science and Technology. Trondheim, Norway; 2010.

[49] Bergman N, Ljung L, Gustafsson F. Terrain navigation using Bayesian statistics. IEEE Control Systems Magazine. 1999;19(3):33–40.

[50] Ånonsen KB, Hallingstad O. Terrain aided underwater navigation using point mass and particle filters. In: Proceedings of the IEEE/ION Position Location and Navigation Symposium (PLANS) 2006. San Diego, CA; 2006. pp. 1027–1035.

[51] Bucy R. Bayes theorem and digital realizations for non-linear filters. The Journal of the Astronautical Sciences. 1969;XVII(2):80–94.

[52] Doucet A. On Sequential simulation-based methods for Bayesian filtering. Cambridge, UK: Department of Engineering, Cambridge University; 1998.

[53] Doucet A, de Freitas N, Murphy K, Russel S. Rao-Blackwellised particle filtering for dynamic Bayesian networks. In: Uncertainty in Artificial Intelligence. Morgan Kaufmann Publishers Inc.; 2000. pp. 176–183.

[54] Karlsson R, Gustafsson F, Karlsson T. Particle filtering and Cramer-Rao lower bound for underwater navigation. In: Proceedings IEEE Conference on Acoustics, Speech and Signal Processing. Vol. 6. Hong Kong; 2003. pp. VI–65.

[55] Hagen OK, Ånonsen KB. Improving terrain navigation by concurrent tidal and sound speed error estimation. In: Oceans-San Diego, 2013. IEEE; 2013. pp. 1–7.

[56] Ånonsen KB, Hagen OK. The HUGIN AUV Terrain Navigation Module. In: Proceedings of the MTS/IEEE Oceans Conference. San Diego, CA, USA: IEEE; 2013. pp. 1–8.

[57] Nygren I, Jansson M. Terrain navigation using the correlator method. In: Position Location and Navigation Symposium, PLANS 2004. Monterey, CA: IEEE; 2004. pp. 649–657.

[58] Ånonsen KB, Hagen OK. An analysis of real-time terrain aided navigation results from a HUGIN AUV. In: Proceedings of the IEEE/ION Oceans 2010 Conference. Seattle, WA, USA; 2010. pp. 1–9.

[59] Meduna DK, Rock SM, McEwen RS. Closed-loop terrain relative navigation for AUVs with non-inertial grade navigation sensors. In: 2010 IEEE/OES Autonomous Underwater Vehicles (AUV). Monterey, CA, USA: IEEE; 2010. pp. 1–8.

[60] Ånonsen KB, Hagen OK. Recent developments in the HUGIN AUV terrain navigation system. In: Proceedings of the MTS/IEEE Oceans Conference. Kona, HI, USA; 2011.

[61] Hagen OK, Ånonsen KB, Sæbø TO. Toward autonomous mapping with AUVs-line-to-line terrain navigation. In: Proceedings of the MTS/IEEE Oceans 15 Conference. Washington, DC, USA: IEEE; 2015. pp. 1–8.

[62] Ånonsen KB, Hagen OK, Berglund E. Autonomous mapping with AUVs using relative terrain navigation. In: Proceedings of the MTS/IEEE Oceans Conference 2017. Anchorage, AK, USA: IEEE; 2017. pp. 1–7.

[63] Ishihara T, Shinohara M, Araya A, *et al.* Development of an underwater gravity measurement system with autonomous underwater vehicle for marine mineral exploration. In: Techno-Ocean (Techno-Ocean). IEEE; 2016. pp. 127–133.

[64] Jircitano A, White J, Dosch D. Gravity based navigation of AUVs. In: Autonomous Underwater Vehicle Technology, 1990. AUV'90, Proceedings of the (1990) Symposium on. IEEE; 1990. pp. 177–180.

[65] Kamgar-Parsi B, Kamgar-Parsi B. Vehicle localization on gravity maps. In: Unmanned Ground Vehicle Technology. Vol. 3693. International Society for Optics and Photonics; 1999. pp. 182–192.

[66] Wang B, Yu L, Deng Z, Fu M. A particle filter-based matching algorithm with gravity sample vector for underwater gravity aided navigation. IEEE/ASME Transactions on Mechatronics. 2016;21(3):1399–1408.

[67] Teixeira FC, Quintas J, Pascoal A. Experimental validation of magnetic navigation of marine robotic vehicles. IFAC-PapersOnLine. 2016;49(23): 273–278. 10th IFAC Conference on Control Applications in Marine Systems CAMS 2016. Available from: http://www.sciencedirect.com/science/article/pii/S2405896316319413.

[68] Teixeira FC, Pascoal AM. Geophysical navigation of autonomous underwater vehicles using geomagnetic information. IFAC Proceedings Volumes. 2008;41(1):178–183.

[69] Kato N, Shigetomi T. Underwater navigation for long-range autonomous underwater vehicles using geomagnetic and bathymetric information. Advanced Robotics. 2009;23(7–8):787–803.

[70] Durrant-Whyte H, Bailey T. Simultaneous localization and mapping (SLAM): Part I: The essential algorithms. IEEE Robotics & Automation Magazine. 2006;13(2):99–110.

[71] Thrun S, Burgard W, Fox D. Probabilistic Robotics. Cambridge, MA/London, UK: MIT Press; 2005.

[72] Willemenot E, Morvan P, Pelletier H, Hoof A. Subsea positioning by merging inertial and acoustic technologies. In: OCEANS 2009-EUROPE; 2009. pp. 1–8.

[73] Midtgaard Ø, Hagen P. Automatic target recognition for the HUGIN mine reconnaissance system. In: Proc. International Conference on Detection and Classification of Underwater Targets; 2007.

[74] Barkby S, Williams SB, Pizarro O, Jakuba MV. A featureless approach to efficient bathymetric SLAM using distributed particle mapping. Journal of Field Robotics. 2011;28(1):19–39. Available from: http://dx.doi.org/10.1002/rob.20382.

[75] Roman C, Singh H. A self-consistent bathymetric mapping algorithm. Journal of Field Robotics. 2007;24(1–2):23–50.

[76] Song Z, Mohseni K. Long-term inertial navigation aided by dynamics of flow field features. IEEE Journal of Oceanic Engineering. 2018;43(4):940–954.

[77] Stanway MJ. Contribution to Automated Realtime Underwater Navigation [PhD thesis]. Dep. Mech. Eng., Massachusetts Institute of Technology. Cambridge, MA, USA; 2012.

[78] Civera J, Davison AJ, Montiel JMM. Inverse depth parametrization for monocular SLAM. IEEE Transactions on Robotics. 2008;24(5):932–945.

[79] Martinelli A. Vision and IMU data fusion: Closed-form solutions for attitude, speed, absolute scale, and bias determination. IEEE Transactions on Robotics. 2012;28(1):44–60.

[80] Roumeliotis SI, Burdick JW. Stochastic cloning: A generalized framework for processing relative state measurements. In: Robotics and Automation, 2002. Proceedings. ICRA'02. IEEE International Conference on. Vol. 2. IEEE; 2002. pp. 1788–1795.

Chapter 6
Ultra-endurance AUVs: energy requirements and terrain-aided navigation

Georgios Salavasidis[1,2], Andrea Munafò[3], Davide Fenucci[1], Catherine A. Harris[1], Thomas Prampart[1], Robert Templeton[1], Micheal Smart[1], Daniel T. Roper[1], Miles Pebody[1], Stephen D. McPhail[1], Eric Rogers[2] and Alexander B. Phillips[1]

6.1 Introduction

Autonomous underwater vehicles (AUVs) have been successfully utilised for a wide range of applications, and examples include (a) oceanographic surveys and research activities, e.g. ocean sampling [1], searching for hydrothermal vent fields in deep oceans [2], seafloor mapping and environmental monitoring [3,4]; (b) commercial applications such as monitoring seafloor assets (e.g. pipeline inspections [5]); and (c) military applications such as mine hunting [6,7]. The endurance of typical AUVs is less than 50 h [8,9] due to the significant energy consumption of the navigation and payload sensors, in combination with propulsion actuators. Therefore, traditionally AUVs are launched and recovered from a support vessel that remains in relatively close proximity to the survey site during the deployment.

However, the prohibitively high cost of scientific ship time (in the order of tens of thousands of UK pounds per day) together with the increasing demand from scientific [10] and industrial [11] communities is driving requirements for flexible autonomous platforms that can conduct subsea measurements on ever larger temporal and spatial scales without the need for persistent and expensive ship support. To meet these needs, platforms such as underwater gliders have become increasingly popular [12,13]. Due to their low energy consumption and efficient gliding behaviour, gliders are able to operate for up to months at a time, periodically contacting remote pilots via a satellite and hence removing the dependence on a costly support ship. However, the saw-tooth gliding behaviour, lack of precise control and navigation as

[1]Marine Autonomous and Robotic Systems, National Oceanography Centre, Southampton, UK
[2]Electronics and Computer Science, University of Southampton, Southampton, UK
[3]SeeByte Ltd. Edinburgh, UK

well as the restricted payload suggest that gliders are not suitable for all application, particularly those requiring operations at a constant depth with sensor suites of a larger volume and of a higher power demand.

In response to this demand, long-range propeller-driven AUVs have been developed. One example of this class of vehicles is Tethys, developed by the Monterey Bay Aquarium Research Institute (MBARI) [14]. Tethys is a relatively small man-portable 300-m depth-rating AUV with multi-week endurance depending on payload. Although Tethys extends what can be achieved with just a glider fleet, it is still limited in depth and range. Thus, there is a remaining requirement for a deeper diving longer-range AUV that is capable of carrying larger sensor payloads. Seeking to combine the expanded capability of a traditional workhorse AUV platform with the vastly increased endurance and remote piloting of a glider, the National Oceanography Centre (NOC) in the UK has been developing a new class of ultra-endurance propeller-driven AUVs, the Autosub Long Range 6000 (ALR6000) [15] and the ALR1500 [16]. These platforms aim to combine the extreme range already achievable using subsea gliders with increased manoeuvrability, position control and payload capacity of large flight-style AUVs. The long endurance of these platforms is achieved via the use of high-energy-density lithium primary cells combined with optimisation in forward speed and energy consumption within all sub-systems. The increasing availability of low-power sensors developed for the glider community [17] coupled with the ample flooded space available has provided the ALR vehicles with a variety of payload options and use cases.

The ALR vehicles and similar long-range platforms open up a world of new AUV applications in the commercial, oceanographic and military domains, including persistent monitoring and data-collection in some of the most remote and inaccessible areas on Earth. The extreme range of these vehicles can facilitate the attempt of previously impossible tasks, including a continuous mid-water column survey from Svalbard (Norway) to Barrow (Alaska) under the Arctic sea-ice [18]. However, operating an AUV during such long deployments comes with its own challenges, as the vehicle is likely to amass a considerable number of pressure cycles. After a long period of time spent at sea, corrosion and biofouling and, most importantly, the navigation drift begin to significantly limit the operational efficiency. Despite the considerable effort directed to underwater navigation, a long-range low-power solution yet remains a challenge precluding multi-month-long applications.

Traditionally, survey-class AUVs rely on simple dead-reckoning (DR) techniques typically coupled with Doppler effect-based velocity sensors [19], such as an acoustic Doppler current profiler (ADCP). Although this provides acceptable accuracy in short oceanographic missions, the error in estimating the vehicle's position is cumulative, as each estimate is relative to the previous one, and hence this approach experiences unbounded error growth. The rate of the error growth depends on the quality of the employed motion sensors, as well as the physical environment where the AUV operates. The typical error rate for survey-class AUVs operating in a straight line is of the order of 0.1–2 per cent of distance travelled (DT) given velocity estimates over the ground [20]. Otherwise, when water-relative (WR) velocity is only available, the error is primarily proportional to the speed of the water currents near the vehicle. Note that the error rate can further be reduced to less than

0.05 per cent of DT by using a high-grade inertial navigation system coupled with an ADCP in continuous bottom-tracking regime. However, such a system can be costly and bulky, as well as power-demanding and thus incompatible with long-range AUVs.

As the DR error grows with time, AUVs are commonly required to regularly surface to obtain global positioning system (GPS) fixes and limit the navigation error. However, such a behaviour can be undesirable for certain applications (e.g. continuous deep-water surveys) or even impossible when operating under a thick layer of sea-ice. Alternative and widely used approaches to place a bound on the navigation error rely on external acoustic-based support systems [21]. Such systems have typically coverage in the range of 100 m to 10 km and positioning accuracy varying from centimetres to a few metres. Various acoustic-based navigation techniques have been developed over the years, although three schemes are the most frequently used: (a) long baseline (LBL); (b) short baseline (SBL); and (c) ultra-short baseline (USBL). The working principle of acoustic navigation lies in measuring the bearing (in case of SBL and USBL) and the time of flight of acoustic signals, with the latter to be transformed to a range measurement. For instance, the standard mode of deploying AUVs is with an acoustic-based support system, e.g. a research ship or an autonomous surface vehicle equipped with a USBL [22,23].

Although acoustic-based navigation techniques are well established and robust, they have the disadvantage of constraining the AUV's operational window, as the vehicle is required to remain within the range of the external support system (which is a major limitation for long-range AUV operations). Furthermore, acoustic-based systems can also be expensive and require a considerable time and effort to be calibrated, deployed, surveyed and recovered (resulting in further cost increases and delays until the mission can begin). If surface support is provided, then the operational cost of a dedicated ship can be high and impractical for some applications. Additionally, assistance from the surface can also be restricted by factors such as limited ship manoeuvrability or inability to access certain areas, e.g. due to a thick layer of sea-ice. Therefore, it becomes apparent that long-range AUVs require a self-contained navigation solution that utilising on-board sensors and efficient estimation algorithms can enable AUV operations of thousands of kilometres without the need for external support or regular surfacing.

Navigation techniques that exploit the geophysical properties of the Earth have appeared in the literature to hold the potential to limit the DR error and meet the requirements mentioned above [24]. In contrast to the acoustic-aided techniques, geophysical navigation can be fully implemented on board a vehicle; hence, this type of navigation is particularly appealing for long-range AUVs. For example, terrain-aided navigation (TAN) is a form of geophysical-based navigation where the terrain morphology is exploited to extract navigational information [19,25]. It has been shown (e.g. in [26]) that the performance of TAN depends on the accuracy of the employed sensors (motion and bathymetric), the number of bathymetric observations available at a time (i.e. whether the vehicle is equipped with multi- or single-beam sonars), as well as the quality of the available bathymetric map (including the terrain variability itself).

However, despite the demonstrated effectiveness of TAN in aerial applications, e.g. as a navigation method for cruise missiles [27], the application of these techniques in the highly unstructured underwater environments is relatively immature [25], with outstanding challenges to be addressed. Real-world implementations are extremely limited in number and the majority of the reported applications concern platforms equipped with power-demanding motion sensors, sonars with high perception ability (i.e. multi-beam echo-sounders), and require the use of high-quality bathymetric maps (with resolutions ranging from sub-metres to only a few metres). As a consequence, the applicability of TAN techniques has been limited to short AUV missions, ranging from hundreds of metres to a few kilometres, and areas that have previously been subjected to detailed bathymetric surveys.

In response to the limitations described above, this work develops a TAN algorithm that relies on basic motion sensors and bathymetric observations obtained by low-power sonars (e.g. a single-beam sounder or a downward-facing ADCP while in bottom-tracking regime) and is sufficiently robust to deal with low-resolution bathymetric maps. The state estimation process is performed by utilising the Rao–Blackwellised particle filter (RBPF) [28]. To make the navigation filter computationally feasible while using low-power processing boards with limited computational resources, the filter estimates the 2D vehicle's position and the 2D speed of the local water currents. Therefore, the proposed navigation solution can enable AUV deployments in remote deep oceans of the order of months, rather than hours or days, without the need for external support or regular surfacing.

This chapter initially presents (Section 6.2) the ALR vehicles and performs a power analysis (Section 6.3) showing that these vehicles are capable of operating for months in a single deployment, even when special missions may require the use of power-demanding sensors (e.g. a gyrocompass to facilitate missions in high-latitude zones). Following, Section 6.4 formulates the localisation task for long-range AUVs as a non-linear Bayesian estimation problem which is solved using an RBPF algorithm.

To evaluate the effectiveness of the TAN algorithm, Section 6.5 uses unique field data that were collected during multi-day long AUV deployments ($\approx$200 km range) at depths greater than 3,000 m in the Southern Ocean. An extensive literature review produced no publications that report TAN results in truly long-range deep-water missions without periodic surfacing to obtain GPS fixes. The algorithm demonstrates sufficient robustness to cope with challenging conditions, such as the use of a sparse terrain map, operating under the influence of strong time-varying water currents, traversing areas with low topographic information (along-track) and lacking bathymetric observations for long periods of time. Using DR navigation in such conditions results in an error of over 40 km after only a 3-day-long mission, whereas TAN appears to be able to limit the localisation error within a few hundreds of metres, despite the use of a small number of low-power navigation sensors.

Following, Section 6.6 investigates whether TAN can be a navigation solution while ALR1500 crosses the Arctic Basin, which is an aspirational science-driven mission for continuous mid-water column survey from Svalbard to Barrow with a range in excess of 3,200 km. The inability to surface in conjunction with the deteriorated performance of the heading sensors and the drift caused by the water circulation make the overall navigation problem extremely challenging. As the determination of the vehicle's direction using a magnetic compass can be unreliable, a power analysis has shown that ALR1500 can also power a gyrocompass (although a power-demanding sensor) to facilitate the heading estimation. Moreover, to enable TAN out of the ADCP's bottom-tracking range, ALR1500 can further be integrated with a bespoke low-power 4,000-m range single-beam echo-sounder specifically developed for the ALR vehicles. For the evaluation of the algorithm, a simulated environment is developed by incorporating a low-resolution bathymetric map of the Arctic Ocean, a model of the water currents in the Arctic and models of the key error sources of the heading sensors at high-latitude zones. While the DR navigation appears to rapidly accumulate an error over the maximum acceptable one for safe transit and surfacing in an ice-free area for recover, numerical results show that TAN is able to estimate the vehicle's position with the required accuracy (given that heading sensor-dependent critical latitudes are avoided), despite the coarseness of the employed terrain model and the use of a sonar with a pinging period of 60 s.

Finally, Section 6.7 summarises this study and provides a short discussion on future considerations.

6.2 ALR1500 and ALR6000: two ultra-endurance AUVs

The NOC in the UK is currently developing a novel flight-style propeller-driven AUV with two configurations: the ALR6000 [15] and the ALR1500 [16]. The first mentioned vehicle, also known as ALR6000, is designed for depths up to 6,000 m and is flexible enough to undertake missions lasting from days to multiple months in a single deployment. In contrast, the second vehicle is a 1,500-m depth-rating version, hence known as ALR1500, with further enhanced endurance and specific design accommodating needs for long-range operations in polar latitudes [18].

The long endurance of these platforms is achieved through the use of high-energy-density lithium cells combined with optimisation of energy consumption within all sub-systems, low hotel load and travelling slowly (0.4–1.0 m/s). Both vehicles can be deployed and recovered either from a ship or directly from a slipway on shore, where the latter model operation does not require towing and eradicates the need for costly ship time. Shipboard launch and recovery will only be necessary for deployments in extremely remote or high-risk locations and conditions, for instance, for under-ice operations in the Arctic/Antarctic, or where the transit would require a significant DT in high-current or high-traffic areas, where the energy cost and risk associated with the transit are substantial.

6.2.1 Mechanical construction and propulsion

ALR6000 is a 3.6-m-long AUV with a dry weight of approximately 660 kg.

The mechanical construction of ALR6000 centres around two cast aluminium alloy pressure spheres (see Figure 6.1). The forward sphere is reserved purely for batteries, whilst the aft sphere contains the mission management computer and additional dry space for payload. The spheres have an outer diameter of 0.711 m and an inner

Figure 6.1 Left: ALR6000 and its construction around two pressure spheres. Right: ALR1500 and its construction around a single dome-ended cylindrical pressure vessel

diameter of 0.684 m, providing 0.335 m^3 of dry volume, with a mean density of 215 kg m^{-3}. The two spheres are mounted in a polypropylene boat frame and covered with a glass-reinforced plastic shell to give a hydrodynamic shape enclosing 1.3 m^3. Within this structure, there is ample flooded space available for wet payload ($\approx$0.6 m^3), enabling multiple capabilities for a wide range of applications, or additional buoyancy in the form of microsphere syntactic foam.

In contrast, although ALR1500 is designed based on the proven architecture of ALR6000, the fundamental difference between these two vehicles is that ALR1500 is constructed around a single dome-ended 700-mm diameter cylindrical aluminium pressure vessel (see Figure 6.1), by replacing the dual spheres used in ALR6000. The ability of ALR6000 to withstand the enormous crushing forces required to operate at 6,000 m comes with a significant volume and weight penalty. As many scientific deployments in the Arctic or Antarctic seas are unlikely that would require a vehicle to dive deeper than 1,500 m, ALR1500 is instead designed to operate at depths not greater than 1,500 m. This has allowed the development of a vehicle with increased available net-buoyancy and dry payload volume (over 0.7 m^3), within the same physical envelope of ALR6000, by reducing the material thickness. As a result, the increased buoyancy and volume can be used to carry a larger battery pack or additional scientific payload, dramatically increasing the capability of ALR1500.

To help overcome the net buoyancy, the ALR vehicles are fitted with high aspect ratio wings, port and starboard, to aid in generating hydrodynamic down force. The propulsion system is located at the aft of the vehicle, comprising of a 600-mm diameter two-bladed slow-speed fixed pitch propeller magnetically coupled to a Maxon brushless DC motor (24 V, 170 W, EC 40) through a two-stage 26:1 ratio planetary gearbox, all dry-mounted in a titanium housing. Furthermore, the motor power is restricted to 80 W for endurance reasons, which is sufficient to drive the hull form to speeds up to 1 m/s. Four control surfaces are present upstream of the propeller in a cruciform arrangement. Manoeuvres are performed using two horizontal stern planes and a single downwards pointing rudder, while the top fin is a fixed control surface containing an independent Iridium beacon and flashing light. The control planes are sized so that the AUVs would be hydrodynamically stable in forward motion, and thus the control planes do not need to be continuously adjusted to maintain heading.

6.2.2 On-board control and autonomy systems

The on-board control system (OCS) of the ALR vehicles relies on a layered component-based architecture [29] that focuses on modularity to support code reuse and rapid development. Following the approach proposed in [30], the system is based on a distributed network of applications, or nodes, using robot operating system (ROS) as main software backbone. The ROS middleware provides the main inter-process communication links and allows to seamlessly interface to simulators or to the real robot (either remotely through the network or on the same machine using interprocess communication).

A further abstraction layer is realised using a set of standard interfaces to simplify integration and decouple the robot hardware from the algorithmic parts. To this aim, the OCS follows the front-seat/back-seat design paradigm [31]. This has two components: a higher-level, behaviour-based autonomy system and an interface to a classical dynamic controller. The latter is responsible for real-time control of the vehicle given the decisions of the autonomy system (e.g. vehicle heading, speed and depth). The front-seat/back-seat driver architecture increases software portability since the autonomy system is completely decoupled from the lower-level details of the vehicle.

From a control perspective, the mission executive layer of the OCS architecture is implemented using behaviour trees (BTs), which is a method to structure the switching between different tasks [32]. One of the main advantages of the BTs relative to a more traditional finite-state machine (FSM) is their modularity. Commands coming from the mission executive layer are then passed on to the vehicle level which is responsible for performing low-level control by interacting reactively with the sensors and actuators. At the lowest level, each actuator is controlled by a proportional-integral-derivative position loop in a distributed embedded electronic system with one board per actuator. In this way, the main controller does not need to continuously control the actuators, instead this task is delegated to actuator-specific micro-controllers. The communication from the main controller to each board is performed using the recommended standard 485 (RS-485), communication protocol. The communication with the sensors is based on serial communication, either transistor-transistor logic (TTL), or recommended standard 232 (RS-232).

6.2.3 Navigation sensor suite and science payload

To minimise the on-board energy consumption and enable long-range missions, the ALR vehicles are equipped with a basic navigation sensor suite. This includes a GPS module to obtain GPS fixes when at the surface; a six-axis PNI TCM XB compass/attitude module; a Sea-Bird SBE 52-MP conductivity, temperature and depth (CTD) probe; and two ADCPs facing upwards and downwards. The downward-facing ADCP is used to estimate the vehicle velocity (bottom-relative (BR) or WR) and has a maximum reliable range of 180 m for bottom lock, whereas the upward-facing ADCP is primarily used to avoid collision with ice while undertaking operations in polar regions. To level the velocity vector in a local horizontal reference frame, the downward-facing ADCP has been integrated with tilt sensors, micro-electro-mechanical system (MEMS) accelerometers, instead of using an IMU (which results in further power savings). In addition, ALR1500 can also be equipped with the Octans north-seeking fibre-optic gyrocompass for finding the true north. Although this is a power-demanding sensor, the gyrocompass can provide reliable heading estimates at high-latitude zones (where ALR1500 is designed to operate).

The current navigation system of the ALR vehicles relies on simple DR techniques, avoiding the use of power-expensive systems and computationally demanding algorithms. However, the uncompensated sensor errors inevitably induce an unbounded growth of the DR error. Shallow-water field experiments in Portland Harbour (UK, October 2015) and deep-water long-range operations in the Southern Ocean (April 2017, see [19]) have revealed that the DR error of the ALR vehicles is of the order of 1–2 per cent of DT in a straight line when using BR velocity estimates, or is otherwise driven by the speed of the water currents.

When undertaking operations near the ocean bottom, range measurements from the downward-facing ADCP can be used to enable TAN and limit the DR error [33]. For supporting (navigationally) operations out of the ADCP's bottom-tracking range (e.g. missions in the mid-water column of deep oceans), the ALR vehicles can further be integrated with a bespoke 4,000 m range 12 kHz single-beam echo-sounder, currently being developed in the NOC (the cylindrical transducer can be seen at the front of ALR1500 in Figure 6.1). The echo-sounder is designed to function in a relatively low repetition rate (of the order of minutes) to conserve on-board energy and has a minimum possible volume to easily fit into the ALR vehicles.

In terms of science-specific payload, the vehicles can be installed with low quiescent power sensors, such as the Rockland's micro-structure turbulence sensor, fluorometers and the NOC's suite of lab-on-chip chemical sensors [34]. For shorter-duration deployments, higher-power sensors can be fitted, such as side-scan sonars. Moderate power applications currently at the trials stage include acoustic and chemical sniffing for gas release from carbon sequestration and storage sites, acoustically coupling the AUVs with an unmanned surface vehicle to provide high-precision multi-week surveys for oil and gas applications [23,35]. Furthermore, the vehicles are already scheduled for use within an integrated long-term measurement monitoring and verification platform for subsea carbon capture and storage sites and long-term continual monitoring of oilfield assets.

6.3 Energy balance for long-range operations

It was first proposed in [36] that a neutrally buoyant flight-style AUV can achieve a cost of transport (J kg^{-1} km^{-1}) comparable to the low transport cost of gliders by having low hotel load as well as low advance speed, thus enabling long-endurance capability [37,38]. Following this rationale, the ALR vehicles maximise the operational range by maximising the on-board energy storage while keeping energy consumption to a minimum.

The power consumption, P_{tot}, of a neutrally buoyant AUV can be assumed to be split between hotel load, P_H, which is invariant to forward speed, and propulsion power, P_P, which is related to propulsion speed raised to the power three [37,38]. For a positively or negatively buoyant vehicle, additional power, P_B, is required to maintain depth. Therefore, the total power consumption can be expressed as

$$P_{tot} = P_H + P_P + P_B \tag{6.1}$$

The endurance for an AUV can be estimated given the total on-board energy budget, the total power consumption and the mean forward speed by the following expression [15]:

$$R = 3.6\frac{E}{P_{tot}} \times \bar{v}^f \tag{6.2}$$

where E is on-board available energy in Wh; $\bar{v}^f$ is the mean forward (advance) speed in m/s; and R is the estimated vehicle range in km. The following subsections demonstrate how terms in (6.1) have been minimised to maximise R.

6.3.1 Battery and maximising energy storage

The ALR vehicles can be powered by Bobin-type lithium thionyl chloride (LTC) primary cells, which have very low self-discharge rates making them suitable for long-term deployments. These cells offer high energy densities, with quoted nominal gravitational energy densities of up to 684 Wh kg^{-1} [39], which is over three times the storage capacity of the best available lithium-ion secondary batteries [40]. They have also very low self-discharge rates making them suitable for long-term deployments and storage. The ALR1500's large dry volume and high available buoyant force allow to float over than 2.8 times the installed batteries currently possible on ALR6000. Alternatively, the installed energy on the ALR6000 vehicle can be matched using lithium polymer secondary batteries, significantly reducing the cost of operation.

It has long been known that both temperature and discharge rates have a significant effect on the performance of the LTC batteries [41]. In AUV power applications, the cells could be exposed to moderate to high power demands and temperatures ranging from −2 °C to 30 °C. It is therefore important to fully characterise the performance of the specific cells being used, in order to determine the vehicle capabilities. Therefore, laboratory testing of these cells has been conducted at the NOC, where cells were discharged under constant and varying load conditions at a variety of temperature and discharge rates [16]. These tests demonstrated that temperature has a significant effect on the available capacity of the cells. Based on these results, an equation for estimating the effective capacity of a cell with respect to cell demand and temperature has been generated:

$$Q = 68 - (239 - 5.3 \times T) \times P \tag{6.3}$$

where Q is the available charge in Wh; T is the temperature in °C; and P is the mean power demand per cell in W.

In order to optimise the energy density of the vehicle battery pack, it is necessary to ensure that the pack contains sufficient cells and that the vehicle load demand can be met whilst keeping the mean load-per-cell low. In order to achieve this, LTC primary cells are placed in a 7s8p battery pack configuration. While ALR6000 is limited to 10 packs and 560 cells providing a nominal voltage of 25.2 V and potential installed energy of approximately 38 kWh, ALR1500 is able to carry 28 packs and 1,568 cells, giving total potential energy over 107 kWh. For example, assuming a total system drain of 30 W, the load per cell will be 0.053 and 0.019 W, for ALR6000 and ALR1500, respectively. This suggests that during a typical deployment of ALR6000, the cells will each achieve approximately 56 Wh capacity (calculated using (6.3)), with a total usable installed energy of 31 kWh. With similar calculations for ALR1500, the corresponding cell capacity will be approximately 63 Wh, with a total usable energy of 100 kWh, which is over three times the installed energy of a typical mid-range electric car [42]. Hence, by increasing the number of cells, the load per cells decreases. From this example of 30 W draw, ALR1500 battery pack will have 3.22 times the available energy of ALR6000 while increasing only 2.8 times the installed batteries.

6.3.2 Propulsion power

For the propulsion power requirements, two ALR6000 vehicles were commissioned and subjected to sea trials in Portland Harbour (UK) in August 2015. Several speed test runs were conducted over a range of propulsion demand set points, by comparing the measured current drawn by the propulsion system combined with recorded battery voltage against the recorded mean forward speed over the period of demand. Following these experiments, it is possible to calculate the relationship between speed and propulsion power demand as

$$P_P = 90 \cdot (\bar{v}^f)^3 \tag{6.4}$$

where P_P refers to the propulsion power demand in W.

Given that ALR1500 has an identical propulsion system and shares the same hull form, it is reasonable to expect for ALR1500 similar propulsion requirements with ALR6000. Note that (6.4) is likely to overestimate the power demand as the performance data were collected with ≈30–35 N of positive buoyancy, which generated additional P_B power demand, to facilitate the vehicle's recovery.

6.3.3 Hotel load

The hotel load, P_H, is the non-propulsive energy consumption. This incorporates the required power to run all on-board computer systems and navigational sensors. To reduce the hotel load of the ALR vehicles, the control and navigation frequency has been set at 1 Hz ($\Delta t = 1$ s), which contrasts with that of Remus 600 at 9 Hz [43] and of Tethys at 2.5 Hz [14]. Field experiments in Portland Harbour have shown that the non-propulsive energy consumption for the ALR vehicles varies from 15 to 20 W [16]. The variation is attributable to the additional consumption of communications devices (such as Wi-Fi, GPS and Iridium) which are active when the vehicle is on the surface.

The above non-propulsive power demand is the base and can increase for applications requiring enhanced navigation capability. To place a bound on the navigation error while operating out of the ADPC's bottom-tracking range, bathymetric observations from the NOC's single-beam echo-sounder can be used to enable TAN. As the echo-sounder is designed to have a range up to 4,000 m, it would thus require a significant portion of the available power budget for pinging frequently the seabed from distance. To minimise the energy consumption, the echo-sounder has been optimised to have a low quiescent power demand (<0.5 W) and pinging time period of the order of minutes. In a 60-s repetition rate, the echo-sounder consumes on average approximately 4 W as recent field experiments in the Whittard Canyon's deep waters have revealed (June 2018). Furthermore, ALR1500 can additionally be equipped with the Octans gyrocompass, which requires approximately 18 W [44], for missions in high-latitude zones where the magnetic compass can be unreliable. By taking into account the power demand of the NOC's echo-sounder and of the Octans gyrocompass, the hotel load grows to approximately 37 W.

It is worth to emphasise that further reduction in the hotel load is possible when operating in the mid-water column of deep oceans. Previous field experiments have shown that ADCPs are likely to provide poor-quality WR velocity estimates, particularly at depths greater than 1,000 m, due to a sporadic lack of particulate matter to provide backscatter in the clear oceanic water. Therefore, for deployments out of the ADCP's bottom-tracking range, the vehicle's forward speed can be computed as a function of the rotation speed of the vehicle's propeller.

6.3.4 Passive buoyancy management

Operating in deep oceans, the buoyancy of the vehicle can significantly vary due to thermal contraction and bulk compression effects on seawater and the vehicle. Consequently, the induced drag on the vehicle varies with depth. While (6.4) shows that the propulsion power varies with velocity to the power three, the power to overcome buoyancy is inversely proportional to speed and thus tends to dominate at low speeds. Considering the low speed that the ALR vehicles are designed to operate, the power to overcome buoyancy can significantly adversely affect range. Therefore, minimising the net buoyancy with respect to depth is crucial to conserve on-board energy.

Various approaches have been investigated to minimise or adjust to buoyancy variation with depth. Techniques relying on hull deflection to match the compressibility of seawater cannot be considered due to the relatively rigid volume of the aluminium pressure vessels. Instead, to accurately and passively match the vehicles' bulk modulus to that of seawater, a medium more compressible than the water is introduced to the vehicles [15]. This brings the mean compressibility of the AUVs closer to that of the seawater, and hence reduces the variability of the buoyancy through the water column. The medium chosen for managing the buoyancy is a polydimethylsiloxane oil, with a typical bulk modulus of 1 GPa. The target is to maintain the net buoyancy between 0 and 20 N and not to go negatively buoyant for safety reasons. However, the volumetric constraint of the vehicles coupled with the necessity for some degree of positive buoyancy at the sea surface (to facilitate ease recovery) precludes from adding the required silicon oil to optimally match the seawater. Therefore, the silicon oil is complemented with controlled surface buoyancy in the form of gas cylinders [15], knowing that this buoyancy will be lost as the vehicle dives. By employing this passive buoyancy mechanism, it is possible to tune the buoyancy to be approximately neutral at the working depth, hence reducing the need to generate lift with the vehicle wings to maintain depth.

6.3.5 Range and endurance calculation

The total power demand for the ALR vehicles can be estimated by using (6.1). The non-propulsion power term, P_H, for both vehicles is approximately 19 W (including the power demand of the NOC's echo-sounder). If there is a need for gyrocompassing, ALR1500 can further be equipped with the Octans gyrocompass that will cause the hotel load to grow to 37 W. Given the passive buoyancy compensation system [15], the power required to maintain depth can be minimised ($P_B = 0$ W). The propulsion power, which is a function of the mean forward speed, can be estimated using (6.4).

By combining (6.3) and (6.4), it is possible to calculate the projected range of the ALR vehicles for a given forward speed at a given water temperature and for a specified hotel load. Three worst case scenarios for the water temperature are considered: (a) 2 °C for deep non-polar waters where ALR6000 may operate; (b) 5 °C for relatively shallower non-polar waters and ALR1500; and (c) −2 °C as the approximately minimum possible water temperature in the Arctic Ocean for ALR1500. Using these temperatures, Figure 6.2(a) shows the obtained range relative to the vehicle's forward speed.

ALR6000 is able to achieve an optimal range of 1,900 km during 51 days with 0.44 m/s optimal forward speed at 2 °C water temperature and using the basic sensor payload (including the echo-sounder). For the same sensor suite but shallower waters (2 °C temperature), ALR1500 extends this range to over 6,000 km in 151 days with 0.46 m/s optimal forward speed. In cold Arctic waters and including the gyrocompass, the corresponding range drops to 3,568 km. Figure 6.2(a) shows that the absolute range is highly sensitive to the magnitude of the hotel load and that as the non-propulsive load increases (e.g. gyrocompass inclusion), the optimum speed also increases.

For deployments where the temporal coverage is prioritised over the spatial coverage, the endurance can be maximised by operating at the minimum speed at which the vehicle can navigate, as can be seen in Figure 6.2(b). In areas with low water flow intensity, where the vehicle's forward speed can be maintained at the minimum value ($\approx$0.4 m/s), the endurance of ALR6000 can exceed 55 days whilst for ALR1500 it is approximately 170 days with the basic sensors and 95 days when including the gyrocompass.

It is expected that the figures presented within this section can be further improved by developing a propulsion system with enhanced propulsion efficiency, by cutting power to non-essential systems whilst underwater, and upgrading to newer battery technologies. Even relatively small savings can result in significant gains in endurance. Through these developments, it is hoped that a total endurance of over 2,000 and 6,500 km, for ALR6000 and ALR1500, respectively, will be demonstrable in the near future.

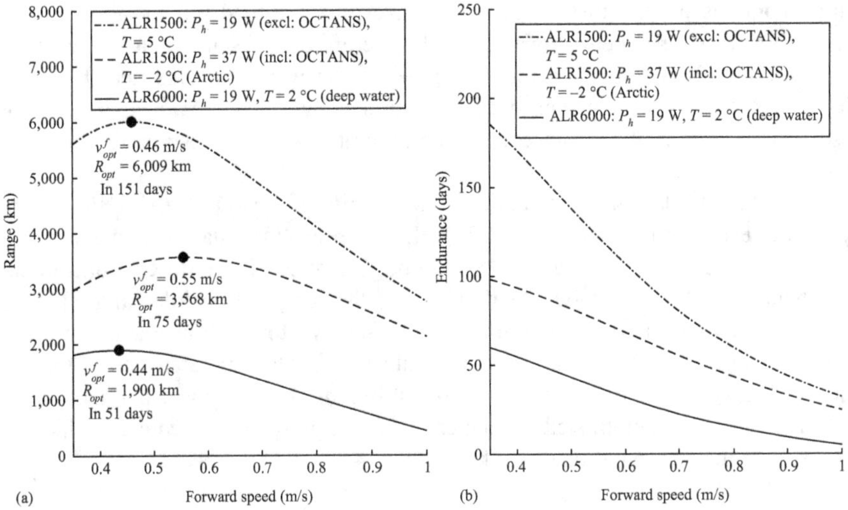

Figure 6.2 Range and endurance of the ALR vehicles: (a) range versus forward speed for the ALR vehicles given hotel load and three water temperature scenarios and (b) endurance versus forward speed for the ALR vehicles given hotel load and three water temperature scenarios

6.4 TAN for ultra-endurance AUVs

The use of a small number of low-power motion sensors will unavoidably cause a rapid error growth in DR navigation. To place an error bound during long-range AUV applications, TAN can be used as a navigational aid. In this work, the navigation task is formulated as a non-linear Bayesian estimation problem. This is then addressed employing the sequential importance sampling with resampling (SISR) particle filter (PF) [19,45]. However, although sample-based approaches can efficiently deal with non-linearities, they may require significant computational power, which is a limiting factor for the applicability of these techniques on AUVs utilising low-power processing boards with limited computational resources [37,38]. The latter is especially true for high-dimensional estimation problems, as PFs would require a larger number of particles to ensure that the state space is sufficiently covered.

To develop a relatively low-complexity filter, the TAN algorithm concentrates at modelling and estimating a minimum number of parameters. Even though an AUV operates in the 3D world, where three parameters are required to fully determine the position of the vehicle, in reality two parameters are unknown and non-directly observable. These unknowns constitute the easting and northing coordinates of some horizontal datum, or the two latitude/longitude (LL) angles of the Earth's spherical coordinate system. The component of the third dimension (operating depth) can be computed by measuring the hydrostatic pressure, using the CTD probe which contains a pressure sensor, and converting it into depth by applying the UNESCO pressure-to-depth formula [46]. Therefore, the state vector consists of the 2D vehicle's position and the 2D speed of the water currents near the vehicle. Given that water currents are characterised by slow dynamics, a common assumption made in the oceanographic community [47], their estimation becomes possible.

To reduce further the computational complexity of the 4D PF, the RBPF can be utilised to estimate the state vector [24,28]. Based on Rao–Blackwell theorem and Rao–Blackwellisation techniques [48], the performance of the SISR algorithm can be enhanced [49] simultaneously with lowering the computational demand. This can be achieved by decomposing/marginalising the state vector into linear and non-linear sub-states. Following, the linear part of the state is optimally estimated using a Kalman filter (KF), whereas a PF is employed for estimating the non-linear parameters. As the dimension of the employed PF decreases, fewer particles are required, which, in turn, results in a computationally lighter filter.

6.4.1 Process model

The ALR vehicles are naturally stabilised in roll and can be further assumed stabilised in pitch during the steady-state phase of maintaining depth. For substantial pitching and/or rolling (e.g. while descending/ascending or manoeuvring), the ADCP body frame velocity can be levelled horizontally using pitch and roll measurements from the tilt sensors (MEMS accelerometers) that the ADCP module is integrated with. Given levelled velocity estimates, the vehicle motion can be modelled as a discrete Markov process which relies on 2D kinematics. This model propagates prior position estimates forward in time by rotating the horizontally levelled body frame velocity (forward and starboard speed) into a velocity expressed in a local North-East (NE) geographical frame (using a rotation matrix parametrised only by yaw), $\boldsymbol{u}_k = [u_k^N, u_k^E]^T$, and then integrating $\boldsymbol{u}_k$ over time. However, to avoid errors associated with the assumption that the Earth is a flat surface (which would be an acceptable assumption for short missions), the vehicle's position is expressed in LL. To include also the water flow in the estimation process, the state vector is augmented with the 2D speed of the water currents in proximity to the vehicle, which are expressed in the local NE reference frame. Assuming that water currents vary slowly with time, they can be modelled as a random walk with a low-intensity Gaussian step. The process model is thus influenced by the vehicle's NE velocity and the speed of the local water currents, as the following discrete-time model shows:

$$\begin{bmatrix} \boldsymbol{x}_{k+1} \\ \boldsymbol{c}_{k+1} \end{bmatrix} = \begin{bmatrix} I_{2\times2} & F(\boldsymbol{x}_k) \\ 0_{2\times2} & I_{2\times2} \end{bmatrix} \begin{bmatrix} \boldsymbol{x}_k \\ \boldsymbol{c}_k \end{bmatrix} + \begin{bmatrix} F(\boldsymbol{x}_k) & 0_{2\times2} \\ 0_{2\times2} & 0_{2\times2} \end{bmatrix} \begin{bmatrix} \boldsymbol{u}_k \\ 0_{2\times1} \end{bmatrix} + \begin{bmatrix} \boldsymbol{\eta}_k^x \\ \boldsymbol{\eta}_k^c \end{bmatrix} \tag{6.5}$$

$$\text{with: } F(\boldsymbol{x}_k) = \Delta t \cdot G(\boldsymbol{x}_k) = \Delta t \cdot \begin{bmatrix} R_e^{-1} & 0 \\ 0 & (R_e \cdot \cos(x_k^{Lat}))^{-1} \end{bmatrix}$$

where $\boldsymbol{x}_k = [x_k^{Lat}, x_k^{Lon}]^T$ corresponds to the Markovian state sub-vector representing the vehicle LL position at time k (expressed in radians); $\boldsymbol{c}_k = [c_k^N, c_k^E]^T$ refers to the state sub-vector containing the speed of the local water currents in NE direction (expressed in m/s); Δt is the sampling time, during which the vehicle's velocity and local water currents are assumed constant; $F(\boldsymbol{x}_k)$ is a matrix used for converting the NE displacement, caused by the vehicle velocity and water currents during Δt, to the respective displacement in LL; R_e is the radius of the Earth in metres; and the vectors $\boldsymbol{\eta}_k^x = [\eta_k^{Lat}, \eta_k^{Lon}]^T$ and $\boldsymbol{\eta}_k^c = [\eta_k^N, \eta_k^E]^T$ compose the process noise. $\boldsymbol{\eta}_k^x$ represents the noise in position, aiming to capture sensor errors and unmodelled vehicle dynamics. $\boldsymbol{\eta}_k^x$ is expressed in radians and follows normal distribution $\mathcal{N}(0, G(\boldsymbol{x}_k)Q_k^x G(\boldsymbol{x}_k)^T)$. As the position noise intensity Q_k^x is expressed in square metres, a conversion to square radians is performed by $G(\boldsymbol{x}_k)$. Similarly, $\boldsymbol{\eta}_k^c$ is the noise accounting for variations in the speed of the water currents and follows $\mathcal{N}(0, Q_k^c)$. The intensity of the noise for the water currents is typically low and expressed in square m^2/s^2. Note that to obtain $F(\boldsymbol{x}_k)$, $G(\boldsymbol{x}_k)$ is scaled by factor Δt. This is required to convert velocities into the associated displacement in radians for a given Δt. For notational simplicity, the matrix parametrisation by latitude is hereafter skipped ($F(\boldsymbol{x}_k) = F_k$, $G(\boldsymbol{x}_k) = G_k$).

Several studies have also included errors in the vertical direction in the estimation process [50]. This allows the navigation filter to compensate for errors in the pressure-to-depth conversion and the errors because of the uncertainty in the tidal estimates. However, to maintain a relatively low-dimension state, which is crucial when designing low-complexity estimation algorithms, these types of errors are assumed to be negligible when compared to the errors introduced by coarse bathymetric maps. For example, the typical tidal range in the open ocean is up to 1 metre [51].

6.4.2 Observation model

One of the factors that significantly affect the performance of TAN is the quality of the utilised bathymetric map, which is a grid model (typically a uniform one) of reality, and the deviation of each grid cell from the true seabed constitutes an error. The size of the grid cells is commonly referred to as resolution of the grid model, with a higher size corresponding to a lower resolution. The accuracy of a grid model has been shown to have a positive correlation with the model's resolution [52]; therefore, high-resolution models can represent the potentially complex underwater terrain with higher fidelity. High-resolution grid models can be obtained by performing extensive surveys, typically in shallow waters, using accurately georeferenced and stable in attitude platforms that are equipped with precise mapping systems. However, it can be challenging and costly, or even impossible, to construct reliable large-scale bathymetric maps in remote deep oceans, as these areas may be inaccessible (e.g. due to a thick sea-ice) or the surface-made maps can be subject to substantial errors due to the large footprint of the ship-mounted sonars [53]. For example, currently available bathymetric maps of deep oceans are gridded in hundreds of metres (per side) because of the sparse and unreliable available depth data [54]. Furthermore, such terrain models may also contain large areas lacking actual depth measurements, which are typically filled via interpolations. As a result, TAN algorithms used during long-range operating in remote waters must be sufficiently robust to deal with sparse* bathymetric maps.

Given the current energy budget of ALR vehicles, a multi-month-long TAN can be enabled only via the use of low-power ranging sonars that measure the distance from the vehicle to the seabed. Such a sensor is the NOC's single-beam echo-sounder or the downward-facing ADCP that the vehicles are already equipped with. Depending on the employed sonar, a measurement can be either a scalar value r_k, when a single-range measurement is available at a time, or a vector containing up to N_r-ranges, $r_k = [r_{k,1}, \ldots, r_{k,N_r}]^T$, obtained by the $\{N_r = 4\}$-beam ADCP. To compensate for the vehicle's attitude and the sonar's beam orientation, each range measurement $r_{k,i}$ can be projected into the 3D space using the following equation:

$$\boldsymbol{\chi}_{r_{k,i}}^{NED} = \begin{bmatrix} \chi^N \\ \chi^E \\ \chi^D \end{bmatrix}_{r_{k,i}} = \mathfrak{R}(\psi_k, \phi_k, \theta_k) \cdot \hat{\boldsymbol{q}}_{r_i} \cdot r_{k,i} \tag{6.6}$$

where $\mathfrak{R}(\psi_k, \phi_k, \theta_k)$ is the 3D rotation matrix parametrised by the vehicle attitude at time k; $\hat{\boldsymbol{q}}_{r_i}$ is the direction unit vector of the i-beam; and $\boldsymbol{\chi}_{r_{k,i}}^{NED}$ is the 3D location where the i-beam hits the seabed with respect to a sonar-attached North-East-Down (NED) reference frame. As $r_{k,i}$ is expressed in metres, $\boldsymbol{\chi}_{r_{k,i}}^{NED}$ will be also in metres.

To compute the absolute location of where the i-beam hits the terrain, the NE component of the projected range needs to be added to the vehicle's position $\boldsymbol{x}_k$. Note that as $\boldsymbol{x}_k$ is expressed in LL, the NE offset needs to be first converted into the corresponding offset in LL:

$$\mathbf{x}_{k,i} = \begin{bmatrix} x_{k,i}^{Lat} \\ x_{k,i}^{Lon} \end{bmatrix} = \begin{bmatrix} x_k^{Lat} \\ x_k^{Lon} \end{bmatrix} + G_k \begin{bmatrix} \chi_{r_{k,i}}^N \\ \chi_{r_{k,i}}^E \end{bmatrix} \tag{6.7}$$

where $\mathbf{x}_{k,i}$ is the LL location where the ith beam hits the terrain and G_k has been defined in (6.5). Moreover, it is possible to calculate a bathymetric observation at $\mathbf{x}_{k,i}$ by adding the AUV's operating depth, computed by applying the pressure-to-depth conversion, to the vertical component $\chi_{r_{k_i}}^D$ of the projected range vector of the i-beam:

$$z_{k,i} = \chi_{r_{k,i}}^D + d_k \tag{6.8}$$

where $z_{k,i}$ is the on-board ith bathymetric observation expressed in metres and d_k denotes the vehicle's operating depth in metres.

On-board bathymetric observations, $z_k = [z_{k,1}, \ldots, z_{k,N_r}]^T$, can be modelled using an observation function that relates z_k to the depth value provided by the terrain model at the location where the sonar beams hit the seabed:

$$z_{k,i} = h(\mathbf{x}_{k,i}) + \zeta_k(\sigma_z) \tag{6.9}$$

where $h(\cdot)$ is a non-linear function representing the map's depth value at a location of interest (e.g. $\mathbf{x}_{k,i}$). To calculate the depth value at inter-grid locations of the grid

*In the content of this work, the term sparsity/coarseness of a bathymetric map refers to both low-resolution and lacking actual depth data over large areas (where the depth values of the grid cells are computed using interpolation techniques).

model, bi-linear interpolation is utilised. The additive noise $\zeta_k \sim \mathcal{N}(0, \sigma_z^2)$ represents the overall impact of the errors in the grid model and the sensors used to form $z_{k,i}$. Therefore, $\zeta_{k,i}$ has typically a spatially variant noise intensity, depending on the map-, sonar- and depth-related uncertainty.

6.4.3 Rao–Blackwellised PF

The state vector in the current formulation consists of four parameters. The position of the vehicle $\boldsymbol{x}_k$, which constitutes the two of all four parameters, is used as an input to the observation function $h(\cdot)$. Given that the terrain below the vehicle can vary in a highly non-linear way (thus $h(\cdot)$ is a non-linear function), no unique mapping always exists between bathymetric observations and the vehicle's position. Because of the non-linearity introduced by $h(\cdot)$, $\boldsymbol{x}_k$ needs to be estimated by a non-linear estimator, such as the SISR PF. PFs are recursive non-parametric non-linear techniques that are able to represent an arbitrary state posterior probability density function and, hence, efficiently deal with non-linearities without requiring strong assumptions, e.g. linearisation. The speed of the water currents $\boldsymbol{c}_k$, which corresponds to the rest two parameters of the state vector, is modelled as a linear function with additive Gaussian noise. Using Bayes' theorem, water currents can be marginalised out from the state vector and can be optimally estimated using a KF. More precisely, the posterior density $p(\boldsymbol{x}_k, \boldsymbol{c}_k|z_k)$ can be decomposed into $p(\boldsymbol{c}_k|\boldsymbol{x}_k, z_k) \times p(\boldsymbol{x}_k|z_k)$, where the linear and the non-linear sub-states are estimated using a KF and a PF, respectively.

6.4.3.1 SISR particle filter

Let assume that N particles are used and $\boldsymbol{x}_k^i$ corresponds to the ith particle with associated weight w_k^i at time k. The SISR PF is a recursive estimation algorithm that consists of two phases: prediction phase and update phase. The implementation of these two phases is given next.

SISR prediction phase
Particles are propagated forward in time using the decomposed state vector derived from (6.5):

$$\boldsymbol{x}_{k+1}^i = \boldsymbol{x}_k^i + F_k^i \boldsymbol{u}_k + F_k^i \boldsymbol{c}_k^i + \boldsymbol{\eta}_k^{i,x} \tag{6.10}$$

where F_k^i is the conversion matrix, defined in (6.5), parametrised by the latitude of the ith particle. Note that the state transition model for the ith particle uses as input the speed of the water currents $\boldsymbol{c}_k^i$, as estimated by ith KF and will be made clear ahead. To compactly represent the noise in position, the uncertainty caused by the water currents is incorporated in $\boldsymbol{\eta}_k^{i,x}$ as

$$\boldsymbol{\eta}_k^{i,x} \sim \mathcal{N}(0, F_k^i P_{k|k-1}^c (F_k^i)^T + G_k^i Q_k^x (G_k^i)^T) \tag{6.11}$$

where $P_{k|k-1}^c$ is the water current error covariance at time k as predicted by the ith KF and G_k^i is a matrix parametrised by the latitude of the ith particle; see (6.5).

SISR update phase
Once a bathymetric observation becomes available, particles are weighted using the likelihood function $p(z_k|\boldsymbol{x}_k^i)$, derived from (6.9), that provides a statistical measure of

conformity (or matching) between bathymetric observations and the *a priori* terrain map. The SISR algorithm weights each particle recursively according to

$$w_k^i = w_{k-1}^i \cdot p(z_k | \boldsymbol{x}_k^i) \tag{6.12}$$

where w_k^i is the weight of the ith particle at time k. This expression is the direct result of using the prior density $p(\boldsymbol{x}_k | \boldsymbol{x}_{k-1})$ as importance density by the SISR algorithm [45].

Resampling

To mitigate the degeneracy effect [55], where after a few iterations only very few particles contribute to the final state estimation, a resampling step has been introduced [49]. The concept behind resampling is to replace particles that have a small weight with others that have a larger weight. However, if resampling is frequently performed it can lead to particle collapse†, where particles are concentrated in a very small region [45]. Therefore, there is a need for a metric that detects degeneracy and enables resampling only when it is necessary. Such a metric is the effective sample size [48]:

$$N_{eff} = \frac{1}{\sum_{i=1}^{N} (w_k^i)^2} \tag{6.13}$$

where w_k^i is the normalised weight of the ith particle. According to [48], severe degeneracy is assumed whenever the effective sample size decreases below $N_{th} = \frac{2N}{3}$. In this work, the systematic resampling (SR) is utilised, as it is computationally more efficient than many other methods [58].

6.4.3.2 Kalman filter

Each particle maintains an estimate of the speed of the water currents local to the vehicle by employing a KF. The KF model is given by

$$\boldsymbol{c}_{k+1} = \boldsymbol{c}_k + \boldsymbol{\eta}_k^c$$

$$\boldsymbol{z}_k^c = F_k \boldsymbol{c}_k + F_k \boldsymbol{u}_k + \boldsymbol{\eta}_k^x$$

where the process model is derived from (6.5) and $\boldsymbol{z}_k^c$ is the KF observation model corresponding to the position displacement during Δt, expressed in LL. According to [28], the predicted $\boldsymbol{x}_{k+1}$ carries information about the speed of the water currents; therefore, it is possible to interpret the residual $\boldsymbol{x}_{k+1} - \boldsymbol{x}_k$ as a KF observation ($\boldsymbol{z}_k^c$ is a fictitious measurement).

KF prediction

KF is a recursive Bayesian estimation technique that alternates between the prediction phase and the update phase. By modelling the speed of the water currents as a Gaussian random walk, the prediction for the ith KF is given by

$$\begin{aligned} \boldsymbol{c}_{k+1}^i &= \hat{\boldsymbol{c}}_{k+1|k}^i \\ &= \hat{\boldsymbol{c}}_{k|k}^i \end{aligned} \tag{6.14}$$

†Note that the particle collapse can be addressed by applying additional noise (so-called jitter) immediately after executing resampling, as reported in [56,57].

and the predicted error covariance by

$$P^{i,c}_{k+1|k} = P^{i,c}_{k|k} + Q^c_k \tag{6.15}$$

KF update

The assumption that each particle provides the true vehicle's position enables the filter with the capability to estimate water currents using a KF, following the Rao–Blackwell theorem adopted in the RBPF. The employed *i*th KF uses as observation z^i_k the LL DT of the *i*th particle during the Δt time interval. This observation is then used to update the predicted speed of the water currents according to the following expressions [24,28]:

$$\begin{aligned} \hat{\boldsymbol{c}}^i_{k|k} &= \hat{\boldsymbol{c}}^i_{k|k-1} + K^i_k \cdot \boldsymbol{y}^i_k \\ P^{i,c}_{k|k} &= (I - K^i_k F^i_k) P^{i,c}_{k|k-1} \end{aligned} \tag{6.16}$$

where $\boldsymbol{y}^i_k$ is the Kalman innovation and K^i_k is the Kalman gain that are computed by (6.17) and (6.18), respectively:

$$\boldsymbol{y}^i_k = \boldsymbol{z}^{i,c}_k - (F^i_k \hat{\boldsymbol{c}}^i_{k|k-1} + F^i_k \boldsymbol{u}_k) \tag{6.17}$$

$$K^i_k = P^{i,c}_{k|k-1} (F^i_k)^T \left[F^i_k P^c_{k|k-1} (F^i_k)^T + G^i_k Q^x_k (G^i_k)^T \right]^{-1} \tag{6.18}$$

In essence, the innovation vector contains the difference between the actual/measured *i*th particle displacement, $\boldsymbol{z}^{i,c}_k$, and the predicted displacement, $\hat{\boldsymbol{z}}^{i,c}_k$, using $\hat{\boldsymbol{c}}^i_{k|k-1}$.

6.4.4 RBPF point estimates

While the RBPF algorithm computes the state posterior probability distribution, an AUV typically needs an estimate of the vehicle's position that, for example, can be used by the control unit. Therefore, the navigation system is required to calculate a single vector, known as a point estimate, which serves as a 'best' estimate of the vehicle's state. Position and water-current estimates can be obtained by averaging the weighted particles (minimum mean square error estimate):

$$\begin{bmatrix} \hat{\boldsymbol{x}}_k \\ \hat{\boldsymbol{c}}_k \end{bmatrix} = \sum_{i=1}^{N} w^i_k \begin{bmatrix} \boldsymbol{x}^i_k \\ \boldsymbol{c}^i_k \end{bmatrix} \tag{6.19}$$

where w^i_k is the normalised particle weight. Covariances associated with the estimates of the vehicle's position and of the speed of the water currents can be computed by (6.20) and (6.21), respectively:

$$P^{\hat{x}}_k = \sum_{i=1}^{N} w^i_k (\boldsymbol{x}^i_k - \hat{\boldsymbol{x}}_k)(\boldsymbol{x}^i_k - \hat{\boldsymbol{x}}_k)^T \tag{6.20}$$

$$P^{\hat{c}}_k = \sum_{i=1}^{N} w^i_k (P^{i,c}_{k|k} + (\boldsymbol{c}^i_k - \hat{\boldsymbol{c}}_k)(\boldsymbol{c}^i_k - \hat{\boldsymbol{c}}_k)^T) \tag{6.21}$$

6.4.5 RBPF initialisation

The recursive nature of Bayesian estimation techniques requires Bayes filters to be initialised. This includes the initialisation of both the PF and the KF. In deep-water missions where long descent is expected, the filter initialisation is typically performed on the arrival of the first bathymetric observation to minimise the use of the on-board computer [19]. However, to properly initialise the filter an estimate of the vehicle's position with the associated uncertainty ($\{\hat{\boldsymbol{x}}_k, P_k^{\hat{x}}\}$) is needed. Therefore, a surface-acquired GPS position and the associated uncertainty are propagated in time using a process equivalent to performing the KF prediction phase:

$$\begin{aligned} \hat{\boldsymbol{x}}_{k+1} &= \hat{\boldsymbol{x}}_k + F_k(\hat{\boldsymbol{x}}_k)\boldsymbol{u}_k \\ P_{k+1}^{\hat{x}} &= S_k P_k^{\hat{x}} S_k^T + G_k Q_k^x G_k^T \end{aligned} \tag{6.22}$$

where $P_{k+1}^{\hat{x}}$ is the time-updated position error covariance and S_k is the Jacobian matrix of the position propagation model with respect to the state vector $\hat{\boldsymbol{x}}$ evaluated at $\hat{\boldsymbol{x}}_k$. Note that the position propagation model above is the simplified version of (6.5), as water-current-related terms are set to zero. Once a bathymetric observation becomes available, the particle weight is initialised to $1/N$ and particles are spread using a normal distribution centred at $\hat{\boldsymbol{x}}_k$ with covariance $P_k^{\hat{x}}$.

In terms of the water currents, KFs can be initialised using a zero-mean normal distribution with either an arbitrary (but realistic) noise intensity or intensity calculated based on *a priori* spatio-temporal data of the speed of the water currents (e.g. using ocean models or historical data).

6.4.6 RBPF simplification

As already discussed in Section 6.4, the number of KFs required to implement the RBPF in the current formulation is equal to the number of particles (N) used in the SISR algorithm. In essence, this dependence occurs from the requirement to track the vehicle's position on a spherical surface during multi-month-long missions, in contrast to short-range AUV applications where typically it is assumed that Earth is locally a flat surface and thus a local static NED reference frame can be used as a navigation origin.

The assumption that the Earth is locally flat introduces a navigation error that is a function of the deviation in latitude (the range between the vehicle's position and the origin of the navigation frame), the length of the circle of latitude (thus the latitude itself) and the vehicle's velocity (which for AUVs travelling slowly can be ignored). For common short-range AUVs, this error can be negligible as the vehicle remains within a short distance from the origin of the navigation frame and typically avoids extreme latitudes. However, the error becomes substantial for long-range AUVs and, therefore, the navigation system needs to take into account the approximately spherical shape of the Earth.

One simple method to tackle this problem is by moving the origin of the NED navigation frame along the vehicle and keeping track of the frame's origin in LL, which would coincide with the vehicle's position. This concept is exploited in (6.10), which propagates each particle using the computed displacement in a local moving NED reference frame attached to its prior position. This propagation is then converted through F_k into the corresponding displacement in LL. Considering a minimum error in latitude (as the origin of the NED frame coincides with prior particle position) and the insignificant displacement during Δt, compared to the radius of the Earth, the error introduced by locally flattening the spherical shape of the Earth can be considered negligible even in high latitudes.

One disadvantage of dealing directly with the approximately spherical shape of the Earth is the non-linearity introduced via the matrix F_k in the process model; see (6.5). This non-linearity leads the RBPF to employ as many KFs as the number of particles the SISR algorithm uses. However, a simplification can be still made. Instead of placing the origin of each particle's reference frame at the prior position of the particle, the origin is now attached to the prior point estimate $\hat{\boldsymbol{x}}_k$. By doing so, all NED reference frames are now collapsed into a single frame. This has an effect over the RBPF computational complexity as it allows the RBPF to use a single KF for all particles (since F_k will be identical for all particles).

The technique described above to avoid the use of multiple KFs inevitably regenerates errors related with the use of an NED navigation frame, although the frame moves along with the particles. This error depends on the particle spatial distribution (particles' deviation from the origin of the navigation frame) and the latitude position of the particles. However, considering that F_k varies slowly with time, the frame-related error will remain insignificant even at high-latitude zones and thus can easily be captured by increasing the process noise. Simple calculations can show that for the extreme scenario of operating at 85° latitude, a particle deviated by 50 km from the origin of the navigation frame will experience a propagation error of less than 4.5 cm in the east direction.

6.4.7 RBPF algorithm

The required steps to perform an iteration of the RBPF are shown in the pseudocode given in Algorithm 1. At every iteration, the algorithm receives as input: the particles (with the corresponding weights) and the speed of the water currents from the preceding iteration; the AUV NE speed $\boldsymbol{u}_k$; and the observation z_k (if available). The number of particles N and the resampling threshold N_{th} are constants and hence not treated as input arguments. Outputs of the algorithm are the posterior particles with the associated weights, the KF estimates and the filter point estimates. All steps listed in the algorithm are as described in this section.

6.5 Long-range TAN using WR velocity

This section evaluates the *a posteriori* performance of the proposed TAN algorithm given a low-resolution bathymetric map while using field data collected during a deep (up to 3,700 m) multi-day-long deployment (up to 195 km in 77 h) of ALR6000 in the Southern Ocean. The results obtained using TAN are compared to the real-time

Algorithm 1: Rao–Blackwellised PF

Inputs: $\{\boldsymbol{x}^i_{k-1}, \boldsymbol{c}^i_{k-1}, w^i_{k-1}\}^N_{i=1}, P^{\boldsymbol{c}}_{k-1|k-2}, \boldsymbol{u}_k, \boldsymbol{z}_k$
Outputs: $\{\boldsymbol{x}^i_k, \boldsymbol{c}^i_k, w^i_k\}^N_{i=1}, P^{\boldsymbol{c}}_{k|k-1}, \hat{\boldsymbol{x}}_k, \hat{\boldsymbol{c}}_k, P^{\hat{\boldsymbol{x}}}_k, P^{\hat{\boldsymbol{c}}}_k$

function RBPF
  if Initialisation needed **then**
    $[\{\boldsymbol{x}^i_{k-1}, \boldsymbol{c}^i_{k-1}, w^i_{k-1}\}^N_{i=1}\}] \leftarrow$ **Initialise RBPF** ▷ Section 6.4.5
  end if
  $[\{\boldsymbol{x}^i_k\}^N_{i=1}] \leftarrow$ **SISR Prediction Phase** ▷ Equation (6.10)
  if $\boldsymbol{z}_k$ available **then**
    $[\{\hat{\boldsymbol{c}}^i_{k-1|k-1}\}^N_{i=1}, P^{\boldsymbol{c}}_{k-1|k-1}] \leftarrow$ **KF Update Phase** ▷ Equation (6.16)
    $[\{w^i_k\}^N_{i=1}] \leftarrow$ **SISR Update Phase** ▷ Equation (6.12)
    $w^i_k \leftarrow \frac{w^i_k}{\sum^N_{l=1} w^l_k}$, for $i = 1 \rightarrow N$ ▷ Weight Normalisation
  end if
  $[\hat{\boldsymbol{x}}_k, \hat{\boldsymbol{c}}_k, P^{\hat{\boldsymbol{x}}}_k, P^{\hat{\boldsymbol{c}}}_k] \leftarrow$ **Point Estimates** ▷ Section (6.4.4)
  $N_{eff} \leftarrow$ **Effective Sample Size** ▷ Equation (6.13)
  if $N_{eff} < N_{th}$ **then**
    $[\{\boldsymbol{x}^i_k, \boldsymbol{c}^i_k\}^N_{i=1}\}] \leftarrow$ **Systematic Resampling** ▷ See, e.g., [58]
    $w^i_k \leftarrow 1/N$, for $i = 1 \rightarrow N$ ▷ Particle Weight Reset
  end if
  $[\boldsymbol{c}^i_k, P^{\boldsymbol{c}}_{k|k-1}] \leftarrow$ **KF Prediction Phase** ▷ Equations (6.14) and (6.15)
end function

position estimates (computed via DR) and USBL measurements, which are treated as the vehicle's baseline position, sporadically recorded by a research ship supporting the mission.

6.5.1 ALR6000 operations in the Southern Ocean

Science expeditions as part of the Dynamics of the Orkney Passage Outflow (DynOPO) research program [19,59] were undertaken in April 2017 at the Orkney passage (see Figure 6.3). This is a submarine valley of the Southern Ocean, NE of Antarctic Peninsula, that connects the Weddell Sea and the Atlantic Ocean with depths in excess of 5,300 m, steep terrain morphology and the presence of strong time-varying water currents (occasionally exceeding 0.6 m/s).

The main objective of the expedition was to map the flow of the deep-water masses and investigate the causes of the warming of the deep ocean. Science data collected by the vehicle, such as seawater properties (e.g. temperature, salinity, etc.) as well as the speed of the water currents and the water turbulence close to the ocean bottom, have provided scientists with invaluable information for understanding better

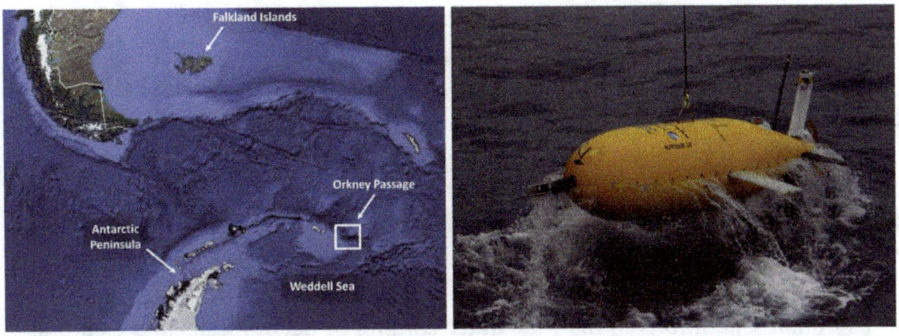

Figure 6.3 Left: ALR6000 operations at the Orkney passage, a narrow gap in the ocean-floor ridge that extends NE of the Antarctic Peninsula [60]. Right: ALR6000 recovery from a deep multi-day-long deployment in the Orkney passage

the deep-water circulation and their role in regulating the Earth's climate system [61]. For the deployment and recovery of ALR6000, Britain's RSS James Clark Ross (JCR) polar ship was used.

To meet the science requirement for data collection near the ocean bottom, ALR6000 employed an altitude following control scheme with 90-m altitude demand and constant 0.7 m/s forward speed over the ground. In terms of navigation, the vehicle was equipped with the basic navigation sensor suite (see Section (6.3.3)) and self-navigated using GPS fixes when on the surface and DR when submerged. The path of the AUV was specified by pre-defined waypoints and the vehicle was guided using a track following algorithm [62]. For safety reasons, mission waypoints were defined along bathymetric contour lines. The maximum achieved depth was approximately 3,700 m while operating for nearly 200 km, a range achieved in about 3 days. Thanks to the on-site research ship, the vehicle's position was intermittently recorded using the ship's USBL head while equipping the AUV with an acoustic transponder.

6.5.2 DR navigation

The entire DynOPO expedition consisted of three deployments. However, this work uses field data collected from the longest and the navigationally most challenging Mission 44 (M44). During this particular deployment, the on-board DR navigation appeared to be very accurate, despite the use of basic low-power motion sensors, because of the successful execution of online calibration and error compensation techniques [19] while navigating in the ADCP's bottom-tracking regime.

The real-time DR navigation accumulated a root mean square error (RMSE) of 1.2 km at the surfacing location after a range of approximately 190 km, which corresponds to an error growth of less than 1 per cent of DT. The largest proportion of the error occurred whilst the vehicle operated under the influence of strong water currents (i.e. while descending to a depth greater than 1,100 and ascending from a depth over 3,500 m), whereas the error increase during the bottom-tracking regime was relatively low [19].

It has already been demonstrated, using the field data from the DynOPO deployments, that TAN can prolong underwater missions to a range of hundreds of kilometres given velocity estimates over the ground and using single altitude measurements [19]. Due to a constant access to BR velocity, the dimension of the navigation filter was reduced to estimate only the vehicle's position via the SISR algorithm. Despite the utilisation of basic low-power sensors and low-resolution bathymetric maps, TAN successfully limited the DR error.

To imitate operations in the mid-water column of deep oceans and examine the TAN performance under the effect of strong water currents, this section removes the available BR velocity during M44 and instead uses WR velocity estimates. One option is to use the WR velocity provided by the ADCP; however, previous field operations have shown that ADCPs are likely to give poor-quality WR velocity due to a sporadic lack of particulate matter to provide backscatter in clear deep oceanic water. An alternative option is to estimate the forward speed of the vehicle from the revolutions per minute (RPM) of the vehicle's propeller. To do so, the propeller RPM has been calibrated against good-quality BT velocity, which was obtained from shallow-water experiments. This has resulted in a first-order mapping function expressed as

$$v_k^f = 0.00742 \cdot Pr_k \cdot \cos(\theta_k) \qquad (6.23)$$

where θ is the pitch displacement of the vehicle; Pr denotes the propeller RPM; and v_k^f corresponds to the horizontally levelled forward speed. Regarding to the speed in starboard direction, this is set to zero due to the lack of actuation in the corresponding direction. In reality, any motion in starboard direction is caused by the local water flow.

By using (6.23) to calculate the vehicle's forward speed, the DR navigation appears to accumulate an error of 41 km (21.38 per cent of DT), when comparing to the last recorded USBL measurement, due to the strong water currents at the Orkney Passage. Table 6.5.2 summarises details of M44 and the performance of the DR navigation. In addition, Figure 6.4 shows the vehicle's path as estimated using DR navigation with and without accessing BR velocity estimates, in comparison to the USBL position estimates.

6.5.3 TAN set-up

To implement the TAN algorithm presented in Section 6.4, the uncertainty in the terrain map and the bathymetric observations (i.e. the likelihood function) as well as other parameters of the RBPF algorithm need to be specified.

6.5.3.1 Likelihood function

Given that ALR6000 operates within the ADCP's bottom-tracking range[‡] during M44, the downward-facing ADCP can be treated as a four-beam sonar, in addition to its

[‡]The ADCP's bottom-tracking range depends on the transducer's working frequency. For the ALR vehicles, the range is approximately 180 m.

Table 6.1 Navigation details of Mission 44

Parameters	Mission 44	Units
Duration	76.7	h
DT	194.3	km
Maximum operating depth	3,691	m
Descent water depth	1,136	m
Ascent water depth	3,505	m
Number of USBL measurements	4,077	–
DT until last USBL	191.7	km
DR RMSE[1] with BR speed	1.2	km
DR RMSE[1] with WR speed	41	km
DR RMSE[1] w.r.t. DT% with BR speed	0.625	DT%
DR RMSE[1] w.r.t. DT% with WR speed	21.38	DT%

[1] Error corresponding to the last recorded USBL measurements.

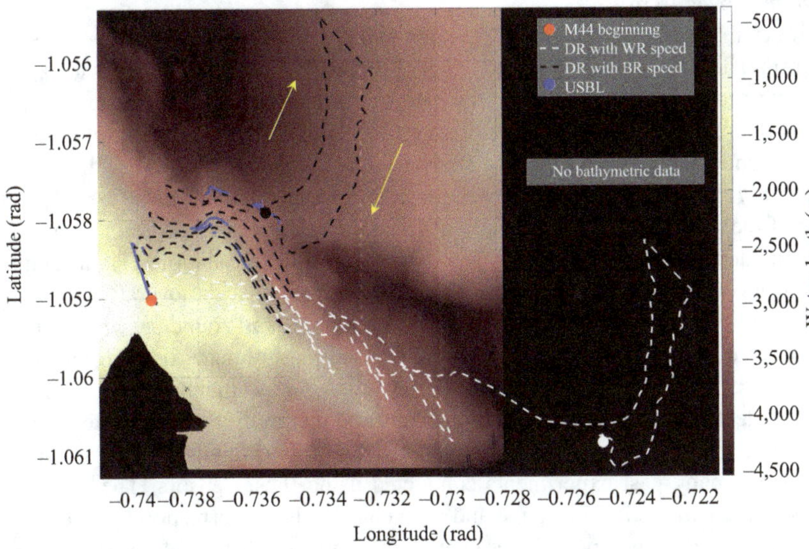

Figure 6.4 ALR6000's position estimated via DR and the ship's USBL system. The black dashed line corresponds to the real-time navigation (using BR velocity when available), whilst the white dashed line is computed using speed obtained from the propeller's RPM. USBL measurements are shown in blue colour

use as a velocity sensor. A common ADCP configuration consists of four beams (four transducers) pointing downwards and displaced by 30° from vertical. Each beam provides the range-to-bottom distance along the beam's direction by measuring the round-trip travel time of the emitted acoustic signal. Provided the known configuration of the sensor, range measurements can be projected into the 3D space, using (6.6), and combined with the vehicle's operating depth to form bathymetric observations. To avoid using erroneous measurements, beam ranges are accepted only when all four beams obtain a return signal from the seabed.

The bathymetric reference map at the Orkney passage was generated by the RSS JCR research ship, which is equipped with a Kongsberg E122 multi-beam echo-sounder, after conducting a series of bathymetric surveys. Due to the steep-terrain morphology and depths sometimes exceeding 5,000 m, a surface-made high-resolution map is not possible [53,54]. The generated reference map first has been gridded into squares of 50 m in longitude and latitude (approximately 0.052° and 0.025°, respectively) with each grid cell to take the mean value of all multi-beam measurements fallen within the cell's footprint. In areas lacking depth measurements, gaps of up to ten grid cells wide have been filled using spline interpolation. Note that the resolution of the constructed map can be sometimes higher of the resolution that the ship's multi-beam sonar can actually provide. For a beam width 1° × 1° (transmit × receive) and depths over 5,000 m, the radius of the approximately circular in shape footprint is over 40 m.

Inevitably, both the terrain model and the bathymetric observations are affected by errors. The additive noise ζ_k used in the observation model in (6.9) aims to capture the combined effect of the errors in the bathymetric map, the range measurements as well as the pressure sensor and the pressure-to-depth conversion (given the statistical independence between the error sources). For the bathymetric map, the vertical uncertainty can be quantified by associating a zero-mean Gaussian noise to each depth value of the grid model. The intensity of the noise is calculated as a function of the depth at each location of interest, according to the S44 International Hydrographic Organization (IHO) standards (Order 2). This assumes that the overall error is the combination of uncorrelated errors in travel time (1 m) and sound speed (expressed as 2.3 per cent of the depth value) of the sonar measurements used to create the bathymetric map [19,57,63]. The error in the range measurements and the computed vehicle's depth is modelled as a zero-mean normal distribution with an intensity that varies proportionally to the measured value. This proportionality is defined empirically from many past experiments conducted using these sensors. The maximum expected error in the ADCP range and the vehicle's depth correspond to 1 per cent of the measured range and 0.1 per cent of the obtained depth. More details about modelling the uncertainty in bathymetric observations and the terrain model can be found in [19].

6.5.3.2 RBPF parameters

To enable TAN, parameters of the RBPF algorithm needs to be specified, e.g. the number of particles N and the intensity of the process noise. It is known that an increased number of particles allows the filter to more accurately represent the posterior density function [45]. However, this comes at the expense of increased on-board power requirements and hence an extremely large number of particles would be infeasible on low-power processing boards with limited computational resources. The computer installed on the ALR vehicles is able to handle over 10,000 particles in real time. However, considering the dimension of the estimation problem (2D), a number of particles beyond a certain value would not result in a significant performance increase [18,57]. To assure filter robustness against estimation fail, $N = 10{,}000$ is selected. This relatively large number of particles allows the filter (a) to effectively cover the state space when the filter is highly uncertain (e.g. while initialising the filter after a long descent); (b) to achieve fast convergence to the true vehicle's position; and (c) to maintain this convergence throughout the entire mission length. It is important to highlight that the filter is assumed to be convergent to a solution when the posterior probability density function becomes unimodal (or nearly unimodal) and the filter's uncertainty satisfies a pre-selected uncertainty threshold [26].

The appropriate selection of the process noise covariance matrix (Q_k^x and Q_k^c) is of high importance for the RBPF algorithm. The intensity of the process noise must be carefully selected to take into account unmodelled system dynamics and to provide a sufficient, but not too large, particle spread. Avoiding an unnecessarily large process, noise intensity is important for mainly two reasons: (a) the number of particles may be insufficient to cover a large search area and (b) potential terrain similarities within the search area may lead the filter to become trapped in local minima. To distinguish the vehicle's operation phase between descending, hence drifting due to the systematic errors introduced by the unknown water currents, with the phase where range measurements become available and TAN has been enabled, two process noise intensities are used: (a) high intensity $\eta_k^x \sim \mathcal{N}(0, Q_k^{x,H})$ while descending (hence used by (6.22)) and (b) low intensity $\eta_k^x \sim \mathcal{N}(0, Q_k^{x,L})$, when TAN has been activated and water currents can be estimated. This distinction allows (6.22) to use a process noise of a sufficient intensity, $Q_k^{x,H}$, so that the time propagation of the vehicle's position, $\{\hat{x}_k, P_k^{\hat{x}}\}$, do not underestimate the long-term effect of the water currents while descending. Following, given an enabled TAN, the noise intensity can drop to a lower value as the employed RBPF algorithm deals explicitly with water currents[§].

During descent, $Q^{x,H}$ depends on the time length required for the vehicle to approach the sea bottom. M44 began in an area with approximately 1,150 m water-depth. Assuming an average drift caused by water currents of the order of 0.15–0.2 m/s (which can be obtained from spatio-temporal historical data) and descent duration of 1.5 h (including the time required for the online compass calibration [19]), the following intensity is assumed sufficient to properly initialise the SISR algorithm:

$$Q_k^{x,H} = \Delta t^2 \begin{bmatrix} \sigma_{N,x}^2 & 0 \\ 0 & \sigma_{E,x}^2 \end{bmatrix} = \begin{bmatrix} 16 & 0 \\ 0 & 16 \end{bmatrix} \tag{6.24}$$

which is expressed in square metres and is used to propagate $P_k^{\hat{x}}$ in time via (6.22).

[§]Recall that despite the bottom-tracking regime, only propeller-based speed estimates are used in order to imitate mid-water column operations.

Once the vehicle approaches the ocean bottom and bathymetric observations become available, TAN is triggered and the process noise intensity reduces to

$$Q_k^{x,L} = \Delta t^2 \begin{bmatrix} \sigma_{N,x}^2 & 0 \\ 0 & \sigma_{E,x}^2 \end{bmatrix} = \begin{bmatrix} 0.25 & 0 \\ 0 & 0.25 \end{bmatrix} \tag{6.25}$$

which is expressed in square metres and is considered sufficient to represent modelling errors as well as errors in the motion sensors.

To enable the estimation of the water currents, the component of the process noise associated with the variation in water currents needs to be defined. Following the common assumption that in overall water currents vary slowly in time, Q_k^c becomes

$$Q_k^c = \Delta t^2 \begin{bmatrix} \sigma_{N,c}^2 & 0 \\ 0 & \sigma_{E,c}^2 \end{bmatrix} = \begin{bmatrix} 10^{-6} & 0 \\ 0 & 10^{-6} \end{bmatrix} \tag{6.26}$$

which is expressed in m^2/s^2.

In terms of the initial position uncertainty, it is assumed that the GPS error is normally distributed around the estimated GPS position with a standard deviation of 5 m in the north and east direction. Similarly, the KF is initialised assuming that water currents are adequately represented by a zero-mean normal distribution and a standard deviation of 0.1 m/s in the north and east direction.

6.5.4 Long-range TAN results

The *a posteriori* TAN performance is numerically evaluated using gathered sensor data from M44 by performing a set of ten simulated runs. The primary aim of this section is to test the performance of the TAN algorithm in terms of (a) the filter accuracy, by comparing TAN position estimates with available USBL data; (b) the filter precision, by reporting the estimation repeatability among all ten runs; and (c) the estimation uncertainty, by comparing the localisation error with the respective 3-σ confidence bounds.

Figure 6.5 shows the vehicle's path by employing TAN, as calculated by averaging the ten simulated runs, in comparison to the USBL estimates. The average RMSE for the entire set of runs when using TAN is 0.23 km, whereas the corresponding DR error is 14.5 km. As the DR navigation is unable to compensate for the drift caused by the water currents, the accumulated DR error is approximately 40 km relative to the last available USBL measurement before ascending, which corresponds to an error of about 21 per cent of DT. At the same time, the TAN algorithm is able to keep the navigation error approximately 0.09 km (averaged over the ten runs), which corresponds to an error lower than 0.05 per cent of DT.

Furthermore, to demonstrate the ability of the filter to re-produce similar position estimates between the simulated runs, Figure 6.5 has included the estimated paths that correspond to the averaged RMSE bounds. These error bounds refer to the highest (upper error bound) and lowest (lower error bound) observed errors among all simulated runs. Despite the challenges for applying TAN, the estimated paths corresponding to the error bounds are almost always convergent to the averaged path. The filter appears imprecise and less accurate only at the bottom right corner of the

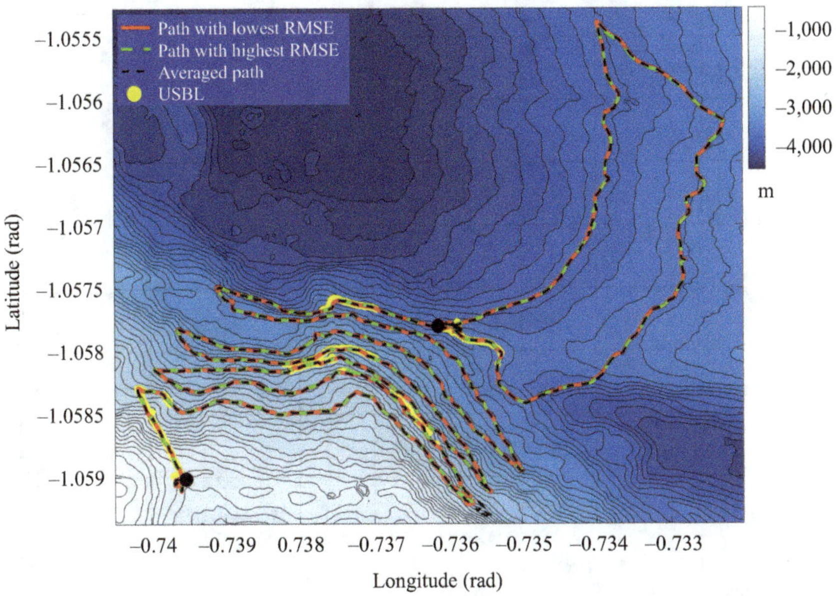

Figure 6.5 *Position estimates using TAN in comparison with the USBL measurements (yellow). For assessing the ability of the filter to provide precise position estimates, the averaged path after performing ten simulated runs (black) is drawn together with the estimated paths corresponding to the lower (red) and upper (green) RMSE bound among all runs*

mission while looking at the vehicle's path in Figure 6.5. The causes for this estimation variability is the combined effect of multiple factors taking place simultaneously: (a) the posterior distribution is mainly multi-modal, hence position estimates fall in between of the modes (however, note that the filter is able to eventually converge to the correct mode); (b) the beams of the ADCP often go out of the functional range due to steep downward slopes (which can be seen by the density of the contour-lines); and (c) due to this steep terrain morphology and the path demand (a turn of almost 180°), the vehicle performs sharp manoeuvrers (high pitching to maintain the demanded distance from bottom and rolling whilst turning). The latter has a direct effect on the accuracy of the bathymetric observations, as the beam projection relies on low-quality tilt sensors.

Figure 6.6 shows the estimated TAN error relative to the USBL measurements in the north and east directions with the respective two-sided 3-σ confidence bounds over the DT.

As USBL measurements are not constantly available, the evaluation of the TAN algorithm is possible only sporadically. Despite the strong assumption (for visualisation purposes) that particles are spatially distributed following a bivariate normal

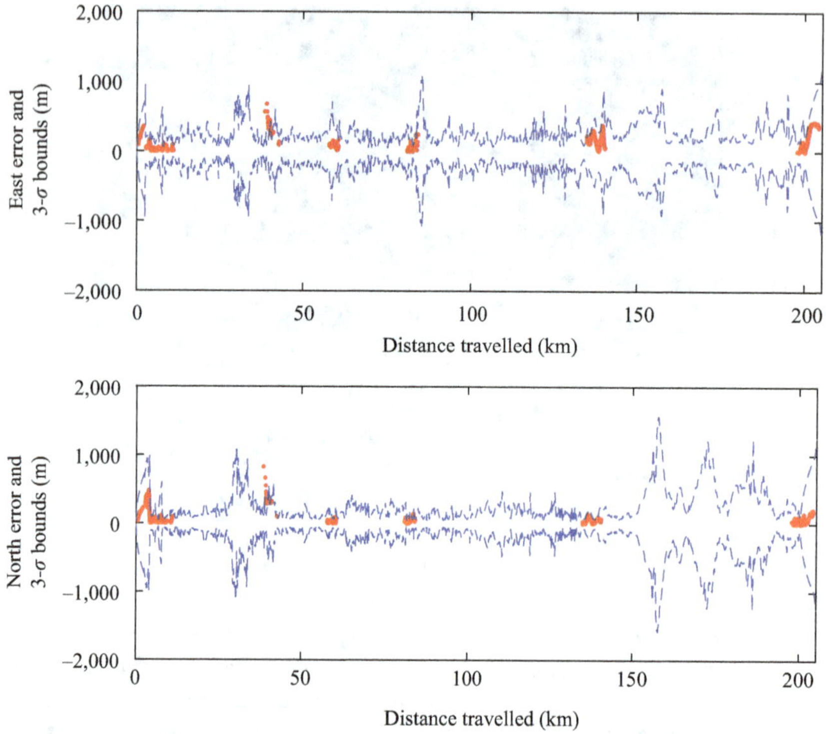

Figure 6.6 Estimates of the TAN error (red) in north and east direction, as calculated by comparing TAN position estimates with the USBL measurements, and the associated approximation of the 3-σ *bounds (blue) over the DT*

distribution uncorrelated in north and east directions, Figure 6.6 shows that the localisation error almost always falls within the calculated uncertainty bounds. Note that although the filter seems to underestimate the positioning error at approximately 40 km of DT (which coincides with the filter's less precise behaviour discussed previously), in reality the assumption that particles are distributed following a normal distribution (to visualise the filter's uncertainty) is invalid as the state posterior probability density function is locally highly multi-modal.

6.6 Crossing the Arctic Ocean with ALR1500

Despite the significant attention that climate change has received over the years, many of its implications remain unclear to the Arctic ecosystem. Currently, the Arctic Ocean is experiencing some of the most rapid transformations on the planet [64]. To be able to take actions in response to these changes, it is necessary to better understand the

Arctic Ocean and the underlying processes. However, the rate that the Arctic Ocean is being sampled is very low, limiting our understanding of the upcoming changes and their fundamental origins. While AUVs have been demonstrated in several Arctic operations (see, e.g., [65,66]), the endurance is typically limited from hours to a few days per deployment. On the other hand, the newly developed ultra-endurance AUVs, such as the ALR vehicles, are flexible platforms that can open doors for research activities at ever larger temporal and spatial scales in remote deep oceans. The use of these platforms as a sensor network for ice-monitoring and data-collection is an example of an application, whilst another is the mid-water column continuous biogeochemical survey from Svalbard to Barrow using ALR1500 [18]. The latter application can provide invaluable insights into the changes occurring as well as the structure and functioning of biogeochemical processes and ecosystems in the Arctic. However, the inability to surface during such a mission and the degraded performance of heading sensors (see later), combined with the need for energy saving (which limits the number of sensors to be used), makes the overall navigation problem extremely difficult.

Following the demonstrated strength of TAN techniques to deal with strong water currents (Section 6.5), this section analyses the potential of ALR1500 to cross the Arctic Ocean using TAN. It has already been shown in Section 6.3.5 that ALR1500 has sufficient energy budget to undertake such a mission. This can be done by either using the basic navigation sensor suite, which will save a significant amount of energy to be used in an emergency, or with the addition of the Octans gyrocompass, which will reduce the risk associated with poor heading estimation but will increase the risk of running out of energy in an emergency. Numerical simulations are used to test the TAN algorithm under the degraded performance of heading sensors and the influence of the water currents while operating out of the ADCP's bottom-tracking range, given a sparse and low-quality bathymetric map. Due to the navigation complexity of such a mission, the localisation requirements are relaxed and an error bounded within 100 km throughout the entire mission is considered acceptable. This error bound allows the AUV to maintain sufficient accuracy during the transit to avoid danger, and to surface in an ice-free area of the Barrow coast for recovery.

6.6.1 Simulation environment of the Arctic Ocean

To investigate whether TAN can enable the crossing of the Arctic Basin, a simulated environment of the Arctic Ocean has been developed. This incorporates a state-of-the-art model of the water flow in the Arctic, allowing to simulate the drift caused by water currents, and a low-resolution bathymetric map of the Arctic Ocean. To take also into account the reduced effectiveness of the heading sensors (whether it is a gyrocompass or a magnetic compass) in determining the vehicle's direction at high-latitude zones, error models have been devised that capture key sensor-dependent error sources. It is assumed that ALR1500 is launched near Svalbard and follows a path of approximately 3,200 km in range, shown in Figure 6.7, to reach the Barrow coast in Alaska. In particular, the path shown in Figure 6.7 is designed to fulfil two basic requirements for safe

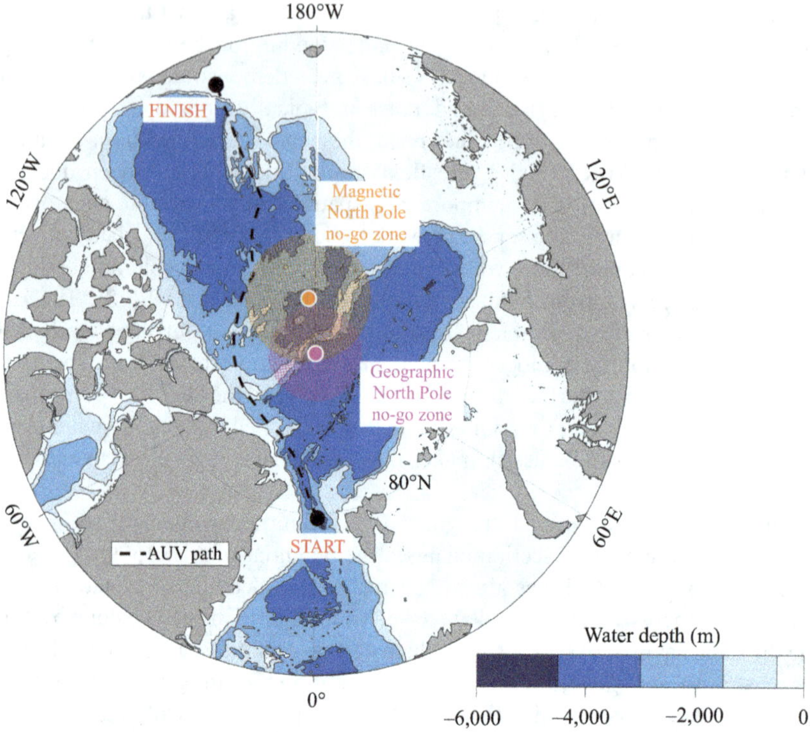

Figure 6.7 One possible route (≈3,200 km) to cross the Arctic Ocean using ALR1500. The designed path aims to maintain maximum possible distance from heading sensor-dependent critical latitudes while avoiding the morphologically flat Canada Basin

transit: maintaining maximum possible distance from heading sensor-dependent critical latitudes (as will be made clear ahead) while simultaneously avoiding traversing areas with low topographic information (see the effect of entering the flat terrain of the Canada Basin in [18]). However, by satisfying the second requirement, the vehicle's path inevitably approaches as close as approximately 500 km to the Magnetic North Pole. Furthermore, it will be explained that the behaviour of heading sensors is not easily predictable in high latitudes (particularly for the magnetic compass). Therefore, the no-go areas drawn in Figure 6.7 are by no means the exact quantification of regions to avoid for safe transit, but rather a visualisation of areas with high risk.

Simulate experiments assume that the vehicle operates with the optimal forward speed, which is dependent on the sensor suite, as calculated in Section 6.3.5 and at a constant 1,000 m depth, unless the ocean depth is shallower and ALR1500 switches to a 100-m altitude following control scheme.

6.6.1.1 Arctic Ocean circulation

To simulate the drift caused by the movement of the water masses in the Arctic, the Nucleus for European Modelling of the Ocean (NEMO) state-of-the-art modelling framework of ocean circulation [67] is used. Among various configurations, NEMO provides a 3D grid of the water flow in the Arctic parametrised by location, depth and time. For the simulation purposes, a spatial resolution of 0.083° in latitude and longitude is selected, which corresponds to the highest available resolution. Following the assumption that ALR1500 operates at 1,000 m depth, water currents at that depth are imported. According to NEMO, peaks in the speed of the water currents are mainly present along the Barents–Kara sea continental slope and are of the order of 0.15–0.2 m/s. While this order of magnitude can cause severe localisation drift, water currents at the Barents–Kara board of the Eurasian Basin are not of particular concern, as these regions are excluded from the potential operating area of ALR1500 (see Figure 6.7). It can be calculated that the navigation drift caused solely by water currents can be lower than 0.01 m/s on average while DR at the particular depth of 1,000 m. However, note that although it appears that strong water currents are not to be expected, particularly if the vehicle's path and/or depth have been optimised against the expected intensity of the water flow, even weak water currents can cause substantial localisation drift to a slow-travelling AUV operating for months underwater. For the following simulations and to avoid underestimating the Arctic water flow, the speed of the water currents obtained from NEMO are scaled by a factor of two.

6.6.1.2 Heading error analysis

Both magnetic compasses and gyrocompasses have degraded performance at high latitudes. Following, short analysis of key factors causing errors in heading estimates is provided for both sensors. However, the subsection does not aim to develop detailed error models, as this can be the subject of a dedicated study. Instead, the objective is to devise error models that are able to represent reasonably well the behaviour of each heading sensor in polar latitudes.

Magnetic compass

The ALR vehicles are fitted as standard with a three-axis magnetometer and three-axis accelerometer for attitude determination. A basic problem for high-latitude operations is the high dip angle for the magnetic field and the low value for the horizontal component of the magnetic field (which is ultimately used for heading determination). For example, the vertical component of the magnetic field near the North Pole at 1,000 m depth is approximately 56 μT, whilst the horizontal field can be about 1 μT [68]. The large ratio between the horizontal and vertical magnetic field causes problems as the vertical field 'leaks' into the compass measurements of the horizontal field via biases in the accelerometers. Simulations suggest that using *in situ* compass calibration techniques [19], it can be possible to operate an AUV successfully in regions with a magnetic horizontal field as low as 1 μT. However, there are still risks by using a magnetic compass.

One risk is due to the effect of lateral acceleration on the sensor performance during manoeuvring, as pitch and roll errors are multiplied up by a large multiplication factor (vertical magnetic field) producing heading errors. For example, the turning (centripetal) acceleration of ALR1500 at 0.7 m/s is 3.2 cm/s^2, which will cause a non-negligible error in heading due to errors in estimating the true roll angle. As the use of a pure magnetic compass can potentially cause a high risk, especially for the vehicle's control unit, the magnetic compass should be combined with a low-power MEMS rate gyros to stabilise the heading output (e.g. a single-yaw axis rate gyro).

The second risk is associated with the rapid change of the magnetic declination with position by approaching the Magnetic North Pole [68], coupled with potential uncertainty in position. Although ALR1500 have a built-in geomagnetic model correction for magnetic declination variation, this will require increasingly more accurate positioning as approaching the Magnetic North Pole. As this cannot be assured *a priori*, one way to mitigate this risk is by keeping a distance from the Magnetic North Pole several times out a maximum acceptable/expected uncertainty in position (see Figure 6.7).

A third risk is associated with the magnetic uncertainties, notable due to solar storms and crustal anomalies. Solar storms can generate anomalies of several hundreds of nano Tesla (which is comparable with the strength of the horizontal magnetic field). Although an initial analysis suggests that these would not pose a large hazard for the navigation system due to their short distance and temporal duration [69], further investigation of the practicality of using a compass given the variation in the magnetic field due to solar storms and crustal anomalies is needed to avoid underestimating the risk.

Overall, it is believed that by using *in situ* compass calibration, keeping a distance from the Magnetic North Pole, and simple addition of a low power MEMS rate gyro, satisfactory performance can be obtained. However, it is important to investigate further for potentially unconsidered risks and hazards.

Gyrocompass

Due to the complexity and the risk of using a magnetic compass at high latitudes, alternative solutions must be also investigated. Such a solution is the use of a gyrocompass, which computes heading by sensing the rotation of the Earth using a combination of triads of inertial grade gyroscopes (rate gyros) and accelerometers. In essence, the accelerometers are used to reference the outputs of the rate gyros to the levelled vehicle axes. Following, the heading is determined based on the resolution of the Earth's rotation vector on the vehicle's levelled horizontal frame.

Similarly to how magnetic sensors have accuracy problems near the Magnetic North Pole, a gyrocompass has degraded performance near the Geographic North Pole (see Figure 6.7), as the Earth's rotation vector has less component in the horizontal plane. Beyond this high-latitude error source, other main sources of error in a gyrocompass is the localisation error in latitude and the error in the north speed. Considering the relatively low errors in north speed (e.g. due to the water currents), the corresponding gyrocompass error can be neglected.

Heading error modelling

State-of-the-art error models for the Octans gyrocompass, developed in conjunction with the sensor manufacturer, suggest that satisfactory performance can be obtained up to 88° latitude [44]. The gyrocompass error model primarily incorporates error sources dependent on the operating latitude and the localisation error in latitude, as follows:

$$e_k^{GC} = \sqrt{\left(\frac{0.05}{\cos(x_k^{Lat})}\right)^2 + \left(\frac{0.022 \cdot e_k^{Lat}}{\cos(x_k^{Lat})}\right)^2} \tag{6.27}$$

where e_k^{GC} is the gyrocompass RMSE expressed in degrees; x_k^{Lat} is the operating latitude in degrees; and e_k^{Lat} is the localisation error in latitude, also expressed in degrees. The first term in (6.27) takes into account errors associated with the operating latitude, whilst the second term corresponds to the heading error due to the localisation error in latitude. Evaluating (6.27) with respect to an extreme latitude of 88° and an error of 100 km in the north direction (maximum acceptable localisation error), it can be seen that the heading RMSE is bounded within 1.6°. However, in reality the localisation error is unknown; therefore, the vehicle needs to operate in lower latitudes to reduce the risk associated with the heading error. The path designed and shown in Figure 6.7 keeps a minimum distance of over 550 km from the Geographic North Pole, which corresponds to an operational latitude below 85°.

An identical error model has been developed for the magnetic compass, though assuming significantly poorer performance, in comparison to the gyrocompass, as the localisation error grows or the distance from the sensor-dependent critical latitudes decreases. As the Magnetic North Pole does not coincide with the Geographic North Pole (see Figure 6.7), the error model requires a latitude angle that is referenced to the Magnetic North Pole. To this end, the Earth's spherical reference frame has been rotated such that the z-axis of the Earth-centred Earth-fixed (ECEF) passes through the Magnetic North Pole (e.g. $\approx${86.5°N, 172.6°W} in 2017). Following this, the error model for the magnetic compass is expressed as

$$e_k^{MC} = \sqrt{\left(\frac{0.3}{\cos(\tilde{x}_k^{Lat})}\right)^2 + \left(\frac{0.15 \cdot \tilde{e}_k^{Lat}}{\cos(\tilde{x}_k^{Lat})}\right)^2} \tag{6.28}$$

where e_k^{MC} is the RMSE for the magnetic compass expressed in degrees; $\tilde{x}_k^{Lat}$ is the operating latitude (in degrees) with respect to a reference frame whose vertical axis (z-axis) passes through the Magnetic North Pole (e.g. a rotated ECEF); and $\tilde{e}_k^{Lat}$ is the positioning error in latitude calculated on the same frame aligned to the Magnetic North Pole, expressed in degrees. The first terms of (6.28) models errors due to operations near the Magnetic North Pole (low horizontal magnetic field), whilst the second term corresponds to heading errors due to errors in localisation and the inability to accurately compensate for the magnetic declination.

By keeping the maximum possible distance from the Magnetic North Pole, the magnetic compass error can be minimised (given a constant localisation error in latitude). Evaluating (6.27) for 2° distance from the Magnetic North Pole and an error of 100 km in the north direction, it can be seen that the heading RMSE is bounded within 3.1°. This shows that the magnetic compass is significantly less reliable compared to the gyrocompass under similar sensor-critical conditions. However, note that the modelling performed in (6.28) is by no means an accurate representation of the error of a magnetic compass, but rather an example of what can be expected at high-latitude zones. As the Earth's horizontal magnetic field can vary unpredictably in polar latitudes, modelling accurately the error of a magnetic compass can be a very challenging task.

In terms of the simulated heading error, the $3\times$RMSE is applied to each heading measurement. In addition, to introduce a stochastic term in the heading error, each measurement is assumed to be further affected by Gaussian noise. This aims to represent unmodelled errors, such as the gyrocompass error caused by inaccuracies in the vehicle's north speed or errors in the magnetic compass due to, e.g., the long-term effect of variations in the Earth's magnetic field because of crustal anomalies and solar storms (while the short-term effect would probably be better described by an error impulse – with a tail over minutes or hours – with Poisson occurrence distribution). This additional noise is assumed to have $\sigma^{GC} = 1/3°$ intensity for the gyrocompass and $\sigma^{MC} = 5/3°$ for the magnetic compass.

6.6.1.3 Arctic terrain model and bathymetric observations

In contrast to the typical TAN applications where the terrain model is constructed by conducting a series of bathymetric surveys [25], good-quality bathymetric maps of the Arctic Ocean are unavailable. The bathymetric map used to enable TAN while crossing the Arctic Ocean is obtained from the International Bathymetric Chart of the Arctic Ocean (IBCAO) multi-national project [70]. Using an accumulated database that contained all available bathymetric data at the time north of 64° and the state-of-the-art oceanographic techniques, IBCAO has generated a terrain model of the Arctic Ocean. The constant increase of bathymetric data, e.g. from icebreaker cruises and declassified echo soundings acquired by submarines, has led to the latest release (Version 3.0) of the grid model with grid cell spacing of 30 × 30 arc seconds.

However, the pace at which the Arctic Ocean is sampled is very slow and only 11 per cent of the Arctic Ocean is covered by multi-beam surveys. Therefore, areas unconstrained by bathymetric data are filled using interpolation techniques. The lack of data inevitably causes errors in the terrain model. As the determination of these errors is a challenging task [71], a previous study in [18] has evaluated the performance of a PF-based TAN algorithm under the effect of a wide range of map errors in the vertical dimension. Following this, the terrain model is assumed to be affected by $\mathcal{N}(0, 100^2\ \text{m}^2)$, which simulates possible bathymetric errors by causing a significant map distortion in the vertical dimension.

To obtain bathymetric observation while operating out of the ADCP's bottom-tracking range, ALR1500 is equipped with the NOC's 4,000 m range single-beam echo-sounder. It is assumed that the echo-sounder has 60 s pinging period. Recent deep-water experiments in Celtic Sea (June 2018) have shown that the error of the echo-sounder is less than 2 per cent of the measured range. Given the statistical independence between the sonar errors, the errors in determining the vehicle's depth (see Section 6.5.3.1) and the terrain model errors, the overall error effect can be represented by the noise term ζ_k of the observation model in (6.9).

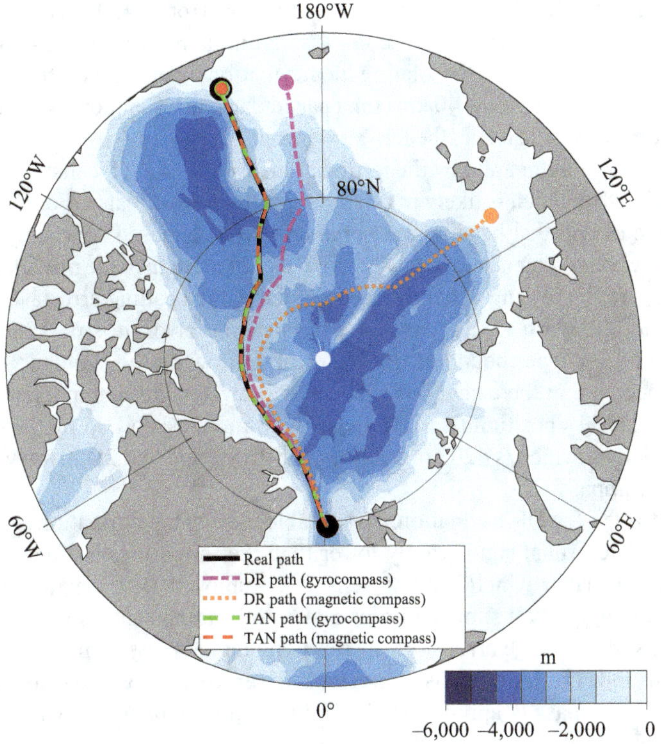

Figure 6.8 Simulated crossing of the Arctic Ocean using ALR1500. The red and the green dashed lines correspond to the averaged vehicle's path as estimated by TAN (while employing a magnetic compass or a gyrocompass, respectively) after performing five simulated runs, in comparison to the true vehicle's path and the estimated path using DR

6.6.2 Arctic crossing navigation results

Using the simulation set-up provided in Section 6.6.1, the TAN algorithm is evaluated in terms of the filter's estimation accuracy and the ability to re-produce identical results when performing multiple simulated crossings of the Arctic Basin. To further investigate the filter's performance while using either a gyrocompass or a magnetic compass, a set of five experiments have been performed for each heading sensor.

Throughout these experiments, the number of particles is set to $N = 5{,}000$. Given that the vehicle uses the 4,000 m echo-sounder, there is no need to dividing the vehicle's operation into two phases (in contrast to Section 6.5.3.2), as the filter is initialised at the very beginning of the mission. The parameters used to initialise the filter and to define the intensity of the process noise have been specified identically as in Section 6.5.3.2 and the DynOPO experiment.

Figure 6.8 shows the estimated vehicle's path by using TAN, as calculated via averaging five simulated runs, in comparison to the true path and the DR path.

It appears that the DR navigation experiences an error of 420 km at the surfacing location (Barrow coast) whilst gyrocompassing and approximately 2,400 km when using a magnetic compass. By isolating the navigation error sources, the drift caused by the water currents is below 40 km (for a particular spatio-temporal sample of water currents from NEMO). Hence, the DR error is in the majority driven by the simulated heading error (predominantly by the term that incorporates the localisation error; see (6.27) and (6.28)), which is likely to be significantly overestimated. For instance, it is unrealistic to expect that heading measurements will be biased constantly towards a specific direction (either it is east or west) for the entire length of the mission, similarly to what the Arctic crossing experiments assume for the heading error. Nevertheless, by simulating the worst case scenario for the heading error, it is possible to obtain an estimate of the upper bound of the DR error. Following the simulation results, it is to be expected that the actual DR solution will always fall within a circle centred at the true vehicle's position (Barrow coast) with radius specified by the calculated localisation error, which is 420 km whilst gyrocompassing and 2,400 km when using a magnetic compass.

In contrast to the DR navigation, TAN is able to keep the navigation error (during all five simulated runs) significantly lower than the initial localisation requirement of 100 km, no matter which heading sensor is employed. The maximum observed TAN error throughout all simulated runs and both heading sensors is approximately 18 km, whilst the average error over the mission length is less than 1.5 km (for both heading sensors). The TAN localisation error at the end of the mission while employing the magnetic compass is approximately 12 km, whereas using the gyrocompass is approximately 10 km.

It is also important to emphasise that the RBPF algorithm never underestimates the localisation error, despite the heavy distortion applied to the terrain model. Even though the state posterior probability density function sometimes is multi-modal, one of the modes always tracks the actual vehicle position. Figure 6.9 shows the mean (over the five runs) TAN error for each direction with the respective two-sided 3-σ confidence bounds over the DT.

Despite the strong assumption (for visualisation purposes) that particles are distributed following a normal distribution uncorrelated in each direction, the TAN error falls always within the 3-σ bounds.

6.7 Conclusion and future work

The ever-increasing demand from scientific and oceanographic communities for conducting research activities in remote rapidly changing oceans has led to the development of ultra-endurance AUVs. Platforms such as the ALR vehicles developed in the NOC open up a world of new AUV applications, including persistent monitoring and data collection in some of the most inaccessible areas on the Earth. The long endurance is achieved through the use of high energy density lithium primary cells combined with optimisation in forward speed and energy consumption within all subsystems. A power analysis has shown that these vehicles are capable of operating for

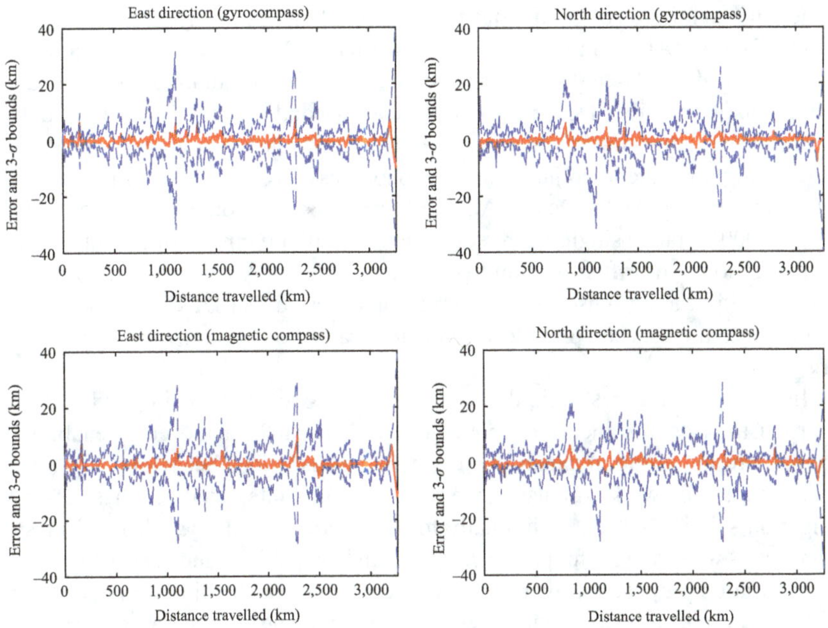

Figure 6.9 TAN estimation error (red) with the associated 3-σ bounds (blue) over DT while using a magnetic compass and a gyrocompass

months, even when unique missions may require power-demanding sensors (e.g. a gyrocompass for operations in high-latitude zones).

However, deploying AUVs for long periods of time leads to various challenges. Even though a considerable effort has been directed towards developing navigation techniques for AUVs, a self-contained low-power solution for multi-month-long missions yet remains a challenge. In response to the presently weak navigation capability, this work has developed a TAN technique that holds the potential to prolong underwater missions without the need for external support or regular surfacing. The proposed TAN algorithm relies on basic low-power sensors, deals with sparse bathymetric maps and is computationally feasible in real time for long-range AUVs utilising processing boards with limited computational resources.

The performance of the algorithm was evaluated using unique field data collected during a deep-water multi-day-long (≈195 km range and 77 h length) mission of ALR6000 in the Southern Ocean. While DR navigation experienced an error of over 40 km due to the presence of strong water currents, the TAN algorithm appeared to be able to limit the DR error within a few hundreds of metres, despite traversing areas with low topographic information (along-track) and using a low-resolution bathymetric map.

As an extension, the algorithm was further evaluated during a simulated crossing of the Arctic Ocean using ALR1500. This is an aspirational example of a science-driven mission of continuous mid-water column survey from Svalbard (Norway) to Barrow (Alaska) under the Arctic sea-ice (≈3,200 km range). A simulated environment of the Arctic Ocean has been developed incorporating a coarse bathymetric map (with grid cell spacing of 30 × 30 arc seconds), state-of-the-art models of the Arctic circulation and key heading error sources while operating in high latitudes (either using a magnetic compass or a gyrocompass). To obtain bathymetric observations in the deep Arctic Ocean, ALR1500 was assumed to be integrated with the NOC's 4,000-m range single-beam echo-sounder with pinging period of 60 s. While DR navigation can drift from hundreds to thousands of kilometres, depending on the employed heading sensor, simulation results showed that TAN can cope with the navigation challenges and provide acceptable localisation accuracy for enabling safe transit.

In terms of future research, this will primarily concentrate on improving the filter's robustness against incorrect convergences. This can be achieved by enabling an automatic re-initialisation capability, e.g. when a suspicious behaviour of the filter is detected. Future work will also move towards a fully integrated geophysical-based navigation solution where, additionally to the ocean bathymetry, other geophysical quantities as well as seawater properties [72] can be combined and used as references for localisation purpose. Furthermore, an interesting consideration for Arctic operations is the use of a hybrid heading estimation technique. In such an approach, the navigation system will alternate between the gyrocompass and the magnetic compass to, for example, safely pass through sensor-dependent critical latitudes, whilst reducing the overall gyrocompassing time.

Acknowledgements

This work was supported by the ROBOCADEMY (FP7 Marie Curie Programme ITN Grant Agreement Number 608096) and NERC Oceanids programmes. Authors would also like to thank the officers, crew and scientific party of RRS James Clark Ross cruise JR16005 (DynOPO) for assistance in acquiring the data used here.

References

[1] Munafò A, Simetti E, Turetta A, *et al.* Autonomous Underwater Vehicle Teams for Adaptive Ocean Sampling: A Data-Driven Approach. Ocean Dynamics. 2011;61(11):1981–1994.

[2] German CR, Yoerger DR, Jakuba M, *et al.* Hydrothermal Exploration with the Autonomous Benthic Explorer. Deep Sea Research Part I: Oceanographic Research Papers. 2008;55(2):203–219.

[3] Kapoutsis AC, Chatzichristofis SA, Doitsidis L, *et al.* Real-time Adaptive Multi-robot Exploration with Application to Underwater Map Construction. Autonomous Robots. 2016;40(6):987–1015.

[4] Salavasidis G, Kapoutsis AC, Chatzichristofis SA, *et al.* Autonomous Trajectory Design System for Mapping of Unknown Sea-floors using a Team of AUVs. In: 2018 European Control Conference (ECC). IEEE; 2018. pp. 1080–1087.

[5] Xiang X, Jouvencel B, Parodi O. Coordinated Formation Control of Multiple Autonomous Underwater Vehicles for Pipeline Inspection. International Journal of Advanced Robotic Systems. 2010;7(1):3.

[6] Yordanova V. Intelligent Adaptive Underwater Sensor Networks. University College London; 2018.

[7] Stokey R, Austin T, Allen B, *et al.* Very Shallow Water Mine Countermeasures Using the REMUS AUV: A Practical Approach Yielding Accurate Results. In: OCEANS, 2001. MTS/IEEE Conference and Exhibition. Vol. 1. IEEE; 2001. pp. 149–156.

[8] Martin B, Tarraf DC, Whitmore TC, *et al.* Advancing Autonomous Systems: An Analysis of Current and Future Technology for Unmanned Maritime Vehicles. Santa Monica, CA: RAND Corporation, 2019. https://doi.org/10.7249/RR2751.

[9] Von Alt C. Autonomous Underwater Vehicles, In Autonomous and Lagrangian Platforms and Sensors Workshop, La Jolla, California, U.S, 31 March–2 April, 2003.

[10] German CR, Jakuba MV, Kinsey JC, *et al.* A Long Term Vision for Long-range Ship-free Deep Ocean Operations: Persistent Presence through Coordination of Autonomous Surface Vehicles and Autonomous Underwater Vehicles. In: Autonomous Underwater Vehicles (AUV), 2012 IEEE/OES. IEEE; 2012. pp. 1–7.

[11] Gilmour B, Niccum G, O'Donnell T. Field Resident AUV Systems-Chevron's Long-term Goal for AUV Development. In: Autonomous Underwater Vehicles (AUV), 2012 IEEE/OES. IEEE; 2012. pp. 1–5.

[12] Bachmayer R, Leonard NE, Graver J, *et al.* Underwater Gliders: Recent Developments and Future Applications. In: Underwater Technology, 2004. UT'04. 2004 International Symposium on. IEEE; 2004. pp. 195–200.

[13] Rudnick DL, Davis RE, Eriksen CC, *et al.* Underwater Gliders for Ocean Research. Marine Technology Society Journal. 2004;38(2):73–84.

[14] Hobson BW, Bellingham JG, Kieft B, *et al.* Tethys-class Long Range AUVs-Extending the Endurance of Propeller-driven Cruising AUVs from Days to Weeks. In: Autonomous Underwater Vehicles (AUV), 2012 IEEE/OES. IEEE; 2012. pp. 1–8.

[15] Furlong ME, Paxton D, Stevenson P, *et al.* Autosub Long Range: A Long Range Deep Diving AUV for Ocean Monitoring. In: Autonomous Underwater Vehicles (AUV), 2012 IEEE/OES. IEEE; 2012. pp. 1–7.

[16] Roper DT, Phillips AB, Harris CA, *et al.* Autosub Long Range 1500: An Ultra-endurance AUV with 6000 km Range. In: OCEANS 2017-Aberdeen. IEEE; 2017. pp. 1–5.

[17] Hayes D, Boyd T, Patterson M. Sensors and Instrument Requirements for Autonomous Underwater Vehicles: Collaborative Autosub Science in Extreme

Environments. In Proceedings: Collins K. and Griffiths G. (eds.) Masterclass in AUV Technology for Polar Science 2007. Society for Underwater Technology, London, pp 39–48.

[18] Salavasidis G, Munafo A, Harris CA, McPhail SD, Rogers E, Phillips AB. Towards Arctic AUV Navigation. IFAC-PapersOnLine, 51(29):287–292. https://doi.org/10.1016/j.ifacol.2018.09.517

[19] Salavasidis G, Munafò A, Harris CA, *et al.* Terrain-Aided Navigation for Long-endurance and Deep-Rated Autonomous Underwater Vehicles. Journal Field Robotics. 2019;35:447–474. https://doi.org/10.1002/rob.21832

[20] Paull L, Saeedi S, Seto M, *et al.* AUV Navigation and Localization: A Review. Oceanic Engineering, IEEE Journal of. 2014;39(1):131–149.

[21] Vickery K. Acoustic Positioning Systems. A Practical Overview of Current Systems. In: Autonomous Underwater Vehicles, 1998. AUV'98. Proceedings of the 1998 Workshop on. IEEE; 1998. pp. 5–17.

[22] Mandt M, Gade K, Jalving B. Integrating DGPS-USBL Position Measurements with Inertial Navigation in the HUGIN 3000 AUV. In: Proceedings of the 8th Saint Petersburg International Conference on Integrated Navigation Systems, Saint Petersburg, Russia; 2001. pp. 28–30.

[23] Phillips AB, Salavasidis G, Kingsland M, *et al.* Autonomous Surface/Subsurface Survey System Field Trials. In: 2018 IEEE/OES Autonomous Underwater Vehicles (AUV). IEEE; 2018.

[24] Teixeira FC. Terrain-aided Navigation and Geophysical Navigation of Autonomous Underwater Vehicles. Dynamical Systems and Ocean Robotics Lab, Lisbon; 2007.

[25] Melo J, Matos A. Survey on Advances on Terrain Based Navigation for Autonomous Underwater Vehicles. Ocean Engineering. 2017;139:250–264.

[26] Meduna DK. Terrain Relative Navigation for Sensor-limited Systems with Application to Underwater Vehicles. PhD thesis, Stanford University; 2011.

[27] Siouris GM. Missile Guidance and Control Systems. New York: Springer-Verlag, 2004, https://doi.org/10.1007/b97614

[28] Schon T, Gustafsson F, Nordlund PJ. Marginalized Particle Filters for Mixed Linear/Nonlinear State-space Models. IEEE Transactions on Signal Processing. 2005;53(7):2279–2289.

[29] Brugali D, Shakhimardanov A. Component-based Robotic Engineering (part II). IEEE Robotics & Automation Magazine. 2010;17(1):100–112.

[30] Quigley M, Conley K, Gerkey B, *et al.* ROS: An Open-source Robot Operating System. In: ICRA Workshop on Open Source Software. Vol. 3. Kobe, Japan; 2009. pp. 5.

[31] Eickstedt DP, Sideleau SR. The Backseat Control Architecture for Autonomous Robotic Vehicles: A Case Study with The Iver2 AUV. Marine Technology Society Journal. 2010;44(4):42–54.

[32] Colledanchise M, Ögren P. Behavior Trees in Robotics and AI: An Introduction, ser. Chapman and Hall/CRC Artificial Intelligence and Robotics Series. CRC Press, Taylor & Francis Group, 2018.

[33] Donovan GT. Development and Testing of a Real-time Terrain Navigation Method for AUVs. In: OCEANS 2011. IEEE; 2011. pp. 1–9.

[34] Taberham A, Kraft M, Mowlem M, *et al.* The Fabrication of Lab-on-chip Devices from Fluoropolymers. Journal of Micromechanics and Microengineering. 2008;18(6):064011.

[35] Salavasidis G, Harris CA, Rogers E, *et al.* Co-operative Use of Marine Autonomous Systems to Enhance Navigational Accuracy of Autonomous Underwater Vehicles. In: Conference Towards Autonomous Robotic Systems. Springer; 2016. pp. 275–281.

[36] Furlong ME, McPhail SD, Stevenson P. A Concept Design for an Ultra-long-range Survey Class AUV. In: OCEANS 2007-Europe. IEEE; 2007. pp. 1–6.

[37] Phillips AB, Haroutunian M, Man SK, *et al.* In: Roberts GN, Sutton R, editors. Nature in Engineering for Monitoring the Oceans: Comparison of the Energetic Costs of Marine Animals and AUVs. 77. The Institution of Engineering and Technology; 2012. pp. 373–405. Available from: http://eprints.soton.ac.uk/336783/.

[38] Phillips AB, Haroutunian M, Murphy AJ, *et al.* Understanding the Power Requirements of Autonomous Underwater Systems, Part I: An Analytical Model for Optimum Swimming Speeds and Cost of Transport. Ocean Engineering. 2017;133:271–279.

[39] Vitzro Cell. Tekcell: Lithium Primary Battery; 2018. Accessed: 6-August-2018. Available from: http://www.vitzrocell.de/catalog_2011.pdf.

[40] Diouf B, Pode R. Potential of Lithium-ion Batteries in Renewable Energy. Renewable Energy. 2015;76:375–380.

[41] Jain M, Nagasubramanian G, Jungst RG, *et al.* Analysis of a Lithium-Thionyl Chloride Battery under Moderate-Rate Discharge. Journal of The Electrochemical Society. 1999;146(11):4023–4030.

[42] Nissan Leaf. Nissan Leaf: A Zero Emission Electric Car. Accessed: 6-August-2018. https://www-europe.nissan-cdn.net/content/dam/Nissan/za/brochures/LEAF.pdf.

[43] Stokey RP, Roup A, von Alt C, *et al.* Development of the REMUS 600 Autonomous Underwater Vehicle. In: OCEANS, 2005. Proceedings of MTS/IEEE. IEEE; 2005. pp. 1301–1304.

[44] iXblue. OCTANS Fiber-optic Gyrocompass; Accessed: 6-August-2018. https://www.ixblue.com/sites/default/files/datasheet_file/octans-datasheet.pdf.

[45] Arulampalam MS, Maskell S, Gordon N, *et al.* A Tutorial on Particle Filters for Online Nonlinear/non-Gaussian Bayesian Tracking. IEEE Transactions on Signal Processing. 2002;50(2):174–188.

[46] Fofonoff NP, Millard Jr R. Algorithms for the Computation of Fundamental Properties of Seawater. UNESCO Technical Papers in Marine Science 44. 1983.

[47] Visbeck M. Deep Velocity Profiling Using Lowered Acoustic Doppler Current Profilers: Bottom Track and Inverse Solutions. Journal of Atmospheric and Oceanic Technology. 2002;19(5):794–807.

[48] Nordlund PJ. Sequential Monte Carlo Filters and Integrated Navigation. PhD thesis, Department of Electrical Engineering, Linköpings University; 2002.

[49] Doucet A, Godsill S, Andrieu C. On Sequential Monte Carlo Sampling Methods for Bayesian Filtering. Statistics and Computing. 2000;10(3):197–208.

[50] Anonsen KB, Hallingstad O. Terrain Aided Underwater Navigation Using Point Mass and Particle Filters. In: Proceedings of the IEEE/ION Position Location and Navigation Symposium; 2006. pp. 1027–1035.

[51] Stammer D, Ray R, Andersen OB, *et al.* Accuracy Assessment of Global Barotropic Ocean Tide Models. Reviews of Geophysics. 2014;52(3):243–282.

[52] Li Z. Variation of the Accuracy of Digital Terrain Models with Sampling Interval. The Photogrammetric Record. 1992;14(79):113–128.

[53] Jensen FB, Kuperman WA, Porter MB, *et al.* Computational Ocean Acoustics. Springer Science & Business Media; 2011.

[54] Mayer L, Jakobsson M, Allen G, *et al.* The Nippon Foundation GEBCO Seabed 2030 Project: The Quest to See the World's Oceans Completely Mapped by 2030. Geosciences. 2018;8(2):63.

[55] Carpenter J, Clifford P, Fearnhead P. Improved Particle Filter for Nonlinear Problems. IEE Proceedings-Radar, Sonar and Navigation. 1999;146(1):2–7.

[56] Salavasidis G, Harris C, McPhail S, *et al.* Terrain-aided Navigation for Long Range AUV operations at Arctic latitudes. In: Autonomous Underwater Vehicles (AUV), 2016 IEEE/OES. IEEE; 2016. pp. 115–123.

[57] Claus B, Bachmayer R. Terrain-aided Navigation for an Underwater Glider. Journal of Field Robotics. 2015;32(7):935–951.

[58] Li T, Bolic M, Djuric PM. Resampling Methods for Particle Filtering: Classification, Implementation, and Strategies. IEEE Signal Processing Magazine. 2015;32(3):70–86.

[59] Naveira Garabato A, Forryan A. RRS James Clark Ross Cruise JR16005: Dynamics of the Orkney Passage Outflow (DynOPO) Cruise Report. University of Southampton, National Oceanography Centre, Southampton, UK; 2017.

[60] Google Earth. Orkney Passage, Antarctic Peninsula, 59° 13′20″S, 49° 57′ 35″W, 3,504.29 km; 2019. Accessed: 6-August-2018. https://earth.google.com/web/@-58.24792076,-54.57702361,-3401.62319381a,3512896.3445127d,35y,0h,0t,0r.

[61] Naveira Garabato AC, Frajka-Williams E, Spingys CP, *et al.* Rapid Mixing and Exchange of Deep-ocean Waters in an Abyssal Boundary Current. Proceedings of the National Academy of Sciences. 2019.

[62] McPhail SD, Pebody M. Navigation and Control of an Autonomous Underwater Vehicle Using a Distributed, Networked, Control Architecture. Underwater Technology. 1998;23(1):19–30.

[63] Marks K, Smith W. An Uncertainty Model for Deep Ocean Single Beam and Multibeam Echo Sounder Data. Marine Geophysical Researches. 2008;29(4):239–250.

[64] Wassmann P, Duarte CM, Agusti S, *et al.* Footprints of Climate Change in the Arctic Marine Ecosystem. Global Change Biology. 2011;17(2):1235–1249.

[65] Bellingham J, Cokelet ED, Kirkwood WJ. Observation of Warm Water Transport and Mixing in the Arctic Basin with the ALTEX AUV. In: 2008 IEEE/OES Autonomous Underwater Vehicles. IEEE; 2008. pp. 1–5.

[66] Kunz C, Murphy C, Camilli R, *et al.* Deep Sea Underwater Robotic Exploration in the Ice-Covered Arctic Ocean with AUVs. In: 2008 IEEE/RSJ International Conference on Intelligent Robots and Systems. IEEE; 2008. pp. 3654–3660.

[67] Madec G. NEMO Ocean Engine. Note du Pole de modelisation, Institut Pierre-Simon Laplace (IPSL), France. 2008;27:1288–1619.

[68] Thébault E, Finlay CC, Beggan CD, *et al.* International Geomagnetic Reference Field: The 12th Generation. Earth, Planets and Space. 2015;67(1):79.

[69] Merrill, RT, McElhinny MW, McFadden PL. The Magnetic Field of the Earth: paleomagnetism, the core, and the deep mantle. San Diego: Academic Press, 1996.

[70] Jakobsson M, Mayer L, Coakley B, *et al.* The International Bathymetric Chart of the Arctic Ocean (IBCAO) Version 3.0. Geophysical Research Letters. 2012;39(12).

[71] Jakobsson M, Calder B, Mayer L. On the Effect of Random Errors in Gridded Bathymetric Compilations, Journal of Geophysical Research: Solid Earth, 107(B12), 2358, doi:10.1029/2001JB000616, 2002.

[72] Munafò A, Fanelli F, Salavasidis G, Storto A, Oddo P. Navigation of AUVs based on Ocean Fields Variability, OCEANS 2019 – Marseille, Marseille, France, 2019, pp. 1–9.

Chapter 7

Data compression, data association and reduced complexity SLAM techniques for UUVs during transit

Ashwin Sarma[1]

In the transiting stage of an unmanned undersea vehicle (UUV) mission, it is of interest to minimize platform localization error with minimal processing. Earlier work [1] derived a simultaneous localization and map building (SLAM)-inspired estimator of platform location and velocity, dubbed "velocity-over-ground" (VOG)-SLAM, that provides virtually identical performance in transit scenarios as conventional SLAM. The method lends itself to simple real-time operation as map building is not required.

The "VOG" simplification was devised based on (a) the observation that the second measurement of a persistent contact was required for potential performance improvement in SLAM and (b) the intuitive idea that SLAM is providing velocity information since contact measurements can only be relative to the platform. We provide here a direct argument by arguing its optimality properties via connection to the maximum likelihood estimator (MLE).

In addition, techniques for sonar data processing, measurement generation and data association methodologies to determine proper assignments between measurements and persistent bottom features are discussed. These further extend concepts found in [2]. The process can be currently completed before the next ping arrives suggesting near real-time SLAM performance in complex undersea environments.

7.1 Introduction

SLAM [3], also sometimes referred to as concurrent mapping and localization [4–6], is a tracking technique with the following goals:

1. construct maps of the locations of strong point scatterers in the sensors' field of view;
2. continually improve these maps as more sensor data arrive;
3. simultaneously achieve improved platform self-localization with respect to scatterer locations.

[1]BAE Systems, Merrimack, USA (formerly of Naval Undersea Warfare Center, Newport, RI 02841, USA)

SLAM's refinement of the platform location estimates is directly linked to its exploitation of the stationary nature of the scatterers of opportunity. However, SLAM is often computationally prohibitive as its architecture attempts to minimize the position error for each scatterer, i.e. "map building." This requires all pairwise correlation information to be maintained and updated leading to very large memory and computation requirements. Clever schemes have been recently devised [7,8] that trade greater algorithmic complexity for ultimately reduced computer operations and near real-time operation on certain computing platforms [9].

The VOG-SLAM method has been shown to provide virtually identical platform position and velocity estimation performance in transit scenarios as conventional SLAM without the need for "map building." In this chapter, a theoretical footing for this behavior is provided by deriving the joint MLE for both the map and platform location and velocity and relating it to the MLE for platform location and velocity alone. This, along with determination that both estimators are efficient, i.e. meet the Cramer–Rao lower bound (CRLB), and equality of the CRLB bounds themselves reveal that no performance is lost. The VOG-SLAM method, closely related to the MLE of platform location and velocity, can then be expected to give observed performance (Section 7.2).

In order for any SLAM-based method to provide maximum performance, correct association of data across scans is required. A UUV with a forward-looking sonar (FLS) along with an inertial navigation system (INS)/Doppler velocity log (DVL) suite is the envisioned platform for this work. Operation in shallow water often results in hundreds of persistent features that must be detected, time-tagged and processed. These persistent features then manifest as measurements relative to the platform in the relative local-level (x–y) plane with associated uncertainty. The generation of measurements that accurately reflect positional uncertainties is an involved process that requires adaptive constant false alarm rate (CFAR) detection at the beam time-series level, connected components clustering and physically reliable injection of positional information and errors. Uncertainty modeling must accurately reflect the nature and resolution of the sonar and the attitude and pointing errors. This must be followed by a robust "scan-to-scan" or ping-to-ping data association process that can reliably detect and associate persistent features, often called landmarks in the robotics literature. These processes build from the work first described in [2] and are described in detail in this chapter (Sections 7.3 and 7.4). Preliminary results are provided in Section 7.5.

7.2 Development of MLE and relation to Kalman update

7.2.1 MLE development

A measurement that mirrors the MLE is certainly a reasonable choice as the maximum a posteriori (MAP) estimator or minimum mean squared estimator (MMSE), as realized by the Kalman Filter, is a weighted combination of the MLE and the mean of the prior pdf [10, p. 91].

As the UUV possesses accurate depth information, the two-dimensional or planar self-localization problem is of ultimate interest. The platform position in an absolute world reference frame (WRF) at time 0 is denoted by $[p_{0x}\ p_{0y}]^T$ and platform velocity, $[\dot{p}_x\ \dot{p}_y]^T$, is considered constant throughout the observation time interval of interest here. This allows the description of the unknown platform state as

$$\theta = \begin{bmatrix} p_{0x} \\ p_{0y} \\ \dot{p}_x \\ \dot{p}_y \end{bmatrix} \tag{7.1}$$

Let us consider the stationary feature (or contact) that was observed by the platform's on-board sensor. The unknown location of this feature in the WRF is denoted $[p_{fx}\ p_{fy}]^T$ and the total vector of interest in the SLAM problem is

$$\Theta = \begin{bmatrix} \theta \\ - \\ p_{fx} \\ p_{fy} \end{bmatrix} \tag{7.2}$$

Consider the estimation of Θ using these available measurements:

$$\vec{z}_p = \begin{bmatrix} z_1 \\ z_2 \end{bmatrix} = \begin{bmatrix} p_{0x} \\ p_{0y} \end{bmatrix} + \begin{bmatrix} n_1 \\ n_2 \end{bmatrix} \tag{7.3}$$

where $[n_1\ n_2]^T$ is zero mean Gaussian noise with covariance $\mathcal{R}$.

$$\vec{z}_A = \begin{bmatrix} z_3 \\ z_4 \end{bmatrix} = \begin{bmatrix} p_{fx} - [p_{0x} + \dot{p}_x(t)] \\ p_{fy} - [p_{0y} + \dot{p}_y(t)] \end{bmatrix} + \begin{bmatrix} n_3 \\ n_4 \end{bmatrix} \tag{7.4}$$

where $[n_3\ n_4]^T$ is zero mean Gaussian noise with covariance R_A.

$$\vec{z}_B = \begin{bmatrix} z_5 \\ z_6 \end{bmatrix} = \begin{bmatrix} p_{fx} - [p_{0x} + \dot{p}_x(t+\Delta)] \\ p_{fy} - [p_{0y} + \dot{p}_y(t+\Delta)] \end{bmatrix} + \begin{bmatrix} n_5 \\ n_6 \end{bmatrix} \tag{7.5}$$

where $[n_5\ n_6]^T$ is zero mean Gaussian noise with covariance R_B. Here $\mathcal{R}$, R_A and R_B need not be diagonal.

The measurement $\vec{z}_p$ may either be a fine (e.g. following the initialization stage of the vehicle INS system when GPS is available) or a coarse (e.g. later on in the mission after accrual of error) estimate of the vehicle position at the initial time 0. The associated measurement noise is reasonably assumed to be uncorrelated with noise sources corrupting $\vec{z}_A$ and $\vec{z}_B$. Measurements $\vec{z}_A$ and $\vec{z}_B$ are measurements of the *relative* location of the observable feature with respect to the sonar face at time t and time $t + \Delta$ and are the result of processing of the sonar beam time-series, connected component clustering, measurement generation and data association (discussed in detail in Sections 7.3 and 7.4). Assume that the feature is first observed on ping k at time t and on ping $k + 1$ at time $t + \Delta$. The measurement noise is reasonably assumed to be uncorrelated across pings.

Define the total vector of measurements as

$$Z = \begin{bmatrix} \vec{z}_p \\ - \\ \vec{z}_A \\ - \\ \vec{z}_B \end{bmatrix} \tag{7.6}$$

Assuming that the noises are zero mean and Gaussian distributed, the vector Z is jointly normal and reflects the parameters of interest via the mean vector. This situation is referred to as the Linear Statistical Model [11, p. 238] and the MLE of Θ, denoted $\hat{\Theta}$ is given by

$$\hat{\Theta} = [H^T R^{-1} H]^{-1} H^T R^{-1} Z \tag{7.7}$$

where R is the covariance matrix for the noise given by

$$R = \left[\begin{array}{c|c|c} \mathcal{R} & 0 & 0 \\ \hline 0 & R_A & 0 \\ \hline 0 & 0 & R_B \end{array}\right] \tag{7.8}$$

and the matrix H connects the elements of Θ to the mean vector of Z via

$$H = \begin{bmatrix} 1 & 0 & 0 & 0 & 0 & 0 \\ 0 & 1 & 0 & 0 & 0 & 0 \\ -1 & 0 & -t & 0 & 1 & 0 \\ 0 & -1 & 0 & -t & 0 & 1 \\ -1 & 0 & -(t+\Delta) & 0 & 1 & 0 \\ 0 & -1 & 0 & -(t+\Delta) & 0 & 1 \end{bmatrix} \tag{7.9}$$

Here H seems to suggest that $\vec{z}_p$, $\vec{z}_A$ and $\vec{z}_B$ could all bring relevant information for optimal estimation of each element of Θ. Thus, it seems reasonable to expect that the MLE for each element of Θ will somehow utilize all six measurements. After direct algebraic manipulation, it can be shown that

$$H^T R^{-1} = \left[\begin{array}{c|c|c} \mathcal{R}^{-1} & -R_A^{-1} & -R_B^{-1} \\ \hline 0_{2x2} & -tR_A^{-1} & -(t+\Delta)R_B^{-1} \\ \hline 0_{2x2} & R_A^{-1} & R_B^{-1} \end{array}\right] \tag{7.10}$$

and the Fisher Information Matrix J which equals $H^T R^{-1} H$ is given by

$$J = \left[\begin{array}{c|c|c} \mathcal{R}^{-1} + R_A^{-1} + R_B^{-1} & tR_A^{-1} + (t+\Delta)R_B^{-1} & -R_A^{-1} - R_B^{-1} \\ \hline tR_A^{-1} + (t+\Delta)R_B^{-1} & t^2 R_A^{-1} + (t+\Delta)^2 R_B^{-1} & -tR_A^{-1} - (t+\Delta)R_B^{-1} \\ \hline -R_A^{-1} - R_B^{-1} & -tR_A^{-1} - (t+\Delta)R_B^{-1} & R_A^{-1} + R_B^{-1} \end{array}\right] \tag{7.11}$$

Postulating

$$\hat{\Theta} = \begin{bmatrix} \vec{z}_p \\ \hline \frac{\vec{z}_A - \vec{z}_B}{\Delta} \\ \hline \vec{z}_p + (1 + \frac{t}{\Delta})\vec{z}_A - \frac{t}{\Delta}\vec{z}_B \end{bmatrix} \tag{7.12}$$

it can be shown that $J\hat{\Theta} = H^T R^{-1} Z$ and thus $\hat{\Theta}$ is indeed the MLE. The optimum estimator, in a maximum likelihood sense, of $[p_{0x}\ p_{0y}]^T$ utilizes only the direct measurement $\vec{z}_p$. Furthermore, the optimal estimator of $[\dot{p}_x\ \dot{p}_y]^T$ merely differences the two measurements of the relative position of the observed feature and scales by the elapsed time, i.e. a direct measurement of average velocity. The quality of $\hat{\Theta}$ will certainly depend on the variances encapsulated in R. However, $\hat{\Theta}$ does not.

The Fisher Information Matrix J is given by $H^T R^{-1} H$ and the CRLB is J^{-1}. As the MLE is unbiased, it meets the CRLB and is efficient. If we were only interested in θ, invariance principles of MLEs can allow us to state that the MLE of θ is given by

$$\hat{\theta} = W(\hat{\Theta}) \tag{7.13}$$

where $W(\Theta) = \theta$. It can be readily shown that $\hat{\theta}$ is unbiased and efficient and that the corresponding CRLB is the 4×4 upper left portion of J^{-1}.

Thus, there is no increase in platform localization error if we choose not to build maps provided we utilize all the available measurements in this fashion. This conclusion was suggested in [1] through the traditional MMSE framework of Kalman filtering used in SLAM [3]. As discussed earlier, the optimum linear combination of the best prior estimate and the maximum likelihood estimate will consider the corresponding error in each. Differencing these vector measurements as in $\frac{\vec{z}_A - \vec{z}_B}{\Delta}$ results in error characteristics that can be accurately handled in a Kalman framework.

It can be shown that if another feature was observed on both pings at the exact times t and $t + \Delta$, the MLE would weight the corresponding platform velocity estimates by their associated errors. However, this case is not exactly physical as every Kalman update step can only process information from a single feature due to the finite wave speed of sound. Thus, the argument here reveals that the method must work when multiple stationary features are ensonified over two pings as considered here. Of interest is the case where three or more pings are considered. The exact slight reduction in optimality of the VOG-SLAM over conventional SLAM may then be determinable from a different point of view than argued in [1] for this case.

7.2.2 VOG-SLAM Kalman update

The VOG-SLAM method differences the position measurements from the *same* persistent feature if observed on consecutive pings. It then interprets this difference in each dimension as the product of a fixed velocity and ΔT, the time elapsed since the arrival of the last measurement associated with that contact. These replace the "contact measurements" of the conventional SLAM method. The Kalman measurement matrix H that relates the measurement to the state vector is of the form

$$H = \begin{bmatrix} 0 & 0 & \Delta T & 0 \\ 0 & 0 & 0 & \Delta T \end{bmatrix} \tag{7.14}$$

We interpret the differenced measurement as providing an observation of platform velocity *at* update time $k+1$.

In [1, Section III], we have derived the recursive MMSE algorithm accounting for correlated measurement noise that arises from the differencing procedure. Note, however, that consecutive updates (e.g. k and $k+1$) will not generally be performed using information derived from the same contact. In fact closely spaced times, each corresponding to an update, will correspond to the re-observation times of distinct persistent contacts. In Figure 7.1 is depicted a scenario in which the FLS has pinged three times in succession. On ping 1, the circle is detected and its position is estimated in the measurement generation process. It is, however, out of the ensonification region on ping 2 and thus cannot be persistent. Note that a feature may be in the ensonified region and still not be detected due to poor detectability or statistical fluctuations. In this case, it would also not be declared persistent. On ping 2, the feature denoted by the square is persistent for pings 1 and 2. On ping 3, the feature denoted by the triangle is persistent for all three pings. Assume we observe the triangle at time t, the time associated with the measurement cluster. We desire the (linear) MMSE estimate of the platform state at time t. This estimate is associated with update $k+1$.

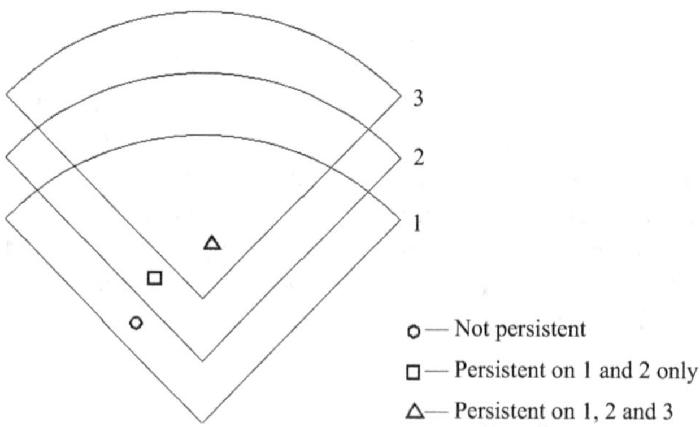

Figure 7.1 Persistence of contacts dependent on geometric opportunity

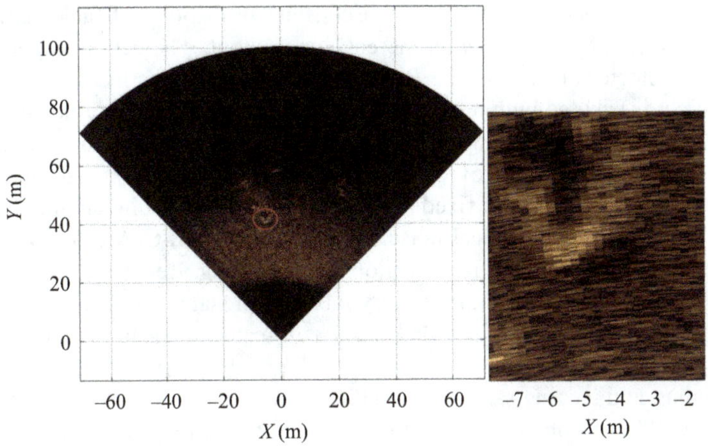

Figure 7.2 Matched filtered beamformed data in x–y reference frame relative to sonar face (in dB)

7.3 Sonar data processing and measurement generation

UUVs are often used to detect and locate stationary objects on the ocean floor. An effective approach for associating multiple active sonar returns from a stationary object is developed. This approach can be used in areas where the seafloor is approximately locally level. Unlike conventional clustering approaches for surveillance systems, the new method directly incorporates estimates of the platform location supplied by an INS/DVL along with the sensor errors. This information is used to eliminate the need for the somewhat arbitrary proximity parameters. In addition, this information allows determination of overlap across multiple pings and accurate distinction between false alarms and stationary objects.

We consider an FLS mounted on the nose of the UUV. High-frequency short sine-wave bursts are transmitted and received on an array with good resolution in the azimuthal dimension and poorer resolution in the vertical dimension. Band-pass filtering, sampling, digitization and beamforming are performed. A representative picture of a generic sonar beam time-series magnitude plot is shown in Figure 7.2.

7.3.1 Detector

Examination of range-beam intensity plots such as Figure 7.2 implies that the data at a fixed (range, beam) point appear, at least visually, as roughly identically distributed in nearby regions. This is due to the slow speed of the UUV platform, the nearly unchanging sound speed characteristics and the similarity of the seafloor and surface interactions. In regions where the seafloor has strong relief, these assumptions may be less reasonable. However, tracking of stationary objects is usually performed in an environment with a relatively flat bottom and these assumptions are quite reasonable.

Most surveillance systems employ a constant probability of false alarm (P_{fa}) or CFAR detector that operates on the envelope output. The cell-averaging (CA)-CFAR [12] detector maximizes the P_d for a fixed P_{fa} when the local noise samples are Rayleigh distributed with unknown variance. The well-known Ordered-Statistic (OS)-CFAR detector of Rohling [13] is a very close approximation to the CA-CFAR detector and has the advantage of robustness to signal contamination of the local noise samples. A bottom feature at a fixed (sample number, beam) point on ping k has the effect of contaminating the local noise sample set. The OS-CFAR detector is thus well-suited to generate the initial detections on each ping. Reliable data association methods (Section 7.4) will eliminate every false alarm that is not spatially consistent over pings. Thus the effective false alarm rate is ultimately much lower than the single ping false alarm rate.

L leading and lagging samples straddling the sample number of the test cell as well as, say, B beams on either side of the beam of the test cell are considered. This leads to a $(2 * L \times B)$ sample set of reference noise samples to be used by the OS-CFAR detector. The test cell is then compared to the threshold based on the 75th-order statistic to declare a detection [13, Equation (14)]. Real-life false alarm performance is often within a factor of two of the design P_{fa}. This implies that the noise distribution and i.i.d. assumptions are somewhat reasonable approximations to reality. Thus, for a user selected design P_{fa} of, say, 10^{-3} and a total of 10^6 time samples across all beams, less than 10^3 usually remain after the detection stage. Figure 7.3 helps visualize the detection output.

An FLS can output beam time-series values at a certain lower-level bit resolution. Methodologies such as [14] that exploit the coarseness of data quantization to reduce overall processing have been considered. Routines that employed both fine histogram (2^b bins for b-bit resolution) and coarser histogram tracking were developed and

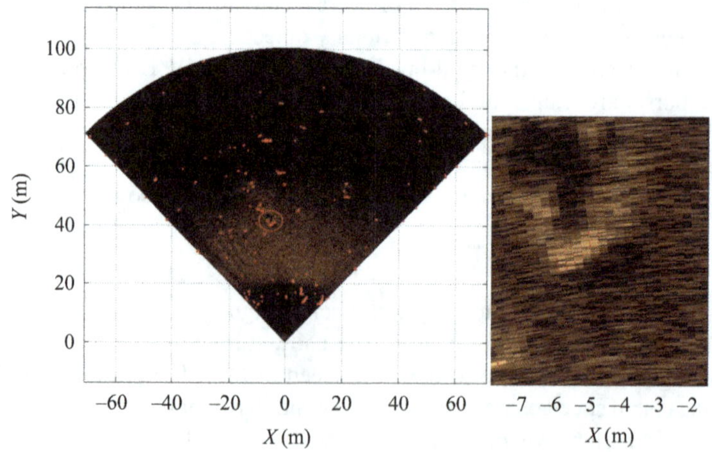

Figure 7.3 Detections overlayed

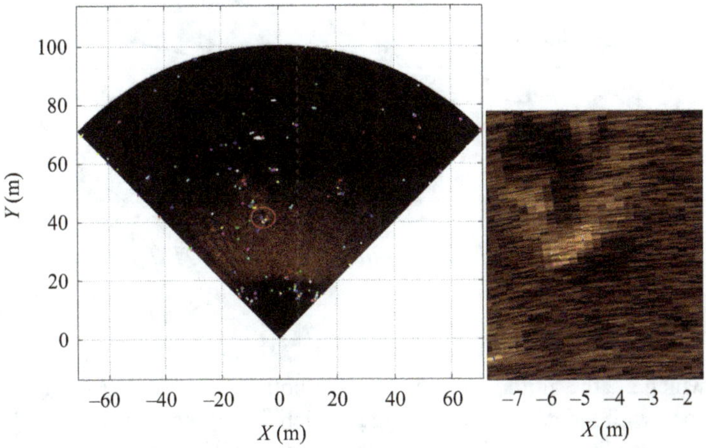

Figure 7.4 Results of connected component clustering and labeling

resulting average run times for the detector stage could be made a small fraction of the time between pings. Ping intervals can be on the order of one second for many FLS systems.

7.3.2 Connected component clustering and labeling

With many detections still flowing from the front-end processing, clustering is an important operation for a sonar-processing system. On each ping, clusters are first formed in the (sample number, beam number) space. This space is viewed as a rectangular binary image with a 1 corresponding to a detection and 0 otherwise. A detection at a certain (sample number, beam number) point is associated to every other detection in an eight-neighborhood. The eight-neighborhood is the eight closest possible detection locations forming a square around the current detection. Clusters formed in this manner can be viewed as regions, often termed eight-connected regions [15, p. 284]. The connected components labeling problem has been extensively studied. Methods related to [16, pp. 28–55] were developed and implemented with run times far less than the time between pings. It is clear that the number of clusters must be less than or equal to the number of detections. Thus if a thousand detections are presented to the clustering module, a few hundred clusters may usually result. In Figure 7.4, each connected cluster is displayed in a different color (cycling through a fixed set) purely to help the reader visually distinguish nearby clusters.

7.3.3 Mapping detections to UTM frame

Each cluster is comprised of a set of connected detections. Each detection must now be associated with a location defined in a WRF. The Universal Transverse Mercator (UTM) frame along with a specific zone is used as our WRF. Figure 7.5 shows a side

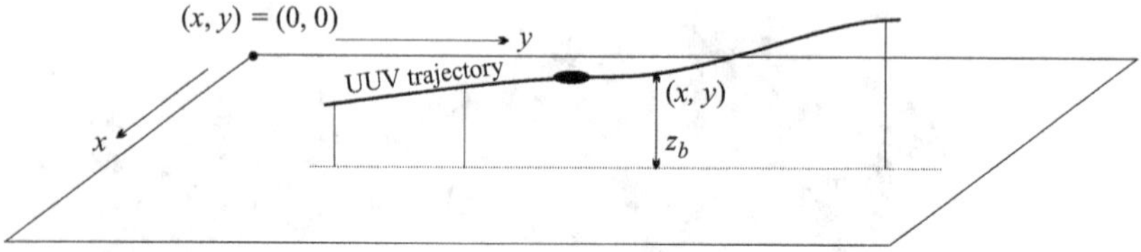

Figure 7.5 Platform motion depicted in UTM frame

view in which z_b, the altitude above the local bottom, is shown. As many vehicles are equipped with Doppler velocity logger which serve to aid the on-board INS, an estimate of z_b is always available especially when the UUV operates close enough to the local bottom.

The sonar we considered here is designed with narrow azimuthal beamwidths and wide vertical beamwidths. At this point, the reader can assume that all detections are from objects on the local bottom. This is a reasonable assumption as persistent features are expected to only arise from bottom features of opportunity. The ping-to-ping data association (Section 7.4) will ensure that only detections from persistent features will be ultimately considered. As the UUV is operating fairly close to the local bottom, a constant speed of sound assumption is reasonable and a straight-line sound path—or pencil beam—from the sonar face to each detection and back can be approximately expected.

7.3.3.1 $\vec{u} \rightarrow \vec{u}_\beta$

Consider the unit length vector $\vec{u} = [1\,0\,0]'$ normal to the face of an FLS array and defined with respect to the body frame. Details of frame conventions are further described in [17]. A pencil beam pointing in a specific vertical and azimuthal direction can be specified by a pair of rotations of $\vec{u}$. Assume that the receive beam is first vertically steered by the vertical depression angle α. The rotation matrix R_α is given by

$$R_\alpha = \begin{bmatrix} \cos\alpha & 0 & \sin\alpha \\ 0 & 1 & 0 \\ -\sin\alpha & 0 & \cos\alpha \end{bmatrix} \tag{7.15}$$

and $\vec{u}_\alpha = R_\alpha \vec{u}$. Next $\vec{u}_\alpha$ is rotated β radians about the body z-axis to the desired azimuthal direction. The rotation matrix R_β is given by

$$R_\beta = \begin{bmatrix} \cos\beta & -\sin\beta & 0 \\ \sin\beta & \cos\beta & 0 \\ 0 & 0 & 1 \end{bmatrix} \tag{7.16}$$

and $\vec{u}_\beta = R_\beta \vec{u}_\alpha$.

7.3.3.2 $\vec{u}_\beta \rightarrow \vec{u}_{NED}$

The beam pointing direction is completely specified with respect to the body frame through $\vec{u}_\beta$. We now wish to specify this direction with respect to the "local" North-East-Down (NED) frame. The INS/DVL system provides attitude measurements—heading (γ), pitch (δ) and roll (ϵ)—describing the orientation of the body with respect to an assumed "local" NED frame.

The "local" NED frame can be rotated to match the body frame. Thus, any vector $\vec{u}_\beta$ defined in the body frame can be redefined in the NED frame as

$$\vec{u}_{NED} = R_\gamma^T R_\delta^T R_\epsilon^T \vec{u}_\beta \tag{7.17}$$

where

$$R_\gamma = \begin{bmatrix} \cos\gamma & \sin\gamma & 0 \\ -\sin\gamma & \cos\gamma & 0 \\ 0 & 0 & 1 \end{bmatrix} \tag{7.18}$$

$$R_\delta = \begin{bmatrix} \cos\delta & 0 & -\sin\delta \\ 0 & 1 & 0 \\ \sin\delta & 0 & \cos\delta \end{bmatrix} \tag{7.19}$$

$$R_\epsilon = \begin{bmatrix} 1 & 0 & 0 \\ 0 & \cos\epsilon & \sin\epsilon \\ 0 & -\sin\epsilon & \cos\epsilon \end{bmatrix} \tag{7.20}$$

7.3.3.3 $\vec{u}_{NED} \rightarrow \vec{u}_{UTM}$

The UTM grid is a projection of the spherical earth onto a cylinder. Each UTM zone is centered on a meridian and defines grid north along it. For points in the zone not along the central meridian, the north and east directions corresponding to the local NED coordinate frame are rotated with respect to grid north and east. The rotation angle η can be readily determined [17, Equation (7)].

The vector $\vec{u}_{NED}$, redefined in the rotated UTM coordinate system, is denoted by $\vec{u}_{UTM}$ and is given by

$$\vec{u}_{UTM} = R_\eta \vec{u}_{NED} \tag{7.21}$$

where

$$R_\eta = \begin{bmatrix} \cos\eta & -\sin\eta & 0 \\ \sin\eta & \cos\eta & 0 \\ 0 & 0 & 1 \end{bmatrix} \tag{7.22}$$

7.3.3.4 $\vec{u}_{UTM} \rightarrow \vec{v}$

Although $\vec{u}_{UTM}$ is defined with respect to the UTM system, we note that the first element of $\vec{u}_{UTM}$ corresponds to the component of the beam pointing ray along grid north, the second element corresponds to the component along grid east and the third element corresponds to the component along the downward z-axis. A final rotation of the coordinate system is required.

$$\vec{v} = R\vec{u}_{UTM} \tag{7.23}$$

where

$$R = \begin{bmatrix} 0 & 1 & 0 \\ 1 & 0 & 0 \\ 0 & 0 & -1 \end{bmatrix} \tag{7.24}$$

Thus, the pencil beam is now coordinatized as "easting, northing, up" in the UTM frame by the vector $\vec{v}$.

At this point, the unit length vector $\vec{v}$ implicitly connects the center of the sonar face to a detection presumed on the local bottom. The slant range r equals $C\tau/2$, where C is the sound speed, f_s is the sampling rate and $\tau =$ sample number$/f_s$. Assuming that all detections are due to bottom returns, the location of the detection in the UTM plane is then given by $(x, y) = (x_0 + rv_x, y_0 + rv_y)$, where (x_0, y_0) correspond to the INS/DVL estimate in UTM coordinates at the time corresponding to the end of the ping transmission. Thus, if $\vec{v}$ is fully specified, the location of detected bottom features projected onto the UTM plane can be estimated without requiring knowledge of local bottom topography.

Bottom returns can arrive over a wide range of vertical angles owing to the poor vertical resolution of the array. For a locally level and smooth seafloor, the angle, ϕ, the receive beam makes with the local tangent plane is fully specified and given by $\tan\phi = z_b/\sqrt{r^2 - z_b^2}$. Note that ϕ is linked to the elements of $\vec{v}$ through

$$\tan\phi = \frac{v_z}{\sqrt{v_x^2 + v_y^2}} \tag{7.25}$$

The left-hand side of (7.25) as well as all rotation matrices except R_α is now specified. Equation (7.25) is nonlinear in α and can be solved in closed form. Solving for α fully specifies $\vec{v}$.

We can now determine the nominal (x, y) UTM location of the detected bottom feature via $(x, y) = (x_0 + rv_x, y_0 + rv_y)$, where (x_0, y_0). In addition, the variability in this location must be studied. This will be accomplished in Section 7.3.4 by accounting for the sources of error.

7.3.4 UTM uncertainty region associated with each cluster

The uncertainty region corresponding to each cluster formed in Section 7.3.2 must somehow reflect each nominal (x, y) location corresponding to each detection in the cluster. The shape and extent of the ultimate uncertainty region for the cluster must also reflect the underlying error in the measurements that were used to determine each nominal location. A process that accurately accounts for these effects is described.

We start with the fact that the INS/DVL system provides estimates of position in geocentric (latitude/longitude) coordinates. Conversion into UTM according to [18] results in (x_0, y_0). The INS/DVL system also provides estimates of the error in terms of variance of these estimates but often in the wander angle frame. The wander angle λ is an arbitrarily chosen reference and is provided by the INS/DVL system [19,20]. This frame is essentially a rotated NED system, or equivalently a rotated UTM system. Rotation is purely about the z (or Down) axis. Assume λ is measured clockwise from true North. The variances along the Wander North Axis and Wander East Axis are denoted by σ^2_{WN} and σ^2_{WE}, respectively. The variance along the z_b dimension is denoted by $\sigma^2_{z_b}$. Define the covariance matrix P_W by

$$P_W = \begin{bmatrix} \sigma^2_{WN} & 0 & 0 \\ 0 & \sigma^2_{WE} & 0 \\ 0 & 0 & \sigma^2_{z_b} \end{bmatrix} \tag{7.26}$$

and the rotation matrix A by

$$A = RR_\eta R_\lambda \tag{7.27}$$

where R_λ is given by

$$R_\lambda = \begin{bmatrix} \cos\lambda & -\sin\lambda & 0 \\ \sin\lambda & \cos\lambda & 0 \\ 0 & 0 & 1 \end{bmatrix} \tag{7.28}$$

Thus, the covariance matrix P of the error about $[x_0\, y_0\, z_b]^T$ is given by

$$P = AP_WA^T \tag{7.29}$$

We assume that these errors are jointly Gaussian distributed. Note that the structure of P will naturally account for any correlation present in P_W.

Estimates of the error in the attitude measurements, C and τ are also utilized. The variance of these errors is denoted by $\sigma^2_\gamma, \sigma^2_\delta, \sigma^2_\epsilon, \sigma^2_C$ and σ^2_τ, respectively. These errors are also assumed jointly Gaussian distributed but need not be mutually uncorrelated. Correlation could also be introduced with the errors about $[x_0\, y_0\, z_b]^T$.

The azimuthal error is defined as uniform over the angular sector defined by the Maximum Response Axes of the adjacent beams, nominally separated by approximately $D°$. This error is assumed statistically independent of the other error sources. Note the vertical angle α is a dependent variable and thus has error dependent on all other sources of error. Thus, the exact statistical characterization of these nine sources of error is elegantly used in a Monte Carlo approach to ultimately define an uncertainty region as follows.

Denote the number of detections in the cluster by M. We will consider each detection $m = 1, \ldots, M$. Considering detection m, our approach is to first generate a realization of the error governed by the joint probability density function (pdf). The joint pdf is the product of the 8×1 multivariate Gaussian pdf and a uniform pdf. This vectorial error realization is then added to the actual values of the nine quantities observed at the time of detection m. The corresponding dithered value of α is derived. Finally, the dithered UTM location $(x, y)_n$ is found according to the process described in Section 7.3.3. The dithered location is indexed by the variable n as this process is repeated N times (i.e. $n = 1, \ldots, N$) to generate a Monte Carlo set of realizations of dithered nominal UTM locations corresponding to the mth detection in the cluster. This is repeated for each of the M detections associated with the cluster to arrive at MN UTM locations.

The sample mean vector and covariance matrix of these MN 2×1 vectors together define the cluster uncertainty region. The corresponding 2σ ellipse is declared as the

UTM Uncertainty Region associated with the cluster to be used in the data association process (Section 7.4).

7.3.5 SLAM measurement and error covariance generation

As stated earlier, the UTM (x, y) location of a detection corresponding to a bottom feature is given by $(x, y) = (x_0 + rv_x, y_0 + rv_y)$, where (x_0, y_0). The corresponding location *relative* to the sonar face in the level plane is then given by $(x_{rel}, y_{rel}) = (rv_x, rv_y)$. The realizations of the error described in the previous section are directly used to generate $(x_{rel}, y_{rel})_n$, $n = 1, \ldots, N$, associated with detection m in a given cluster. The sample mean vector of these MN 2×1 vectors defines a measurement to be used in the SLAM and VOG-SLAM processes. The sample covariance of the MN 2×1 vectors defines the measurement error matrix. Specifically, in the VOG-SLAM process, two such vectors are differenced to act as a measurement to be used for the $k + 1$ update. The corresponding time, denoted t, for this update is the time associated with the later of the two vectors. These vectors are chosen according to the data association process (Section 7.4). This differenced measurement is viewed as noisy observation of the platform velocity at time t scaled by the elapsed time between the two measurements. The corresponding error of the differenced measurement is a function of the two sample covariances as explained in [1, Section III].

7.4 Data association to determine persistent features

The process of properly associating pairs of measurements obtained in Section 7.3.5 to be used in SLAM and VOG-SLAM is as follows:

1. Form clusters on ping 1 according to the procedures in Section III.
2. Form clusters on ping 2. We compare each ping 2 cluster with every ping 1 cluster. We can readily observe that each ping 2 cluster can either intersect no ping 1 clusters or intersects at least one. A ping 2 cluster that does not intersect any ping 1 clusters is placed in another set denoted 2_2. Thus if a ping 2 cluster intersects K ping 1 clusters, the data association code associates the ping 1 cluster with maximum overlap. Ping 2 clusters with intersection(s) are accounted for in set 2_1. If we denote the number of associated pairs in 2_1 as N_2, the VOG-SLAM processing stage will process each such pair sequentially in accordance with the update time associated with the ping 2 cluster in the pair. Ping 2 clusters with no intersection are accounted for in set 2_2.
   We define a persistent bottom feature as possessing a non-empty intersection region over at least 2 contiguous pings. The 2_1 clusters correspond to persistent bottom features that are "on" for *at least* pings 1 and 2. The 2_2 clusters correspond to *potentially* persistent bottom features, i.e. features that are "on" for ping 2 and possibly beyond.
3. Form clusters on ping 3. Note we compare each ping 3 cluster against every ping 2 cluster. Note, a ping 3 cluster could possess a non-empty intersection with an element of 2_1 and an element of 2_2. Again the mapping is chosen according to

maximum overlap. If a ping 3 cluster best overlaps with an element of 2_1, it is accounted for in set 3_1. Similarly if it best overlaps with a cluster in 2_2, it is accounted for in set 3_2. If we denote the number of associated pairs in 3_1 and 3_2 as N_3 the VOG-SLAM processing stage will process each such pair sequentially in accordance with the update time associated with the ping 3 cluster in the pair. Ping 3 clusters that do not overlap any ping 2 clusters are accounted for in set 3_3.

4. The process is repeated until the final ping. Note that the number of sets after each ping grows linearly with ping number. However, the number of elements in each set reduces rapidly. In fact, it can be the case that many of the sets are eventually empty. For example, the set 15_1 corresponds to all persistent features that possess a non-empty intersection over each of the pings 1 through 15. As this event is unlikely, the number of such features is expected to be small. Moreover, any bottom feature can only be visible for a maximum number of pings determined by the UUV speed, the receive sonar coverage and location of the bottom feature. Figure 7.6 depicts these effects for the first ten pings of a VOG-SLAM process.

The description above in terms of sets is meant to reveal the inherent variability in the types of persistence features that can be observed in a shallow water environment. It is important to note that only the mapping from each ping $p+1$ cluster to the best ping p cluster (if any) must be stored for the SLAM or VOG-SLAM processes. This requires a much simpler data structure than described above. Run times for the ping-to-ping data association process amount to a small fraction of the ping interval. The

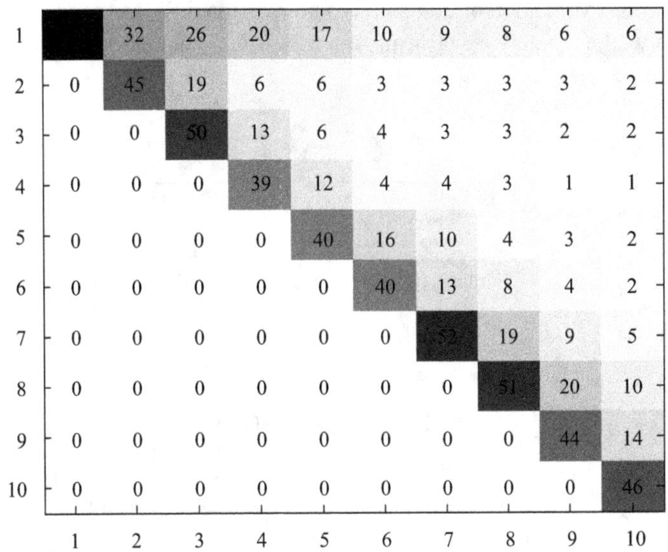

Figure 7.6 Elements in each set P_q after data association. Here N_2 equals 32, N_3 equals 45 (26 + 19)

combination of steps performed in Sections III and IV can still be completed before the arrival of the next ping.

7.5 Performance example

It is important to note that a short transit when the vehicle is most likely to maintain constant course and speed is ideal for studying the achievable platform localization performance of the VOG-SLAM method as compared to the results from an INS/DVL unit on board the UUV (see Figure 7.7). Note that the INS/DVL estimates are not to be used in the VOG-SLAM process. Only the initial position information at the start of the event is used similar to z_p in Section 7.2.1. Specifically, z_p is the MAP estimate of the platform location prior to any sonar-derived measurements. The prior estimate of platform velocity is assumed as zero and the prior covariance of the platform state is a compound covariance matrix with errors given by the INS/DVL system in position and velocity errors assumed unbounded. Thus, velocity information during the run must be learned purely from the sonar-derived measurements. A total of 4,266 updates were performed over a 78-s interval with a ping rate of 1 Hz. This can be obtained by the sum $1 + N_2 + N_3 + \cdots + N_{78}$. This number may seem high but similar results are commonly observed in a bay environment in which accidental (e.g. lobster pots, debris, refuse and relief) landmarks can serve as persistent features. The number of persistent features of opportunity can vary across environments and drives the complexity of the method. However, even in a rich feature environment the implementation can allow for real-time operations. Both the simplicity and optimality of the VOG-SLAM method as compared to conventional SLAM (and even its faster variants) as well as the efficient means to process and data associate the FLS beam data play a part.

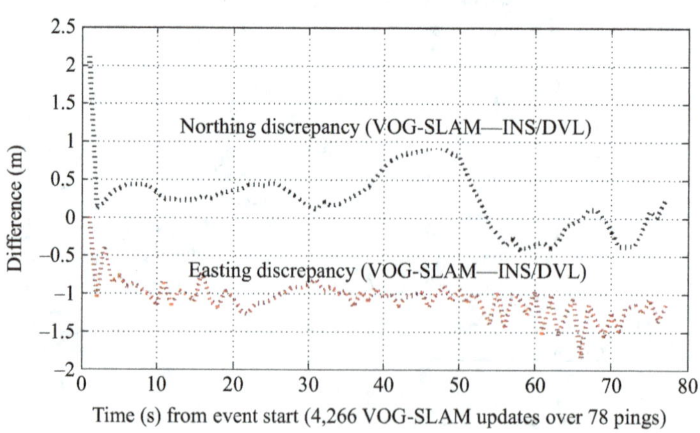

Figure 7.7 Results of VOG-SLAM for short transit run in Narragansett Bay, RI

Results reveal that position estimates from the VOG-SLAM approach can closely track those from a DVL-aided INS system used here as ground truth. Using the DVL-aided INS as ground truth so early after GPS-initialization (or mission start) is a reasonable choice. Note, however, that although a DVL-aided INS system will help to retard the positional error growth rate as compared to an unaided INS system, the positional error will grow over the mission. This is not only due to the errors in the local speed estimates but also in the bias and noise of the accelerometers and gyroscopes and initial heading.

7.6 Conclusions and future work

In [1], an argument was made that the VOG-SLAM method could approximately match the power of conventional SLAM with a fraction of the complexity when the vehicle is performing constant course and speed behavior—the desired behavior in long transits. Here we argue the near-optimality under that behavior from a simpler maximum likelihood perspective. We also suggest a possible sonar signal-processing method for rapid data compression and data association for use in all SLAM systems.

One observation is that, unlike a DVL-aided INS system, the VOG-SLAM method can utilize a group of features, perhaps deliberately emplaced along a known track, to result in potentially lower error growth rates provided the vehicle attempts constant course and speed behavior. Specifically, observation and data association of these emplaced contacts can allow the vehicle to more quickly converge to accurate estimates of its own speed along the assumed heading when traveling over a level ocean floor.

Although the spatially localized clusters are here determined by accounting for the inherent errors of the sensors, their finite extent and implicit dependence on the DVL-aided INS outputs in their creation must lead to incorrect declarations of persistence and data association. Due to the careful CFAR design of the signal processing, this is expected to have a weak effect when the vehicle obeys the constant course and speed behavior. However, this can lead to a biased velocity estimate when the vehicle deviates from this behavior. Note, DVL systems can ultimately also only most reliably provide speed along the current vehicle body axis and not velocity. Note that neither system can provide velocity with respect to a WRF. Thus, VOG-SLAM suffers from a similar drawback to DVL when the vehicle deviates from the constant course and speed behavior or travels over non-level surfaces. Future work is concerned with a means to analyze and compartmentalize the errors in VOG-SLAM and DVL that most contribute to false data association and incorrect persistence declaration when the vehicle deviates from the preferred behavior. It is crucial to note that conventional SLAM is not immune and must also reflect a loss every time vehicle behavior does not match any of the kinematic expectations that govern the assumed Kalman platform state evolution equation.

In addition, future work is also concerned with optimal ways to bound the positional error growth rate of VOG-SLAM and DVL systems. This is important as, even after GPS-initialization, the initial estimate of vehicle heading will be biased and no "velocity"-aiding can remove this bias. This bias will lead to positional error growth even assuming error-free outputs from the accelerometers, gyroscopes and derived estimates of speed either along the body axis when under the constant course and speed behavior. This growth is akin to a cross-range error.

References

[1] A. Sarma, "CML/SLAM studies and 'velocity-over-ground' SLAM," in *Proceedings of the IEEE/ION Position Localization and Navigation Symposium (PLANS), Monterey CA*, 2014.

[2] A. Sarma, A. Brown, A. Palmer and D. Prowse, "Novel detection, localization and registration methods for persistent contacts," in *Proceedings of the Underwater Acoustic Measurements (UAM) Conference, Kos Greece*, 2011, pp. 323–330.

[3] R. Smith, M. Self and P. Cheeseman, "Estimating uncertain spatial relationships in robotics," in *Autonomous Robot Vehicles*, I. Cox and G. Wilfong, Eds. Springer-Verlag, 1990, pp. 167–193.

[4] J. J. Leonard and H. F. Durrant-Whyte, *Directed Sonar Sensing for Mobile Robot Navigation*. Boston: Kluwer Academic Publication, 1992.

[5] J. J. Leonard, R. N. Carpenter and H. J. S. Feder, "Stochastic mapping using forward look sonar," *Robotica*, vol. 19, pp. 467–480, 2001.

[6] R. N. Carpenter and M. R. Medeiros, "Concurrent mapping and localization and map matching on autonomous underwater vehicles," in *Proceedings of the MTS/IEEE Oceans Conference, Honolulu, HI*, 2001.

[7] M. Kaess, A. Ranganathan and F. Dellaert, "iSAM: Incremental smoothing and mapping," *IEEE Transactions on Robotics*, vol. 24, pp. 1365–1378, 2008.

[8] V. Nguyen, A. Harati, A. Martinelli, R. Siegwart and N. Tomatis, "Orthogonal SLAM: A step toward lightweight indoor autonomous navigation," in *Proceedings of the IEEE/RSJ International Conference on Intelligent Robots and Systems (IROS) 2006*, 2006.

[9] L. Paull, M. Seto and J. J. Leonard, "Decentralized cooperative trajectory estimation for autonomous underwater vehicles," in *Proceedings of the IEEE/RSJ International Conference on Intelligent Robots and Systems (IROS) 2014*, 2014.

[10] Y. Bar-Shalom and X.-R. Li, *Estimation and Tracking: Principles, Techniques and Software*. Artech House: Boston, 1993.

[11] L. L. Scharf, *Statistical Signal Processing: Detection, Estimation and Time Series Analysis*. Addison-Wesley: Reading Massachusetts, 1991.

[12] H. M. Finn and R. S. Johnson, "Adaptive detection mode with threshold control as a function of spatially sampled clutter-level estimates," *RCA Review*, vol. 29, pp. 414–464, 1968.

[13] H. Rohling, "Radar CFAR thresholding in clutter and multiple target situations," *IEEE Transactions on Aerospace and Electronic Systems*, vol. 19, pp. 608–621, 1983.

[14] T. Huang, G. Yang and G. Tang, "A fast two-dimensional median filtering algorithm," *IEEE Transactions on Acoustics, Speech and Signal Processing*, vol. 27, no. 1, pp. 13–18, 1979.

[15] R. Duda and P. E. Hart, *Pattern Classification and Scene Analysis*. John Wiley and Sons: New York, 1973.

[16] R. M. Haralick and L. G. Shapiro, *Computer and Robot Vision—Vol. 1*. Boston: Addison-Wesley Longman Publishing Co., Inc. 1992.

[17] A. Sarma, "Maximum likelihood estimates and Cramer–Rao bounds for map-matching based self-localization," in *Proceedings of the MTS/IEEE Oceans Conference, Vancouver CA*, 2007.

[18] Department of Army, *Universal Transverse Mercator Grid, U. S. Army Technical Manual TM 5-241-8*. U.S. Army, 1973.

[19] M. P. Grewal, L. Weill and A. Andrews, *Global Positioning Systems, Inertial Navigation and Integration*. New York: Wiley-Interscience, 2000.

[20] C. Jekeli, *Inertial Navigation Systems with Geodetic Applications*. Walter de Gruyter: Berlin, 2000.

Part III

Toolset

Chapter 8

Toolchain for autonomous systems and its use in real-life scenarios: from science to defence

Paulo Sousa Dias[1], José Pinto[1] and João Borges Sousa[1]

8.1 Introduction

The ocean is an essential support system for most if not all Earth ecosystems, but in spite of this, it still remains for the most part unexplored. It covers more than 70 percent of the Earth's surface [1] and has a big impact, for example, in the global weather patterns by distributing the absorbed sun's heat around the world via its ever-moving ocean currents [1]. The ocean depths still hold lots of mysteries that we are still getting knowledge of. Pockets of life appear in extreme environments like around hydrothermal vents where chimney-like structures spew gases and mineral-rich water from beneath the Earth's crust. This becomes a feeding place for organisms like tube worms, clams, and mussels that feed on heat-loving bacteria [1].

These environments pose some challenges for its exploration. Traditional tools involve long cruises on-board research vessels with seasonal expeditions where the amount of time, assets, crew, and scientists escalate the costs to a very large sum. Also, some characteristics may be eluded from scientists by not having a continuum phenomenon observation.

We see an increased usage of robotic exploration in these cruises. A vessel like the R/V Falkor from Schmidt Ocean Institute [2] is a good example. Robotic exploration allows easier access to difficult places; the deployment of complex sensors; the sample collection; and long periods of exploration; among others. Despite a decreasing number of available research ships, the new generation ships, like the future REV from Norwegian patron Kjell Inge Røkke or OceanX's M/V Alucia2 from patron Ray Dalio, offer more and more robotic facilities [3].

One of the most versatile robotic assets for oceanographic exploration is the autonomous unmanned vehicle. Autonomous and unmanned are the keywords. These vehicles can be launched from ships or from shore and operate autonomously spanning from a few hours to some months. These vehicles operate on the surface of the water

[1]LSTS – Laboratório de Sistemas e Tecnologia Subaquática (Underwater Systems and Technology Laboratory), University of Porto, Faculty of Engineering, Porto, Portugal

(ASVs*), underwater (AUVs†, gliders, or crawlers), and on the air (UAVs‡). These robotic vehicles can be used to complement by extending the reach of the research vessels, or in some situations to replace them altogether. This means that in some cases these vehicles operate for long periods away from human reach. So they need to have some autonomy to react to observations that its sensor measure.

Nowadays there are lots of options in the market, with plenty of configurations of capabilities and payloads, where the manufacturers provide a software interface where one can programme the vehicle and command and control what it is doing. Once people get a hand on these systems and start to operate it, they realise that perhaps they need more sensors and are able to buy another vehicle, or it is interesting to operate with other people with a slightly different vehicle. Great idea! Unfortunately, the vehicles are from different vendors, and we cannot see the data and position of one another. This is for the most part what happens because each vendor has his own interfaces, both for hardware and software. Fortunately, throughout the years, this situation has been somewhat mitigated since vendors started providing APIs§ to access to the vehicles as back-seat drivers. But for the most part, there are still per vendor APIs.

As things stand, we felt a need to start an effort to build a toolchain that was able to interface with such a vast variety of systems. The LSTS (Laboratório de Sistemas e Tecnologia Subaquática (Underwater Systems and Technology Laboratory)) Toolchain for Autonomous Systems addresses the following problems:

- a modular toolchain that was able to be used as a whole or just parts;
- an on-board control software to fully control the vehicle or to plug-in as a back-seat driver;
- a message interface definition with flexible serialisation options;
- a command and control software that
  - can be extended through plug-ins (visual, data processing, or communications);
  - can display vehicles and other assets on a geo-referenced map;
  - can be used to plan a mission to a vehicle, with feasible manoeuvres and payload;
  - can adjust to display telemetry and data arriving from vehicles;
  - has a display where visualisations and data processors can be flexibly laid out;
  - can be used to display mission collected data, as well as extendable data exporters.

8.2 The toolchain

The LSTS Toolchain for Autonomous Systems [4–6] started its development at the end of 2004 with the start of the command and control software Neptus. This development

*ASV – Autonomous Surface Vehicle
†AUV – Autonomous Underwater Vehicle
‡UAV – Unmanned Aerial Vehicle
§API – Application Programming Interface

started with the need to reduce errors on the planning of the missions of, at the time, Woods Hole Oceanographic Institution's REMUS AUV. This REMUS was acquired in 1997 and named Isurus [7]. The planning interface was very rudimentary, with a text interface for mission plans creation. It was sometimes easy to introduce errors while typing or copying and pasting coordinates.

Around that time, LSTS was already in the process of creating an AUV from the ground up [8]. This provided freedom to modify, try new sensors, or experiment with vehicle navigation and control software. To accomplish that, an on-board control software was in need, and that was when in 2006 the toolchain on-board control software was born.

With the LSTS's creation of the LAUV (Light Autonomous Underwater Vehicle), a light and flexible AUV [8], a complete product line was established. The LAUV went to market with an LSTS's spin-off company around 2007 (Oceanscan MST), which imposed a stable product of the toolchain. Despite the focus on the LAUV as a product, we never closed or restricted the toolchain development direction. We somewhat used it as a test bed.

The LSTS Toolchain for Autonomous Systems is composed of the following components:

- IMC (Inter-Module Communication) that is the definition of a message protocol;
- DUNE (DUNE Uniformed Navigational Environment) that is the on-board control software;
- Back-seat driver control that allows the control of a vehicle through a message API;
- Neptus Command and Control software, to command and monitor the autonomous vehicles operation;
- Ripples that is a web interface and REST[‖] API for command and monitor autonomous vehicles.

8.2.1 Communication

The IMC [9] is a message-based protocol that defines standard messages to interconnect the different on-board subsystems, the vehicles with the command and control software, and also vehicle to vehicle. IMC has similarities with other protocols like the Google's Protocol Buffers or Apache's Thrift. The main difference of IMC is that it was conceived to target communication-challenged networks with heterogeneous transports.

IMC is currently being used for communication between nodes using IP protocol working with both UDP (User Datagram Protocol) and TCP (Transmission Control Protocol). Additionally, it is used with acoustic modems, RF[¶] telemetry radios, GSM[**], and Iridium. The availability of these communication transport protocols depends on the vehicle and on which mean the vehicle is (underwater, water surface, or air) at any given time. So one or more can be used to transmit and receive IMC messages.

‖REST – REpresentational State Transfer

¶RF – Radio Frequency

**GSM – Global System for Mobile Communications

```
<message id="251" name="Voltage" abbrev="Voltage" source="vehicle"
  ↪  flags="periodic">
  <description>
    Report of electrical voltage.
  </description>
  <field name="Measured Voltage Value" abbrev="value" type="fp32_t"
   ↪  unit="V">
    <description>
      The value of the internal electrical voltage as measured by the
    ↪  sensor.
    </description>
  </field>
</message>
```

Listing 8.1 Voltage IMC message XML definition

The definition of the messages is by an XML†† file. The rules for its definition are described in an XSD‡‡ file where the structure is defined and checked. These files are exposed on LSTS's GitHub [10]. IMC definition is composed of the following parts:

- field types;
- a description of the serialisation (mainly for documentation);
- the header and footer definition;
- the allowed units for the fields;
- global enumerations and bit fields;
- definition of message groups; and
- definition of the messages.

Listing 8.1 shows the voltage message definition. As said above, the header and footer are declared in another section, with the listing showing just the body. It defines a message ID and name, and for each field its type and name. The descriptions are used for the documentation that is generated from the definition XML.

IMC has an addressing scheme that is independent of the transports, making it possible to establish the communication session. For example, messages from a system can be arriving by Wi-Fi and switch to acoustics, and because of the source persistence, the receiver can know from whom it came and keep track of the best mean to connect with the system.

The serialisation format is for the most part binary (LSF§§ format described in the specification) but can also be JSON‖‖, or other custom transport dependent. There are a couple of official bindings for Java/Android, and C++ programming languages, but bindings for Python or Dart/Flutter can also be found.

††XML – eXtensible Markup Language
‡‡XSD – XML Schema Definition
§§LSF – LSTS Serialisation Format
‖‖JSON – JavaScript Object Notation

The current version of IMC has 305 messages defined, distributed by several groups that match the layered control architecture we have for the toolchain. Adding custom messages is the way to extend the base messages. These custom messages have IDs higher than 1,000 with no guarantees that all systems support them. They serve the purpose of testing for latter submission for integration into the IMC base message set.

8.2.2 DUNE – on-board control software

DUNE Unified Navigation Environment is the *on-board control software* and responsible for *logging*, *control and navigation*, and *sensor interfacing* of autonomous platforms. It is written in C++ and composed of a *core set of libraries* that makes external dependencies to a minimum allowing a better *portability*. For the moment it complies with support for C++14. DUNE is portable but maintains its flexibility. It may be set up to control a simple communications gateway (Figure 8.1) that works as a bridged for several communications means (Wi-Fi, acoustics, GSM, Iridium, etc.), or set up to control an autonomous vehicle.

DUNE supports *several CPU architectures* with the most tested being the Intel x86 compatible and ARM, but others like Sun SPARC, PowerPC, and MIPS can be used. It is also *OS independent* with the possibility to use in Linux, Apple macOS, Microsoft Windows 2000 or above, and less used ones Sun Solaris, FreeBSD, NetBSD, OpenBSD, eCos, RTEM, and QNX Neutrino.

The core logic is supported by the notion of a *Task*. These tasks are organised into several modules that serve several purposes spanning from logging, control and navigation, guidance, mission control, sensor interface, and communication transports. In essence, a *task can be viewed as a thread*, either periodic or event triggered. The idea is to follow the philosophy of the UNIX [11] where one task does one job only and does it well. The *output of one task can be the input of potential other tasks* for further processing. DUNE bases itself on an *asynchronous message bus* mechanism,

Figure 8.1 Manta Gateway with visible Wi-Fi antenna and acoustic modem

which allows IMC messages to be produced and consumed from the bus. By the *composition* of these simple tasks, *complex systems are easy to build.*

DUNE follows a *layered control architecture.* Close to the hardware, we find the *sensors and actuators tasks*. These connect the hardware to the upper layers. The next level comes with the *guidance tasks* where the desired path is translated to actuation commands. *Navigation tasks* consumes the information from the sensors and output the vehicle's location and attitude. *Manoeuvre tasks* are responsible for controlling the vehicle in higher-level behaviours. They operate on top the guidance tasks generating the desired path. Above, the *plan control tasks* trigger the execution of manoeuvres according to a mission plan. On top of all, the *vehicle monitor and supervisor tasks* monitor the sanity and safety of the vehicle for operation. A mission plan typically, in DUNE, is composed of a sequence of manoeuvres, ran by the plan control tasks. Figure 8.2 depicts the control layer for an AUV focusing on plan execution.

The upper layers modules (plans, manoeuvres, guidance, autopilot, allocator, and actuators) are *parametrisable* to run specific implementations. These configurations are created by INI files[¶¶] that can import other INI files for easy reuse. DUNE starts with an *INI configuration and a profile*. The INI describes the vehicle, indicating the tasks to start and their configurations. A task has a profile associated, meaning that it only starts if the appropriate profiles are at DUNE start command. Profiles are used, for example, to start tasks only if the hardware is connected and others if in simulation. This way, hardware tasks are replaced with their simulated equivalent enabling a *simulated vehicle with little effort* both for SIL[***] or partial HIL[†††] simulations.

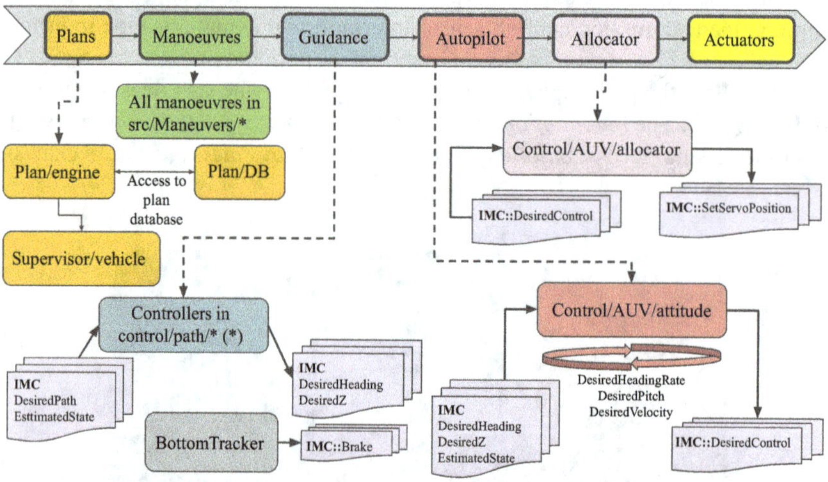

Figure 8.2 DUNE layered control architecture (AUV)

[¶¶]INI files are text-based configuration common format with sections, properties, and values.
[***]SIL – Software in the Loop
[†††]HIL – Hardware in the Loop

DUNE has dedicated transport for logging messages passing the communication bus. As stated, the messages passing the bus are IMC, and so the log is a serialisation of those messages in sequence. Which messages to log, and their rate, is a matter of task configuration. A new log is created with a new plan start or by request with a specific IMC message. The log is usually compressed and has the name of *Data.lsf.gz*. Additionally, other sensors use the log-related IMC messages to be able to create their specific data logging in the same folders.

8.2.3 Extensions to the toolchain

DUNE already provides an expressive planning framework, where manoeuvres are sequenced and actions can be specified during the activation of manoeuvres (e.g. to change payload configuration). However, there are times when more interesting or advanced behaviours are required. DUNE provides support for safe high-level behaviour customisation via safety supervisors and a back-seat driver API.

The back-seat access is done by the implemented FollowReference manoeuvre and by consuming its state and of the vehicle. The back seat, when activating this manoeuvre, puts DUNE in external control mode. The back seat sends desired reference messages, which makes the front seat to track that moving reference.

The safety supervisor can be configured with a set of operational limits that restricts what external controllers are able to do. If the safety supervisor detects anything outside, these limits will force the controller to stop and bring the vehicle back to idle state. Currently the operational limits include minimum altitude (distance to sea bottom), maximum depth, and area boundaries. Whatever happens inside this 'sandbox' is allowed, but as soon as the vehicle gets out, it will be stopped.

8.2.4 Neptus – command and control framework

Neptus is the command and control software of the toolchain. Neptus is built as a framework and so has the flexibility to adapt to different systems, even if they don't speak the same language.

Neptus is built in Java and supported to run on Linux and Microsoft Windows 2000 and above. Apple macOS is not fully supported but is functional (apart from some external libraries dependent code).

This C2 (Command and Control) framework supports all typical mission life cycles: planning, simulation, execution, and review and analysis. Different interfaces serve parts of the life cycle. The two main interfaces are the operational Console (or simply Console) and the Mission Review & Analysis, described in the following sections.

Different components are provided to operate in these two interfaces. Plug-ins are used to extend both main interfaces. These plug-ins can be visual components, build support for other types of communication layers, or connectors to an external source of data, for instance.

Table 8.1 Example of Services from Announce IMC Message

Service	Description
dune://0.0.0.0/uid/7389082285	Unique ID for detecting reboots
dune://0.0.0.0/version/2018.07.00-rc1	Version info for DUNE
ftp://127.0.0.1:30021/	FTP service for logs download
http://127.0.0.1:8080/dune	Web page for DUNE status
imc+info://0.0.0.0/version/5.4.11	Info of IMC version
imc+udp://127.0.0.1:6002/	IMC UDP IP and port used
imc+udp://127.0.0.1:6002/acoustic/operation	IMC ACOMMS[1] service through UDP
imc+udp://127.0.0.1:6002/iridium	IMC SAT[2] service through UDP

[1]Acoustic communication.
[2]Satellite communication.

8.2.4.1 Communications

Like in DUNE, Neptus also uses IMC as the main communication protocol but is easily extensible to support others. It implements IMC transports for TCP and UDP based on generic TCP and UDP transports. The IMC message manager takes care of the connection with the vehicle and keeps track of the capabilities that each system announces.

This mechanism of announcing is done by periodically sending an Announce IMC message. It is expected to arrive by a UDP multicast or broadcast message (usually with a periodicity of 10 seconds). The Announce message informs of the system name and type, its location, and its available services. These services are in the form of URLs[‡‡‡] and publicise the system's communication mean that are available, e.g., IMC over UDP or TCP with IP address and port number. Table 8.1 shows examples of such URLs. You can see not only ways to connect to the system but also services that allow their use by others. Every system can then extend its capabilities without the need for extra hardware. The Manta Gateway from Figure 8.1 is used to do such capabilities extension for the operator consoles. Neptus, which usually runs on a laptop computer in the field, can use Manta's acoustic, GSM, or Iridium modems without having the laptop USB ports full of devices hanging around.

8.2.4.2 Console

The *Neptus Console* is the interface used for interacting with autonomous systems. See it as an empty canvas where console panels, interactions, or layers are added and laid out. At its base, the console offers a communication bus that allows *publish-subscribe-style communication*. IMC messages travel on the bus, but any POJO[§§§] can be published and subscribed. For *IMC, a message manager* takes care of all communication and addressing details, publishing on the communication bus the received messages. This manager is available for all console elements to use for

[‡‡‡]URL – Uniform Resource Locator

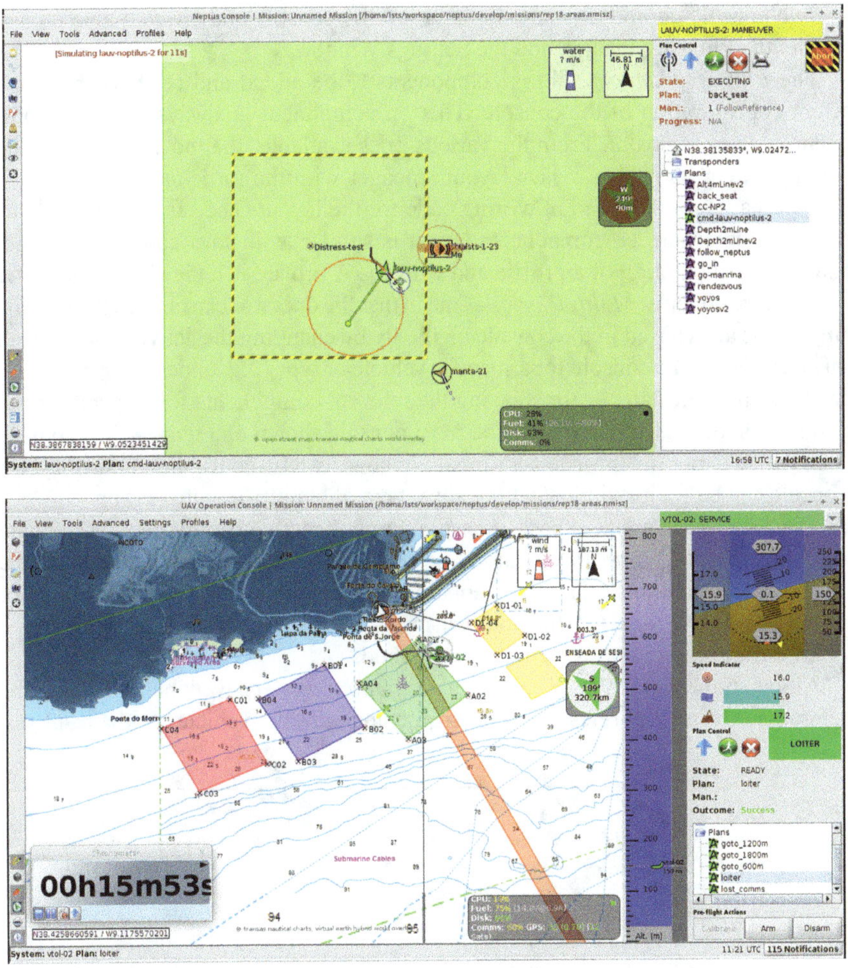

Figure 8.3 Console configurations examples (AUV and UAV operations)

sending messages. The *communication bus* is also utilised to pass some internal events, and it can be used to introduce additional communication protocols other than IMC.

XML files *store consoles configuration and layout*. Out of the box, Neptus already offers some consoles customised to different operations. In Figure 8.3, two different examples of console configurations for use on underwater and aerial operations is shown.

§§§POJO – Plain Old Java Object

Two console components are the base for any console and are available on any console: the Plug-in Manager and the Settings Window.

The *Plug-in Manager* allows components to be added and removed from the current configuration of the console. The top component is always a special component, a *console layout.* From the various ones available, the most versatile is the MigLayout. On its base is a Java layout manager with the same name [12], but an XML format is used instead of writing code for the layout. The same rules apply as the Java counterpart, so components layout is similar as in Java coding. The main additions are the notion of a profile and window. A profile is a view with a specific layout of components. *Multiple profiles* mean that the operator can change the way the components are viewed in the console on the fly by changing the active profile. Each profile can adapt the console for a specific vehicle or type (e.g. underwater vs aerial), vehicle or mission state, or for showing the relevant components for a specific task. Listing 8.2 shows a MigLayout for the AUV console shown on Figure 8.3. A console is one window, but in a profile additional windows can be configured, meaning that a console can be multi-window taking advantage of a multi-monitor set-up.

The *Settings Window* exposes console components configurations on a single window to be changed by the operator. The developers of the components expose the available configurations either by annotating a Java field with *@NeptusProperty* or overriding the *getProperties* method. Customisation or aggregation of fields into a single configuration implies the override method.

All components on a console have a *notification system* available. This way, the operator is informed of relevant events happening with the vehicles or the results of

```
<profiles>
  <profile name="Normal">
    <container layoutparam="ins 0" param="w 100%, h 100%">
      <child name="Map Panel" param="w 100%, h 100%"/>
    </container>
    <container layoutparam="ins 3" param="w 300px!, h 100%">
      <container layoutparam="ins 0, filly" param="w 100%, h 100%">
        <child name="Plan Control" param="w 80%, h 50px!"/>
        <child name="Abort Button" param="w 20%, h 50px!, wrap"/>
        <child name="Plan Control State" param="w 100%!, h 100px!,
         ↪  span, wrap"/>
        <child name="Mission Tree" param="w 100%, h 100%, span"/>
      </container>
    </container>
  </profile>
  <profile name="Map">
    <container layoutparam="ins 0" param="w 100%, h 100%">
      <child name="Map Panel" param="w 100%, h 100%"/>
    </container>
  </profile>
</profiles>
```

Listing 8.2 Console MiGLayout XML definition example

tasks that run on the background. The notifications sent are revealed on the bottom right of the console as a pop-up dialogue.

Vehicles send their positions which calls for a map component. In Neptus, this component is the *Map Panel*. It provides a canvas with a map layer that allows the choice of various world maps, or world map layers (used on top of another map). Examples of world maps are OpenStreetMaps, GoogleMaps, VirtualEarth, and S57 nautical charts, among others. For the layers, examples are Open Sea Map or nautical charts. Plug-ins are the mechanism to add additional maps or layers. The Map Panel also allows other console elements to paint themselves on the map, or interact with it with the mouse clicks. This Map Panel is the base for any map base component such as

- *Map editor* to add map elements, e.g. marks, polygons, paths, and images;
- *Plan editor* to edit mission plans directly on the map;
- *Depth Soundings Interaction* to add soundings data from a file or S57 Charts;
- *Control interactions* to send motion commands to the vehicles by clicking on the map. Examples are the following moving reference, or the command simple plans (e.g. go to position, loiter here, dive here);
- *Overlay painters* that show geo-reference data on the map. Examples are the systems list with vehicles present and past positions, vehicle status, Automatic Identification System (AIS), or GPS information; and
- Other can be customised through the plug-in interface.

8.2.4.3 Mission review and analysis

The *Mission Review & Analysis (MRA)* is the interface to interact with the collected data. The interface (Figure 8.4) has two panels: the data tree and visualisations list; and the active visualisation panel. The DUNE logs are the main source of data for this

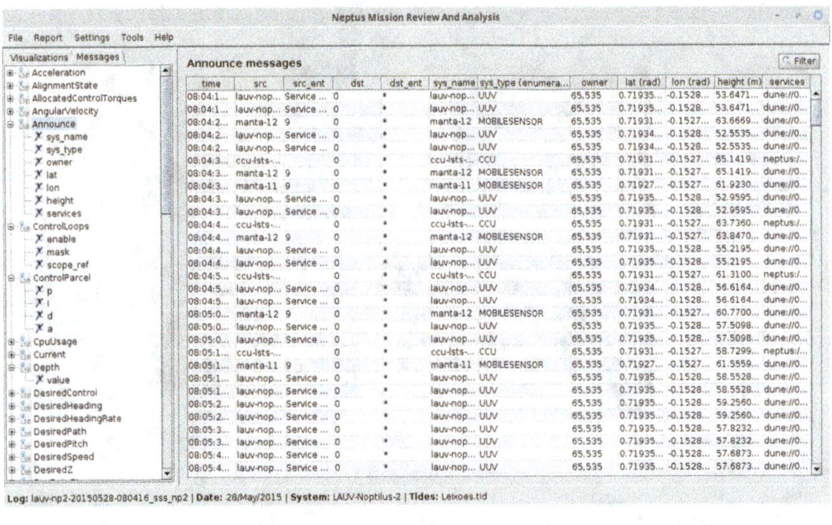

Figure 8.4 MRA data tree

application. As explained in the DUNE section, the logs are stored in a new folder as the mission plan changes. Stored on these folders, we find serialised logged IMC messages and other sensors output vendor data.

The *data tree and visualisations list* has two tabs: the messages and the visualisations. The messages tab exposes the *messages data tree*. This tree (left side of Figure 8.4) shows all the IMC messages stored by the vehicle during the log duration. This tree structure has messages on the branches and fields on the leaves. A double click on a message shows a table with all the messages of that type (right side of Figure 8.4). By selecting one or more fields (either from the same or different messages), different plots can be created.

The second tab is the *visualisations* (left part of Figure 8.5). This tab lists the available data visualisations for the data accessible from the log (and active on the settings), as well as the open data tables and plots.

Markers are a feature shown in Figure 8.5. Operators can create markers on the visualisations, or be automatically generated. A marker is a time label that all open visualisations can utilise to display on themselves in a suitable matter.

The *active visualisation panel* displays the current visualisation. A *map panel* is available in the same way as in the console. Several map layers are activated dependent on accessible data. The MRA map panel has the same world maps as the ones on the Neptus Console.

Several *data exporters* can be used to transform the data into other formats. Out of the box, several exporters are offered, e.g.: CSV; MATLAB®; netCDF; JSON; KML (trajectory and sensor image tiles); and TID (tides format).

A pool of visualisations is already available for the most common data that autonomous vehicles collect. This pool consists of photos or videos visualisers,

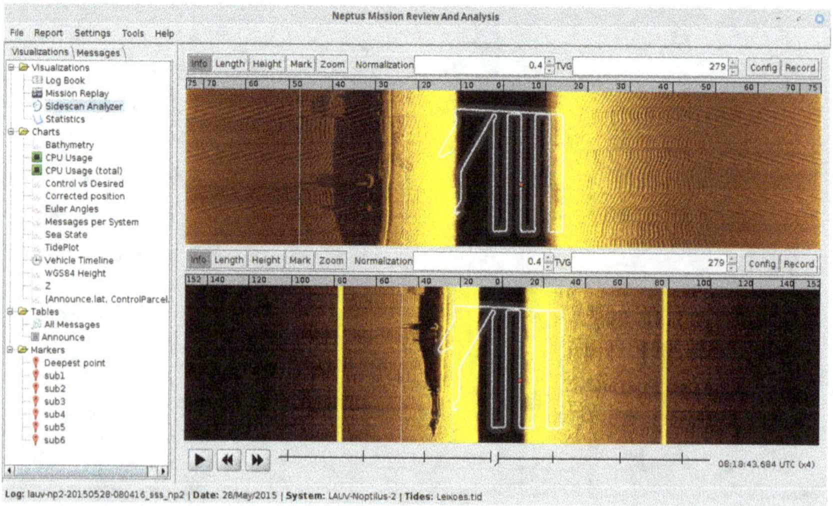

Figure 8.5 MRA Edgetech SSS Visualisation of the World War II German submarine U-1277

colourmaps data overlays, multibeam sonar data (Imagenex DeltaT), or side-scan sonar (e.g. Imagenex YellowFin, Edgetech JSF format, or Klein SDF) like the one shown at Figure 8.5.

At the end of the data review, a small report creation with a summary of the mission plan, data collection, and mission statistics is possible.

8.2.4.4 Plug-in interface

Neptus offers a plug-in interface that enables the functionality extension. A plug-in materialises as an autonomous JAR (Java ARchive) that placed on the plug-in folder is loaded and available for use. A plug-in can implement different functionality as follows:

- Console plug-ins:
  - *ConsolePanel* to create panels (visible or not) for the console, it can also be a layer and/or interaction;
  - *ConsoleLayer* to create a layer painter on the Map Panel;
  - *ConsoleInteraction* to create a mouse interaction on the Map Panel;
  - *IPlanFileExporter* to create a plan exporter to different format;
- MRA plug-ins:
  - *MRAVisualization* to create a visualisation panel;
  - *MRAExporter* to create a data exporter;
  - *LogReplayLayer* to create a layer or replay layer on the Map Panel. A replay means data being fed to the component by time flowing;
  - *LogReplayPanel* to create a similar component as the previous, but opening on a dialog window; and
- World Map plug-ins:
  - *MapTileProvider* to create a map tile provider, or a map painter for the World Map layer on the Map Panels.

If, for example, you want to create a component able to interact with other message protocol other than IMC, creating a ConsolePanel plug-in is the way to go. You can use the communication bus to pass along the received messages.

At the moment, a plug-in should be self-contained, meaning it may only depend on the core code, and not on any other plug-in. The plug-in JAR must have a file named *plug-ins.lst*, in which it's listed all the available plug-in components it provides. These components must extend one of the available functionality on the previous list.

8.2.4.5 Data structures

The following data structures are used to store and hold data related to mission and plans, map elements, or vehicles configurations. These data structures are XML files.

The *mission file* is the entry point for the information on a console, holding relevant data for the operation, starting with the home location. Mission files are opened by the consoles and the MRA (in this case as read-only). The map uses this location as a centre and is the default location for added objects. The map editor, previously mentioned, adds extra elements to maps. *Map files*, part of the mission file, store these elements.

Mission plans consist of sequences of manoeuvres with extra actions on manoeuvres entry points. Mission files save these plans. *Manoeuvres* are movement specifications, all supported on the IMC definition. Examples of manoeuvres are goto (go to a waypoint); loiter or station keeping on an area; rows (lawnmower pattern); elevator (ascending or descending on a small area); follow trajectory or path (this has some predefined patterns for easy edition like RI pattern or cross-hatch pattern); and others. New manoeuvres are added by extending the base *Maneuver* Java class, providing a preview (a painter on a map), and an additional interaction editor if needed map-based edition.

A *vehicle file* contains configuration information of the vehicle, like ID (unique), name, type, model, size, vehicle presentation images, and icon colour. Each vehicle has its feasible manoeuvres. On this file, each compatible manoeuvre is stored, with default values. Additional information regarding connectivity and protocol support is also found, allowing initial connection with the vehicle. This information is overridden or complemented by the information on the IMC Announce message discussed in Section 8.2.4.1.

The *vehicle parameters* files contain vehicle internal parameters. Sending entity parameters IMC messages to a vehicle makes changes to these parameters and so the vehicle configuration. DUNE creates these files which are loaded by Neptus. These files are not exclusive for IMC usage. For non-IMC vehicles, these files can hold configurations of the vehicle and payload. The format of these files describes configurations divided into sections (corresponding in DUNE to tasks). Each section has parameters with the following information: name, type (e.g. numeric, string, enumeration), visibility (normal or advanced), scope (all, plan, or manoeuvre), a default value (for reset purposes), units, and description. Figure 8.6 depicts the parameters

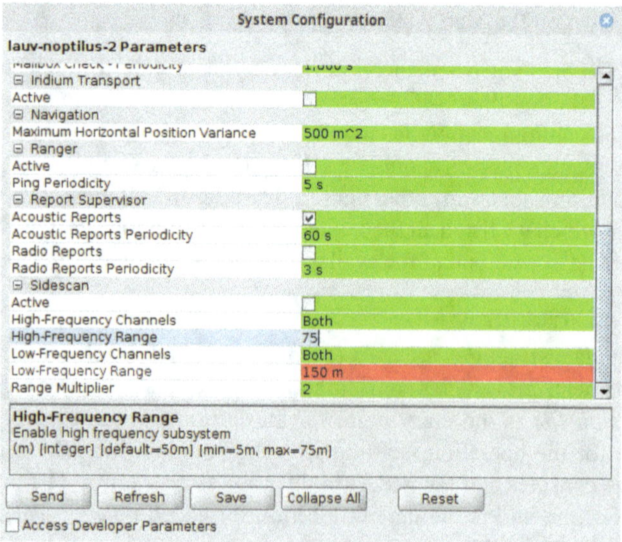

Figure 8.6 LAUV-Noptilus-2 vehicle params

of LAUV-Noptilus-2 vehicle and the interface for configuring the parameters for all scopes. These configurations with plan and manoeuvre scopes can be activated and configured on a plan or manoeuvre levels, respectively, in the plan editor.

8.3 Ripples

Considering the distributed nature of many of the deployments of the LSTS toolchain, it makes sense to consider networks that span many kilometres and potentially several disconnected clusters. To address this type of networks, we use a cloud-based infrastructure named Ripples to aggregate and disseminate data coming from multiple remote locations, while also providing different visualisations of these data. To begin with, Ripples provides APIs to post and consult the state of any deployed systems. The state includes not only the position and health (fuel, hardware failures) but also the configuration and planned behaviour of the systems. This allows Ripples to run a simulation of the systems' behaviour used to provide an estimate of what each system is doing between communications.

To allow the integration of other systems not using the LSTS toolchain, Ripples encodes plans as a set of timed waypoints. This allows systems like ships or external robots to send their planned tracks for multi-robot coordination schemes.

Ripples also receives information for ships, provided via AIS. These data are used to visualise, in real time, the estimated position of ships and generate future-timed waypoints (based on provided speed and heading). The fact that ship behaviour can be compared/integrated with robot's behaviour improves the awareness of the operators and even allows automatic detection of potential collisions, using simple iterative algorithms.

Ripples serves not only as a repository of information. Ripples will also push commands and relevant information between remote locations. Ripples provides an Iridium API that allows sending data to Ripples via satellite. This is especially useful for oceanographic missions where vehicles and ships remain mostly disconnected.

Whenever a message is posted and addressed to a system, Ripples will consult the available communication means for that system and push it over the available communication mean, such as Iridium. Nonetheless, all commands sent through Ripples are stored and can be consulted. Another use of the Iridium API is when a console is disconnected from the Internet but still requires updates of the deployed systems. For this, Ripples allows a console to subscribe to 'Iridium updates'. As a result, whenever relevant information is sent to Ripples such as a change of state or new commands being sent to the vehicles, this information also gets pushed to the consoles that have subscribed via Iridium. This effectively allows operation in the open sea without any connection to the Internet.

For visualising all deployed assets, Ripples provides a simplified web interface based on HTML5 and Leaflet (Figure 8.7). This interface shows a map with all active vehicles, planned behaviours, and transmitted data such as water column profiles. The user can also select from a plethora of different map layers including currents, wind, sea surface temperature, traffic density, etc.

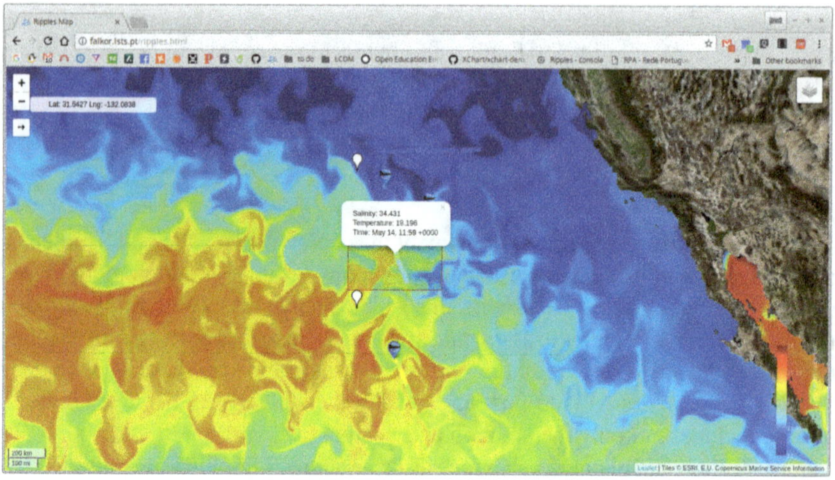

Figure 8.7 Ripples web interface showing collected (in-situ) data overlaid with satellite data

8.4 Applying the toolchain

8.4.1 Mine countermeasures

Mine countermeasures, or MCMs, operations are a typical military application. The objective of these operations is to clear safe pathways for ships to pass by unharmed. Side-scan sonar is the most pivotal sensor for sweeping potential minefields.

Every year ever since 2010, we are organising with the Portuguese Navy and more recently with CMRE, the REP Atlantic[‖‖] exercise. Each year we invite participants from leading industries, navies and armed forces, and universities from Europe and the USA. The objectives of these exercises are two-fold. First, it's a showcase for new technologies and methods experimentation in real operational scenarios. Second, by having several participants from research to industry and end users creates a fortuitous environment for sharing experiments and knowledge.

The vehicles used in this exercise were equipped with side-scan sonar from Edgetech and Klein. In one of the vehicles was installed a self-compensating magnetometer (SCM) system from Ocean Floor Geophysics (OFG). The scenario in which the tests ran was prepared by the Portuguese Navy and consisted on a site with dummy mines and magnetic targets.

Alongside with our vehicles, namely the LAUVs, other vehicles like Iver 3 from US Naval Undersea Warfare Center, and also Gavias and LAUVs from the Portuguese Navy, were also present (Figure 8.8). In the exercise was evaluated the operation

[‖‖]Recognized Environmental Picture, Atlantic, since 2020 Robotic Experimentation and Prototyping – Maritime Unmanned Systems

Figure 8.8 MCM vehicles used on REP Atlantic exercise

Figure 8.9 MCM scenario

procedures to do these kind of surveys from the different teams and equipment. Additionally, tests for acquisition of data for building detection probability curves for some of the sonars used.

The vehicles rotated around four areas. Three areas had one mine-like contact each, and a smaller area with multiple contacts including contacts with magnetic signatures. We focused our operations on the area with magnetic signatures. This area had multiple diving bottles on it, and the different side-scan sonars could detect the small air bottles (17 × 60 cm). On the top left of Figure 8.9 are displayed a side-scan sonar images and a captured photo of these contacts. Figure 8.9 depicts the scenario where after side-scan sonar surveys and identification of potential contacts, second runs with an optical camera and an SCM were executed.

Magnetometers are extremely sensitive to magnetic disturbances like the electric motors of an AUV, and so towing these sensors is the usual choice. The sensor we used from OFG is installed on the AUV body instead. To overcome the AUV disturbances and detect the self-magnetic anomalies, prior to the survey, we execute a compensating manoeuvre with the vehicle travelling at a similar speed and payload configuration. The top right of Figure 8.9 displays the plot with raw (red) and compensated (blue) magnetometer data. The compensated value distinguishes from the raw at the end of the compensating manoeuvre (around 9:25, where the blue line is distinct from the red).

The small diving bottles were not detected using just the magnetometer data. There was no correlation between the diving bottles and anomalies, with the data collected being quite flat. This was the first time with this sensor on such a scenario, and because of the compensation manoeuvre conditions, the battery of the AUV depleted quickly. The compensation manoeuvre took 25 minutes to complete, and due to the operation slots and wait time with the payloads on, all this didn't allow many tries. Because of the fishnets, we needed to operate at 5-metre altitude from the bottom, which proved not to be the ideal altitude to operate.

8.4.2 Digital acoustic communications in submarine distress

An emergency event with a manned submarine while submerged poses serious challenges for its rescue (DISSUB[¶¶]). Current protocols for the rescue in the event of a problem uses acoustic telephony. Underwater acoustics is a challenging technology, but nowadays it has evolved a lot, and digital communication is a valid option for data transmission.

We have been experimenting [13] on these scenarios in collaboration with NATO CMRE[****]. CMRE is pushing JANUS [14] as a standard for digital underwater communication protocol, and its use in DISSUB events [15].

Figure 8.10 shows a depiction of a DISSUB event. Surface ships deploy buoys with acoustic modems and connect through Wi-Fi and radio networks. Acoustic communication flows with the distressed submarine, buoys, and the deployed AUVs. Distressed submarines send distress messages informing of the predicted location, internal environmental conditions, and the number of crew members. Additional messages can flow between the parts.

The LAUV and manned submarine on-board situation awareness used the LSTS toolchain. Upon the distress operations start, the search vessel launches one or more LAUVs, which starts a waiting pattern in the vicinity, listening for the distress messages. When the message arrives, the LAUV extracts location and attitude information of the submarine. This reception triggers a search pattern around the predicted submarine location, surveying with side-scan sonar. On the situation awareness side, both on board the submarine and on the ship, we used the submarine's position to show on the Neptus map display. Neptus also displays the distress messages on the map.

[¶¶]DISSUB – DIStressed SUBmarine

[****]CMRE – Centre for Maritime Research and Experimentation, in La Spezia, Italy

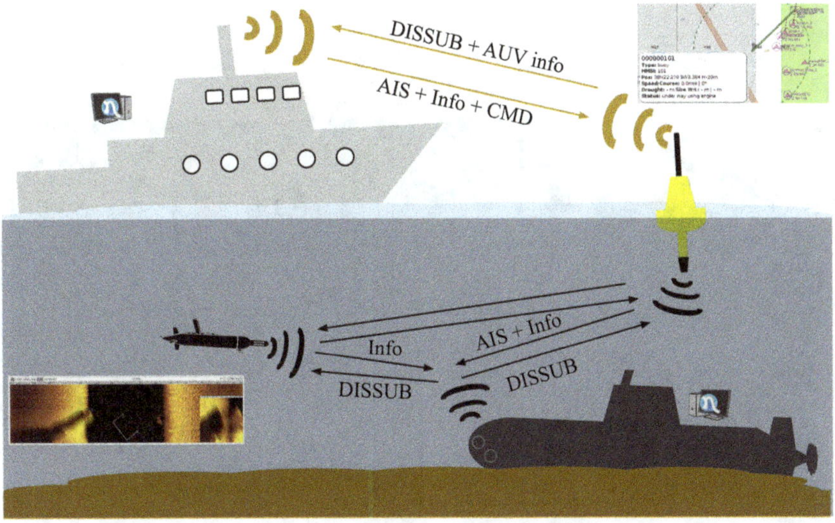

Figure 8.10 Distressed manned submarine (DISSUB) scenario

8.4.3 Multi-robot coastal oceanography

Physical oceanography bases a lot of its results on remote sensing, using satellite imagery. However, the availability of the satellite information has shortcomings such as follows: (1) it samples only the data from the surface of the water; (2) the data has low resolution given the long distances between the sensor and the object of interest; (3) clouds may block the sensing; and (4) satellite passes can be infrequent (in the order of days). As such, the models currently used are found many times to be imprecise when collecting in-situ data. In September 2018, as part of the REP exercise, a team of researchers used different types of robots coordinated together to collect different types of data from the Sado estuary, in the south of Portugal (Figure 8.11). Coordination of the robots is important for correct data correlation and data augmentation with complementary information.

In the first scenario, a team of two AUVs continuously monitor the river outflow (denoted by lower salinity) by travelling back and forth in front of the river mouth (in a 3-km transect). The vehicles travelled coordinately by having both vehicles reaching each side of the transect at the same time. We did it by creating temporal plans and having the vehicles adapting the execution (speed, time to communicate) to the perceived delays. Since the AUVs were travelling far away from the base station, we control them over Iridium, using Ripples as described in Section 8.3. Whenever the AUVs surfaced to communicate and receive new commands, they would also upload collected data over Iridium.

Simultaneously with the oceanographic AUVs, we deployed another AUV close to the base station to use an ADCP (Acoustic Doppler Current Profiler) and measure the current profile in front of the river. This AUV was close by so we have received

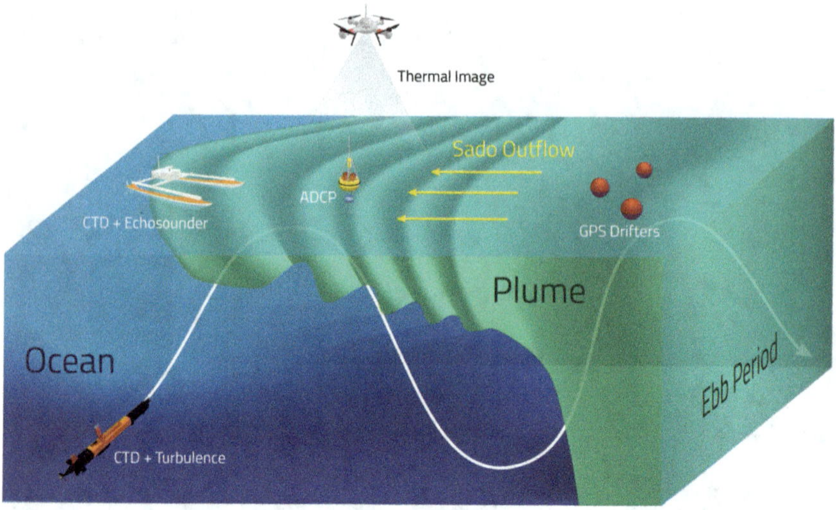

Figure 8.11 Operational concept for the coastal oceanography mission, undertaken in September 2018

all data in real time using Wi-Fi. Moreover, an aerial vehicle capable of taking off and landing on the ship (base station location) was used to monitor water temperature at the surface using an infrared radiometric camera. The UAV flew multiple times throughout the mission to acquire data at different tidal times.

8.4.4 Large-scale open ocean exploration

In May 2018, a cruise took place aboard R/V Falkor from Schmidt Ocean Institute for 20 days in the Pacific Ocean, 800 nautical miles off the coast of California. The location for the cruise was selected based on the estimated location of Pacific's Subtropical Front. This is a well-known but understudied oceanic front where fresher, colder waters from the Arctic meet those of the tropics which are warmer and more saline. Due to its sheer size (the front itself can have a cross section of up to 100 km), a multitude of assets was coordinated to locate, track, and map the front. This cruise introduced a novel approach to observe the ocean by using multiple underwater, surface, and aerial vehicles and the scientific ship as the base and control centre.

Before departing to the estimated location of the front, a team of unmanned surface vehicles was deployed to more precisely locate it. This team was composed of two Sail Drones and one Wave Glider SV3. Both platform types allow supervision using web APIs, which were integrated into the Ripples cloud infrastructure. Every minute, Ripples would poll new data coming from these platforms and disseminate it to other software applications such as Neptus and web clients. This data overlaid with salinity models from Copernicus Marine Environment Monitoring Service (CMEMS). This allowed not only to visualise, in real time, the difference between model and in-situ measurements but also detect any steep changes which would hint at the front location.

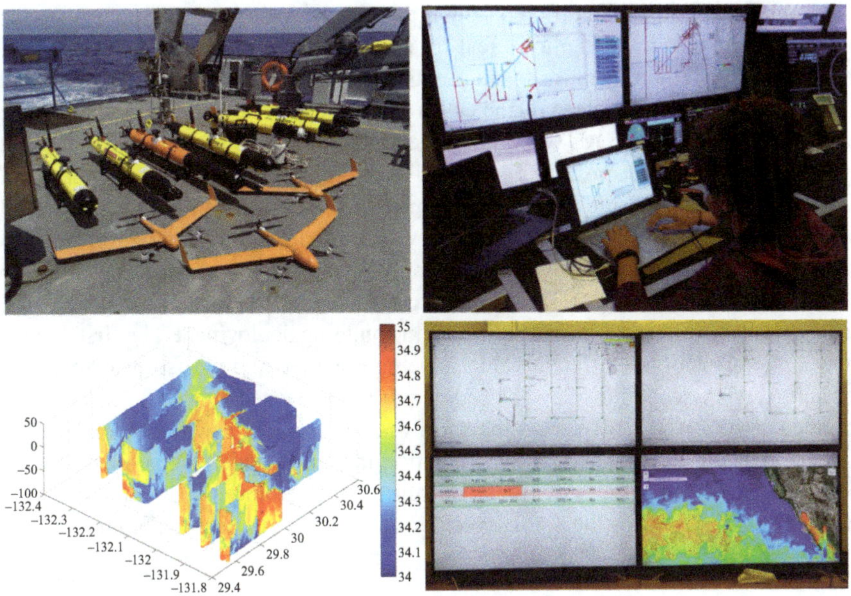

Figure 8.12 Top-left: vehicles used on-board R/V Falkor, top-right: science control room (base station) on-board R/V Falkor, bottom-left: 3D map of the front acquired using six AUVs, bottom-right: OSC in Porto, at the time of the cruise

During the 3-day transit from San Diego to the front location, scientists on board the ship have selected the methodology to map the front based on the in-situ data that was arriving in real time. The initial survey was a high-resolution mapping of the front using three AUVs at a time and having the ship separate from this deployment. This resulted in the mapping of a front jet in unprecedented detail which took 6 days to complete. After this initial survey, we conducted another large-scale survey using both the ship and AUVs, to show the possibility of augmenting ship data resolution with more assets (Figure 8.12). Together with AUVs and ASVs, we also deployed multiple times a vertical take-off and landing UAV (Figure 8.12). We have used this platform with different sensors according to the different surveys.

During the entire expedition, we have controlled the AUVs using just satellite communications. To support this type of operation, we devised a specific on-board executive that would adapt and execute a plan autonomously and exchange data/commands using small Iridium messages (Short Burst Data). With every plan generation or change on board the vehicle, it synchronised with the base station via Iridium (flowing through Ripples) informing all operators of the current system configuration and planned waypoints, including the time of arrival for each. This approach has allowed the operators to maintain a high degree of awareness while the vehicles remained disconnected and very far (over 10 miles) for most of the time.

Operation of the UAVs differed from the AUVs as, in this case, we controlled the vehicles using Wi-Fi which allows real-time visualisation of data such as DMS†††† and surface temperature. However, since both AUVs and UAVs were being controlled using Neptus, it was possible to fly the UAVs while keeping an accurate estimate of the locations of all deployed AUVs (while underwater and disconnected). This allowed flying the UAVs over the positions of the AUVs to build a synoptic image of the front.

We continuously deployed the AUVs for over 30 hours at a time, which required different operators' shifts. To help with supervision, other scientists that stayed in Porto followed the execution and interacted via Ripples using an Ocean Space Center (OSC) system in Porto. This OSC was a computer connected to four displays (Figure 8.12), each containing different information about the ongoing deployment. The first display showed the plans of the different vehicles being executed and the estimated position of each AUV. A second display showed acquired data together with historical data obtained by the AUVs. We used the third display to send new commands to the vehicles (change the plans) and, by simulating the commanded plans, predict what the vehicles will do in the near future. A fourth display showed a simple table that, for each vehicle, informs the operator of reported failures or potential collisions.

8.5 Integration with other frameworks

In the preceding sections, we presented the toolchain for autonomous systems created and used by the LSTS. This toolchain has also room to interact with other systems that do not speak the same language. This need comes from operating with people that have other systems that are not by design compatible with our toolchain.

One example is ROS (Robot Operating System). It has a lot of attention but lacks standardisation of data structures. It is easy to introduce IMC to ROS and at the same time making an ROS vehicle available to Neptus. With ROS IMC Broker and the adapter node, all IMC messages and a base communication protocol are available for use, where we can consume and produce IMC messages. We need some implementation for the base functionality, but this exposes an ROS vehicle to the toolchain, allowing Neptus command over it.

The same can be done for MOOS-IvP (Mission Oriented Operating Suite — Interval Programming). The moos-ivp-dune is a piece of software that exposes IMC state to MOOS.

Both bridges are available in the GitHub (either LSTS's [10] or its collaborators). We helped or integrated vehicles to work on the toolchain using these bridges or a translator approach.

Additionally other vehicles, some commercially available, already use our toolchain. Companies like, e.g., Oceanscan MST, Evologics, or Atlas Elektronik, to name a few, already ship their vehicles with LSTS's toolchain (full or partially).

††††DMS – Dimethyl sulphide

8.6 Conclusion

In recent years scientists perceived the advantage of unmanned vehicles in ocean exploration. They can extend the time of observation or reach areas that are difficult to access with conventional means. The market offers a diversity of options of unmanned vehicles, but mostly each system is self-contained, and it is difficult to operate in collaboration with others from different vendors. A flexible tool that enables the interconnection and command of all assets independent of the vendor is a must if we want to integrate diverse assets from multiple vendors.

LSTS toolchain for autonomous systems has been in the development for several years and is presented here. For greater flexibility, the toolchain is divided into several components:

- IMC that defines a message protocol independent of the communication means that allows the passing of the information between vehicles and command and control interfaces;
- DUNE that is on-board control software that fully controls an autonomous vehicle or interfaces with a front seat to control the vehicle;
- back-seat driver control that exposes an API from DUNE in order to allow the creation of more complex controllers without compromising the front-seat operation;
- Neptus Command and Control software allowing the command of autonomous vehicles, being flexible enough by adding a plug-in framework for easy functionality or protocol extension allowing connection to non-IMC vehicles. It also provides a data review and analysis tool with plug-in extensions for data visualisation and export; and
- Ripples, a REST web interface, allowing the monitor and command of autonomous systems through a web browser interface, and at the same time exposing functionality and data persistence for command and control interfaces.

Several examples of real-world use of the toolchain in scenarios spanning from military to science were shown depicting the flexibility of the toolchain. Additionally, we discussed the integration of the toolchain with other frameworks. The toolchain is used as the main interface but can also be used to expose IMC vehicles to other frameworks.

The development of the toolchain is now in a mature phase, with some adoption not only by other universities but also from commercial companies that use the full or partial toolchain.

References

[1] Climate 101: Oceans [homepage on the Internet]. National Geographic [cited 2018 Nov 14]. Available from: https://www.nationalgeographic.com/environment/habitats/ocean/.

[2] Research vessel R/V Falkor [homepage on the Internet]. Schmidt Ocean Institute [cited 2018 Nov 14]. Available from: https://schmidtocean.org/rv-falkor/.

[3] Norwegian billionaire funds deluxe deep ocean research ship [homepage on the Internet]. Science Magazine, AAAS [updated 2018 Nov 19; cited 2018 Nov 19]. Available from: http://www.sciencemag.org/news/2018/11/norwegian-billionaire-funds-deluxe-deep-ocean-research-ship.

[4] Pinto J, Dias PS, Martins R, *et al.* The LSTS toolchain for networked vehicle systems. In: 2013 MTS/IEEE OCEANS - Bergen; 2013. pp. 1–9.

[5] Pinto J, Calado P, Braga J, *et al.* Implementation of a control architecture for networked vehicle systems. IFAC Proceedings Volumes. 2012;45(5):100–105. 3rd IFAC Workshop on Navigation, Guidance and Control of Underwater Vehicles. Available from: http://www.sciencedirect.com/science/article/pii/S1474667016305870.

[6] LSTS's Home Page [homepage on the Internet]. FEUP-LSTS [cited 2018 Dec]. Available from: https://www.lsts.pt/.

[7] Cruz N, Matos A, Martins A, *et al.* Estuarine environment studies with Isurus, a REMUS class AUV. In: Oceans '99. MTS/IEEE. Riding the Crest into the 21st Century. Conference and Exhibition. Conference Proceedings (IEEE Cat. No.99CH37008). vol. 3; 1999. pp. 1205–1210.

[8] da Silva JE, Terra B, Martins R, *et al.* Modeling and simulation of the LAUV autonomous underwater vehicle. In: 13th IEEE IFAC International Conference on Methods and Models in Automation and Robotics. Szczecin, Poland; 2007.

[9] Martins R, Dias PS, Marques ERB, *et al.* IMC: A communication protocol for networked vehicles and sensors. In: OCEANS 2009-EUROPE; 2009. pp. 1–6.

[10] LSTS's GitHub [homepage on the Internet]. FEUP-LSTS [cited 2018 Dec]. Available from: https://github.com/LSTS/.

[11] McIlroy MD, Pinson EN, Tague BA. UNIX time-sharing system: Foreword. Bell System Technical Journal. 57(6):1899–1904. Available from: https://onlinelibrary.wiley.com/doi/abs/10.1002/j.1538-7305.1978.tb02135.x.

[12] MigLayout – Java Layout Manager for Swing, SWT and JavaFX [homepage on the Internet]. MiG InfoCom AB; [cited 2019 Jan]. Available from: http://www.miglayout.com/ | https://github.com/mikaelgrev/miglayout.

[13] Dias PS, Costa M, Sousa JB. REP18 Atlantic: A large scale exercise using unmanned systems. Instrumentation Viewpoint. 2018; p. 52. Available from: http://hdl.handle.net/2117/126290.

[14] NATO. NATO Standardization Office STANAG 4748 Ed. A Ver. 1, Digital underwater signalling standard for network node discovery & interoperability. 2017. Available from: https://nso.nato.int.

[15] Petroccia R, Alves J, Zappa G. JANUS-based services for operationally relevant underwater applications. IEEE Journal of Oceanic Engineering. 2017; 42(4):994–1006.

Chapter 9
Telemetry for AUVs

João Alves[1], Konstantinos Pelekanakis[1] and Roberto Petroccia[1]

9.1 Introduction

In this chapter, we introduce the topic of telemetry for autonomous underwater vehicles (AUVs) and draw considerations on different techniques and approaches to be used when establishing underwater communications.

In broad terms, telemetry refers to automated measurements and transmission of such acquired data from a remote location. This wide definition tends then to be tailored to specific areas of use. In the context of AUVs, telemetry is typically referred to as the transmission of the vehicle status or sensor data. Within the context of this chapter, we also add the transmission of commands or queries by a control station or a peer node. This way, and for the purposes of this chapter, we define AUV telemetry as the underwater wireless data exchanges that support any AUV-based mission. We deliberately left out of this definition any above-water communications for the simple fact that terrestrial radio networks have been vastly studied and are commercially mature in a wide variety of forms. Hence, the focus of this chapter is on underwater wireless communications, covering from the propagation physics of the links up to the establishment of networks.

9.1.1 The requirement for telemetry

In the early days of AUV operations, the underwater links were used to exchange a fairly limited amount of information. Operators would not expect more than simple 'OK/NOT OK' status report or 'START/STOP' commands to instruct a single AUV from a user control station. For the most part, the underwater link (mostly acoustic as will be shown later in the chapter) was assumed as non-existent or non-dependable. This was in fact the fundamental motivation for the development of autonomy in underwater vehicles.

With the technological developments of the last few decades, AUVs have matured and gained increasing levels of autonomy. This advancement motivated the

[1]Research Department of the NATO Centre for Maritime Research and Experimentation (CMRE), La Spezia, Italy

requirement to use them cooperatively in teams or fleets and contributed to a sympathetic development of underwater communications equipment that became more available from a wide range of manufacturers.

In fact, the interest in underwater communications has grown immensely in the last few decades and such technology currently plays a critical enabling role. The requirement went from very basic, single-vehicle monitoring and command to the establishment of underwater ad-hoc networks including all networking support protocols.

Underwater acoustic networks (UANs) are currently an enabling technology in a wide spectrum of different collaborative scenarios that find application to science, security, and commercial activities. Such scenarios of interest include (among others) environmental monitoring, prediction of and reaction to natural disasters, surveillance for defence applications and port safety, offshore oil and gas, aquaculture, and geological and oceanographic science [1,2].

9.2 Underwater communications modalities

Underwater wireless communications can explore different propagation channels. Optical, electromagnetic radio frequency (RF) and acoustics are described in the following. Acoustics is the modality of choice for most open sea applications and so we present expanded sections on underwater acoustic (UWA) communications.

9.2.1 Underwater optical communications

For a long time, there has been a significant interest on the optical properties of water, focusing on geophysical, bio-optical, and remote sensing applications. Recently, underwater optical communication has witnessed a growing interest motivated mainly by the potential to provide high-speed data links that can reach the order of gigabit per second [3]. However, there are some disadvantages for wireless optical communications through water (also commonly referred to as free space optical communications as opposed to the case where optical wave guides provide a path for the light to travel from one node to the next). The first is that optical signals are rapidly absorbed in water (through the interaction of photons with water molecules), limiting their operational range. Another source of losses is optical scattering caused by the redirection of the photons. Ocean scattering is caused by suspended particles and marine life in general and may occur in the forward or backward directions (forward scattering or backscattering). Additionally, the ambient light close to the surface is a source of high ambient noise, making shallow water optical communications particularly challenging. Communication can typically be carried out over ranges of a few hundred metres, depending on water clarity and the availability blue-green band (400–500 nm). Due to the aforementioned absorption and scattering, optical waves deliver the best performance in very clear water, and (because of the typically high beam directionality) with well-aligned nodes. This requirement for line-of-sight when using optical communications imposes additional practical (engineering) constraints. In terms of energy, the requirements of optical data transmission are much lower than for acoustics when considered on a bit-per-joule basis. A factor of 300 is mentioned in [4].

The viability of optical laser transmission at Mbps rates over ranges up to a few hundred metres in clear waters was demonstrated at the Woods Hole Oceanographic Institution [5] and preceded a range of other efforts such as the ones in [6,7]. This is clearly seen as a major breakthrough for specific applications such as contactless broadband data exchange between AUVs and cabled observatories or other fixed infrastructures.

Underwater optical communication using focused light sources is typically considered point-to-point because of the high directionality. Point-to-point links require strict transmitter–receiver alignment and tracking, especially on mobile platforms. This is reasonable in systems that use collimated laser links and are typically large enough to accommodate dedicated gimbal systems. Adopting optical communications for localised underwater networking of several assets in close proximity seems like an obvious technological choice. Interesting applications of underwater optical communications for data muling using AUVs can be found in [8,9]. The usage of optical communications to remotely operate underwater is documented in [10,11].

9.2.2 Underwater RF communications

Underwater RF propagation is considerably different from air. The main driving parameters for such difference are the high permittivity and electrical conductivity of water [12]. Underwater RF models are described in detail in [13,14].

Propagation Speed c (m/s) can be expressed by [12]:

$$c = \sqrt{\frac{f \times 10^7}{\sigma}} \tag{9.1}$$

showing the dependence on frequency, f(Hz) and conductivity σ(S/m). It is, in any case, always faster than acoustic propagation by at least one order of magnitude in the worst case (low frequency, highly conductive water) to five orders of magnitude in the best case (high frequency, low-conductivity water). This has important advantages in terms of latency, an important consideration when the channel needs to be shared in a network. The wavelength, λ (m), of an electromagnetic wave is given by

$$\lambda = \frac{1}{\sqrt{f\sigma \times 10^{-7}}} \tag{9.2}$$

which immediately shows that for the same frequency, the wavelength in freshwater will be longer than that in seawater, resulting in the need for bigger antennas. The key problem for electromagnetic communication in seawater is the strong attenuation of the signals. RF bandwidth only exceeds the counterpart acoustic systems at very short range.

Acoustic and RF technologies behave in a reversed way when in water and air. In air, acoustic signals are highly attenuated (the reason why several groups of people can hold independent conversations in a crowded room). Radio waves have low attenuation in air, so communications are separated in frequency through spectrum management. In water, the two roles reverse themselves, and as pointed out in [12], there are potential benefits of high RF spatial attenuation in a multi-user environment. On the other hand, significantly longer acoustic propagation requires dedicated channel access techniques.

An interesting property of electromagnetic waves is the ability to cross the air–water boundary. This has been explored in the past by submarines towing large antennas to communicate to shore. RF is susceptible to electromagnetic interference which may pose a challenge for installation in small and highly integrated unmanned underwater vehicles. In the recent years, electromagnetic underwater communications efforts were oriented to AUV short-range communications including AUV docking stations [15]. The authors in [16] test various IEEE 802.11 standards for the 2.4 GHz ISM frequency band. Results show a maximum rate of 11 Mbps at 16 cm range. AUV applications based on RF are described in [17].

9.2.3 UWA communications

In applications where AUVs need to establish links for more than hundred metres of range, underwater communications systems need to rely on acoustic waves. In this case, it is important to understand the distortions any sound signal suffers as it travels from the transmitter to the receiver. An excellent review of the impact of acoustic channels on communications systems is described in [18–23]. Here, we revisit the acoustic channel impairments from an AUV telemetry perspective.

The operational bandwidth is severely limited due to high-frequency signal absorption by seawater and low-frequency signal contamination by ambient noise. The typical bandwidth for long-range (>50 km) communications is in the order of a kilohertz. For medium-range (1–50 km) communications, the bandwidth ranges from tens of kilohertzs to few kilohertzs while for small-range communications (<1 km), the bandwidth is in the order of tens or hundreds of kilohertzs. Using the range-rate product as a metric to understand performance, most of the systems (with some exceptions) are below the curve $range \cdot rate = 40$ km·kbps [21]. For example, a system that operates close to 20 km range is expected to achieve less than 2 kbps bit rate. Furthermore, the carrier frequency is typically more than 20% of the bandwidth classifying these systems as ultrawideband [24]. Modelling and simulating ultrawideband UWA channels to compute realistic Bit Error Rates (BERs) is tedious because it requires to account for frequency-dependent fluctuation rates. In addition to the limits imposed by the channel physics, platform-specific constraints may affect the choice of operational frequencies. The permissible size of the equipment that can be hosted in an AUV obviously depends on the AUV size. For example, if ranges of hundreds of kilometres are desired (e.g. in Arctic communications), then the operational frequencies must be close to 100 Hz to account for absorption power loss. Since the size of the sound source is proportional to the acoustic wavelength, at very low frequencies the transducers may become prohibitively large for an AUV. In this case, the carrier frequency will be dictated by the platform constraints.

Multipath propagation, i.e. multiple sound paths between the transmitter and the receiver, is more prominent in horizontal than vertical underwater links. It is governed by three mechanisms: refraction due to the inhomogeneous sound speed at sea, reflection off physical boundaries (e.g. sea surface), and scattering off objects (e.g.

school of fish). Multipath propagation results in multiple replicas of the transmitted signal at various delays, amplitudes, and phases. The constructive and/or destructive combination of these replicas at the receiver gives rise to *frequency selective fading* – i.e. some frequencies are enhanced (constructive interference) while some other are suppressed (destructive interference). Maintaining good communication is challenging since rapid power fluctuations of the received signal occur as the AUV moves by a few wavelengths. This issue is exacerbated in the scenario where an AUV receives the direct path and the phase-inverted path from the sea surface. At moderate ranges, these two paths add destructively and the signal loss decreases with the radial distance to the fourth power (Lloyd's Mirror effect [25]). The remedy to multipath fading is to use wideband signals so to exploit *frequency diversity* techniques. For single-carrier wideband signals, the channel generates intersymbol interference (ISI), namely, the delayed replicas of the previously transmitted symbols interfere with the decision of the current symbol. The ISI is measured as $L = R \cdot T_d$, where R and T_d denote the baud rate and multipath spread, respectively, and it usually ranges between tens and hundreds of symbol intervals. Assuming an AUV with a single hydrophone, the optimum maximum likelihood sequence estimation (MLSE) receiver has computational complexity $O(M^L)$ [26], where M is the symbol constellation size. Consequently, the MLSE receiver is prohibitively complex for an AUV and suboptimal receivers that yield similar performance are needed. Considering the widely used decision feedback equaliser (DFE), its computational complexity is $O(T_d \cdot R^2)$ flops [20] excluding the decoding complexity. Hence, a careful hardware implementation is needed given that an AUV may not be able to host a power-hungry signal processing unit.

The relative motion between the transmitter and the receiver induces a different timescale from the transmitted signal. Specifically, different parts of the transmitted signal may be either dilated or compressed. The magnitude of compression/dilation is equal to $\alpha = u/c$, where u is the line-of-sight speed and $c \approx 1{,}500$ m/s is the sound speed in water. For typical AUV speeds of a couple of m/s, α lies close to 10^{-3}. Theoretically, signal resampling is needed to achieve bit synchronisation. Practically, α is compared against the time-bandwidth product, $T \cdot W$, of the transmitted signal. A rule of thumb indicates that if $T \cdot W \geq 1/\alpha$ (e.g. a binary phase-shift keying (BPSK)-modulated signal carrying 1,000 bits at 1 kBaud rate), then resampling is required. On the other hand, if $T \cdot W \ll 1/\alpha$ (e.g. a BPSK-modulated signal carrying 100 bits at 1 kBaud rate), then resampling can be neglected. Note though that the demodulation is typically performed in baseband as it is less power hungry for an AUV. Regardless the signal time-bandwidth product, carrier-phase synchronisation to account for the Doppler effect is necessary.

Time variability of multipath formation occurs even without AUV motion. This variability happens at different timescales. Slow-time variations are due to changes of large-scale spatial features of the environment (e.g. changes in sound speed due to daily cooling and heating). Fast time variations are due to rapid, small-scale environmental fluctuations (e.g. changes in sea surface motion). For example, considering the tone reflection off a smooth surface wave (ignoring scattering effects), the transmitted tone undergoes frequency spreading (or Doppler spread) due to translation of the reflection point of a moving surface. If both the transmitter and the receiver are at a depth much greater than the surface wave amplitude, a transmitted tone at carrier frequency f_c is received as a phase-modulated signal:

$$\cos\left(2\pi f_c\left(t-\tau_0-\frac{2s(t)\cos(\theta)}{c}\right)\right) \tag{9.3}$$

where τ_0 is the time of arrival of the beginning edge of the tone, $s(t)$ is the wave amplitude, and θ is the incident angle. Assuming wind generated surface waves following a Pierson–Moskowitz wave amplitude spectrum, the surface wave modal frequency is determined by [27]

$$f_w = 0.877\frac{g}{2\pi u_{19.5}} \tag{9.4}$$

where g is the gravitational acceleration (m/s^2) and $u_{19.5}$ is the wind speed (m/s) at height 19.5 m above the sea surface. The significant amplitude of the waves at that frequency is given by

$$h = 0.21\frac{u_{19.5}^2}{2g} \tag{9.5}$$

If the sea surface elevation is approximated as $s(t) = h\sin(2\pi f_w t)$, then the received signal is not appeared as a single tone any more but exhibits a frequency spread (Doppler spread) which is approximated by using Carson's rule [28]:

$$B_d = 2f_w\left(1+\frac{2\pi f_c 2h\cos(\theta)}{c}\right) = 0.877\frac{g}{\pi u_{19.5}} + 0.368\frac{\cos(\theta)f_c u_{19.5}}{c} \tag{9.6}$$

If $u_{19.5} = 7$ m/s (sea state 3) is assumed, then it follows that $f_w = 0.2$ Hz and $h = 0.5$ m from (9.4) and (9.5), respectively. Assuming $f_c = 10$ kHz, $c = 1{,}500$ m/s, and $\theta = \pi/4$, (9.6) results in 12.5 Hz Doppler spread. The authors in [29] suggest that the ratio of Doppler spread over the baud rate (or symbol rate) should be less than 10^{-2} for phase coherent communications. Hence, for 12.5 Hz Doppler spread one would require more than 1 kBaud rate. Obviously, this is possible only if the operational bandwidth is more than 1 kHz. Higher baud rates increase the ISI, which ultimately limits the operational data rate as it was observed in several shallow water environments [30,31].

Finally, special attention must be given to shadow zones. These are regions in the water column where the transmitted signal has little penetration [25]. This may happen either due to obstructions, e.g. a sea mount, or due to sound refraction. In these zones, acoustic communications is practically impossible. Current research predicts these areas by running acoustic propagation models on board an AUV using environmental data from its sensors [32].

9.2.3.1 Acoustic noise

Noise signals received in an AUV hydrophone can be categorised into self-noise and ambient noise. Self-noise can be produced by the on-board systems (propulsion, buoyancy control systems, and other vehicle actuators) and by the interaction of the hull with the medium (flow noise). Self-noise depends on the vehicle size and is mainly generated by the AUV propulsion and actuators and may reach up to 10 kHz [33].

At high speeds, self-noise may become more severe than ambient noise and thus a limiting factor. Ambient noise is generated by natural sources (such as rain, breaking waves, turbulence, and biologics, to name a few) or man-made (such as distant ships and industrial activity) [34]. In deep water, the ambient noise soundscape is roughly divided into four spectral regions. Turbulent currents dominate frequencies up to 10 Hz. Distant shipping dominates the frequency range of 10–500 Hz. Bubbles associated with breaking waves dominate the frequency range of 500 Hz to 100 kHz. Finally, thermal noise (random motion of water molecules) dominates the soundscape above 100 kHz. The frequency region from 1 Hz to 100 kHz can be roughly characterised by a spectral slope of -18 dB/decade and Gaussian statistics is typically assumed all year long. On the other hand, the spectral levels in shallow waters depend on time and location and significant differences exist, for example, between bays and harbours. In addition, Gaussian statistics are not valid in warm shallow waters with snapping shrimp [35] or polar regions with ice cracking [36], as ambient noise has impulsive characteristics. For instance, symmetric α-stable (SαS) distributions have been used to model snapping shrimp noise in Singapore waters. This results in unique implications on the design of communications systems [37].

9.2.3.2 Acoustic channel modelling

The defining characteristic of UWA channels is time-varying multipath, which leads to spreading of the propagating signal in both time and frequency. Following the line of thought of terrestrial wireless communications [26], UWA channels are modelled as a system with a time-varying impulse response, $h(\tau, t)$; t corresponds to the time variations of the impulse response due to motion (platforms and/or physical boundaries) while the variable τ represents the channel multipath delay for a fixed value of t. The transmitted signal, $x(t)$, can be of any modulation type such as frequency-shift keying (FSK) or phase-shift keying (PSK). The received signal, $y(t)$, is related to $x(t)$ as follows [21]:

$$y(t) = \int_{-\infty}^{+\infty} h(\tau, t)x(t - \tau)d\tau + w(t) \tag{9.7}$$

where $w(t)$ denotes the ambient noise. Assuming $x(t)$ has bandwidth W and applying the sampling theorem, then the discrete-time model of (9.7) is given as

$$y[m] = \sum_{\ell=0}^{L-1} x[m - \ell]h_\ell[m] + w[m] \tag{9.8}$$

where $h_\ell[m]$ denotes the ℓth channel coefficient (or tap) at discrete time m and L is the ISI span. The value of $h_\ell[m]$ is determined by all sound paths whose arrival times (delays) fall within the window $[\ell/W - 1/2W, \ell/W + 1/2W]$ and assuming that the fastest path is at zero delay. Equation (9.8) indicates that the doubly spread channel can be modelled as a tapped-delay line filter with time-varying taps. Figure 9.1 shows the block diagram of the model.

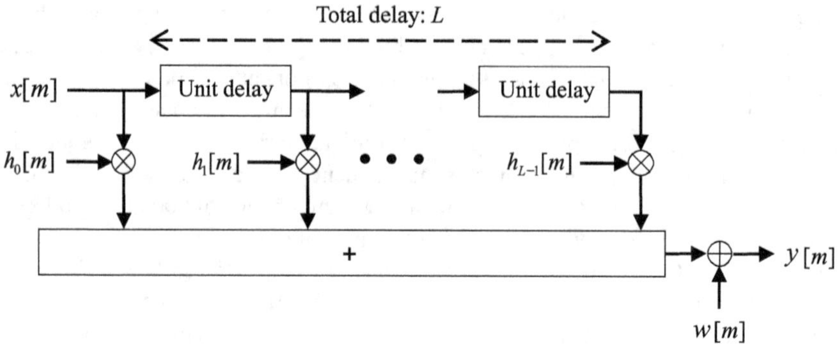

Figure 9.1 Block diagram of discrete-time channel model

Estimating each channel tap $h_\ell[m]$ is of paramount importance for phase coherent communication systems. Classical adaptive algorithms such as the least mean squares or the recursive least squares may not be suitable to track $h_\ell[m]$ in many acoustic environments [38,39]. Due to this issue, the research focus has shifted to formulating additional structure into $h_\ell[m]$. The correlation between taps via autoregressive models has been studied in [38,40,41]. Channel sparseness – i.e. the fact that the energy of the channel impulse response is concentrated in a small fraction of its duration – has gained considerable attention since the late 2000s. Current efforts include matching pursuit (MP) algorithms [42,43], a Bayesian models [44,45], a mixture of L_2 and L_1 norms [46], and a mixture of natural gradient and L_0 norm in impulsive [47] and non-impulsive noise [39].

A statistical model of $h_\ell[m]$ is necessary for computing the maximal data rate the channel can support (channel capacity) and choosing error correction codes that approach capacity. The capacity studies in [48,49] predict 2–4 bits/s/Hz at an average Signal-to-Noise Ratio (SNR) of 15–20 dB based on Rician fading statistics. Other studies propose two-path Rayleigh fading [50] with SαS ambient noise, K-distribution fading [51] and Rician fading [52,53]. The authors in [30] challenge the Gaussian fading assumption (i.e. the existence of a large number of scatterers that lead to the central limit theorem) as intermittent distinct arrivals were identified in a surf zone environment. The aforementioned studies were established at different environments, different experimental set-up and signal frequencies. Hence, it is plausible to claim that one universal fading model cannot fit all environments. The issue of not having standardised channel models leads to the choice of robust coded modulation schemes that can work efficiently under different channel fading statistics. For instance, the work in [54] proposes traditional coding schemes (i.e. trellis coded modulation (TCM) or bit-interleaved coded modulation) that have both large *free*

For incoherent systems, channel estimation is not important.

Table 9.1 Advantages and disadvantages of different underwater communications modalities

	Advantages	**Disadvantages**
Acoustic	• Established technology • Full networking support • Supports ranges of 10s of kilometres	• Sensitive to pressure and temperature • Performance degrades in shallow water • Limited bandwidth • Sensitive to biofouling
Optical	• Low power • Small hardware • High bandwidth	• Requires line of sight • Requires alignment of end points • Prototype technology • Sensitive to suspended particles and turbidity • Short range
Electromagnetic	• Unaffected by turbidity, biofouling, or acoustic noise • Crosses the air–water boundary • High bandwidth	• Very short effective ranges • Loop antennas are far from ideal for small AUV integration • Sensitive to salinity

Euclidean and *Hamming* distances. Other choices include non-binary low-density parity-check (LDPC) codes [55] and Reed–Solomon block turbo codes [56].

9.2.4 Modalities recap

Table 9.1 offers an overview of advantages and disadvantages of the different underwater communications modalities described above. The effective communication range achievable by practical systems capable of being installed on AUVs vary from a few metres for RF, up to 100 m for optical and up to tens of kilometres for acoustic systems. Since UWA communications is the most used technology for AUV telemetry and can effectively enable a wide range of possible applications and scenarios, the remainder of the chapter focuses on the acoustic modality.

9.3 Physical layer

As in terrestrial communications, an underwater communications system is typically described in terms of layers based on the Open Systems Interconnection (OSI) standard [57]. In this section, we address communication techniques pertained to the physical layer of the OSI. Techniques related to the Data Link and Network layers are described in Section 9.4.

The aim of the physical layer is to transmit and receive signals that contain digital information, i.e. bits. The choice of the acoustic waveform depends on how receiver demodulation copes with time-varying ISI. For AUV telemetry, two demodulation techniques are typically considered: non-coherent and coherent demodulation. Next, we describe these techniques and then we cite examples of representative AUV telemetry systems.

9.3.1 Non-coherent systems

FSK modulation has been playing a pivotal role since the early developments of UWA communications [18,21]. It is well known for its robustness in the harsh UWA propagation environment since phase information is not needed for demodulation. Moreover, the basic pulse duration is typically much larger than the expected multipath spread and so the receiver neglects the ISI. Obviously, this receiver robust design comes at the expense of low bandwidth efficiency, which does not exceed 0.5 bits/s/Hz. Another desired trait is that FSK signals have a constant envelope and so they are less sensitive to amplitude distortions from power-efficient non-linear amplifiers. Frequency-hopped-binary FSK (FH-BFSK) is an extension of FSK and uses a sequence of short-duration tones (chips) to transmit bits. The transmitter produces only a single tone at any time but this tone may be one of two adjacent frequencies in a block. The important idea behind FH-BFSK is that the location of the block in the wider available operating band is determined by an algorithm that pseudo-randomly finds distinct locations. That is to avoid any interference caused by multipath propagation and Doppler. In addition, FH-BFSK is robust to signal collisions, supporting a degree of multiple simultaneous access. The receiver is synchronised with the frequency hopping pattern and uses an energy detector to retrieve the transmitted bit transmitted at every block.

A characteristic FSK system is described in [58]. In this system, a convolutional encoder followed by bit interleaver and binary FSK modulation is used. The bit interleaver ensures that the coded bits are transmitted across different coherent bands of the channel. To account for ISI, the system uses a guard interval bigger that the channel delay spread before transmission of data. The achieved rate was 250 bps and the reported BER was on the order of 10^{-3} over nine different short-range, shallow-water channels. Perhaps, the most well-known FH-BFSK system is the NATO standard JANUS [59], which provides 80 bps and uses the 9,400–13,600 Hz band with centre frequency of 11,520 Hz. Robustness to multipath fading and noise is provided by a 1/2-rate convolutional encoder followed by an interleaver. The JANUS standard has been used in various AUVs configurations to broadcast situational awareness messages (a.k.a. underwater AIS) or as a first contact and language switching protocol [60].

9.3.2 Coherent systems

The perpetual need for high data rates drives the community's interest to research systems with high bandwidth efficiencies (higher than 1 bit/s/Hz). In these systems, the basic pulse duration is smaller than the expected ISI and so dealing with the resulting

ISI is essential. As mentioned above, the optimal structure to deal with the ISI is the MLSE receiver. However, its computational complexity grows exponentially with the ISI length; therefore, suboptimal signal processing techniques such as DFE, orthogonal frequency-division multiplexing (OFDM) and direct-sequence spread spectrum (DSSS) are typically used.

A DFE consists of a feedforward and a feedback filter. The feedforward filter operates directly on the received signal and tries to recover the transmitted symbol while the feedback filter operates on previously detected symbols and tries to remove the ISI. A DFE outperforms a typical linear equaliser, yet, degradation in the DFE performance occurs when erroneous feedback is applied due to incorrect decisions. The DFE was initially demonstrated at sea in [61]. The essential element of the receiver structure was the combination of a fractionally spaced adaptive DFE with a second-order phase-locked loop (PLL). Excellent performance at 10 kbps using uncoded four-PSK signals over a short-range, shallow-water channel was achieved. The authors also reported excellent performance at rates up to 1,000 bps using uncoded 4-PSK/8-PSK signals over long-range, deep- and shallow-water channels. In the last 20 years, a plethora of systems based on the DFE-PLL receiver structure have been developed. One example is the ASIMOV system, which used a four-state TCM code for eight-PSK modulation and was able to achieve 30 kbps at a near vertical configuration between an AUV and an autonomous surface craft [62]. Another representative example is the implementation of the WHOI Micromodem originally developed for use on small AUVs in shallow water [63]. The fastest data rate approaches 5 kbps in horizontal channels based on PSK signals.

OFDM is a multi-carrier modulation scheme that distributes the channel symbols over a number of subcarriers that are spaced apart at precise frequencies. In addition, the sub-carriers are overlapped in an orthogonal manner to result in a bandwidth efficient frequency division multiplexing transmission. OFDM may be viewed as transmitting many parallel low-rate narrowband signals rather than one fast-rate wideband signal. In practice, OFDM signals are generated using the fast Fourier transform algorithm, which is straightforward to implement on a chip. However, OFDM systems suffer high peak-to-average power ratio, which reduces the transmission range. In addition, a time-varying channel entails large Doppler spreads which yield intercarrier interference (ICI) among the OFDM subcarriers. During the last decade, a surge of research has been addressing ICI suppression in OFDM systems studies [64–68].

In DSSS modulation, the transmitter spreads the data symbols over a larger bandwidth by multiplying each symbol with a pseudo-random code. This code is known at the receiver in order to achieve message synchronisation. The original data symbol is recovered by multiplying the blocks of the received signal with the spreading code followed by an integrator. Essentially, the DSSS demodulator becomes a correlator, which is very efficient to build in hardware. The spreading of data over a wider band automatically achieves frequency diversity and long codes can render the ISI negligible at the expense of lower data rates. In addition to achieving frequency diversity, DSSS systems provide anti-jamming capability and message privacy in the presence of other listeners [26]. In UWA communications, coherent DSSS modulation is used

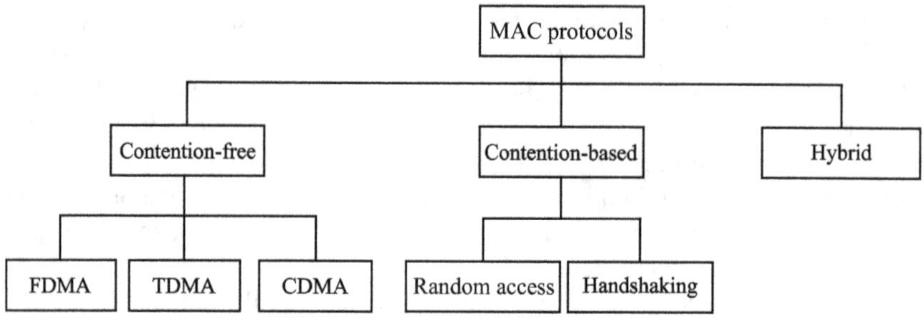

Figure 9.2 Categories for MAC protocols

in conjunction with a RAKE receiver [69], a DFE receiver [70], or with OFDM [71]. A communication system between an AUV and a vessel is described in [72]. Two communication modes are designed: (a) variable rate DSSS modulation resulting in 70–400 bps; (b) variable rate OFDM modulation resulting in 600–2,400 bps.

9.4 Underwater acoustic networking

The unique characteristics of underwater environments make the ability to communicate and to build a network very challenging. The challenge increases when moving assets are deployed in the network introducing changes in the topology and links reliability over time.

As described in Section 9.2.4, in order to cover large areas of deployment, which is typical of underwater networks, acoustic transmissions have to be employed. Solutions for terrestrial wireless networks cannot be directly applied to build reliable and robust UANs. Hence, various novel solutions have been proposed addressing both single-hop and multi-hop communication scenarios. In what follows, special attention is given to the challenges of sharing the communication medium and of routing messages in the network.

9.4.1 Data link layer

This section discusses some of the strategies and protocols proposed to share the communication medium for UANs. Multiple surveys on medium access control (MAC) protocols for UANs have been published in the open literature in the recent past [73–76]. As described in [74], the MAC protocols can be classified in three main categories: (1) contention-free; (2) contention-based; and (3) hybrid (see Figure 9.2).

Here, we do not try to review the state of the art in MAC protocols for UANs, instead we aim at describing advantages and disadvantages of the different strategies when considering the deployment of mobile vehicles. In particular, we highlight the impact on the protocol performance under different networking

Networking solutions designed for terrestrial wireless networks assume the usage of radio communications resulting in a higher bit rate and negligible propagation delays with respect to UANs.

scenarios: (i) a reduced number of vehicles with respect to the number static nodes; (ii) a similar number of static and mobile nodes; and (iii) a fully mobile network.

9.4.1.1 Contention-free MAC protocols

When using contention-free MAC protocols, the nodes of the network do not compete to access the channel. Communication channel resources are statically allocated to the network nodes and collisions are avoided by ensuring that each node uses its allocated resources exclusively. The usage of acknowledgment packets may still be required to cope with possible errors in the delivery of the messages due to the communication channels. Static allocation of resources is typically performed before the deployment of the network nodes, regardless the fact that nodes may require a different usage of resources over time. The usage of such static allocation strategies may introduce inefficiency in the communications. To overcome this limitation, dynamic approaches has been explored for UANs, and are presented in Section 9.4.1.3.

Next, we discuss the three major contention-free multiple access techniques considered for UANs, i.e. frequency-division multiple access (FDMA), time-division multiple access (TDMA), and code-division multiple access (CDMA). Table 9.2 summarises the main advantages and disadvantages for these contention-free strategies along with the most suitable networking scenarios for each strategy.

Frequency-division multiple access

FDMA divides the available frequency bandwidth into sub-bands and assigns each sub-band to an individual user. At any given time, each sub-band is used only by the user until it is released. It is possible to have multiple users transmitting simultaneously in the different sub-bands. Guard bands are usually required to avoid interference between close sub-bands, thus reducing the actual occupation of the available bandwidth.

The FDMA strategy is largely used for terrestrial radio networks and a few protocols have also been proposed for UANs [77–79]. Given the narrow bandwidth in UWA channels and the vulnerability of limited band systems to fading and multipath, the FDMA strategy should be used in small-scale single-hop scenarios with a limited number of users. In multi-hop scenarios or when the number of users increases, the reduction in bandwidth usage (due to guard bands) and in the transmission capability at each node may severely impair the network performance in terms of achievable throughput, end-to-end delays and energy consumption. The presence of mobile nodes does not impact the resources allocation.

Time-division multiple access

When using a TDMA approach, each node of the network has an assigned time slot and it is allowed to perform a transmission only in its own time slot, while being idle in the others. Guard times between slots are needed to avoid overlapping of transmissions and receptions. The duration of the guard time depends on the signal propagation delay. The sum of different time slots plus the guard times composes the TDMA frame. TDMA assumes some level of synchronisation among the network nodes. In case of poor synchronisation, longer guard times are needed to cope with the mismatch in the clock of the nodes.

The TDMA strategy is largely adopted for terrestrial networks [80] and for satellite systems [81]. Many solutions have also been proposed for UANs [82–88]. Avoiding any transmission/reception and reception/reception collision, TDMA-based protocols have the potential to obtain a high channel usage. However, UWA channels suffer from long propagation delays which impact the guard time duration. Traditional TDMA-based solutions require guard times to be long enough in order to ensure that the transmission and reception of a message occur within the same slot. Therefore, the guard time has to account for the maximum propagation delay in the network, which could be long in UANs. In the presence of unreliable communication channels, acknowledgement (ACK) packets may be required to confirm the reception of a message. In this case, every slot has to be long enough to enable the transmission of the data packet and the reception of the corresponding ACK message within the same slot. When the number of nodes increases, the frame duration increases as well. Each node may need to wait for long time before transmitting even if high-priority messages are generated, e.g. urgent mission updates for mobile nodes as well as request/notification from the vehicles. Additionally, when considering multi-hop networks or when transmissions are not meant to be received by all the other nodes in the network, it could be possible to have node transmitting/receiving at the same time without interfering with each other or without actually affecting the delivery of the messages to the intended nodes. In these cases, traditional TDMA-based protocol may result in poor network performance.

The usage of this kind of MAC protocols should be therefore preferred for small-scale networks (i.e. limited number of nodes and limited maximum distance) and in single-hop scenarios where all the nodes (or a large set of them) need to communicate with each other. Additionally, it should be preferred in scenarios where all nodes have always data to transmit in their slots, thus avoiding the waste of resources. Since guard times are needed for TDMA-based solutions, the presence of mobile nodes may introduce changes on the maximum expect propagation delay (depending on the network geometry) and guard times definition. One way to overcome this limitation is to impose constraints on the operational area of the vehicles. This makes possible to respect the guard time limits but it may impair the vehicle mission requirements.

In order to reduce the effect of long propagation delays, the use of centralised mathematical models has been explored to produce an optimal slot allocation for TDMA-based protocols [89–91] (Opt TDMA in Table 9.2). However, these models are suitable for static scenarios. The presence of a single mobile vehicle in the network would in fact introduce changes over time affecting the optimal slot allocation. This would require the need to recompute the optimal solution at each change and to distribute it to the network nodes, thus making this approach not feasible especially in the presence of an unreliable communication channel.

Code-division multiple access

CDMA allows multiple users to operate simultaneously over the entire frequency band. Signals from different users are distinguished employing different coding

Table 9.2 Summary of advantages and disadvantages for the three major contention-free multiple-access techniques (discussed in Section 9.4.1.1) in UANs

	Advantages	**Disadvantages**	**Suitable for**
FDMA	Multiple users can access simultaneously the communication channel	Narrow bandwidth; needs of guard bands; vulnerability of limited band systems	Small-scale single-hop scenarios with a limited number of users
TDMA	No collisions	Limited channel utilisation efficiency in large-scale networks	Small-scale single-hop scenarios where all nodes need to communicate with each other and all nodes have always data to transmit
Optimised TDMA	No collisions and improvement in channel utilisation with respect to traditional TDMA	Centralised solutions which assume the knowledge of propagation delays in the network	Static networks with a known traffic pattern
CDMA	Multiple users operating simultaneously over the entire frequency band; robust to frequency-selective fading	Near-far problem reduces the performance; need of link quality estimation to adjust the transmission power and code length	Static networks or with limited mobility

sequences. These coding sequences are used for spreading the user messages. There are two basic spreading techniques, namely direct sequence (DS) and frequency-hopped (FH) spread spectrum. In the DS-CDMA, information signals are linearly modulated using wideband codes, whereas in FH-CDMA the carrier frequencies of the nodes are changed according to a pattern obtained from codes. CDMA transmission techniques are robust to frequency-selective fading caused by underwater multipaths.

Various CDMA-based solutions have been proposed for UANs [70,92–95] exploring the usage of different codes. These solutions have the potential to achieve a high network throughput, low channel access delay, and low energy consumption. However, CDMA-based protocols suffer the near-far problem, which is a condition where a receiver captures a strong signal and thereby makes it impossible for the receiver to detect a weaker signal. This can significantly reduce the network performance. In terrestrial radio systems, this problem is commonly solved by dynamic output power adjustment of the transmitters. In [93–95], a similar approach is proposed for UANs. A closed-loop distributed algorithm is employed to jointly set the optimal transmit power and code length to use. In [94,95], the combination of a CDMA technique with a hybrid automatic repeat request strategy is instead explored. Nodes estimate the quality of the links with the neighbour nodes and cooperate in order to increase the achievable throughput and reduce the number of packet retransmissions and energy consumption.

According to the CDMA strategies currently proposed in the open literature, it is fair to say that CDMA is more suitable for static UANs (or with limited mobility). The presence of mobile nodes makes this kind of solution more complex due to Doppler issues. Node mobility results in changes in the link quality over time, complicating the estimation required for dynamic power management. Additionally, in many applications, mobile nodes are used for data muling, thus moving close to other nodes for data offloading. This may increase the near-far problem effects.

9.4.1.2 Contention-based MAC protocols

Contention-based MAC protocols are designed assuming that multiple users need to access the communication channel without pre-coordination. With respect to contention-free solutions, no channel resources are assigned to the nodes, which need to contend to access the medium. Additionally, each node accesses the channels only when it needs to transmit data, thus solving the problem of under/over allocation of resources to a node over time. This approach could lead to increased channel utilisation, but it may cause collisions among nodes. Contention-based protocols can be divided into random access and handshaking protocols.

Table 9.3 summarises the main advantages and disadvantages for these contention-based strategies along with the most suitable networking scenarios for each strategy.

Random access

Pure random access MAC protocols do not make any use of control messages to reserve the channel. A typical example is the ALOHA protocol [96], which has been largely studied for terrestrial radio networks. In pure ALOHA, when a node has a packet to send, it transmits the information immediately, without performing any channel sensing before accessing the medium. An ACK message is sent back by the receiver if the packet is received without errors. In case of errors in delivering the data or the ACK, the sender retransmits the same packet again. A version without ACK can be also considered if no delivery confirmation is required. Many different flavours of the ALOHA protocol have been proposed for UANs [97–99] trying to enhance the protocol performance with respect to pure ALOHA. The usage of carrier sensing (CS) before transmission (CS-ALOHA) has been considered as well as the usage of a probability value to decide if performing a transmission or not (probabilistic ALOHA). CS [100] is largely used in radio networks since it prevents collisions with ongoing transmissions at the transmitter side. When using CS, a node with a packet to transmit first listens to the medium for a certain amount of time. If it does not hear a transmission from another node, it is then allowed to begin its own transmission. However, due to the long propagation delay of UWA channels, when CS is used, the channel may be sensed idle while a transmission is ongoing, since the signal may not have reached the receiver yet. In UANs, CS most of the time means checking

Table 9.3 Summary of advantages and disadvantages for the two classes of contention-based MAC protocols (discussed in Section 9.4.1.2) in UANs

	Advantages	**Disadvantages**	**Suitable for**
Random access	Limited use of control messages and introduced delays; easy to implement; suffer less the presence of unreliable channels	Possibility of collisions	Large-scale and multi-hop scenarios with a low-medium traffic load; both static and mobile networks
Handshaking	Can reduce the number of data/data collisions	Use of multiple control messages; reduce channel utilisation in the presence of long propagation delays	Small-scale scenarios with a reliable communication channel; limited presence of mobile nodes in case of protocol optimisation

if a reception is currently ongoing before starting a transmission. In [98], a version of the ALOHA protocol with collision avoidance (ALOHA-CA) and the other with advance notification (ALOHA-AN) have been proposed. ALOHA-based solutions are typically very simple and easy to implement. Although these kinds of solutions are able to achieve limited performance in terms of throughput for terrestrial radio networks, they have the potential to better perform with respect to more complex solutions (requiring the transmissions of multiple control messages) in UANs [101]. Limiting the usage of control messages makes possible to reduce the delays and overhead introduced in the network. This can have a positive impact when deploying networks at sea.

ALOHA-based protocols should be preferred in multi-hop scenarios, where parallel communications can occur in different areas of the network, with a low-medium traffic load. Additionally, these solutions seems to be suitable in the presence of significant unreliable channels, where the usage of many control messages can introduce long delays and a large overhead. Not requiring any channel estimation or similar mechanism, ALOHA-based solutions can be employed in the presence of mobile nodes, without additional constraints with respect to fully static deployments.

Handshaking

Handshaking MAC protocols represent another class of contention-based techniques where control messages are exchanged between the sender and receiver to reserve the channel, before the actual data transmission. This class of protocols introduces handshaking by adding two control packets called request-to-send (RTS) and clear-to-send (CTS) [102]. The RTS is transmitted by the sender to reserve the communication channel. Once the receiver correctly decodes the RTS, it replies with a CTS in case the data message can be received, otherwise no CTS is transmitted and the sender will have to restart the process again after some back-off time. The usage of ACK packets can be introduced as well to confirm the reception of the data, if needed. The usage of the RTS and CTS messages allows the sender and receiver to inform their neighbours about the coming data message, thus reducing/avoiding the presence of possible interfering nodes. This mechanism solves the hidden terminal problem but does not solve the exposed terminal problem. Additionally, it relies on channel symmetry. A version combining handshaking with an Automatic Repeat ReQuest (ARQ) strategy has also been proposed [103] in order to improve the performance and reliability of the protocol. This class of protocols is largely used for terrestrial radio networks with solutions also proposed for UANs [104–106]. In the underwater domain, however, the long propagation delays have a significant impact on the protocol performance. Long time may be required to complete the handshaking, thus spending more time in acquiring the channel rather than in transmitting data [101]. Some of the protocol proposed for UANs try to estimate the actual propagation delay based on the RTS–CTS round trip time, thus avoiding the use of long timeout values. This has however an impact when mobile nodes are deployed, since the propagation delay changes over time, depending on the position of the node and on the channel. Moreover, in UANs it is possible to have packets already transmitted in the channel and not yet received at the sender and receiver nodes. This may results in possible interference in the neighbourhoods if additional control or data messages are transmitted. In [105], the authors try to address this problem introducing a warning time before the actual data transmission. This can help in reducing the probability of control/data collisions, at the cost of increasing the time require to reserve the channel. Other solutions have explored the usage of different channels for control and data messages [107–110]. In [111], an handshaking protocol using the long propagation delays to improve the channel usage is explored. Similarly to the optimised TDMA solutions, nodes can perform overlapping transmissions of control and data packets as long as there is no interference at the destination nodes. An accurate knowledge of the propagation delays and of the packets to be transmitted is required.

Similarly to the case of TDMA-based solutions, handshaking protocols should be preferred for small-scale scenarios, thus reducing the impact of long propagation delays. Additionally, these solutions should be preferred in the presence of reliable communication channels, thus reducing the effect of many retransmissions of RTS/CTS messages. In case nodes try to estimate the propagation delays to their neighbours to improve the protocol performance, static or quasi static networks should be preferred.

9.4.1.3 Hybrid MAC protocols

Hybrid MAC protocols explore the use of different medium access techniques and different type of MAC strategies to be able to adapt to the need of the network over time and to improve the overall performance. This approach can help in improving the

The hidden terminal problem occurs when a node can communicate with a neighbour node x, but cannot directly communicate with neighbours of x.

The exposed node problem occurs when a node is prevented from sending packets to other nodes because of a neighbour transmitter.

Table 9.4 Summary of advantages and disadvantages for the hybrid strategies discussed in Section 9.4.1.3

	Advantages	**Disadvantages**	**Suitable for**
ALOHA+ CDMA [112]	More flexible adaptation of transmission power and code length	Less effective in the presence of mobile nodes	Static networks or with limited number of mobile nodes
TDMA+ CDMA [113]	Spatial reuse of channel resources	Need of a hierarchical organisation of the network	Static network
ALOHA+ TDMA [114, 115]	Dynamic allocation of data slots	Suffer errors occurring during the request phase and changes in the network topology	Static networks or with limited number of mobile nodes

usage of the channel resources with respect to static allocation, as for contention-free MAC protocols. Different strategies have been proposed for UANs, mostly focusing on (1) combining CDMA with random access or TDMA protocols and (2) combining TDMA and random access strategies.

In [112], a distributed hybrid MAC protocol combining both random access and CDMA is presented. Each data packet is composed of two parts: header and payload. Nodes simultaneously access the channel using an ALOHA-based MAC for the header and a CDMA scheme for the payload. Adaptive power control and code length is employed. The header part contains information required to decode the data part. This solution however still suffers the problem of correct adaptation of transmission power and code length in the presence of mobile nodes. In [113], a combined TDMA–CDMA strategy is explored for clustered UANs. TDMA is used for the intra-cluster communication and CDMA for inter-cluster communication, thus achieving a spatial reuse of channel resources. Nevertheless, maintaining a hierarchical organisation of the network can be challenging, especially when mobile nodes are employed. In [114,115], the combination of TDMA and ALOHA protocols is considered. The proposed MAC protocol divides a time frame into two portions. In the first portion, ALOHA is used to transmit the requests to acquire slots in the second portion. In the second portion, which is much larger than the first one, nodes use TDMA to transmit data according to the already acquired slots. A guard time is required between the two portions to avoid collisions between request and data packets. By estimating the propagation delays among node, the use of guard times can be reduced in the second portion. This approach enables to achieve dynamic resource allocation but significantly suffers of errors occurring in the first portion of the frame. Additionally, the presence of mobile nodes and varying propagation delays require the introduction of guard times between data slots, thus reducing the channel utilisation.

Table 9.4 summarises the main advantages and disadvantages for the hybrid strategies described above along with the most suitable networking scenarios for each strategy.

9.4.2 Network layer

This section presents some of the strategies and protocols proposed to route messages in an UAN. Various surveys on routing solutions for UANs have been published in the recent past [73,116–123]. Both regular networks and delay-tolerant networks have been considered. Similar to the section about MAC protocols, in what follows we try to highlight advantages and disadvantages of different routing strategies when applied to UANs in the presence of mobile nodes.

As discussed in [73], routing protocols for terrestrial radio networks are traditionally divided into three categories:

- *Proactive protocols*. These protocols attempt to minimise the message latency by maintaining up-to-date routing information at all times from each node to every other node. Every time a change in the network occurs, the information about the whole network needs to be updated and delivered to the nodes. Some proactive routing protocols have been proposed for UANs assuming a tree-based hierarchical organisation of the network [124,125]. A central manager is used to compute the tree structure and the routing tables which are then distributed to the network nodes. These protocols are not suitable for dynamic underwater networks since a high overhead and long delays may be introduced to compute and propagate the required information. To reduce the overhead and delays introduced by a tree-based organisation of the network, various cluster-based routing protocols have also been investigated [126–129]. Different techniques have been explored to build, reorganise, and interconnect the clusters. However, in the presence of mobile nodes, the network topology may change over time as well as the node neighbourhoods. Many network updates can be therefore introduced over time which have to be distributed in the network, thus incurring in a significant overhead.
- *Reactive protocols*. Network nodes initiate a route discovery process only when a route to a destination is required. Once a route has been established, it is maintained by a route maintenance procedure until it is no longer desired. Multiple reactive routing protocols have been proposed for UANs [130–132]. However, for this class of protocols, the full route to the destination has to be found and maintained. This makes them not very suitable for dynamic underwater networks since long delays and a high overhead may be introduced to establish and maintain the route. Additionally, most reactive protocols assume the presence of stable symmetrical links which may be not the case in UANs, especially with mobile vehicles.
- *Geographical protocols*. These protocols use the information about the position of the nodes to deliver packets in the network. Each node selects its next hop based on its own position, the position of its neighbours, and the position of the destination node. These techniques are very promising for their scalability and limited signalling requirements with solutions proposed for UANs [133–135]. However, without the availability of a Global Positioning System (GPS), obtaining and maintaining accurate position data in an UANs is a challenging task, especially in the presence of mobile nodes.

Table 9.5 Summary of advantages and disadvantages for the three categories of routing protocols usually employed for terrestrial radio networks when applied to UANs

	Advantages	**Disadvantages**	**Suitable for**
Proactive	Routing information always available; minimum delay and overhead	Large overhead and delay; scalability is a major issue	Small static or semi-static networks
Reactive	Route is only determined when actually needed	The full path has to be established; long delays and a high overhead; need of symmetric links	Static or semi-static networks
Geographical	Scalable with limited signaling requirements	Obtaining, and distributing accurate position data	Static networks or with limited number of mobile nodes

Table 9.5 summarises the main advantages and disadvantages for these three categories of routing protocols when applied to UANs along with the most suitable networking scenarios for each category.

According to the challenges described above, most of the solutions proposed so far for dynamic UANs (with the presence of mobile nodes) are based on broadcasting replicas of the messages in the network (flooding) and on hop-by-hop strategies. In some cases, limited knowledge of position data is also assumed. Additionally, one of the main constraints of UANs is the energy budget since nodes are usually battery powered. Mobile nodes are even more impaired by this constraint since a large part of the energy is required for the motion of the node. Adding more batteries is not always possible since it increases the size and weight of the node and can impact the vehicle navigation capability. Special attention should be therefore placed in selecting the best routes to use, depending on the energy budget of the nodes and the type of the nodes (static or mobile).

All these aspect are discussed in what follows. Tables 9.6 and 9.7 summarise the main advantages and disadvantages for the various considered routing solutions along with the most suitable networking scenarios for each solution. As expected, routing solutions that tries to use topological information to reduce the introduced overhead may be less effective in the presence of mobile nodes, as presented in Table 9.7.

9.4.2.1 Flooding-based routing

Flooding is a simple routing solution where every new incoming packet is broadcasted in the network. When a node identifies an incoming packet as one it already forwarded, that packet is discarded. Traditional flooding does not assume the presence of any common data collection point (sink) in the network or link symmetry and it works for both static and mobile networks. Flooding-based protocols usually increase the robustness of the network in terms of packet delivery, introducing however a large number of message replicas transmitted at each node. This usually leads to a high overhead and energy consumption in the network. In [136], a flooding-based solution for UANs is presented where a desynchronisation mechanism is introduced to avoid that several packets forwarded at the same time result in collisions at the receiver. This can impact UANs where the bit rates of the communication devices are low and the packet transmission time and data collision window can be long.

Restricted-flooding based on geographical information

Various restricted-flooding strategies have been investigated to reduce the number of replicas transmitted in water. Some of these strategies assume the knowledge of the position of the destination node in order to control the packet flow towards the right area. Only nodes within a given area [137–139] or a given angle [140,141] are allowed to forward the received messages. However, as mentioned above, determining geographic information under water can be difficult and it could require high cost/overhead. To address this issue, some protocols have been designed to use only partial geographical information, i.e. node depth, when forwarding packets in the network [143–145]. Although all these restricted flooding protocols, based on geographical data, can work in the presence of asymmetric links and mobile nodes, they limit the selection of relay nodes to the ones providing a positive advancement (i.e. distance reduction) towards the destination. This may incur in the possibility to drop the packets at dead-end nodes, even if a stable path between source and the destination exists, passing through node(s) not providing such a positive advancement. Additionally, all the solutions presented so far are designed for networks where there is a sink node receiving all the generated data and do not support the possibility to route packets between any pair of nodes in the network. This restricts the usage of these solutions to specific scenarios and can be a problem if mission updates (or other control messages) have to delivered to the mobile nodes, e.g. to change/adjust the current ongoing operations.

Restricted-flooding based on link and neighbourhood information

To overcome the limitation introduced by a geographical approach, other restricted-flooding strategies have been proposed for UANs. In [136], a multipoint relay protocol is proposed for UANs where historical information on link quality and topology status is used to select the best set of relay nodes. The node selection is performed as a trade-off between the reduction in the number of replicas and the maximisation in the probability of delivering the message. The presence of mobile nodes is taken into account and messages can be delivered to any other node in the network – i.e. there is no assumption on the presence of a common data collection point. In order to estimate the required link quality information, control messages are periodically broadcasted in the network or piggybacked to regular data packets. This introduces an additional overhead in the network, especially in scenarios where the transmission of

Depth estimation can be easily determined with greater accuracy using on-board sensors [142].

Table 9.6 Summary of advantages and disadvantages for the different flooding-based routing strategies for UANs

	Advantages	**Disadvantages**	**Suitable for**
Traditional flooding	Robust; no assumption on link symmetry and presence of a sink node; possibility to deliver messages to any nodes in the network	Multiple replicas of each data message are broadcasted in the network; high overhead; scalability issues	Small-medium size networks ranging from static to fully mobile
Restricted-flooding with position information	Reduced number of replicas with respect to traditional flooding	Need of geographical data; possible dead ends in the route	Static network or with limited mobility; a static sink node needs to be deployed
Restricted-flooding with depth information	Reduced number of replicas with respect to traditional flooding	Possible dead ends in the route	A sink node must be deployed at the surface
Restricted-flooding with link and neighbourhood information	Reduced number of replicas with respect to traditional flooding; no need of a sink node	Estimation of data required in the neighbourhood which may introduce additional overhead	Networks with medium-high traffic loads to increase the estimation accuracy

few data packets is required over long time periods, such as environmental monitoring applications. This approach is enhanced in [146] where control data are only added in piggyback to regular data messages while a traditional flooding solution is used when no updated information is available. In [147], nodes try to estimate the number of neighbour nodes in order to reduce the number of relays. Each node waits for back-off time before broadcasting the message. This back-off time is adjusted according to the number of duplicates heard of each packet. If the number of duplicates is larger than a given threshold, the node will not broadcast the message. This approach reduces the number of message replicas transmitted in the network, but it also reduces the robustness of the flooding strategy since no feedback is provided to estimate the correctness of the decision. The selection of the correct threshold value to use depends in fact on the topology configuration and may vary at each node or over time, due to node mobility.

Table 9.6 summarises the main advantages and disadvantages for the various flooding-based routing solutions considered in this section along with the most suitable networking scenarios for each solution.

9.4.2.2 Hop-by-hop routing

Hop-by-hop protocols are usually designed to reduce the number of packets broadcasted in the network with respect to the flooding-based approach. Handshaking mechanisms are typically adopted to find the next hop relay and to tune the parameters of the data exchange. Short control messages are exchanged between the transmitting node and the possible relay nodes to find the best available option [148–151]. Although reducing the number of replicas broadcasted in the network, this class of protocols usually requires the presence of symmetric links over the paths connecting the source node(s) to the destination node(s). Additionally, the presence of a common data collection point is typically required. Due to the exchange of control information, hop-by-hop solutions can impose long latency and a reduction in the protocol robustness. To reduce the introduced overhead and improve the network performance, approaches based on cross-layering and reinforcement learning techniques have been proposed.

Cross-layering

Recently, it has been shown that cross-layer techniques can impact protocol performance positively, especially in networks with limited resources and/or deployed in challenging environments, like UANs [152,153]. Various cross-layer solutions have been proposed for UANs combining MAC and hop-by-hop routing tasks to reduce the introduced overhead [154–157]. Link quality information is usually estimated and transmission power or coding modulation schemes are tuned to increase the reliability and energy efficiency of the communications. Although cross-layering can improve the overall network performance, hop-by-hop solutions are still limited to scenarios where a sink node is employed. In case the sink is mobile, the task of routing messages in the network becomes more challenging.

Reinforcement learning

Nodes learn about the status of the network over time in order to promptly and intelligently adapt to topological changes without any offline training. Various strategies have been proposed for UANs [158–161], mostly focusing on energy consumption (or residual energy) criteria. Each node establishes a reinforcement function based on the collected information and selects the best relay node to use. Although reinforcement learning techniques can help in reducing the energy consumption in the network and improve the overall performance, the presence of many topological changes (multiple mobile vehicles or a mobile data collection point) makes the learning process slower and more challenging.

Table 9.7 summarises the main advantages and disadvantages for the hop-by-hop routing solutions considered in this section along with the most suitable networking scenarios for each solution.

9.4.3 Cognitive communications

AUV networks are faced with the bottleneck of communications because there is not a universal waveform/protocol that can cope with all different environments and network topologies [162]. For this reason, there is a need to abandon monolithic integrated modems and develop software-defined open-architecture modems (SDOAMs) that can adjust their waveform and protocol parameters depending on channel conditions and network requirements. The main characteristic of a SDOAM is the ability

Table 9.7 Summary of advantages and disadvantages for the different hop-by-hop routing strategies for UANs

	Advantages	**Disadvantages**	**Suitable for**
Traditional hop-by-hop	Single relay node and reduced number of packets broadcasted in the network	Use of handshaking mechanisms; need of symmetric links; need of a sink node	Static networks or with reduced mobility with a static sink node
Hop-by-hop with cross-layering	Single relay node and improved uses of available communication links with respect to traditional hop-by-hop solutions	Use of handshaking mechanisms; need of a sink node	Static networks or with reduced mobility with a static sink node
Hop-by-hop with reinforcement-learning	Nodes learn and adjust the selection; reduction in the energy consumption	Many topology changes make the learning process slower and more challenging; need of a sink node	Static networks or with reduced mobility with a static sink node

Note: General observation: none of this methods are really efficient for teams of AUVs.

to learn the channel dynamics and adapt at all layers of the communication stack. For instance, a cognitive modem may have the capability to provide physical layer information in the upper layers for adjusting AUV control messages or selecting the most appropriate AUV node for data delivery. Similarly, upper layers of the communication stack may share quality of service data, such as message priority and maximum delivery time, to make the best choice of coded modulation, encryption, and communication modality. An excellent review of current industrial and academic SDOAMs is found in [163] and references therein.

SDOAMs that allow flexibility and modularity have been the research focus for a decade. In [164], a unified software for both simulation and implementation of MAC protocol development is proposed. Both the modem and simulator use the same C code to implement the logic of the selected MAC protocol; therefore, performance comparisons between simulated and real environments are possible. An extension of [164] is proposed in [165] where the UNET-2 framework is presented. The paper describes a software-defined modem and a communication protocol stack that could investigate new coding schemes and to evaluate networking protocols in simulation and at sea. In [166], a novel open framework, coined SUNSET, is described. SUNSET supports the design and evaluation of networking protocols via simulations, in-lab and at-sea testing. It is based on two well-known and largely used simulation tools for wireless sensor networks: ns-2 [167] and ns2-Miracle [168]. By extending the ns2-Miracle tool, SUNSET can make use of the Bellhop ray tracing software [169] to model the UWA channel. SUNSET supports commercial modems and has been interfaced with several AUVs. Similarly to SUNSET, the DESERT framework [170] is a simulation/emulation software based on ns-2 and ns2-Miracle, focusing on the design of MAC and routing solutions. SUNSET and DESERT are interoperable and both of them are compatible with the SDOAM design presented in [171]. They provide, therefore, a layered structure with the support for multi-protocol co-existence at each layer and cross-layering functionalities. A similar approach has been recently proposed by the NATO Centre for Maritime Research and Experimentation (CMRE) with the development of a Cognitive Communications Architecture (CCA) [172]. Differently from SUNSET and DESERT, this architecture is specifically designed for in-field experimentation. The usage of the CCA has been successfully demonstrated during various at-sea trials [173,174] on board of different platforms.

Understanding the channel at the physical layer is the first step towards achieving cognitive communications. Hence, an integral part of the SDOAM is adaptive modulation and coding (AMC). AMC refers to intelligent selection of the bit rate to meet user requirements under various channel conditions. As in radio channels, the conventional approach to AMC in underwater is to maximise a throughput-related objective function subject to a set of link reliability constraints. These constraints are appropriately defined in terms of link quality metrics. In [175], both the SNR and the channel Doppler spread are used as metrics to switch between non-coherent FSK and adaptive equalisation. The authors in [176] use closed-form BER expressions for OFDM under the assumption of Rayleigh fading. From these expressions, a set of thresholds determine the modulation and/or power level needed on each sub-carrier of the OFDM system. Moreover, the authors have addressed the issue of feedback delay through linear prediction of channel state information. A method that adapts the code rate of Reed–Solomon codes based on the guard interval needed to accommodate interference free TDMA links in an UAN network is proposed in [177]. The adaptation is based on range information incorporated into the packet exchange of the ARQ protocol. Recently, data-driven machine learning techniques for AMC have been applied. In particular, multi-armed bandit techniques have been applied by the researchers in [178–180] while BER prediction based on decision trees have been applied in [181,182]. Real-time AMC implementation for OFDM acoustic modems has been demonstrated in [183–186].

At upper layers, adaptive routing [157] and strategies that propose a smart switch between acoustic and optical signals have been studied [187–189]. Acoustic signals are typically used for long-range low-bit rate data transmissions, while optical signals are used for short-range high-bit rate data delivery. In [190], the use of optical and acoustic communications is combined with the problem of optimising the path of an AUV, acting as a data mule. The objective is to maximise the value of the information retrieved by the AUV, which decays in time, while using fast data transfer over short ranges and low data transfer over long ranges.

9.5 Security in underwater communications

In terrestrial radio networks, secure solutions and standards are well established and security aspects are nowadays considered as key metrics during the design and implementation of novel protocols. The properties of acoustic channels and underwater networks exposes UANs against a large array of malicious attacks. Attackers can very easily jam or intercept underwater communications and this applies to both static and mobile nodes. Additionally, current technology of UANs does not consider security as a performance indicator. Aspects related to securing the communications and the network operations have been marginally investigated (or not considered at all), leaving room for potential attackers, which might make a UAN unusable [191]. Given to the unique characteristics of UANs, existing security solutions in terrestrial networks cannot be directly applied and novel strategies have to be explored.

To secure the communication link, the use of sophisticated modulation techniques employing spread spectrum (SS) signals have been investigated in [192,193]. These SS signals have the inherent ability to provide covertness, i.e. low probability of intercept/detection characteristics due to their low-level transmit power densities. To ensure confidentiality and integrity of the communication, data encryption protocols have been addressed in [194,195]. Encryption keys handled by upper network layers are typically used for security of information. Unfortunately, updating and distributing these keys typically induce an overhead in the network, especially in the presence of mobile nodes that can join and leave the network over time. In [196,197], the concept of physical layer security [198] has been explored to use specific properties of the communication channel to generate secret keys. Received signal strength is explored in [196], while channel frequency response is used in [197].

Although physical layer security and data encryption solutions can be used to ensure confidentiality, integrity, and authentication on the communication link, various denial of service attacks can be still conducted to impair the performance of a network of collaborative nodes. Jamming, spoofing, and packet manipulation and redirection (just to name a few) can be effectively applied to impair the navigation and localisation capabilities of mobile robots. These are critical capabilities which have to be ensured in order to correctly accomplish the requested tasks. Various strategies against denial of service attacks (mostly jamming and wormhole) have been proposed in the open literature, along with secure routing solutions and reputation-based models [199–207]. However, these efforts focus only on attacks conducted at a particular layer of the protocol stack, without considering any coordinated approach spanning over multiple layers. It is known that attacks at different layers require different countermeasures. Therefore, independent (layered) countermeasures could result in conflicting requirements leading to network performance degradation. One recent attempt to combine data encryption and routing has been proposed in [195]. In [208,209], the use of a software-defined approach with intense use of cross-layer messages is instead proposed. The objective of this approach is to enable the different layers of the protocol stack (from the physical layer up to the application layer) to work in a coordinated way to adapt and react to changes and attacks occurring in the network. Special attention is given to the use of node cooperation and mobility in order to react to possible attacks by adapting the network geometry similarly to what has been described in [210]. Different strategies can be in fact explored to allow mobile nodes to cover connectivity holes in the network, to increase the quality of the communication links and to avoid the presence of single points of failure. In same cases, the use of mobile nodes and a data muling approach, combined with a multi-modal communication system (e.g. acoustic and optical links), could be the only effective solution against jamming.

9.6 Interoperability and standards

Our twenty-first-century experience of connectivity is built on standards. Such standards support interoperability at planetary scale and provide unprecedented connectivity in a range of environments through diverse technologies (Wi-Fi, Bluetooth, cell networks, etc.). A user of today's communications technology is offered the convenience of being able to interconnect devices or plug a gadget into a communications infrastructure regardless of its manufacturer. This happens in stark contrast with the world of underwater communications that, until recently, was manufacturer-specific generally using proprietary modulation and coding technologies. Such manufacturer dependence introduces procurement and obsolescence management risks and prevents the development of scalable and truly ad-hoc AUV networks. To fill this gap, the NATO CMRE developed and brought to the community the first-ever standard for digital underwater communications – JANUS. JANUS is, in a nutshell, a modulation and coding scheme that also incorporates some additional mechanisms to render it practical in real use. A FH-FSK modulation scheme is employed as the base of JANUS. The specification of JANUS can be found in NATO's Standardisation Office Website (www.nso.nato.int) under the reference STANAG4748. An academic publication with a draft specification was made available in [59] and reference implementations are freely available online [211] so that anyone can construct a transmitter/receiver to communicate via JANUS with any other compliant platform.

The key motivation for JANUS was to provide a simple baseline mode that manufacturers could add to their modem products as an enabler for interoperability, where JANUS would be the *lingua franca* of underwater communications systems – nodes could use JANUS to establish a first contact, discover capabilities, and eventually switch to another common and more performant modulation [212]. JANUS also offers basic data exchange capabilities for cases where no other common modulations are present.

The main design drives behind JANUS were simplicity of implementation and robustness to challenging channels. The simple implementation means that manufacturers willing to experiment or even implement JANUS in their hardware will not require latest generation digital signal processors or any type of advanced hardware. The robustness aspect typically comes at the cost of reduced data throughput, something that can well be accommodated in JANUS.

The following list enumerates some of the key features of JANUS

- **Common frequency band(s):** The JANUS waveform is fully parameterised based on centre frequency, providing flexibility in both definition and implementation details. For the initial JANUS specification, a centre frequency of 11,520 Hz was chosen, resulting in a frequency band between 9,400 and 13,600 Hz.

- **Openly available:** JANUS is open and free to all. Even though born from the efforts of NATO and CMRE, the requirement for global underwater interoperability was embraced from the outset and helped attract the broad international underwater communications community that has participated so strongly in JANUS development. Additionally, JANUS definition does not have any classification mark and is open to be used by NATO, non-NATO, military, and civilian.
- **Flexible packet definition:** The JANUS specification defines a flexible, nested bit allocation scheme under which digital packets can be composed. A baseline JANUS packet consists of 64 bits of information. Besides control and specification bits, the 'core' of the packet is primarily defined by the User Class ID, Application Type, and Application Data Block. The logic behind this arrangement is that for each of the 256 available User Class IDs (Generic applications and user groups), different application types can be specified along with their respective application data blocks.
- **Community driven:** JANUS was built on community consensus and its strength resides precisely in that community engagement. There are no technical limitations to the establishment on an underwater digital communications standard but the barriers can be immense if the different interests and requirements are not, at least, analysed and brought to the table for discussion.
- **Designed to live alongside proprietary solutions:** Lastly it may be important to underline that JANUS was never intended to replace existing industrial or even academic efforts. The fact that there is no 'one-fits-all' solution to solve the underwater communications problem opens the door for multiple approaches that can and should be employed whenever best suited. In this context, JANUS plays the role of 'lingua franca'.

In the recent past, at least two additional standardisation efforts in the field of digital UWA communications have been identified. From an industrial perspective, the Subsea Wireless Interest Group (SWIG) has proposed a JANUS-based standard to support offshore oil and gas exploration activities [213].

From a multi-national perspective, the International Organization for Standardization (ISO), together with the International Electrotechnical Commission (IEC) are developing the ISO/IEC 30140 [214] series of Standards. These aim at providing general requirements and guidance to assist the development of UWA sensor networks.

9.7 Future trends

Today, it is difficult to identify the one dominant application of underwater communications that may drive the market and push the technology. Civilian and military applications alike are taking advantage of the developments in AUV technology and

www.subseawirelessgroup.com

exploring the increased size of their fleets. This trend in AUV employment may well set a future landscape where underwater seamless connectivity will become a need. A subsea world where sensors, instruments, and vehicles are able to connect and build awareness of each other, just as we do now with our 'smart' gadgets, seems to still be a far-fetched proposition. However, under the concept of the 'Internet of Things' (IoT) one can certainly extrapolate and dream of a world of underwater connectivity.

The realisation of underwater IoT is contingent upon the ability to send digital information via acoustic signals. However, the combined effect of platform mobility with multipath variability is still the major limitation for achieving bandwidth-efficient high-speed acoustic communications. Hence, research on advanced receiver structures will not cease until the theoretical capacity limits are achieved. Given the lack of universal UWA channel models, the community is rapidly moving towards machine learning approaches for adaptive modulation. Receiver feedback is essential and future research will manifest the type of channel characteristics that are invariable during the two-way travel time. In addition, future mobile platforms will include transmit arrays to allow spatial division multiplexing strategies in multi-user networks and prevent energy propagating to an eavesdropper for security purposes.

Networking protocols will play an increasingly important role as the numbers of assets increase and the applications explore the those numbers to gain efficiency. Standards will be required addressing the format of the messages exchanged in the network and the encoding/decoding of the information, thus enabling interoperability beyond what currently done at the waveform level (i.e. JANUS).

The investigation of smarter strategies will require to take full advantage of multi-modal and heterogeneous communication and sensing solutions available at the node. This will make possible to truly explore software-defined and cognitive systems. Additionally, a coordinated integration of the communication, sensing localisation and navigation capabilities will allow mobile nodes to complete the required tasks in a more efficient way and to enhance their cooperation ability. This coordinated integration will make possible to effectively use AUV mobility to cover connectivity holes in the network, increase the quality of the communication links, and avoid the presence of single points of failure, just to name a few.

Security for underwater communications will certainly be an important feature in future applications. Novel protocols that will establish cryptographic keys without sharing any data between nodes combined with state-of-art cryptography will lead to substantial security gains. Additional security enhancements will be provided by distributed and cross-layer networking solutions that allow cooperation between the nodes of the network.

References

[1] N. Meratnia, P. J. M. Havinga, P. Casari, *et al.*, 'CLAM – Collaborative embedded networks for submarine surveillance: An overview', in *Proceedings of IEEE/OES OCEANS 2011*, 6–9 June, 2011.

[2] J. Heidemann, M. Stojanovic, and M. Zorzi, 'Underwater sensor networks: Applications, advances, and challenges', *Royal Society*, vol. 370, no. 1958, pp. 158–175, 2012.

[3] J. Xu, A. Lin, X. Yu, *et al.*, 'Underwater laser communication using an OFDM-modulated 520-nm laser diode', *IEEE Photonics Technology Letters*, vol. 28, no. 20, pp. 2133–2136, 2016.

[4] N. Farr, A. Bowen, J. Ware, C. Pontbriand, and M. Tivey, 'An integrated, underwater optical/acoustic communications system', in *OCEANS 2010 IEEE – Sydney*, May 2010, pp. 1–6.

[5] C. Pontbriand, N. Farr, J. Ware, J. Preisig, and H. Popenoe, 'Diffuse high-bandwidth optical communications', in *OCEANS 2008*, Sept 2008, pp. 1–4.

[6] M. Doniec, I. Vasilescu, M. Chitre, C. Detweiler, M. Hoffmann-Kuhnt, and D. Rus, 'Aquaoptical: A lightweight device for high-rate long-range underwater point-to-point communication', in *OCEANS 2009, MTS/IEEE Biloxi – Marine Technology for Our Future: Global and Local Challenges*, Oct 2009, pp. 1–6.

[7] J. Simpson, W. Cox, J. Krier, B. Cochenour, B. Hughes, and J. Muth, '5 Mbps optical wireless communication with error correction coding for underwater sensor nodes', in *OCEANS 2010*, Sept 2010, pp. 1–4.

[8] M. Dunbabin, P. Corke, I. Vasilescu, and D. Rus, 'Data muling over underwater wireless sensor networks using an autonomous underwater vehicle', in *IEEE International Conference on Robotics and Automation 2006*, Orlando, Florida, 2006, pp. 2091–2098.

[9] M. Doniec, 'Autonomous underwater data muling using wireless optical communication and agile AUV control', Ph.D. dissertation, Massachusetts Institute of Technology, 2013.

[10] N. Farr, J. Ware, C. Pontbriand, T. Hammar, and M. Tivey, 'Optical communication system expands CORK seafloor observatory's bandwidth', in *OCEANS 2010 MTS/IEEE SEATTLE*, Sep. 2010, pp. 1–6.

[11] M. Doniec, C. Detweiler, I. Vasilescu, and D. Rus, 'Using optical communication for remote underwater robot operation', in *2010 IEEE/RSJ International Conference on Intelligent Robots and Systems*, Oct 2010, pp. 4017–4022.

[12] X. Che, I. Wells, G. Dickers, P. Kear, and X. Gong, 'Re-evaluation of RF electromagnetic communication in underwater sensor networks', *Communications Magazine, IEEE*, vol. 48, no. 12, pp. 143–151, 2010.

[13] C. Uribe and W. Grote, 'Radio communication model for underwater WSN', in *2009 3rd International Conference on New Technologies, Mobility and Security*, Cairo, Egypt, December 2009, pp. 1–5.

[14] A. Zoksimovski, D. Sexton, M. Stojanovic, and C. Rappaport, 'Underwater electromagnetic communications using conduction – Channel characterization', *Ad Hoc Networks*, vol. 34, pp. 42–51, 2015.

[15] X. Che, I. Wells, G. Dickers, P. Kear, and X. Gong, 'Re-evaluation of RF electromagnetic communication in underwater sensor networks', *IEEE Communications Magazine*, vol. 48, no. 12, pp. 143–151, 2010.

[16] J. Lloret, S. Sendra, M. Ardid, and J. J. P. C. Rodrigues, 'Underwater wireless sensor communications in the 2.4 GHz ISM frequency band', *Sensors*, vol. 12, no. 4, pp. 4237–4264, 2012.

[17] A. Palmeiro, M. Martin, I. Crowther, and M. Rhodes, 'Underwater radio frequency communications', in *Oceans 2011 IEEE*, Santander, Spain, June 2011, pp. 1–8.

[18] A. Baggeroer, 'Acoustic telemetry – An overview', *IEEE Journal of Oceanic Engineering*, vol. 9, no. 4, pp. 229–235, 1984.

[19] M. Stojanovic, 'Recent advances in high-speed underwater acoustic communications', *IEEE Journal of Oceanic Engineering*, vol. 21, no. 2, pp. 125–136, 1996.

[20] D. Brady and J. C. Preisig, 'Underwater acoustic communications', in *Wireless Communications: Signal Processing Perspectives*, H. V. Poor and G. W. Wornell, Eds. Prentice-Hall, Inc., Upper Saddle River, 1998, vol. 8, pp. 330–379.

[21] D. B. Kilfoyle and A. B. Baggeroer, 'The state of the art in underwater acoustic telemetry', *IEEE Journal of Oceanic Engineering*, vol. 25, no. 1, pp. 4–27, 2000.

[22] M. Chitre, S. Shahabudeen, and M. Stojanovic, 'Underwater acoustic communications and networking: Recent advances and future challenges', *Marine Technology Society Journal*, vol. 42, no. 1, pp. 103–116(14), 2008.

[23] T. Melodia, H. Kulhandjian, L.-C. Kuo, and E. Demirors, *Advances in Underwater Acoustic Networking*. John Wiley and Sons, Ltd, 2013, ch. 23, pp. 804–852.

[24] P. v. Walree and R. Otnes, 'Ultrawideband underwater acoustic communication channels', *IEEE Journal of Oceanic Engineering*, vol. 38, pp. 678–688, 2013.

[25] F. B. Jensen, W. A. Kuperman, M. B. Porter, and H. Schmidt, *Computational Ocean Acoustics*, 2nd ed. New York: Springer, 2011.

[26] D. Tse and P. Viswanath, *Fundamentals of Wireless Communication*. Cambridge University Press, 2005, ISBN 978-0521845274.

[27] R. Stewart, *Introduction to Physical Oceanography*. University Press of Florida, 2009, ISBN 978-1616100452.

[28] J. Proakis and M. Salehi, *Communication Systems Engineering*, ser. Prentice-Hall International editions. Prentice Hall, 1994, ISBN: 9780131589322.

[29] M. Stojanovic, J. G. Proakis, and J. A. Catipovic, 'Analysis of the impact of channel estimation errors on the performance of a decision-feedback equalizer in fading multipath channels', *IEEE Transactions on Communications*, vol. 43, no. 2/3/4, pp. 877–886, 1995.

[30] J. C. Preisig and G. B. Deane, 'Surface wave focusing and acoustic communications in the surf zone', *The Journal of the Acoustical Society of America*, vol. 116, no. 4, pp. 2067–2080, 2004.

[31] A. Song, M. Badiey, H. C. Song, W. S. Hodgkiss, M. B. Porter, and K. Group, 'Impact of ocean variability on coherent underwater acoustic communications during the Kauai experiment (KauaiEx)', *The Journal of the Acoustical Society of America*, vol. 123, no. 2, pp. 856–865, 2008.

[32] T. Schneider and H. Schmidt, 'Model-based adaptive behavior framework for optimal acoustic communication and sensing by marine robots', *IEEE Journal of Oceanic Engineering*, vol. 38, pp. 522–533, 2013.

[33] R. Zimmerman, G. L. D'Spain, and C. D. Chadwell, 'Decreasing the radiated acoustic and vibration noise of a mid-size AUV', *IEEE Journal of Oceanic Engineering*, vol. 30, no. 1, pp. 179–187, 2005.

[34] L. Bjørnø, 'Chapter 6 – Ambient noise', in *Applied Underwater Acoustics*, T. H. Neighbors and D. Bradley, Eds. Elsevier, 2017, pp. 363–401, Paperback ISBN: 9780128112403.

[35] A. Mahmood and M. Chitre, 'Ambient noise in warm shallow waters: A communications perspective', *IEEE Communications Magazine*, vol. 55, no. 6, pp. 198–204, 2017.

[36] E. H. Roth, J. A. Hildebrand, S. M. Wiggins, and D. Ross, 'Underwater ambient noise on the Chukchi Sea continental slope from 2006–2009', *The Journal of the Acoustical Society of America*, vol. 131, no. 1, pp. 104–110, 2012.

[37] M. Chitre, J. Potter, and S. Ong, 'Viterbi decoding of convolutional codes in symmetric α-stable noise', *IEEE Transactions on Communications*, vol. 55, no. 12, pp. 2230–2233, 2007.

[38] R. Nadakuditi and J. C. Preisig, 'A channel subspace post-filtering approach to adaptive least-squares estimation', *IEEE Transactions on Signal Processing*, vol. 52, no. 7, pp. 1901–1914, 2004.

[39] K. Pelekanakis and M. Chitre, 'New sparse adaptive algorithms based on the natural gradient and the L_0-norm', *IEEE Journal of Oceanic Engineering*, vol. 38, no. 2, pp. 323–332, 2013.

[40] T. H. Eggen, A. B. Baggeroer, and J. C. Preisig, 'Communication over Doppler spread channels. Part I: Channel and receiver presentation', *IEEE Journal of Oceanic Engineering*, vol. 25, no. 1, pp. 62–71, 2000.

[41] S. H. Huang, J. Tsao, T. C. Yang, and S.-W. Cheng, 'Model-based signal subspace channel tracking for correlated underwater acoustic communication channels', *IEEE Journal of Oceanic Engineering*, vol. 39, no. 2, pp. 343–356, 2014.

[42] W. Li and J. C. Preisig, 'Estimation of rapidly time-varying sparse channels', *IEEE Journal of Oceanic Engineering*, vol. 32, no. 4, pp. 927–939, 2007.

[43] C. R. Berger, S. Zhou, J. C. Preisig, and P. Willett, 'Sparse channel estimation for multicarrier underwater acoustic communication: From subspace methods to compressed sensing', *IEEE Transactions on Signal Processing*, vol. 58, no. 3, pp. 1708–1721, 2010.

[44] J. Ling, K. Zhao, J. Li, and M. L. Nordenvaad, 'Multi-input multi-output underwater communications over sparse and frequency modulated acoustic channels', *The Journal of the Acoustical Society of America*, vol. 130, no. 1, pp. 249–262, 2011.

[45] W.-J. Zeng and W. Xu, 'Fast estimation of sparse doubly spread acoustic channels', *The Journal of the Acoustical Society of America*, vol. 131, pp. 303–317, 2012.

[46] A. S. Gupta and J. C. Preisig, 'A geometric mixed norm approach to shallow water acoustic channel estimation and tracking', *Physical Communication*, vol. 5, no. 2, pp. 119–129, 2012.

[47] K. Pelekanakis and M. Chitre, 'Adaptive sparse channel estimation under symmetric alpha-stable noise', *IEEE Transactions on Wireless Communications*, vol. 13, no. 6, pp. 3183–3195, 2014.

[48] F.-X. Socheleau, M. Stojanovic, C. Laot, and J.-M. Passerieux, 'Information-theoretic analysis of underwater acoustic OFDM systems in highly dispersive channels', *Journal of Electrical and Computer Engineering*, vol. 2012, pp. 1–15, 2012.

[49] Y. M. Aval, S. K. Wilson, and M. Stojanovic, 'On the achievable rate of a class of acoustic channels and practical power allocation strategies for OFDM systems', *IEEE Journal of Oceanic Engineering*, vol. 40, pp. 785–795, 2015.

[50] M. Chitre, 'A high-frequency warm shallow water acoustic communications channel model and measurements', *The Journal of the Acoustical Society of America*, vol. 122, pp. 2580–2586, 2007.

[51] W.-B. Yang and T. C. Yang, 'High-frequency channel characterization for M-ary frequency-shift-keying underwater acoustic communications', *The Journal of the Acoustical Society of America*, vol. 120, pp. 2615–2626, 2006.

[52] F.-X. Socheleau, C. Laot, and J.-M. Passerieux, 'Stochastic replay of non-WSSUS underwater acoustic communication channels recorded at sea', *IEEE Transactions on Signal Processing*, vol. 59, pp. 4838–4849, 2011.

[53] P. Qarabaqi and M. Stojanovic, 'Statistical characterization and computationally efficient modeling of a class of underwater acoustic communication channels', *IEEE Journal of Oceanic Engineering*, vol. 38, pp. 701–717, 2013.

[54] K. Pelekanakis and A. B. Baggeroer, 'Exploiting space–time–frequency diversity with MIMO–OFDM for underwater acoustic communications', *IEEE Journal of Oceanic Engineering*, vol. 36, no. 4, pp. 502–513, 2011.

[55] J. Huang, S. Zhou, and P. Willett, 'Nonbinary LDPC coding for multicarrier underwater acoustic communication', *IEEE Journal on Selected Areas in Communications*, vol. 26, no. 9, pp. 1684–1696, 2008.

[56] A. Goalic, J. Trubuil, C. Laot, and N. Beuzelin, 'Underwater acoustic communication using Reed Solomon Block Turbo Codes channel coding to transmit images and speech', in *OCEANS 2010 MTS/IEEE SEATTLE*, Sep. 2010, pp. 1–6.

[57] H. Zimmermann, 'OSI Reference Model - The ISO model of architecture for open systems interconnection', *IEEE Transactions on Communications*, vol. 28, no. 4, pp. 425–432, 1980.

[58] S. Coatelan and A. Glavieux, 'Design and test of a multicarrier transmission system on the shallow water acoustic channel', in *IEEE OCEANS'94*, vol. 3, Sep. 1994, pp. III/472–III/477.

[59] J. Potter, J. Alves, D. Green, G. Zappa, I. Nissen, and K. McCoy, 'The JANUS underwater communications standard', in *Proceedings of the Second IEEE OES Underwater Communications Networking Conference*, ser. UComms'14. IEEE, 3–5 September 2014.

[60] J. B. de Sousa, J. Pereira, J. Pinto, *et al.*, 'Rapid environmental picture Atlantic exercise 2015: A field report', in *MTS/IEEE OCEANS'16*, Monterey, CA, USA, Sep. 2016, pp. 1–6.

[61] M. Stojanovic, J. Catipovic, and J. Proakis, 'Phase-coherent digital communications for underwater acoustic channels', *IEEE Journal of Oceanic Engineering*, vol. 19, no. 1, pp. 100–111, 1994.

[62] J. Gomes, V. Barroso, G. Ayela, and P. Coince, 'An overview of the ASIMOV acoustic communication system', in *MTS/IEEE OCEANS 2000*, Providence, RI, USA, 2000.

[63] L. E. Freitag, M. Grund, J. Partan, K. Ball, S. Singh, and P. Koski, 'Multiband acoustic modem for the communications and navigation aid AUV', in *OCEANS 2005 MTS/IEEE*, Sep. 2005, pp. 1080–1085, Vol. 2.

[64] B. Li, S. Zhou, M. Stojanovic, L. Freitag, and P. Willett, 'Multicarrier communication over underwater acoustic channels with nonuniform Doppler shifts', *IEEE Journal of Oceanic Engineering*, vol. 33, no. 2, pp. 198–209, 2008.

[65] J. Huang, S. Zhou, J. Huang, C. R. Berger, and P. Willett, 'Progressive intercarrier interference equalization for OFDM transmission over time-varying underwater acoustic channels', *IEEE Journal of Selected Topics in Signal Processing*, vol. 5, no. 8, pp. 1524–1536, 2011.

[66] S. Yerramalli, M. Stojanovic, and U. Mitra, 'Partial FFT demodulation: A detection method for highly Doppler distorted OFDM systems', *IEEE Transactions on Signal Processing*, vol. 60, no. 11, pp. 5906–5918, 2012.

[67] Z. Wang, S. Zhou, J. Catipovic, and J. Huang, 'Factor-graph-based joint IBI/ICI mitigation for OFDM in underwater acoustic multipath channels with long-separated clusters', *IEEE Journal of Oceanic Engineering*, vol. 37, no. 4, pp. 680–694, 2012.

[68] Y. M. Aval and M. Stojanovic, 'Differentially coherent multichannel detection of acoustic OFDM signals', *IEEE Journal of Oceanic Engineering*, vol. 40, no. 2, pp. 251–268, 2015.

[69] C. C. Tsimenidis, B. S. Sharif, S. Y. L. Goff, O. R. Hinton, and J. A. Neasham, 'Adaptive spread-spectrum based RAKE receiver for shallow-water acoustic channels', in *OCEANS 2007 – Europe*, June 2007, pp. 1–6.

[70] M. Stojanovic and L. Freitag, 'Multichannel detection for wideband underwater acoustic CDMA communications', *IEEE Journal of Oceanic Engineering*, vol. 31, no. 3, pp. 685–695, 2006.

[71] G. Leus and P. v. Walree, 'Multiband OFDM for covert acoustic communications', *IEEE Journal on Selected Areas in Communications*, vol. 26, no. 9, pp. 1662–1673, 2008.

[72] H. Zhang, Z. Yue, Z. Xie, and Z. Wang, 'Design and testing of underwater acoustic communications for an AUV', in *Underwater Acoustics and Ocean Dynamics, Proceedings of the 4th Pacific Rim Underwater Acoustics Conference*, Singapore, 2016.

[73] T. Melodia, H. Khulandjian, L.-C. Kuo, and E. Demirors, 'Advances in underwater acoustic networking', in *Mobile Ad Hoc Networking: Cutting Edge Directions*, S. Basagni, M. Conti, S. Giordano, and I. Stojmenovic,

Eds. Hoboken, NJ: John Wiley & Sons, Inc., March 5 2013, ch. 23, pp. 804–852.
[74] K. Chen, M. Ma, E. Cheng, F. Yuan, and W. Su, 'A survey on MAC protocols for underwater wireless sensor networks', *IEEE Communications Surveys & Tutorials*, vol. 16, no. 3, pp. 1433–1447, 2014.
[75] S. S. Shahapur and R. Khanai, 'Underwater sensor network at physical, data link and network layer – A survey', in *Proceedings of the IEEE International Conference on Communication and Signal Processing*, ser. ICCSP'15. Melmaruvathur, India: IEEE, 2–4 April 2015.
[76] S. Jiang, 'State-of-the-art medium access control (MAC) protocols for underwater acoustic networks: A survey based on a MAC reference model', *IEEE Communications Surveys & Tutorials*, vol. 20, no. 1, pp. 96–131, 2018.
[77] M. Hayajneh, I. Khalil, and Y. Gadallah, 'An OFDMA-based MAC protocol for underwater acoustic wireless sensor networks', in *Proceedings of the 5th International Conference on Wireless Communications and Mobile Computing*, ser. IWCMC'15, Leipzig, Germany, 21–24 June 2009, pp. 810–814.
[78] F. Bouabdallah and R. Boutaba, 'A Distributed OFDMA medium access control for underwater acoustic sensors networks', in *Proceedings of the IEEE International Conference on Communications*, ser. ICC'11. Kyoto, Japan: IEEE, 5–9 June 2011, pp. 1–5.
[79] I. Khalil, Y. Gadallah, and M. H. Khreishah, 'An adaptive OFDMA-based MAC protocol for underwater acoustic wireless sensor networks', *IEEE Sensors Journal*, vol. 12, no. 7, pp. 8782–8805, 2012.
[80] S. Lam, 'Delay analysis of a time division multiple access (TDMA) channel', *IEEE Transactions on Communications*, vol. 25, no. 12, pp. 1489–1494, 1977.
[81] R. M. Huiltberg, F. H. Jean, and M. E. Jones, 'Time division access for military communications satellites', *IEEE Transactions on Aerospace and Electronic Systems*, vol. AES-1, no. 3, pp. 272–282, 1965.
[82] A. Syed, W. Ye, and J. Heidemann, 'Understanding spatio-temporal uncertainty in medium access with ALOHA protocols', in *Proceedings of the third ACM International Workshop on UnderWater Networks*, ser. WUWNet'07. Montréal, Quebec, Canada: ACM, 14 September 2007, pp. 41–48.
[83] C.-C. Hsu, K.-F. Lai, C.-F. Chou, and C.-J. K. Lin, 'ST-MAC: Spatial-Temporal MAC scheduling for underwater sensor networks', in *Proceedings of the 28th IEEE International Conference on Computer Communications*, ser. INFOCOM'09. Rio de Janeiro, Brazil: IEEE, 19–25 April 2009, pp. 1827–1835.
[84] R. Diamant, M. Pinkhasevich, and I. Achrak, 'A novel spatially shared TDMA protocol and quality measure for ad-hoc underwater acoustic networks', in *Proceedings of 2009 International Conference on Advanced Information Networking and Applications Workshops*, Bradford, UK, 26–29 May 2009, pp. 1160–1165.
[85] K. Kredo, P. Djukic, and P. Mohapatra, 'STUMP: Exploiting position diversity in the staggered TDMA underwater MAC protocol', in *Proceedings*

of the 28th IEEE International Conference on Computer Communications, ser. INFOCOM'09. Rio de Janeiro, Brazil: IEEE, 19–25 April 2009, pp. 2961–2965.

[86] Y. Guan, C.-C. Shen, and J. Yackoski, 'MAC Scheduling for high throughput underwater acoustic networks', in *Proceedings of Wireless Communications and Networking Conference*, ser. WCNC'11. Cancun, Mexico: IEEE, 28–31 March 2011, pp. 197–202.

[87] N.-Y. Yun, Y.-P. Kim, S. Muminov, J.-Y. Lee, S.-Y. Shin, and S.-H. Park, 'Sync MAC protocol to control underwater vehicle based on underwater acoustic communication', in *Proceedings of the 9th International Conference on Embedded and Ubiquitous Computing*. Melbourne, VIC, Australia: IEEE, 24–26 October 2011, pp. 452–456.

[88] W.-H. Liao and C.-C. Huang, 'SF-MAC: A spatially fair MAC protocol for underwater acoustic sensor networks', *IEEE Sensors Journal*, vol. 12, no. 6, pp. 1686–1694, 2012.

[89] M. Chitre, M. Motani, and S. Shahabudeen, 'Throughput of networks with large propagation delays', *IEEE Journal of Oceanic Engineering*, vol. 37, no. 4, pp. 645–658, 2012.

[90] R. Diamant, G. N. Shirazi, and L. Lampe, 'Robust spatial reuse scheduling in underwater acoustic communication networks', *IEEE Journal of Oceanic Engineering*, vol. 39, no. 1, pp. 32–46, 2014.

[91] S. Lmai, M. Chitre, C. Laot, and S. Houcke, 'Throughput-efficient super-TDMA MAC transmission schedules in ad hoc linear underwater acoustic networks', *IEEE Journal of Oceanic Engineering*, vol. 42, no. 1, pp. 156–174, 2017.

[92] J.-P. Kim, J.-W. Lee, Y.-S. Jang, K. Son, and H.-S. Cho, 'A CDMA-based MAC protocol in tree-topology for underwater acoustic sensor networks', in *Proceedings of 2009 International Conference on Advanced Information Networking and Applications Workshops*, Bradford, UK, 26–29 May 2009, pp. 1166–1171.

[93] D. Pompili, T. Melodia, and I. F. Akyildiz, 'A distributed CDMA medium access control for underwater acoustic sensor networks', in *Proceedings of Mediterranean Ad Hoc Networking Workshop, Med-Hoc-Net 2007*, Corfu, Greece, 13–15 June 2007.

[94] M. Rahmati, D. Pompili, and R. Petroccia, 'Collaborative hybrid ARQ for CDMA-based reliable underwater acoustic communications', in *Proceedings of the Fourth Underwater Communications and Networking Conference*, ser. UComms'18. Lerici, Italy: IEEE, 28–30 August 2018, pp. 1–5.

[95] M. Rahmati, R. Petroccia, and D. Pompili, 'In-network collaboration for CDMA-based reliable underwater acoustic communications', *IEEE Journal of Oceanic Engineering (Early Access)*, pp. 1–14, May 2019.

[96] N. Abramson, 'The Aloha system – Another alternative for computer communications', in *Proceedings of the American Federation of Information Processing Societies Conference*, ser. AFIPS'70, Atlantic City, NJ, USA, 5–7 May 1970, pp. 295–298.

[97] L. Vieira, J. Kong, U. Lee, and M. Gerla, 'Analysis of ALOHA protocols for underwater acoustic sensor networks', in *Proceedings of the First ACM International Workshop on UnderWater Networks, WuWNet 2006*. Los Angeles, CA: ACM, September 25 2006.

[98] N. Chirdchoo, W.-S. Soh, and K. C. Chua, 'ALOHA-based MAC protocols with collision avoidance for underwater acoustic networks', in *Proceedings of the 26th IEEE International Conference on Computer Communications*, ser. INFOCOM'07, Anchorage, Alaska, USA, 6–12 May 2007, pp. 2271–2275.

[99] C. Petrioli, R. Petroccia, and J. Potter, 'Performance evaluation of underwater MAC protocols: From simulation to at-sea testing', in *Proceedings of IEEE/OES OCEANS 2011*, Santander, Spain, 6–9 June 2011.

[100] A. Colvin, 'CSMA with collision avoidance', *Computer Communications*, vol. 6, no. 5, pp. 227–235, 1983.

[101] R. Petroccia, C. Petrioli, and J. Potter, 'Performance evaluation of underwater MAC protocols: At-sea experiments', *IEEE Journal of Oceanic Engineering*, vol. 43, no. 2, pp. 547–556, 2018.

[102] P. Karn, 'MACA: A new channel access protocol for packet radio', in *ARRL/CRRL Amateur Radio 9th Computer Networking Conference*, London, Ontario Canada, 22 September 1990.

[103] V. Bharghavan, A. Demers, S. Shenker, and L. Zhang, 'MACAW: A media access protocol for wireless LAN's', in *Proceedings of ACM Special Interest Group on Data Communication*, ser. SIGCOMM. London, Ontario Canada: ACM, 31 August–2 September 1994.

[104] M. Molins and M. Stojanovic, 'Slotted FAMA: A MAC protocol for underwater acoustic networks', in *Proceedings of MTS/IEEE OCEANS 2006*. Singapore: IEEE, 16–19 May, 2006.

[105] B. Peleato and M. Stojanovic, 'Distance aware collision avoidance protocol for ad-hoc underwater acoustic sensor networks', *IEEE Communications Letters*, vol. 11, no. 12, pp. 1025–1027, 2007.

[106] H.-H. Ng, W.-S. Soh, and M. Motani, 'MACA-U: A media access protocol for underwater acoustic networks', in *Proceedings of IEEE GLOBECOM*, New Orleans, LA, USA, 30 November–4 December 2008, pp. 1–5.

[107] L. Tracy and S. Roy, 'A reservation MAC for ad-hoc underwater acoustic sensor networks', in *Proceedings of the 14th Annual International Conference on Mobile Computing and Networking*, ser. MobiCom. San Francisco, CA, USA: ACM, 14–19 September 2008.

[108] S. Basagni, C. Petrioli, R. Petroccia, and M. Stojanovic, 'Multiplexing data and control channels in random access underwater networks', in *Proceedings of MTS/IEEE OCEANS 2009*. Biloxi, Mississippi, USA, October: IEEE, 26–29 October 2009.

[109] Z. Peng, Y. Zhu, Z. Zhou, Z. Guo, and J.-H. Cui, 'COPE-MAC: A contention-based medium access control protocol with parallel reservation for underwater acoustic networks', in *Proceedings of IEEE OCEANS 2010*. Sydney, Australia: IEEE, 24–27 May 2010.

[110] C.-M. Chao, Y.-Z. Wang, and M.-W. Lu, 'Analyzing multi-channel MAC protocols for underwater acoustic sensor networks', *IEEE Transactions on Systems, Man, and Cybernetics: Systems*, vol. 43, no. 1, pp. 128–138, 2013.

[111] X. Guo, M. Frater, and M. Ryan, 'An adaptive propagation-delay-tolerant MAC protocol for underwater acoustic sensor networks', *IEEE Journal of Oceanic Engineering*, vol. 34, no. 2, pp. 170–180, 2009.

[112] D. Pompili, T. Melodia, and I. F. Akyildiz, 'A CDMA-based medium access control for underwater acoustic sensor networks', *IEEE Transactions on Wireless Communications*, vol. 8, no. 4, pp. 1899–1909, 2009.

[113] R. Diamant and L. Lampe, 'A hybrid spatial reuse MAC protocol for ad-hoc underwater acoustic communication networks', in *Communications Workshops (ICC), 2010 IEEE International Conference on*, Cape Town, South Africa, 23–27 May 2010, pp. 1–5.

[114] K. B. Kredo, II and P. Mohapatra, 'A hybrid medium access control protocol for underwater wireless networks', in *Proceedings of the third ACM International Workshop on UnderWater Networks (WUWNet '07)*, Montréal, Quebec, Canada, 14 September 2007, pp. 33–40.

[115] J. Yackoski and C. Shen, 'UW-FLASHR: Achieving high channel utilization in a time-based acoustic MAC protocol', in *Proceedings of the third ACM International Workshop on UnderWater Networks*, ser. WUWNet'08. San Francisco, California, USA: ACM, 15 September 2008, pp. 59–66.

[116] A. Wahid and K. Dongkyun, 'Analyzing routing protocols for underwater wireless sensor networks', *International Journal of Communication Networks and Information Security*, vol. 2, no. 3, pp. 253–261, 2010.

[117] M. Ayaz, I. Baig, A. Abdullah, and I. Faye, 'Review: A survey on routing techniques in underwater wireless sensor networks', *Journal of Network and Computer Applications*, vol. 34, pp. 1908–1927, 2011.

[118] G. Han, J. Jiang, N. Bao, L. Wan, and M. Guizani, 'Routing protocols for underwater wireless sensor networks', *IEEE Communications Magazine*, vol. 53, no. 11, pp. 72–78, 2015.

[119] M. Fazeli and S. N. Basharzad, 'A survey on underwater wireless sensor networks routing algorithms', in *2017 IEEE 4th International Conference on Knowledge-Based Engineering and Innovation (KBEI)*. Tehran, Iran: IEEE, 22 December 2017, pp. 373–378.

[120] E. Liou, C. Kao, C. Chang, Y. Lin, and C. Huang, 'Internet of underwater things: Challenges and routing protocols', in *2018 IEEE International Conference on Applied System Invention (ICASI)*. Chiba, Japan: IEEE, 13–17 April 2018, pp. 1171–1174.

[121] H. Cho, C. Chen, T. K. Shih, and H. Chao, 'Survey on underwater delay/disruption tolerant wireless sensor network routing', *IET Wireless Sensor Systems*, vol. 4, no. 3, pp. 112–121, 2014.

[122] R. Amirthavalli, R. Dhaya, and M. S. Chandrasoodan, 'A survey of routing algorithms in delay tolerant networks', in *2016 2nd International Conference on Applied and Theoretical Computing and Communication Technology (iCATccT)*. Bangalore, India: IEEE, 21–23 July 2016, pp. 469–473.

[123] N. Li, J.-F. Martínez, J. M. Meneses Chaus, and M. Eckert, 'A survey on underwater acoustic sensor network routing protocols', *Sensors*, vol. 16, no. 3, p. 414, 2016.

[124] E. M. Sozer, M. Stojanovic, and J. Proakis, 'Underwater acoustic networks', *IEEE Journal of Oceanic Engineering*, vol. 25, no. 1, pp. 72–83, 2000.

[125] C.-J. Huang, Y.-W. Wang, H.-H. Liao, C.-F. Lin, K.-W. Hu, and T.-Y. Chang, 'A power-efficient routing protocol for underwater wireless sensor networks', *Applied Soft Computing*, vol. 11, no. 2, pp. 2348–2355, 2011.

[126] M. C. Domingo and R. Prior, 'A distributed clustering scheme for underwater wireless sensor networks', in *Proceedings of the 2007 IEEE 18th International Symposium on Personal, Indoor and Mobile Radio Communications*. Athens, Greece: IEEE, 3–7 September 2007, pp. 1–5.

[127] P. Wang, C. Li, and J. Zheng, 'Distributed minimum-cost clustering protocol for underwater sensor networks (UWSNs)', in *Proceedings of the 2007 IEEE International Conference on Communications*. Glasgow, UK: IEEE, 24–28 June 2007, pp. 3510–3515.

[128] M. Ayaz, A. Abdullah, and L. T. Jung, 'Temporary cluster based routing for underwater wireless sensor networks', in *Proceedings of the 2010 International Symposium on Information Technology*. Kuala Lumpur, Malaysia: IEEE, 15–17 June 2010, pp. 1009–1014.

[129] M. Faheem, G. Tuna, and V. C. Gungor, 'LRP: Link quality-aware queue-based spectral clustering routing protocol for underwater acoustic sensor networks', *International Journal of Communication Systems*, vol. 30(12), 2017. Available at https://onlinelibrary.wiley.com/doi/abs/10.1002/dac.3257, accessed August 2017

[130] G. Toso, R. Masiero, P. Casari, O. Kebkal, M. Komar, and M. Zorzi, 'Field experiments for dynamic source routing: S2C EvoLogics modems run the SUN protocol using the DESERT Underwater libraries', in *Proceedings of MTS/IEEE OCEANS 2012*. Hampton Roads, VA, USA: IEEE, 14–19 October 2012, pp. 1–10.

[131] Y. Wei and D. Kim, 'Reliable and energy-efficient routing protocol for underwater acoustic sensor networks', in *Proceedings of the 2014 International Conference on Information and Communication Technology Convergence*, ser. ICTC. IEEE, 22–24 October 2014, pp. 738–743.

[132] E. Kim, J. Kang, P. K. Chong, S. Yoo, and D. Kim, 'Energy-efficient local area source routing protocol of underwater sensor networks in the deep ocean', in *Proceedings of the 2007 International Symposium on Communications and Information Technologies*. Sydney, Australia: IEEE, 17–19 October 2007, pp. 948–953.

[133] E. A. Carlson, P. Beaujean, and E. An, 'Location-aware routing protocol for underwater acoustic networks', in *Proceedings of MTS/IEEE OCEANS 2006*, Boston, MA, USA, 28–21 September 2006, pp. 1–6.

[134] F. Al-Salti, N. Alzeidi, and B. Arafeh, 'A new multipath grid-based geographic routing protocol for underwater wireless sensor networks', in *Proceedings of the 2014 International Conference on Cyber-Enabled*

Distributed Computing and Knowledge Discovery. Shanghai, China: IEEE, 13–15 October 2014, pp. 331–336.

[135] J. Jiang, G. Han, H. Guo, L. Shu, and J. J. Rodrigues, 'Geographic multipath routing based on geospatial division in duty-cycled underwater wireless sensor networks', *Journal of Network and Computer Applications*, vol. 59, no. C, pp. 4–13, 2016. [Online]. Available: http://dx.doi.org/10.1016/j.jnca.2015.01.005

[136] J. Alves, R. Petroccia, and J. R. Potter, 'MPR: Multi-point relay protocol for underwater acoustic networks', in *Proceedings of the 9th ACM International Conference on Underwater Networks and Systems*, ser. WUWNet'14, Rome, Italy, 12–14 November 2014.

[137] P. Xie, J. H. Cui, and L. Lao, 'VBF: Vector-based forwarding protocol for underwater sensor networks', in *Proceedings of the International Conference on Research in Networking*, Coimbra, Portugal, 15–19 May 2006.

[138] X. Xiao, X. P. Ji, G. Yang, and Y. P. Cong, 'LE-VBF: Lifetime-extended vector-based forwarding routing', in *Proceedings of the International Conference on Computer Science*, ser. CSSS'12, Nanjing, China, August 2012, pp. 1201–1203.

[139] E. Isufi, H. Dol, and G. Leus, 'Advanced flooding-based routing protocols for underwater sensor networks', *EURASIP Journal on Advances in Signal Processing*, vol. 2016, 2016.

[140] N. Chirdchoo, W. Soh, and K. C. Chua, 'Sector-based routing with destination location prediction for underwater mobile networks', in *Proceedings of the 2009 International Conference on Advanced Information Networking and Applications Workshops*. Bradford, UK: IEEE, 26–29 May 2009, pp. 1148–1153.

[141] D. Shin, D. Hwang, and D. Kim, 'DFR: An efficient directional flooding-based routing protocol in underwater sensor networks', *Wireless Communication and Mobile Computing*, vol. 12, pp. 1517–1527, 2012.

[142] B. Jalving, 'Depth accuracy in seabed mapping with underwater vehicles', in *Proceedings of MTS/IEEE OCEANS 1999*, Seattle, WA, 13–16 September 1999, pp. 973–978.

[143] H. Yan, Z. J. Shi, and J.-H. Cui, 'DBR: Depth-based routing for underwater sensor networks', in *Proceedings of the 7th international IFIP-TC6 NETWORKING 2008*, 5–9 May 2008, pp. 72–86.

[144] U. Lee, P. Wang, Y. Noh, L. Vieira, M. Gerla, and J.-H. Cui, 'Pressure routing for underwater sensor networks', in *Proceedings of IEEE INFOCOM 2010*. San Diego, CA, USA: IEEE, 15–19 March 2010, pp. 1–9.

[145] Y. Noh, U. Lee, P. Wang, B. S. C. Choi, and M. Gerla, 'VAPR: Void-aware pressure routing for underwater sensor networks', *IEEE Transactions on Mobile Computing*, vol. 12, no. 5, pp. 895–908, 2013.

[146] R. Petroccia and J. Alves, 'A hybrid routing protocol for underwater acoustic networks', in *Proceedings of the 17th Annual Mediterranean Ad Hoc Networking Workshop (Med-Hoc-Net)*. Anacarpi, Italy: IEEE, 20–22 June 2018, pp. 1–8.

[147] R. Otnes and S. Haavik, 'Duplicate reduction with adaptive backoff for a flooding-based underwater network protocol', in *Proceedings of MTS/IEEE OCEANS 2013*, Bergen, Norway, 10–13 June 2013.

[148] P. Xie, Z. Zhou, N. Nicolaou, A. See, J.-H. Cui, and Z. Shi, 'Efficient vector-based forwarding for underwater sensor networks', *EURASIP J. Wirel. Commun. Netw.*, vol. 2010, pp. 1–13, 2010.

[149] Z. Zhou, Z. Peng, J. Cui, and Z. Shi, 'Efficient multipath communication for time-critical applications in underwater acoustic sensor networks', *IEEE/ACM Transactions on Networking*, vol. 19, no. 1, pp. 28–41, 2011.

[150] L. Kuo and T. Melodia, 'Tier-based underwater acoustic routing for applications with reliability and delay constraints', in *Proceedings of the IEEE International Workshop on Wireless Mesh and Ad Hoc Networks*, ser. WiMAN'11. Maui, HI, USA: IEEE, 31 July 2011.

[151] M. Ayaz, A. Abdullah, I. Faye, and Y. Batira, 'An efficient dynamic addressing based routing protocol for underwater wireless sensor networks', *Computer Communications*, vol. 35, no. 4, pp. 475–486, 2012. [Online]. Available: http://dx.doi.org/10.1016/j.comcom.2011.11.014

[152] T. Melodia, M. C. Vuran, and D. Pompili, 'The state of the art in cross-layer design for wireless sensor networks', in *Wireless Systems and Network Architectures in Next Generation Internet*, M. Cesana and L. Fratta, Eds. Springer: Berlin, Heidelberg, 2006, pp. 78–92.

[153] R. Jurdak, A. G. Ruzzelli, and G. M. P. O'Hare, 'Design considerations for deploying underwater sensor networks', in *Proceedings of SensorComm 2007*, 14–20 October 2007, pp. 227–232.

[154] J. M. Jornet, M. Stojanovic, and M. Zorzi, 'Focused beam routing protocol for underwater acoustic networks', in *Proceedings of the third ACM International Workshop on UnderWater Networks (WUWNet'08)*, San Francisco, California, USA, 15 September 2008, pp. 75–82.

[155] D. Pompili and I. F. Akyildiz, 'A multimedia cross-layer protocol for underwater acoustic sensor networks', *IEEE Transactions on Wireless Communications*, vol. 9, pp. 2924–2933, 2010.

[156] S. Basagni, C. Petrioli, R. Petroccia, and D. Spaccini, 'CARP: A channel-aware routing protocol for underwater acoustic wireless networks', *Ad Hoc Networks*, vol. 34, pp. 92–104, 2015.

[157] R. Petroccia, K. Pelekanakis, J. Alves, S. Fioravanti, S. Blouin, and S. Pecknold, 'An adaptive cross-layer routing protocol for underwater acoustic networks', in *Proceedings of the Fourth Underwater Communications and Networking Conference*, ser. UComms'18. Lerici, Italy: IEEE, 28–30 August 2018, pp. 1–5.

[158] T. Hu and Y. Fei, 'QELAR: A Machine-learning-based adaptive routing protocol for energy-efficient and lifetime-extended underwater sensor networks', *IEEE Transactions on Mobile Computing*, vol. 9, no. 6, pp. 796–809, 2010.

[159] T. Hu and Y. Fei, 'An adaptive routing protocol based on connectivity prediction for underwater disruption tolerant networks', in *Proceedings of*

the 2013 IEEE Global Communications Conference, ser. GLOBECOM'13. Atlanta, GA, USA: IEEE, 9–13 December 2013, pp. 65–71.

[160] R. Thangarajan, T. Siva, and R. Boopalachakaravarthy, 'Adaptive energy efficient routing protocol with extended lifetime in underwater sensor networks', in *Proceedings of the 2013 International Conference on Communication Systems and Network Technologies*. Gwalior, India: IEEE, 6–8 April 2013, pp. 322–326.

[161] R. Plate and C. Wakayama, 'Utilizing kinematics and selective sweeping in reinforcement learning-based routing algorithms for underwater networks', *Ad Hoc Networks*, vol. 34, no. C, pp. 105–120, 2015. [Online]. Available: https://doi.org/10.1016/j.adhoc.2014.09.012

[162] A. Caiti, K. Grythe, J. M. Hovem, *et al.*, 'Linking acoustic communications and network performance: Integration and experimentation of an underwater acoustic network', *IEEE Journal of Oceanic Engineering*, vol. 38, no. 4, pp. 758–771, 2013.

[163] H. S. Dol, P. Casari, T. van der Zwan, and R. Otnes, 'Software-defined underwater acoustic modems: Historical review and the NILUS approach', *IEEE Journal of Oceanic Engineering*, vol. 42, no. 3, pp. 722–737, 2017.

[164] S. Shahabudeen, M. Chitre, M. Motani, and A. L. Y. Siah, 'Unified simulation and implementation software framework for underwater MAC protocol development', in *MTS/IEEE OCEANS'09*, Biloxi, MS, 26–29 October 2009, pp. 1–9.

[165] M. Chitre, I. Topor, and T. Koay, 'The UNET-2 modem – An extensible tool for underwater networking research', in *2012 Oceans – Yeosu*, Yeosu, South Korea, May 2012, pp. 1–7.

[166] C. Petrioli, R. Petroccia, J. R. Potter, and D. Spaccini, 'The SUNSET framework for simulation, emulation and at-sea testing of underwater wireless sensor networks', *Ad Hoc Networks*, vol. 34, pp. 224–238, 2015.

[167] The VINT Project, 'The ns manual', http://www.isi.edu/nsnam/ns/.

[168] N. Baldo, F. Maguolo, M. Miozzo, M. Rossi, and M. Zorzi, 'ns2-MIRACLE: A modular framework for multi-technology and cross-layer support in network simulator 2', in *Proceedings of the 2nd International Conference on Performance Evaluation Methodologies and Tools, ValueTools 2007*, Nantes, France, 23–25 October 2007, pp. 1–8.

[169] M. Porter *et al.*, 'Bellhop code', http://oalib.hlsresearch.com/Rays/index.html.

[170] F. Campaganro, R. Francescon, F. Favaro, *et al.*, 'The DESERT underwater framework v2: Improved capabilities and extension tools', in *IEEE Third Underwater Communications and Networking Conference (UComms)*, Lerici, Italy, 30 August–1 September 2016.

[171] J. Potter, J. Alves, T. Furfaro, *et al.*, 'Software defined open architecture modem development at CMRE', in *Proceedings of the 2nd IEEE OES International Conference on Underwater Communications and Networking*, ser. UComms14, Sestri Levante, Italy, September 2014.

[172] R. Petroccia, G. Zappa, T. Furfaro, J. Alves, and L. D'Amaro, 'Development of a software-defined and cognitive communications architecture at CMRE', in *Proceedings of MTS/IEEE OCEANS 2018*, Charleston, USA, 22–25 October 2018.

[173] R. Petroccia, J. Śliwka, A. Grati, *et al.*, 'Deployment of a persistent underwater acoustic sensor network: The CommsNet17 experience', in *Proceedings of MTS/IEEE OCEANS 2018*, Kobe, Japan, 28–31 May 2018.

[174] G. Ferri, R. Petroccia, G. D. Magistris, *et al.*, 'Cooperative autonomy in the CMRE ASW multistatic robotic network: Results from LCAS18 trial', in *Proceedings of MTS/IEEE OCEANS 2019*, Marseille, France, 17–20 June 2019.

[175] A. Benson, J. Proakis, and M. Stojanovic, 'Towards robust adaptive acoustic communications', in *Proc. MTS/IEEE OCEANS 2000*, vol. 2, 2000, pp. 1249–1249.

[176] A. Radosevic, R. Ahmed, T. M. Duman, J. G. Proakis, and M. Stojanovic, 'Adaptive OFDM modulation for underwater acoustic communications: Design considerations and experimental results', *IEEE Journal of Oceanic Engineering*, vol. 39, no. 2, pp. 357–370, 2014.

[177] R. Diamant and L. Lampe, 'Adaptive error-correction coding scheme for underwater acoustic communication networks', *IEEE Journal of Oceanic Engineering*, vol. 40, no. 1, pp. 104–114, 2015.

[178] S. Shankar, M. Chitre, and M. Jayasuria, 'Data driven algorithms to tune physical layer parameters of an underwater communication link', in *IEEE OCEANS'10*, Sydney, Australia, 2010.

[179] S. Shankar and M. Chitre, 'Tuning an underwater communication link', in *MTS/IEEE OCEANS'13*, Bergen, Norway, 2013.

[180] P. Anjangi and M. Chitre, 'Model-based data-driven learning algorithm for tuning an underwater acoustic link', in *Fourth Underwater Communications and Networking Conference (UComms)*, 2018, pp. 1–5.

[181] K. Pelekanakis, L. Cazzanti, G. Zappa, and J. Alves, 'Decision tree-based adaptive modulation for underwater acoustic communications', in *IEEE Third Underwater Communications and Networking Conference (UComms)*, Lerici, Italy, 2016.

[182] K. Pelekanakis and L. Cazzanti, 'On adaptive modulation for low SNR underwater acoustic communications', in *MTS/IEEE OCEANS'18*, Charleston, USA, 2018.

[183] L. Wan, H. Zhou, X. Xu, *et al.*, 'Adaptive modulation and coding for underwater acoustic OFDM', *IEEE Journal of Oceanic Engineering*, vol. 40, no. 2, pp. 327–336, 2015.

[184] D. Nams, M. L. Seto, and J. J. Leonard, 'On-line adaptation of underwater acoustic transmission rates to optimize communications for collaborative AUV missions', in *MTS/IEEE OCEANS 2015*, Washington, DC, USA, 2015.

[185] M. Sadeghi, M. Elamassie, and M. Uysal, 'Adaptive OFDM-based acoustic underwater transmission: System design and experimental verification', in *2017 IEEE International Black Sea Conference on Communications and Networking (BlackSeaCom)*, June 2017, pp. 1–5.

[186] E. Demirors, G. Sklivanitis, G. E. Santagati, T. Melodia, and S. N. Batalama, 'A high-rate software-defined underwater acoustic modem with real-time adaptation capabilities', *IEEE Access*, vol. 6, pp. 18 602–18 615, 2018.

[187] T. Hu and Y. Fei, 'MURAO: A multi-level routing protocol for acoustic-optical hybrid underwater wireless sensor networks', in *Proceedings of the 9th Annual IEEE Communications Society Conference on Sensor, Mesh and Ad Hoc Communications and Networks*, ser. SECON'12, Seoul, Korea, 18–21 June 2012.

[188] F. Campagnaro, F. Guerra, P. Casari, R. Diamant, and M. Zorzi, 'Implementation of a multi-modal acoustic-optical underwater network protocol stack', in *Proceedings of MTS/IEEE OCEANS 2016*, Shanghai, China, 10–13 April 2016.

[189] F. Campagnaro, R. Francescon, P. Casari, R. Diamant, and M. Zorzi, 'Multimodal underwater networks: Recent advances and a look ahead', in *Proceedings of the 12th ACM International Conference on Underwater Networks and Systems*, ser. WUWNet'17. Halifax, NS, Canada: ACM Press, 6–8 November 2017.

[190] P. Gjanci, C. Petrioli, S. Basagni, C. A. Phillips, B. Ladislau, and D. Turgut, 'Path finding for maximum value of information in multi-modal underwater wireless sensor networks', *IEEE Transactions on Mobile Computing*, vol. 17, no. 2, pp. 404–418, 2018.

[191] E. Souza, H. C. Wong, I. Cunha, A. A. F. Loureiro, L. F. M. Vieira, and L. B. Oliveira, 'End-to-end authentication in under-water sensor networks', in *Proceedings of the 18th IEEE International Symposium on Computers and Communications*, ser. ISCC'13, Split, Croazia, 7–10 July 2013, pp. 299–304.

[192] J. H. Park, 'LPI techniques in the underwater acoustic channel', in *Proceedings of the IEEE Military Communications Conference - Communications-Computers: Teamed for the 90's*. Monterey, CA, USA: IEEE, 5–9 October 1986.

[193] P. A. van Walree and G. Leus, 'Robust underwater telemetry with adaptive turbo multiband equalization', *IEEE Journal of Oceanic Engineering*, vol. 34, no. 4, pp. 645–655, 2009.

[194] G. Ateniese, A. Capossele, P. Gjanci, C. Petrioli, and D. Spaccini, 'SecFUN: Security framework for underwater acoustic sensor networks', in *Proceedings of MTS/IEEE OCEANS 2015*. Genova, Italy: IEEE, 18–21 May 2015, pp. 1–9.

[195] G. Dini and A. L. Duca, 'A secure communication suite for underwater acoustic sensor networks', *Sensors*, vol. 12, no. 11, pp. 15 133–15 158, 2012.

[196] Y. Luo, L. Pu, Z. Peng, and Z. Shi, 'RSS-based secret key generation in underwater acoustic networks: Advantages, challenges and performance improvements', *IEEE Commun. Mag.*, vol. 54, no. 2, pp. 32–38, 2016.

[197] Y. Huang, S. Zhou, Z. Shi, and L. Lai, 'Channel frequency response-based secret key generation in underwater acoustic systems', *IEEE Trans. Wireless Commun.*, vol. 15, no. 9, pp. 5875–5888, 2016.

[198] L. Lai, Y. Liang, and H. V. Poor, 'A unified framework for key agreement over wireless fading channels', *IEEE Transactions on Information Forensics and Security*, vol. 7, no. 2, pp. 480–490, 2012.

[199] M. Zuba, Z. Shi, Z. Peng, J.-H. Cui, and S. Zhou, 'Vulnerabilities of underwater acoustic networks to denial-of-service jamming attacks', *Security and Communication Networks*, vol. 8, no. 16, pp. 2635–2645, 2015.

[200] R. Zhang and Y. Zhang, 'Wormhole-resilient secure neighbor discovery in underwater acoustic networks', in *Proceedings of 29th IEEE International Conference on Computer Communications*, ser. INFOCOM'10, San Diego, CA, USA, 15–19 March 2010, pp. 1–9.

[201] G. Dini and A. L. Duca, 'SeFLOOD: A secure network discovery protocol for underwater acoustic networks', in *Proceedings of the 16th IEEE International Symposium on Computers and Communications*, ser. ISCC'11, Kerkyra, Corfu, Greece, 28 June–1 July 2011, pp. 636–638.

[202] M. Zuba, M. Fagan, Z. Shi, and J.-H. Cui, 'A resilient pressure routing scheme for underwater acoustic networks', in *Proceedings of the 57th IEEE Global Communications Conference*, ser. GLOBECOM'14, Austin, Texas, USA, 8–12 December 2014.

[203] A. Capossele, G. De Cicco, and C. Petrioli, 'R-CARP: A reputation based channel aware routing protocol for underwater acoustic sensor networks', in *Proceedings of the 10th ACM International Workshop on Underwater Networks*, ser. WUWNet'15, Washington DC, USA, 22–24 October 2015.

[204] G. Han, J. Jiang, L. Shu, and M. Guizani, 'An attack-resistant trust model based on multidimensional trust metrics in underwater acoustic sensor network', *IEEE Transactions on Mobile Computing*, vol. 14, no. 12, pp. 2447–2459, 2015.

[205] G. Han, J. Jiang, N. Sun, and L. Shu, 'Secure communication for underwater acoustic sensor networks', *IEEE Communications Magazine*, vol. 53, no. 8, pp. 54–60, 2015.

[206] H. Li, Y. He, X. Cheng, H. Zhu, and L. Sun, 'Security and privacy in localization for underwater sensor networks', *IEEE Communications Magazine*, vol. 53, no. 11, pp. 56–62, 2015.

[207] M. C. Domingo, 'Securing underwater wireless communication networks', *IEEE Transactions on Wireless Communications*, vol. 18, no. 1, pp. 22–28, 2011.

[208] C. Lal, R. Petroccia, M. Conti, and J. Alves, 'Secure underwater acoustic networks: Current and future research directions', in *Proceedings of the 3rd IEEE OES International Conference on Underwater Communications and Networking*, ser. UComms16, Lerici, Italy, 30 August–1 September 2016.

[209] C. Lal, R. Petroccia, K. Pelekanakis, M. Conti, and J. Alves, 'Towards the development of secure underwater acoustic networks', *IEEE Journal of Oceanic Engineering*, vol. 42, no. 4, pp. 1075–1087, 2017.

[210] A.Munafò, J. Sliwka, and J. Alves, 'Dynamic placement of a constellation of surface buoys for enhanced underwater positioning', in *Proceedings of MTS/IEEE OCEANS 2015*. Genova, Italy: IEEE, 18–21 May 2015, pp. 1–9.

[211] NATO STO CMRE, 'JANUS Wiki', http://www.januswiki.org/.

[212] R. Petroccia, J. Alves, and G. Zappa, 'JANUS-based services for operationally relevant underwater applications', *IEEE Journal of Oceanic Engineering*, vol. 42, no. 4, pp. 994–1006, 2017.

[213] A. Smerdon, F. Bustamante, and M. Baker, 'The SWIGacoustic standard: An acoustic communication standard for the offshore energy community', in *2016 IEEE Third Underwater Communications and Networking Conference (UComms)*, Aug 2016, pp. 1–4.

[214] ISO Central Secretary, 'Underwater acoustic sensor network UWASN', International Organization for Standardization, Geneva, CH, Standard ISO/IEC 30140, parts 1 to 4, 2018. [Online]. Available: https://www.iso.org/standard/53260.html

Chapter 10

Pressure-tolerant and pressure-neutral systems for AUV

Gunnar Brink[1] and Gerhard Körner[2]

In almost all industries, savings in raw materials, costs and energy play an increasingly important role in the manufacturing, use and recycling of products. This means that the overall device, propulsion engine system and fuel supply can be made smaller and lighter. The savings in operating the optimized product are often offset by additional development, manufacturing and assembly costs.

Today, lightweight construction is important in all industries, especially in vehicle construction, shipbuilding and aircraft construction. This is not different with subsea systems. A submersible vehicle, which is lighter and smaller than previous models for the same performance figures, is easier to launch and recover from the sea. A smaller cross section leads to less water resistance and so reduces the energy consumption of the engine. This increases the possible speed, reduces the battery capacity or increases the range of the vehicle.

In underwater vehicles, the pressure vessels offer a particularly worthwhile starting point for reducing vehicle weight and vehicle size. Pressure vessels are needed because on-board batteries, engines, optics and, above all, electronic components like computers have been developed for operation in air and must be protected from corrosion and ambient pressure in the sea.

Although the cavities in pressure vessels provide a certain amount of buoyancy, the thick-walled vessel walls made of glass, aluminium, titanium, stainless steel or other materials have to become thicker and thicker as the vehicles' operating depth increases, increasing their weight.

Since many sensors need access to the environment, special cables and plugs are needed to carry electricity and data through the vessel walls, but need to withstand the pressure difference. Such plugs and cables are expensive and heavy. Feedthroughs are made of metal and glass and pose a permanent risk. At the slightest defect, the vessel would fill with seawater, which reduces the buoyancy of the vehicles and increases the risk of sinking.

To ensure that the vehicles do not sink to the bottom of the sea like stones, the downforce is compensated by pressure-resistant buoyancy elements. The larger and heavier the pressure vessels, the more buoyancy foam is required. In a weight spiral,

[1]Fraunhofer Institute of Optronics, System Technologies and Image Exploitation IOSB, Karlsruhe, Germany
[2]Meelve, Rövershagen, Germany

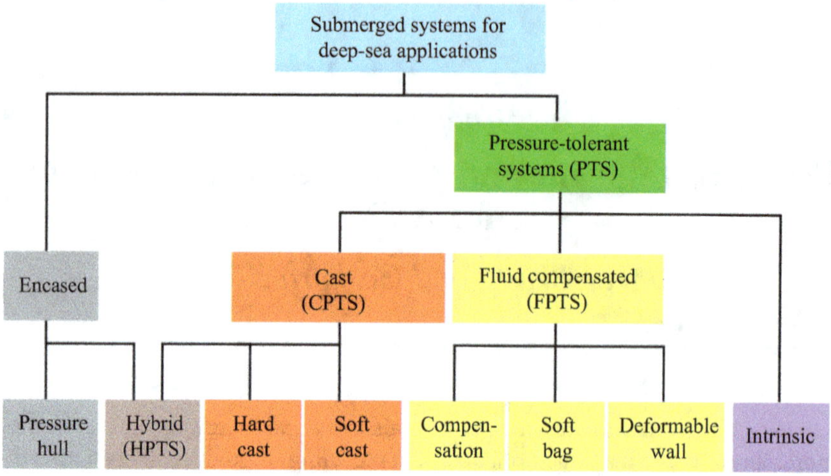

Figure 10.1 Taxonomy of pressure vessel and pressure-tolerant electronic deep-sea technology proposed by Kampmann et al. [2]

larger vehicles require more drive energy, larger engines and more batteries, which in turn need more buoyancy foam which increases their size and weight.

In addition, the construction of electronic assemblies and drive systems in pressure vessels becomes problematic with increasing diving depth and requires a great deal of effort for the pressure vessel and the necessary seals. The electrical feedthroughs in particular pose a problem if, for example, high currents or high-frequency signals have to be routed through feedthroughs in the vessel wall which simultaneously insulate electronically, have low impedances and nevertheless have to absorb pressure differences of several hundred bar. The construction of electronic assemblies and propulsion engines in pressure vessels becomes problematic with increasing diving depth and requires a great deal of effort for the pressure vessel and the necessary seals.

For this reason, pressure-compensated or pressure-neutral systems are increasingly being used. Australian-American pressure-resistant electronics expert Nic Bingham even used a pressure-tolerant approach to implement the power distribution systems, vehicle control networks and pressure-tolerant engine control that power all the engines and hydraulics of the manned dive vehicle used in James Cameron's world-renowned dive to the bottom of the Marianas Trench at Challenger's depth of nearly 11,000 m [1].

In pressure-compensated systems and pressure-neutral systems, all internal components are subjected to the ambient pressure and must withstand it. The difference between the two systems lies in the housing forms and the pressure-transmitting materials.

Kampmann *et al.* [2] have proposed a taxonomy of pressure vessel and pressure-tolerant and pressure-neutral electronic deep-sea technology (see Figure 10.1).

Pressure-compensated electronics has been a scientific and technological topic at least since the 1970s. In a period of 2 years, Holzschuh [3] had examined more than 3,000 active semiconductor devices and came to the following conclusions:

1. Many commercially available electronic components can be operated under deep-sea pressure conditions.
2. High pressures have negligible effects on the electrical properties of semiconductors.
3. Most pressure-related failures are mechanical in nature and are due to poor manufacturing or packaging techniques.

A summary of electronic components that are usable in the use of pressure-tolerant circuits, can already be found in Sutton (1976). The work of Barnes *et al.* [4] gives an overview of the companies active in this field at that time and lists different fluids, their characteristics and their usability for pressure-compensated technology.

The pressure compensation takes place either via a compensator, which is connected to the fixed tank via a pipe system, or via a membrane, which is integrated directly into a housing wall. The choice of fluid depends on which components are to be protected against the surrounding seawater. Dielectric fluids are used for electrical assemblies. Due to the risk of fire and environmental incompatibility, synthetic fluids are increasingly replacing used previously used mineral oils (Pittini *et al.*, 2011).

An important milestone in this development was the funding of competence-building projects in this field of underwater energy supply at SINTEF Energy Research by the Research Council of Norway (RCN). The projects included more or less the entire energy supply chain from the production site to motor drives, pumps, compressors, etc., and consisted of the following components:

1. electric energy systems for deep-water treatment and the transport of oil and gas;
2. feasible power electronics for demanding deep-water applications;
3. electrical insulating materials and insulation systems for underwater high-voltage installations.

One example of this is the use of insulated gate bipolar transistor (IGBT) electrodes in Pittini *et al.* (2010). In this research project, a pressure-tolerant circuit was developed to investigate the influence of high ambient pressure on IGBTs.

In pressure-compensated systems, the system components, including the cabling, are enclosed in watertight housings, which are filled with oil or a protective fluid. Compensators achieve pressure compensation. For repairs, the compensation fluid must be drained and the housings opened. The electronic modules must be thoroughly cleaned before repairs are carried out. After the repair, the housings must be closed again and filled with the compensation liquid. All system parts are vented gas-free with great effort to make sure that the working ranges of the compensators are not exceeded. Such an approach was tested, for example, in the Norwegian funding project 'Subsea Power Supply' for high-voltage electronics, with eight partners.

In pressure-neutral cast systems, housing-like skin casts with membrane function replace the housings. In most cases, the casing-like skin castings are prefabricated and are only open on the mounting side. They form a watertight shell after the assembly has been closed. Cable trees and underwater connectors can be integrated into the skin casting. In pressure-neutral systems, the non-liquid embedding cast replaces the compensation liquid in pressure-compensated systems. Depending on the module type, the embedding cast has varying hardness from gelled (almost liquid) to relatively solid but still soft and pressure compensating.

Skin cast and embedding cast are carried out in a vacuum during component manufacture to avoid air inclusions. Depending on the scope of a repair, the assemblies are cast in vacuum or with degassed but not hardened casting compound.

In many cases, pressure-neutral components are enclosed in non-pressure-resistant plastic housings for mechanical mounting and assembly.

During repair work, the skin casts are partially cut open; the soft embedding cast is removed; it forms hardly any residue when removed. After the repair, the embedding casting is carried out at the repair site and then the skin casting is carried out. This can also be done at the same place as often as desired.

Advantages of pressure-neutral cast systems are:

- Volume and weight are reduced due to the lack of pressure vessels.
- Reduced leakage problems due to lack of pressure difference.
- Almost any diving depth.
- Can be repaired, changed and extended as often as required.
- Reduction of chemical and galvanic corrosion.
- Very good long-term behaviour, continuous underwater operation.
- Integration of antifouling measures in the casting process.
- Good insulation properties even in a water-saturated state provided the appropriate grouting properties be maintained.
- Mechanically resilience, low corrosion due to seawater and low galvanic corrosion.
- Pressure-neutral vehicles are one-half to one-third as heavy as comparable pressure-vessel vehicles.
- Accessibility, mountability and repair are no worse than for pressure vehicles or oil-compensated systems.
- There is no danger of loss of buoyancy or complete destruction in the event of a fault or implosion.

The successful testing of pressure-neutral technology prompted the German Ministry of Economics to fund several projects on pressure-neutral systems for the deep sea. The findings described in the following are largely derived from these projects.

With some cast systems, a combination of solid and locally liquid casts is also possible. For example, assemblies with silicone embedding, castings or silicone gels and local areas with silicone oils are possible. In this case, the skin casting in its membrane function takes over the function of the compensators.

10.1 Basic design of pressure-neutral systems

10.1.1 Mechanical construction

In early-stage deep-sea tests, we found several casting technology-related failures in electronic assemblies. It was found that the simple cast of a pressure-neutral circuit in a built plastic housing with embedding cast and subsequent one-sided skin cast is not suitable as a membrane and failed after the second to third immersion test. After conversion of the design and the cast technology to a closed skin cast, we were able to manufacture and operate reliable electronic assemblies without failures.

10.1.2 Casting systems

Silicone cast systems have proved their worth as cast materials. They showed the most favourable properties. For pressure-neutral cabling, we assembled and cast cables with silicone cores with a minimum cross section of 0.5 mm^2. These cables have also proved themselves in combination with PUR sprues in the connector area.

Electrical measurements with special measuring needles through skin and embedding cast are possible. After the measurement, the pierced channel closes automatically due to the sticky embedding cast and the sliding material of the needles.

The transfer of this concept to submersible motors has also achieved significant performance and reliability improvements.

Typical resins used for castings are silicones, polyurethanes and epoxy casting compounds. Silicones as insulating materials require appropriate insulation distances in the water-saturated state. They can be designed for high reliability and long-term operation.

10.2 Pressure-neutral electrical and electronic assemblies

10.2.1 Suitable components

For diving depths up to 6,000 m and oceanic conditions, there is a sufficient selection of suitable components to realize all necessary circuit structures.

Many suitable components and construction elements can be used at pressures above 600 bar. However, the suitability limits above 600 bar are largely unknown due to a lack of long-term testing.

For some component groups, use in pressure-neutral circuit structures is critical and depends not only on the pressure but also on the electrical loads in the circuit.

For example, some of our circuit structures were successfully tested several times in pressure tanks up to 700 bar, but under real conditions in the deep sea, they almost systematically failed.

These component groups include all functionally self-compressing components such as quartz crystals and accelerometers as well as pre-impregnated or poorly pre-cast wound components with ceramic cores.

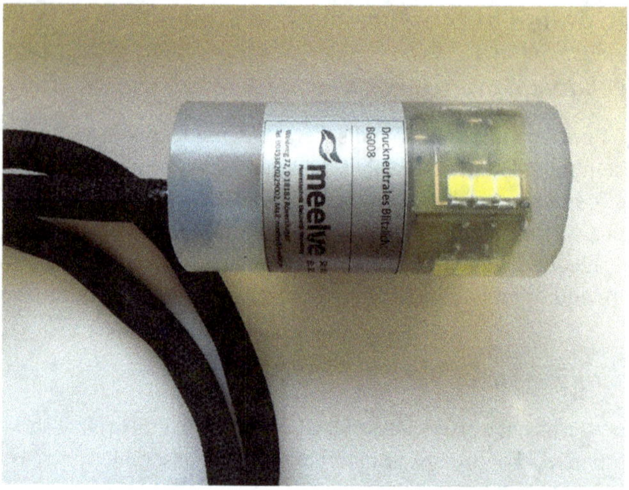

Figure 10.2 Pressure-neutral deep-sea flashlight

Capacitors in particular are critical components of which only a few subgroups are suitable. Electrolytic capacitors, tantalum capacitors, some film capacitors and some ceramic capacitors have not proved themselves reliable because of air inclusions caused by their specific design and production processes. A few types of film capacitors that have not been pre-cast have been found to be suitable. They can be reliably operated with a high operating voltage; especially in high current, circuit structures such as energy converters and drive controllers, but unfortunately have relatively large sizes in relation to their capacitance and operating voltage.

10.2.2 Optical systems and displays

Due to the transparent cast, important system states can be indicated optically but can also be controlled by optical receivers in the cast assemblies. Cast optical serial interfaces have also been successfully tested.

Furthermore, powerful flashlights, optical on/off switches as well as optical underwater modems for data transmission have been built and tested (see Figure 10.2). Crystal-clear silicone areas for optical sensors with favourable optical properties can be integrated into the skin and embedding castings.

10.3 Pressure-neutral power supplies and battery systems

10.3.1 Battery systems

Modern lithium polymer battery systems with a very high energy density are best accommodated in the water in a pressure-neutral battery systems. The expandability of pressure-neutral battery systems in water is given. The water gives ideal cooling. There is no danger of explosion of a battery pressure body.

10.3.1.1 Suitable battery cells

After extensive tests in the joint projects on pressure-neutral systems and pressure-neutral deep-sea systems, it was established that lithium polymer cells are suitable as pressure-neutral components of energy supply systems. Due to the high gravimetric and volumetric energy density, light and small energy supply systems without pressure bodies can be constructed for underwater vehicles. For these applications, the cells must have a robust cell membrane system and a low residual gas content. They must not have a tendency to outgas over a long period and must be as uniform as possible with regard to self-discharge capacity and temperature behaviour due to mechanical mass production. The current carrying capacity plays an important role in these applications.

The capacity of a battery block is determined by the capacity of the cells used and their parallel connection. There is a mechanical limit to the capacity of the cells in terms of their manageability and weight, which enables reliable embedding and skin casting without endangering the mechanical support of the cells. Pressure-neutral battery blocks with a capacity of 2 kWh have a weight around 20 kg. In addition to the cells, the battery block must include battery management, protection against destruction, cell fuses for levelling outputs and, in some cases, a circuit breaker and cabling.

The cell groups connected together to form higher voltage stacks must be cast in a special form in order to compensate for the change in volume caused by the residual gas content of the cells at high pressure without creating larger differential pressures inside the blocks. Also for the battery stacks, closed, extra-robust skin casts, special soft embedding casts and membrane surfaces on all sides by means of a mounting and fastening housing with water breakthroughs have proved themselves. In addition to the main current connections, the electronically fused cell connections for external levelling, the safety resting circuits and the connections for maximum cell voltage control are all connected via underwater connectors.

The 24 V battery blocks usually have isolated battery management system (BMS) outputs, so that they can be interconnected to battery banks and entire power supply systems with different voltage levels via further pressure-neutral modules. Depending on the discharge power, 24, 48 and 110 V are used as the primary voltage levels.

10.3.1.2 Battery management and safety systems

A battery management system (see Figure 10.3) with safety circuits for overtemperature and overvoltage during charging and overtemperature and undervoltage in discharge mode for each cell group must be integrated into the battery block. In addition, protection against destruction (short-circuit protection) must be provided in the battery block. In some cases, monitoring functions with communication connection and a pressure-neutral electronic circuit breaker are integrated into the block.

The battery management of a 24 V battery block must have the following minimum characteristics:

- low self-consumption when switched on;
- undervoltage monitoring of each cell;

Figure 10.3 Ready-to-cast pressure-neutral battery management system for 48 V battery

- temperature monitoring for seven cells;
- maximum cell voltage control of each cell;
- avoidance of a cell current load due to measuring currents in the switched off state (<0.01 μA) by isolating switch;
- safety rest circuits for undervoltage monitoring and max. cell voltage control;
- potential-separated structure of the safety rest circuits for charging and discharging.

10.3.1.3 Destruction protection: levelling protections

Pressure-neutral cast fuses provide destruction protection in the battery blocks. Silicone embedding cast is used as pressure-neutral cast. The silicone embedding cast ensures good spark quenching, is not flammable or only slightly flammable and changes the tripping values only insignificantly.

10.3.1.4 Charge levelling

Chargers for lithium polymer battery blocks should have the following charging functions:

- constant current charging;
- constant voltage control with 29.05 V, +0 V, −0.5 V;
- at a cell voltage of 4.1 ± 0.05 V of any cell group, a maximum cell voltage regulation (MZR safety resting circuit) takes place;
- if the MZR signal voltage is permanently too low (no regulation), the charger will continue to switch off after approx. 30 s;
- status displays;

- start, stop button;
- measuring sockets for battery voltage and the MZR signal.

Charging can be stopped at any state of charge without damaging the battery. The power supply system can be operated during charging.

If the battery is stored for a longer period or is not used, it should be charged to a battery voltage of approx. 25 V. Before transport, we recommend setting a battery voltage of 24 V in order to keep the energy content of the battery to a minimum. The power supply system must be switched off during transport. If the battery is stored for a longer period of time, it should be checked every 6 months to prevent deep discharge by self-discharge.

10.3.1.5 Levelling battery blocks

The easiest way of cyclically levelling battery blocks over time or levelling a battery block with excessive cell voltage differences is to charge all cells to the same voltage in parallel via the levelling inputs.

To do this, the battery block is discharged to a specific voltage (e.g. 27 V) and then brought to the same voltage (e.g. 4.00 V) for each cell in the battery block via a parallel constant voltage charge with subordinate constant current control (e.g. 10 A). The levelling takes place in the almost full state of the cells, so that in this state the cells have as equal voltages and charges as possible. This means that differences in capacity between the cells only have an effect during discharge when the deep discharge protection of the worst cell responds.

When the largest cell charge current is less than 0.5 A, the levelling process can be terminated. The individual cell voltages should differ by less than 0.02 V. The voltage of the individual cell voltages should be less than 0.02 V. The battery voltage is in the range of 28.0–28.2 V with a constant voltage charge of 4.00 V for each cell. The individual cell voltages can be precisely measured and controlled via the currentless levelling inputs.

10.3.1.6 Circuit-breaker assemblies

Circuit-breaker assemblies are used for the safe operation of a pressure-neutral lithium polymer battery. They contain all safety circuits and safety circuits for overtemperature and overcharge protection as well as the safety shutdown in the event of overload, short-circuit, overtemperature and imminent deep discharge in discharge mode. The battery can be switched on and off via the circuit breaker.

A circuit breaker which has been opened or tripped by overload or short-circuit can be switched on and off at will via the on/off signal. In the open state, the system is de-energized during operation and can be stored for a longer period.

In addition to overload protection, the circuit breaker also provides safe behaviour in the event of a short-circuit and as safe a behaviour as possible in the event of a short-circuit in the closed state. For this purpose, the circuit breaker must have sufficient short-circuit resistance.

Furthermore, when operating power controllers, e.g., for drive controllers to be supplied with large backup capacitors, the inrush current must be limited and a time limit set for charging the capacitors in which the charging currents reach normal load current values.

The charging input is protected by a charging diode. The switched-on state and special states are indicated optically.

10.3.1.7 Power distributions

Power distribution systems are needed to interconnect different battery blocks and battery banks, to build emergency and auxiliary supplies and to distribute the energy to different consumers and consumer groups, often controllable.

Power distributions are mainly implemented via electronic circuit breakers that supply individual consumers or groups of consumers.

Supply outlets can also be protected via pressure-neutral cast fuses, with the disadvantage that if a fuse is tripped, it must be replaced by opening the casting. A tripped electronic circuit breaker, on the other hand, can be re-closed if it has not been destroyed by exceeding the permissible short-circuit strength. Since pressure-neutral batteries can drive very high currents, everything depends on the short-circuit resistance and the type of short-circuit. For this reason, pressure-neutral fuses should only be used as destruction protection to limit consequential damage and to achieve the selectivity of a distribution.

An important function in distributions is overvoltage protection. When loads are switched off or when overload cut-offs are necessary, overvoltage always occurs which can lead to the failure of electronic modules without overvoltage protection. It should be installed decentralized near switching elements. An overload of an overvoltage protection should not lead to a short-circuit and cause the entire distribution to fail.

Auxiliary power supply levels of 24 or 48 V can be buffered and electrically isolated via energy converters and buffer batteries. Any power distribution system can be set up using pressure-neutral circuit breaker and distribution modules as well as electronic on/off switches. The transparent cast optically indicates important system states. Electrical measurements with special measuring needles through the skin and embedding cast are also possible to a limited extent during troubleshooting. The insertion channel closes automatically after the measurement due to the soft, sticky embedding casting and the sliding material of the needles.

10.3.1.8 Energy converters, charging converters and energy transformers

Pressure-neutral energy converters are used to convert electrical energy to another voltage level. The associated battery is charged or buffered.

Pressure-neutral energy transmitters are used for contactless transmission of electrical energy to another underwater system or vehicle. The energy transfer only takes place if both systems are mechanically coupled, whereby one system can move away only if the energy transfer is interrupted.

All converters should transmit the energy with the highest possible efficiency and, in most cases, should handle mechanical separation situations.

We have built pressure-neutral converters for all voltage levels from 5 V to approx. 2 kV and powers in the range of 5 W to approx. 2 kW.

In addition to energy conversion, the converters are designed to measure input voltage, output voltage or battery voltage and current consumption or charge current, as well as operating states such as ready for operation, voltage regulation, current regulation, maximum cell voltage regulation, docked or fault messages. The working voltage ranges are adapted to the battery systems used.

Since the converters contain switched power electronics, they must be equipped with EMC filters, anti-corrosion fuses and surge protection devices on both sides of the converter.

The pressure-neutral design of the converters requires that the pressure on the core of the converter transformer and storage chokes be taken into account and that the storage capacitors on the primary and secondary sides of the converter be accurately dimensioned using suitable power capacitors.

At high converter powers, heat must be dissipated. For this purpose, heat exchangers via the skin cast to the ambient water must be provided.

In the case of pressure-neutral energy exchangers, the transformer can be mechanically separated and must be designed accordingly. It is not possible to nest or arrange the primary and secondary windings one above the other to improve efficiency. In addition, both core halves must be insulated from the seawater. As a result, the transducer must necessarily be designed for relatively large water gap. Therefore, a completely different core and winding construction and different switching frequencies than usual are necessary.

If one side of the transformer is removed, the water gap is enlarged to infinity. The entire transducer must be designed for these transition states and protected from undefined states and loads.

10.4 Pressure-neutral drives

10.4.1 Submersible motors and drives

The following drive motors are used underwater:

– Drives in pressure vessels with pressure retaining shaft seal; these operate in air or in an inert gas. For large immersion depths and/or high speeds, they suffer from low efficiencies due to necessary shaft seals.
– Pressure-compensated drives operate; these in protective fluids. Internal pressure and external pressure are adjusted via compensators or membranes. Housing and shaft seal must be seawater tight.
– Pressure-neutral drives; operated in seawater.

Development focus for pressure-neutral submersible motors:

– High efficiency from battery to mechanical output.
– Flow-optimized internal design for motor parts running in seawater or protective fluid.

Large ‘air gap’ (1.5–3 mm) for motors running in seawater for casting the magnetic circuit. For high efficiencies, these motors must be designed and developed completely differently from ‘land motors’.

In the development of pressure-neutral underwater drives, high energy densities with high efficiency are to be aimed at. For double-strand, electronically commutated, permanently excited main drive motors with position sensing and trapezoidal package power supply, an overall efficiency of motor with actuator in water of over 73% is obtained with an output between 1 and 4 kW. Further increases in efficiency seem possible.

The efficiency is distributed approximately as follows:

- controller approx. 94%;
- motor in air approx. 89%;
- fluid filling of the motor approx. 88%.

In addition, we have built and used a large number of pressure-neutral actuators and rudder drives, winch motors, control thrusters and binary drives. Since these drives are only operated for a short time, they are not optimized in terms of efficiency, but in terms of size, weight and energy density.

In addition to the efficient design of the submersible motors, the development of a corresponding robustness and reliability is necessary. This can only be done by means of many tests.

In particular, our further development from double-strand electronically commutated, permanently excited motors to robust submersible motors with a high degree of efficiency was successful. This could only be achieved by an efficient design of the bidirectional actuator-motor energy conversion at the relatively large water gaps between 1 and 2 mm. Furthermore, the use of dynamic water slide bearings was made possible by the compensation of the tightening forces of rotors on both sides of the stator. These motors can be operated at very low speeds in electronically commutated stepper motor operation. This ensures good controllability. After many failures with underwater coatings and thin casting layers in the water gap area, a satisfactory robustness and reliability could only be achieved by designing the stator and rotor as a closed casting system with water-impermeable plastic membranes that can be integrated into the casting.

10.4.1.1 Differences to motors in air

It is almost impossible to convert brushless land motors for underwater use. Land motors are primarily optimized for efficiency by means of a small air gap. If the air gap is increased to provide corrosion protection in the millimetre range between stator and rotor by grinding or turning, the magnetic circuit of the motor is considerably changed and has completely different properties and much poorer efficiencies.

10.4.1.2 Water gap

The water gap between rotor and stator is determined by the need for reliable corrosion protection of the motor components of the magnetic circuit. The magnetic circuit of the motor consists of a dynamo plate and permanent magnets with unfavourable corrosion properties. These components must be protected from corrosion even in a water-saturated state by protective layers that are as thin as possible; this determines the water gap width. Reliable protective coatings allow a water gap of 1.2–1.8 mm.

The magnetic flow of the submersible motor must be optimized for these water gap sizes and the necessary permanent magnets.

10.4.1.3 Rotating rotors in fluids

The rotating rotor of the submersible motor in a fluid (seawater or oil) causes considerable hydrodynamic losses depending on the speed, radius, surface and fluid, which considerably reduces the efficiency of the motor. For this reason, the lowest possible speeds are required with favourable hydrodynamic surfaces of the rotor. Normal motors designed for use in air and operated in low-viscosity oil-protective fluids have considerable efficiency losses. At high pressures and low temperatures between 0°C and 4°C, these motors can come to a standstill due to these losses.

10.4.1.4 Pressure-neutral casting of coil systems

If wound coils are used in pressure-neutral systems, the windings must be pre-cast separately. Otherwise, there is a risk that the casting material, which only enters the coil at approx. 1 bar during casting and contains residual air inclusions, will be pressed into the coil at high operating pressure (e.g. 600 bar). When the pressure is returned to one bar, the compressed air pushes the grouting material out of the roll again.

The coil systems 'breathe' with every dive. This can have serious consequences, especially with fluid-filled motors, if the compensating membranes can no longer compensate for the difference in volume and seawater enters the fluid. In addition, the embedding cast can be considerably damaged.

10.4.1.5 Cooling

The waste heat in pressure-neutral systems must be transferred to the surrounding seawater via the pressure-neutral casting. Cooling in pressure-neutral modules is highly dependent on the casting system.

With silicone cast systems, the thermal conductivity of the cast material is better than that of air.

In the case of high dissipated power losses (e.g. in energy converters or power controllers for motors), heat exchangers, usually made of aluminium, are used which lie over a large surface on the skin casting and transfer the waste heat to the water. For this purpose, the skin casting can be thinner at this point or made of special, highly heat-conducting silicone material in order to achieve better thermal conductivity.

10.5 Elastic buoyancy foams in pressure-neutral technology

In the joint publicly funded German project 'Pressure Neutral Systems for the Deep Sea', a technology for the production of mechanically compressible, cut and processable hollow silicone-sphere foams was developed for the first time. The produced silicone foam with a density of 0.68–0.7 kg/dm^3 was successfully tested in the deep sea in autumn 2010. The approach proved to be very robust and reliable. The density values can be significantly improved once better spheres are available.

10.5.1 Recap and outlook

The seemingly easiest way to use electronics for deep-sea applications is to use the same circuits and electronic components as in air but to place them in a thick-walled pressure vessel. But these pressure vessels are expensive and heavy. Connectors and penetrators that transmit current and signals with reasonable impedance and withstand several hundred bar pressure from the outside to the inside are expensive, and can sometimes only be procured with unpredictable delivery times. When the pressure vessels are closed, human errors can easily occur which lead to water penetrating into the pressure vessel during operation. In the event of an implosion, not only is the content destroyed, but the weight of the vehicles also increases with the risk of sinking and total loss. If, on the other hand, electronic circuits containing only incompressible materials are used, it is possible to avoid the pressure housings and, instead to expose the circuits directly to the isotropic external pressure. The trick is to allow components and circuit boards to be completely compressed by the water outside and inside the vehicle. Some of the commonly used components are unsuitable because they cannot withstand the external pressure or no longer function properly in fluids or plastics. These include aluminium electrolytic capacitors and oscillating quartz crystals. If these components are avoided and replaced by other components such as ceramic capacitors and solid-state oscillators, almost any circuit can be designed to be pressure-tolerant. The deeper you dive, the more pronounced the advantages of pressure-neutral technology become. Therefore, it is no wonder that James Cameron used pressure-neutral technology on his famous dive to Challenger Deep of almost 11 km in the sea. It is to be expected that due to such prominent examples, pressure-neutral technology will soon gain new followers and become more widespread.

References

[1] Bingham, N. (2017). Building pressure tolerant electronics (PTE) for deep ocean vehicle applications. Presentation at Hardware Developers Didactic Galactic meetup held on April 20th 2017 in San Francisco, available: https://www.youtube.com/

[2] Kampmann, P., Lemburg, J., Hanff, H., & Kirchner, F. (2012). Hybrid pressure-tolerant electronics. Oceans'2012 (pp. 1–5).

[3] Holzschuh, J. E. (1974, August). Pressure tolerant electronics. Oceans'74, MTS/IEEE, Halifax, Nova Scotia.

[4] Barnes, H. E., & Gennari, J. J. (1976). A review of pressure-tolerant electronics (PTE) (No. NRL-MR-3317). Naval Research Lab, Washington DC.

Further Reading

[1] Bingham, N. (2013). Designing pressure-tolerant electronic systems. Unmanned Underwater Technology, White Paper, 2013. [Online]. Available: http://uutech.com/ptepaper

[2] Degel, C., Becker, F. J., Heinz, M., *et al.* (2011, June). Concept and demonstrator for modular pressure tolerant autonomous underwater vehicle. In Oceans, 2011 IEEE-Spain (pp. 1–5). IEEE.
[3] Gilbert, D., Manes, E., & Allard, F. (1980, September). Analysis of the performance of a pressure-tolerant light-emitting diode/fiber optic cable interface. Oceans'80 (pp. 440–444). IEEE.
[4] McBride, L. R., & Scholfield, J. T. (2007). Solid- state pressure-tolerant illumination for MBARI's underwater low-light imaging system. Journal of Display Technology, 3(2), 149–154.
[5] Morash, J., *et al.* (2013). Systems and methods for pressure tolerant energy systems. U.S. Patent Application No. 10/000,260.
[6] Nellessen, V. Jr. (2013). Pressure tolerant battery. Schlumberger Technology Corporation, Sugar Land, TX, USA, Patent WO 2 013 103 679 A1.
[7] Nellessen Jr, P. (2017). Pressure tolerant battery. U.S. Patent No. 9,637,994.
[8] Pittini, R., Hernes, M., & Ljøkelsøy, K. (2012). Pressure tolerant power electronics: IGBT gate driver for operation in high pressure hydrostatic environment. Journal of Energy and Power Engineering, 6(9), 1500.
[9] Pittini, R., & Hernes, M. (2012). Pressure-tolerant power electronics for deep and ultradeep water. Oil and Gas Facilities, 1(01), 47–52.
[10] Spillman, D. M., Ebel, S. J., & Clark, W. D. K. (1992). Pressure tolerant lithium cells. IEEE 35th International Power Sources Symposium, Cherry Hill, NJ, USA, pp. 84–87.
[11] Sutton, J. (1979, September). Pressure tolerant electronic systems. OCEANS'79 (pp. 460–469).
[12] Thiede, C., Buscher, M., Lück, M., *et al.* (2009). An overall pressure tolerant underwater vehicle: DNS Pegel. Oceans'09, IEEE Conference, Bremen.
[13] Vedachalam, N., Babu, S. M., Ramadass, G. A., & Atmanand, M. A. (2014). Review of maturing multi-megawatt power electronic converter technologies and reliability modeling in the light of subsea applications. Applied Ocean Research, 46, 28–39.
[14] Wilson, R. A., & Bales, J. W. (2006, September). Development and experience of a practical, pressure-tolerant, lithium battery for underwater use. Bluefin Robotics Corporation, Oceans'2006, MTS / IEEE, Boston, MA, USA.
[15] Wilson, R. A., & Somlyody, S. A. (2009). Pressure-tolerant lithium polymer batteries. Sea Technology, 50(7), 31–35.
[16] Woolsey, M., Adams, L. D., & Jarnagin, R. (2017, September). Integration of pressure-tolerant lithium ion battery modules within the Mola Mola AUV. In Oceans' 2007, Anchorage (pp. 1–7). IEEE.

Part IV

Cooperation

Chapter 11

Optimization of waterspace management and scheduling for multiple unmanned vehicles

Matthew J. Bays[1] *and Thomas A. Wettergren*[2]

We describe an approach for accomplishing the high-level mission planning required for a heterogeneous team of autonomous vehicles performing surveys in multiple areas, such as required for mine countermeasure (MCM) missions. The high-level mission scheduling and waterspace management require sequencing the order and location of lower-level tasks to be completed by each vehicle in the heterogeneous team: unmanned surface vessels (USVs) and unmanned underwater vehicles (UUVs). In this context, the USVs serve as transport vehicles while the UUVs perform the actual surveys and execute any other local actions. We develop a solution to this complex sequencing operation by leveraging unique information processing, communication, refueling, and planning windows that form constraints within the system within a formal scheduling optimization framework. We use mixed-integer linear programming (MILP) as a solution method for the associated optimization problem. By using such a standard optimization approach, we both take advantage of optimality guarantees on the solution and extensive commercial numerical solvers. In addition to developing the numerical optimization problem, we also include methods to account for risks due to schedule slip for the individual tasks. We conclude with an example of joint USV/UUV planning using the optimization algorithm.

11.1 Introduction

The use of autonomous systems to perform increasingly complex and coordinated tasks has necessitated creating heterogeneous teams of agents, where different agent types specialize in different parts of an operation [1–3]. One such heterogeneous team operation is the heterogeneous unmanned system MCM scenario, when one type of mobile agent, typically a UUV, is tasked with performing the direct surveys and reacquires/identifies tasks for the operation, while another type of larger, faster-moving, or longer-range agent, a USV, is responsible for transporting the UUVs

[1]Naval Surface Warfare Center, Panama City Division, Panama City, FL 32407, USA
[2]Naval Undersea Warfare Center, Newport Division, Newport, RI 02841, USA

between jobs for faster completion. Within manned systems, there are numerous examples of the USV/UUV concept such as aircraft carriers and their respective aircrafts, garbage trucks and accompanying garbage workers, or mail delivery and their respective postmen. While this form of close interaction between unmanned systems is still far from common, the underlying hardware and guidance infrastructure to allow autonomous docking and deployment between unmanned systems is actively researched [4–8].

We present a high-level planner that takes the lower-level vehicle tasks required to complete a typical underwater survey mission and sequences them in a near-optimal manner to minimize objective functions such as mission time while accomplishing mission objectives. We use MILP as an optimization technique for our approach [9]. MILP has previously been used for probabilistic vehicle motion control, and is capable of quickly finding near-optimal solutions to constrained optimization problems involving linear and integer variables [10]. Other recent studies [11] have shown that a carefully constructed constraint programming approach can be superior to MILP approaches in multiagent scheduling and allocation problems, but we restrict our attention to the MILP approach in order to take advantage of well-established commercial solvers.

The principal theory developed in this research was how to pose the multiple unmanned vehicle task-scheduling problem in a formal mathematical manner such that it can be input into an off-the-shelf MILP solver, while allowing for the inherent uncertainty in the system. The use of an MILP solver requires two fundamental components: a matrix of mathematical constraint equations that capture both the independent variables of optimization, as well as linear and binary variables that are dependent on those optimization variables. The second component is the linear cost function that is dependent on all independent and dependent variables used in the constraint equations. If these components are posed successfully to capture the dynamics of the unmanned vehicle mission, solutions can be found relatively quickly. While our work predominantly considers heterogeneous teams of unmanned vehicles for the development of the optimization constraints, it is possible that this approach could be applied to teams of manned systems provided similar constraints on communication, refueling, or other mission characteristics exist.

11.1.1 Related work

The problem of scheduling and task allocation has been extensively studied with a variety of techniques in multiple different forms.* Koes *et al.* developed a coordination architecture for modeling multirobot coordination and task allocation in [12]. Our framework complements this methodology by expanding the types of scenarios under

*Excerpts of the literature review related to Koes, Korsah, Gombolay, and Lin are reprinted from *Robotics and Autonomous Systems*, vol. 85, M. J. Bays and T. A. Wettergren, "Service agent-transport agent task planning incorporating robust scheduling techniques," pp. 15–26, Copyright 2015, with permission from Elsevier.

which a constraint optimization framework can be used, namely to problems where there are strongly coupled constraints on vehicles such as transportation actions.

Korsah *et al.* create a framework to explore optimal vehicle routing with cross-schedule constraints with applications to robotic assistance in [13]. However, there are several notable differences to our work. First, the principal element of optimization is an agent *route*, not the collection of agent actions comprising the route. Additionally in Korsah's problem, the individuals being transported along the routes are static; there is no mechanism within the optimization framework for the individuals themselves to move about the area. As such [13] is similar to the traditional vehicle routing problem [14]. Similarly, Mathew *et al.* address finding the shortest route for an unmanned ground vehicle (UGV) and unmanned aerial vehicle (UAV) pair to transit and make deliveries to locations only reachable by the UAV in [15]. However, their work focuses primarily on the overall path planning for the two vehicles, and not on fuel limitations and the higher-level scheduling aspects required when there are an arbitrary number of UAVs and UGVs. Our work takes a scheduling-centric approach, formally incorporates fuel constraints, and is generalized for an arbitrary number of UUVs and USVs.

Gombolay *et al.* develop a centralized algorithm to efficiently schedule manufacturing processes using robotic teams in [16]. However, they primarily focus on the development of a task sequencing heuristic that allows the generalized problem, a simple temporal problem [17], coupled with simple spatiotemporal constraints, to be tractable for large numbers of tasks and robots to be assigned to a fixed set of tasks. Our principal contribution is in the unique modeling and constraint-based logic required for the docking, transport, and deployment of unmanned vehicles throughout the environment. The development of the nontrivial constraints within our framework allows for applications where a list of required tasks is not given, but instead a variety of different methods are available to the vehicles to collectively complete a set of higher-level objectives by the vehicles' interaction.

Previous work specifically for MCM scheduling has typically focused on the task allocation problem [1,18]. For example, several efforts have been undertaken to determine optimal partitioning of search areas for undersea search with various assumptions on sensor modeling and false alarm information in [19,20]. Our work extends these results by focusing on the *when* and *how* to optimally have vehicles prosecute different mission areas, as opposed to simply *where* vehicles should be tasked. More recently, work by Molieneaux *et al.* have developed algorithms to aid in schedule repair for previously developed MCM operation schedules using a cognitive architecture framework [21]. With their framework, the authors attempt to repair user-designed MCM schedules that have become infeasible due to unforeseen circumstances such as asset breakdown or tasking delays. In our work, we attempt to solve the initial step by providing a formal, automated method to plan an overall MCM schedule from a mathematical model of generalized MCM operation constraints.

General robust scheduling techniques have been studied using a variety of methods. Lin, Janak, and Floudas develop complimentary approaches to formally incorporate uncertainty in scheduling problems in [22,23]. In their work, Lin *et al.* develop formal methodologies for converting MILP scheduling problems where model parameters are uncertain under either a bound or known probability distribution into another MILP or MINLP-type problem where the uncertainty is explicitly incorporated into the optimization framework, yielding a robust solution that still meets the problem constraints with a fixed probability. In a similar manner, Bertsimas and Sim create a method to adjust the conservatism of robust MILP solutions in [24] while keeping the robust problem linear. Other work on robust scheduling and integer programming problems in the general sense may be found in [25–27]. However, to our knowledge there is no work explicitly incorporating Bayes risk to schedule slips nor for the application of MCM scheduling. Our work has the similar goal for the specific problem of MCM scheduling, but instead of robustly optimizing with a fixed uncertainty of meeting time constraints or an acceptable number of constraints that may be violated, we incorporate a cost function derived from Bayes risk using an adjustable cost of a schedule slip and the potential increase in buffer time for a task.

11.1.2 Overview of the chapter

The remainder of this chapter is outlined as follows: in Section 11.2, we discuss the general setup of the multiple UUV task-scheduling problem, including the definition of the theoretical base problem we call the service agent transport problem (SATP). In Section 11.3, we outline the mathematical constraints that make up the framework. In order to account for necessary uncertainties in task duration, we discuss robust scheduling extensions in Section 11.4. These extensions provide a modification to the cost function that is based on Bayes risk; this modification adds buffer times in between tasks into the system while maintaining the MILP framework. Finally, simulation results showing the performance of various implementations of the UUV task scheduler are discussed in Section 11.5.

11.2 Multiple UUV logistics scheduling using MILP

We now outline the general multiple UUV scheduling problem by developing a formal mathematical framework for optimizing a vehicle schedule.† Figure 11.1 shows a notional illustration of the UUV scheduling problem for a survey application. Let there exist a set of autonomous vehicles $\mathscr{A} = \{1, \ldots, A\}$ that are tasked with surveying a number of survey areas $\mathscr{S} = \{1, \ldots, S\}$. The survey areas are connected by a directed graph $\langle \mathscr{D}, \mathscr{V} \rangle$. The nodes $\mathscr{D} = \{o, 1, \ldots, D\}$ represent locations at which an agent $a \in \mathscr{A}$ may enter a survey area $s \in \mathscr{S}$ in order to perform operations. The node labeled o is the origin node which represents the starting point for all valid paths in the graph. The agents may move from node i to node j along an edge of the graph

†The general task-scheduling problem formulation is reprinted from *Robotics and Autonomous Systems*, vol. 85, M. J. Bays and T. A. Wettergren, "Service agent-transport agent task planning incorporating robust scheduling techniques," pp. 15–26, Copyright 2015, with permission from Elsevier.

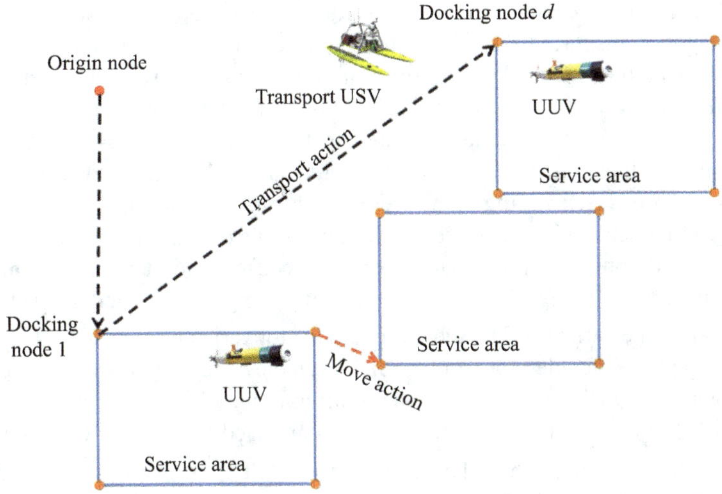

Figure 11.1 Notional illustration of an MCM mission utilizing unmanned vehicles, including USVs to transport UUVs

$v = (i,j) \in \mathscr{V}$. The subset $\mathscr{D}_s \subset \mathscr{D}$ represents nodes from which agents may survey area s.

The use of graph representations of scheduling problems has been historically used effectively due to the ability of the methods to directly lead to effective heuristic decomposition procedures [28] that can greatly reduce the computational complexity of the optimization. Structural simplifications of the graph structure may be used to create heuristics on the survey scheduling; for example, the resulting optimized schedule is expected to be from the subset of acyclic paths within the graph structure. Thus, reducing the problem size by decomposing the graph structure into a directed acyclic graph structure is a reasonable complexity reduction strategy; such an approach is being considered as a future extension to this work.

The vehicles are separated into two mutually exclusive and collectively exhaustive subsets $\mathscr{A}_u$ and $\mathscr{A}_s$ consisting of UUVs to perform survey actions (with potentially additional post-survey task execution) and transport USVs, respectively. A survey UUV $m \in \mathscr{A}_u$ is capable of directly surveying the survey areas. Additional post-survey task executions are actions requiring additional effort on areas previously surveyed and in which objects of interest have been found. A USV $n \in \mathscr{A}_s$ may collect a fixed number of the UUVs and transport them between nodes. The UUVs may move between nodes with a time cost c_{ij}^{move}; however , these are significantly slower than USVs with transportation cost of $c_{ij}^{trans} < c_{ij}^{move}$. The USVs also require a fixed amount of time to collect and deploy the survey UUVs. The collection and deployment costs are denoted c^{dock} and c^{deploy}, respectively. The vehicles initially start at an origin node o, with each USV n containing A_n UUVs. This problem is a variant of the SATP which we first described in [29].

We outline the general logistical constraints for an operation where multiple UUVs may be transported throughout the field via an USV based largely on [29]. We then expand the logistical constraints with the specific constraints developed for UUVs performing both general survey tasks as well as additional tasks (so-called "post-survey actions"), the conduct of which is dependent on the survey results (such as examining a location where the survey detected an object). We also include an extension for robust scheduling in the presence of task duration slippage and develop a strategy to reduce the impact of schedule slips on the overall mission.

The primary optimization variables for our specific solution to this variant of the SATP, which we call *decision variables*, fall into two groups: scheduling indicator variables and time variables. The scheduling indicator variable $\bar{\mathbf{I}}^{k}_{a,p,\mathscr{L}} \in \{0, 1\}$ denotes that task type k is scheduled for completion by vehicle a during phase p at location set $l \in \mathscr{L}_k$, where $\mathscr{L}_k$ is the set of all feasible locations required for execution of task type k and $\bigcup_{k\in\mathscr{K}} \mathscr{L}_k = \mathscr{L}$. Phases in this problem are used as slots for the scheduling of the tasks to be accomplished. The timing variables $\bar{\mathbf{T}}^{start}_{a,p} \in \mathbb{R}$ and $\bar{\mathbf{T}}^{end}_{a,p} \in \mathbb{R}$ denote the times at which the task of vehicle a during phase p is scheduled to start and end, respectively. The type of tasks that are considered in this model include deploying, docking, moving, waiting, surveying, and post-survey tasks (which are additional tasks to be conducted based upon the results of a successful survey). In contrast to the decision variables, additional variables are needed for the optimization problem in order to develop the necessary constraints to formulate the optimization problem. While the additional variables, which we denote as *helper variables*, are technically variables used in the optimization, they are not of primary concern for the optimization problem. We discuss the helper variables as the details of the optimization problem are formulated.

11.3 Constraint formulation

We now turn to defining the constraints required for the formal mathematical formulation of the multiple UUV scheduling problem pertaining to the survey constraints and logistics management. Many of the following constraints were originally characterized within the original SATP formulation in [29].[‡]

11.3.1 Docking and deployment constraints

The first constraint within the SATP is that a UUV must only complete certain tasks when they are docked or deployed, as appropriate. For enforcing this constraint, we

[‡]The constraints related to docking/deployment, fuel, goal/capabilities, time dependency/waiting, and transit/transport are reprinted from *Robotics and Autonomous Systems*, vol. 85, M. J. Bays and T. A. Wettergren, "Service agent-transport agent task planning incorporating robust scheduling techniques," pp. 15–26, Copyright 2015, with permission from Elsevier.

create the helper variables to track whether a survey UUV m is docked or deployed from a USV n. We denote this variable

$$\bar{\mathbf{D}}_{mn,p} = \begin{cases} 1 & \text{UUV } m \text{ is docked with USV } n \text{ in phase } p \\ 0 & \text{Otherwise} \end{cases} \tag{11.1}$$

Within an integer programming framework, we create (11.1) with the formal constraint

$$\begin{aligned} &\forall m \in \mathscr{A}_u, \quad \forall n \in \mathscr{A}_s, \quad \forall p \in \mathscr{P} \\ &\bar{\mathbf{D}}_{mn,p} = \sum_{p' \leq p, d \in D} \left[\bar{\mathbf{I}}^{dock}_{mn,p',d} - \bar{\mathbf{I}}^{deploy}_{mn,p',d} \right] \end{aligned} \tag{11.2}$$

where $\bar{\mathbf{I}}^{dock}_{mn,p',d}$ and $\bar{\mathbf{I}}^{deploy}_{mn,p',d}$ are indicator variable docking and deploying, respectively, at particular dock/deploy nodes d. We note that initially all vehicles are docked at the origin node o. We bound $\bar{\mathbf{D}}_{mn,p}$ with the constraint

$$\forall m \in \mathscr{A}_u, \quad \forall n \in \mathscr{A}_s, \quad \forall p \in \mathscr{P} \qquad 0 \leq \bar{\mathbf{D}}_{mn,p} \leq 1 \tag{11.3}$$

With constraints (11.2) and (11.3), $\bar{\mathbf{D}}_{mn,p}$ equals 1 if UUV m is currently docked with USV n during phase p, and 0 otherwise. Constraint (11.2) and the lower bound in (11.3) also enforces the additional requirement that in order for a survey UUV to be deployed, it must first have been docked. It is also necessary to track if a UUV is docked or deployed in general. This is performed using the helper variable $\bar{\mathbf{D}}_{mn,p}$ and constraints

$$\forall m \in \mathscr{A}_u, \quad \forall p \in \mathscr{P} \qquad \bar{\mathbf{D}}_{m,p} = \sum_{n \in \mathscr{A}_s} \bar{\mathbf{D}}_{mn,p} \tag{11.4}$$

and

$$\forall m \in \mathscr{A}_u, \quad \forall p \in \mathscr{P} \qquad 0 \leq \bar{\mathbf{D}}_{m,p} \leq 1 \tag{11.5}$$

Additionally, we limit the total number of UUVs that can be docked at a given time by a USV to D_{max} using the constraint

$$\forall n \in \mathscr{A}_s, \quad \forall p \in \mathscr{P} \qquad \sum_{m \in \mathscr{A}_{ur}} \bar{\mathbf{D}}_{mn,p} \leq D_{max} \tag{11.6}$$

With constraints (11.4)–(11.6), we can now limit the feasible actions of the individual vehicles when deployed or docked as appropriate.

11.3.2 Fuel limitation and charging constraints

Fuel consumption is an additional variable that must be tracked for each vehicle during all phases of the mission. For tracking fuel, we define the variable $\bar{\mathbf{F}}_{m,p} \in \mathbb{R}$ as the amount of fuel survey UUV m has available during the start of phase p. The variable $\bar{\mathbf{F}}_{m,p}$ is assigned a value using the constraints

$$\forall m \in \mathscr{A}_u, \quad \forall p \in \mathscr{P} \tag{11.7}$$

$$\bar{\mathbf{F}}_{m,p} = \sum_{p' \leq p} \Bigg[\bar{\mathbf{F}}^{charge}_{m,p'} - \sum_{s \in \mathscr{S}} f^{survey}_{m,s} \bar{\mathbf{I}}^{survey}_{m,p',s} - \sum_{d \in \mathscr{D}} \left(f^{dock}_{m,d} \bar{\mathbf{I}}^{dock}_{m,p',d} + f^{deploy}_{m,d} \bar{\mathbf{I}}^{deploy}_{m,p',d} \right) - \sum_{v \in \mathscr{V}} f^{move}_{m,p',v} \bar{\mathbf{I}}^{move}_{m,p',v} \Bigg]$$

where $f^{survey}_{m,s}$, $f^{dock}_{m,d}$, $f^{deploy}_{m,d}$, and $f^{move}_{m,v}$ are fixed parameters representing the fuel usage rate for executing each of the respective tasks, and $\bar{\mathbf{F}}^{charge}_{m,p}$ is a floating variable indicating the amount a vehicle is charged during a phase. The variable $\bar{\mathbf{F}}^{charge}_{m,p}$ is fixed to its value using the linear survey UUV charge rate c^{charge}_m and the constraints

$$\forall m \in \mathscr{A}_u, \quad \forall p \in \mathscr{P} \tag{11.8}$$

$$\bar{\mathbf{D}}_{m,p} = 1 \Rightarrow \bar{\mathbf{F}}^{charge}_{m,p} \leq c^{charge}_m \left(\bar{\mathbf{T}}^{end}_{m,p} - \bar{\mathbf{T}}^{start}_{m,p} \right)$$

$$\bar{\mathbf{D}}_{m,p} = 0 \Rightarrow \bar{\mathbf{F}}^{charge}_{m,p} = 0$$

Additionally, we limit the fuel capacity with the constraints

$$\forall m \in \mathscr{A}_u, \quad \forall p \in \mathscr{P} \tag{11.9}$$

$$\bar{\mathbf{F}}_{m,p} \leq F_{max}$$

$$\forall m \in \mathscr{A}_u, \quad \forall p \in \mathscr{P} \tag{11.10}$$

$$0 < \bar{\mathbf{F}}_{m,p}$$

11.3.3 Goal and capability constraints

In our formulation, each vehicle can only prosecute one goal during a phase of the mission. As such, we have the constraint

$$\forall a \in \mathscr{A}, \quad \forall p \in \mathscr{P} \qquad \sum_{k \in \mathscr{K}} \sum_{l \in \mathscr{L}_k} \bar{\mathbf{I}}^k_{a,p,l} \leq 1 \tag{11.11}$$

As not every vehicle is capable of completing every task, this requires additional constraints to force only vehicles capable of completing a task to be scheduled for the task. As such, we have the constraint

$$\forall a \notin \mathscr{A}_k, \quad \forall p \in \mathscr{P}, \quad \forall l \in \mathscr{L}_k \qquad \bar{\mathbf{I}}^k_{a,p,l} = 0 \tag{11.12}$$

where $\mathscr{A}_k$ is the set of vehicles that are capable of action k. We note that this constraint may also be implemented for tractability by limiting both the scheduling variables $\bar{\mathbf{I}}^k_{a,p,l}$ and constraints specific to a certain capability that are created in the optimization problem to those allowed by the individual vehicle's capabilities.

11.3.4 Time dependency and wait constraints

We now turn to the constraints required for determining the specific times at which phases are scheduled for starting and ending. Without loss of generality, we assume that the start time of the initial phase is at $t = 0$. As such we have the constraint

$$\forall a \in \mathscr{A} \qquad \bar{\mathbf{T}}^{start}_{a,p=0} = 0 \tag{11.13}$$

In order to adhere to the boundary conditions between the end of the last phase and the start of the next phase, we have the constraint

$$\forall a \in \mathscr{A}, \quad \forall p \in \mathscr{P} \mid p \neq 0 \qquad \bar{\mathbf{T}}^{start}_{a,p+1} \geq \bar{\mathbf{T}}^{end}_{a,p} \tag{11.14}$$

With constraints (11.13) and (11.14), we have defined all variables with the exception of the end time of each phase, $\bar{\mathbf{T}}^{end}_{a,p}$. The end time may be defined by taking advantage of (11.11) with only one task being scheduled for a vehicle during a phase

$$\forall a \in \mathscr{A}, \quad \forall p \in \mathscr{P} \qquad \bar{\mathbf{T}}^{end}_{a,p} = \bar{\mathbf{T}}^{start}_{a,p} + \sum_{k \in \mathscr{K}} \sum_{l \in \mathscr{L}_k} c^k_{a,l} \bar{\mathbf{I}}^k_{a,p,l} \tag{11.15}$$

where $c^k_{a,L}$ is the duration of task k if performed by vehicle a on location set L. Additionally, the dual relationship between UUV and USV performing docking and deployment constraints necessitates that we must ensure that those particular operations occur at the same time for both vehicles. Thus, we have the time-dependency constraints

$$\forall m \in \mathscr{A}_u, \quad \forall m \in \mathscr{A}_s, \quad \forall p \in \mathscr{P} \tag{11.16}$$
$$\sum_{d \in \mathscr{D}} \bar{\mathbf{I}}^{dock}_{mn,p,d} = 1 \Rightarrow \bar{\mathbf{T}}^{start}_{m,p} = \bar{\mathbf{T}}^{start}_{n,p}$$

$$\forall m \in \mathscr{A}_u, \quad \forall m \in \mathscr{A}_s, \quad \forall p \in \mathscr{P} \tag{11.17}$$
$$\sum_{d \in \mathscr{D}} \bar{\mathbf{I}}^{deploy}_{mn,p,d} = 1 \Rightarrow \bar{\mathbf{T}}^{end}_{m,p} = \bar{\mathbf{T}}^{end}_{n,p}$$

Some scheduling situations necessitate an vehicle loitering or otherwise performing no meaningful actions during a given phase so that it may properly coordinate with the actions of other vehicles. For example, a USV may need to wait at a docking point during a phase in which a survey UUV is servicing an area with which it will dock in the proceeding phase. In this case, we require an additional variable indicating that a vehicle is idle during that phase at a given location. We define the $\bar{\mathbf{I}}^{wait}_{a,p,d}$ indicator variable by the constraints

$$\forall a \in \mathscr{A}, \quad \forall p \in \mathscr{P}, \quad \forall d \in \mathscr{D} \tag{11.18}$$
$$\sum_{k \in \mathscr{K}} \left[\sum_{l \in \mathscr{L}_k} \bar{\mathbf{I}}^k_{a,p,l} \right] + \sum_{d' \in \mathscr{D} | d' \neq d} \bar{\mathbf{I}}^{wait}_{a,p,d'} = 1 \Rightarrow \bar{\mathbf{I}}^{wait}_{a,p,d} = 0$$

$$\forall a \in \mathscr{A}, \quad \forall k \in \mathscr{K}, \quad \forall p \in \mathscr{P}, \quad \forall d \in \mathscr{D} \tag{11.19}$$

$$\sum_{k \in \mathscr{K}} \left[\sum_{l \in \mathscr{L}_k | d \notin \mathscr{L}} \bar{\mathbf{I}}^{k}_{a,p-1,l} \right] + \sum_{d' \in \mathscr{D} | d' \neq d} \bar{\mathbf{I}}^{wait}_{a,p-1,d'} = 1 \Rightarrow \bar{\mathbf{I}}^{wait}_{a,p,d} = 0$$

$$\forall a \in \mathscr{A}, \quad \forall p \in \mathscr{P} \sum_{k \in \mathscr{K}} \sum_{l \in \mathscr{L}_k} \bar{\mathbf{I}}^{k}_{a,p,l} = 0 \Rightarrow \sum_{d \in \mathscr{D}} \bar{\mathbf{I}}^{wait}_{a,p,d} = 1 \tag{11.20}$$

Constraint (11.18) forces $\bar{\mathbf{I}}^{wait}_{a,p,d}$ to equal zero if the vehicle is otherwise tasked or waiting at another location. Constraint (11.19) forces $\bar{\mathbf{I}}^{wait}_{a,p,d}$ to equal zero if the vehicle performed an action at a different location d' during the previous time step, thereby prohibiting the vehicle from waiting at location d. Constraint (11.20) forces $\bar{\mathbf{I}}^{wait}_{a,p,d}$ to one if the vehicle is not otherwise tasked.

11.3.5 Transit and transport constraints

We now turn to the constraints required for transiting vehicles and transporting them. In order for a vehicle to transit from one location to the next, it must have performed an action at the previous location

$$\forall m \in \mathscr{A}_u, \quad \forall n \in \mathscr{A}_s, \quad \forall p \in \mathscr{P}, \quad \forall d_0 \in D \tag{11.21}$$

$$\sum_{d \in \mathscr{D}} \bar{\mathbf{I}}^{move}_{m,p,(d_0,d)} = 1$$

$$\Rightarrow \sum_{d' \in \mathscr{D}} \bar{\mathbf{I}}^{move}_{m,p-1,(d',d_0)} + \sum_{n \in N} \bar{\mathbf{I}}^{deploy}_{mn,p-1,d_0} + \sum_{s \in \mathscr{S} | d_0 \in \mathscr{D}_s} \bar{\mathbf{I}}^{survey}_{m,p-1,s} + \bar{\mathbf{I}}^{wait}_{m,p-1,d} = 1$$

where $(d_0, d), (d', d_0) \in \mathscr{V}$. Additionally, we have the constraint that for a vehicle to either move from one area to the next or to survey an area, the vehicle must be deployed

$$\forall m \in \mathscr{A}_u, \quad \forall p \in \mathscr{P} \qquad \sum_{v \in \mathscr{V}} \bar{\mathbf{I}}^{move}_{m,p,v} = 1 \Rightarrow \bar{\mathbf{D}}_{m,p} = 0 \tag{11.22}$$

For a UUV m to be docked with USV n, both vehicles must have been located previously at the location of the capture. Thus, we have

$$\forall m \in \mathscr{A}_u, \quad \forall n \in \mathscr{A}_s, \quad \forall p \in \mathscr{P}, \quad \forall d \in \mathscr{D} \tag{11.23}$$

$$\bar{\mathbf{I}}^{dock}_{mn,p,d} = 1 \Rightarrow \sum_{d_{0_1} \in D} \bar{\mathbf{I}}^{move}_{m,p-1,(d_{0_1},d)} + \bar{\mathbf{I}}^{wait}_{m,p-1,d} + \sum_{s \in \mathscr{S} | d \in \mathscr{D}_s} \bar{\mathbf{I}}^{survey}_{m,p-1,s} = 1$$

$$\bar{\mathbf{I}}^{dock}_{mn,p,d} = 1 \Rightarrow \sum_{d_{0_1} \in D} \bar{\mathbf{I}}^{move}_{n,p-1,(d_{0_1},d)} + \bar{\mathbf{I}}^{wait}_{n,p-1,d}$$

$$+ \sum_{m' \in \mathscr{A}_{ur} | m' \neq m} \left[\bar{\mathbf{I}}^{dock}_{m'n,p-1,d} + \bar{\mathbf{I}}^{deploy}_{m'n,p-1,d} \right] \geq 1 \tag{11.24}$$

where $(d_{0_1}, d), (d_{0_2}, d) \in \mathscr{V}$. Likewise, we have the constraint for deploying UUVs as

$$\forall m \in \mathscr{A}_u, \quad \forall n \in \mathscr{A}_s, \quad \forall p \in \mathscr{P}, \quad \forall d \in \mathscr{D}$$

$$\bar{\mathbf{I}}^{deploy}_{mn,p,d} = 1 \Rightarrow \bar{\mathbf{D}}_{mn,p} = 1 \tag{11.25}$$

$$\bar{\mathbf{I}}^{deploy}_{mn,p,d} = 1 \Rightarrow \sum_{d_{0_1} \in D} \bar{\mathbf{I}}^{move}_{mn,p-1,(d_{0_1},d)} + \bar{\mathbf{I}}^{wait}_{n,p-1,d} + \sum_{m' \in \mathscr{A}_{ur} | m' \neq m} \left[\bar{\mathbf{I}}^{dock}_{m'n,p-1,d} + \bar{\mathbf{I}}^{deploy}_{m'n,p-1,d} \right] \geq 1$$

11.3.6 Survey constraints

The first constraint for a servicing action is that a survey area can only be surveyed if it is scheduled after a deployment action at a node $d \in \mathscr{D}_s$ or move action. Thus, we have the constraint

$$\forall m \in \mathscr{A}_u, \quad \forall n \in \mathscr{A}_s, \quad \forall p \in \mathscr{P}, \quad \forall d \in \mathscr{D}, \quad \forall s \in \mathscr{S}$$

$$\bar{\mathbf{I}}^{survey}_{m,p,s} = 1 \Rightarrow \sum_{n \in \mathscr{A}_s} \bar{\mathbf{I}}^{deploy}_{mn,p-1,d} + \sum_{d' \in D} \bar{\mathbf{I}}^{move}_{m,p-1,(d',d)} = 1 \tag{11.26}$$

where $(d', d) \in \mathscr{V}$. Next, we have the constraint that a survey action can only be performed if it is scheduled while the UUV is deployed

$$\forall m \in \mathscr{A}_u, \quad \forall p \in \mathscr{P} \qquad \sum_{s \in \mathscr{S}} \bar{\mathbf{I}}^{survey}_{m,p,s} = 1 \Rightarrow \bar{\mathbf{D}}_{m,p} = 0 \tag{11.27}$$

Additionally, every survey goal should be completed exactly once by the group of vehicles. Thus,

$$\forall s \in \mathscr{S} \qquad \sum_{m \in \mathscr{A}_u} \sum_{p \in \mathscr{P}} \bar{\mathbf{I}}^{survey}_{m,p,s} = 1 \tag{11.28}$$

11.3.7 Additional post-survey task execution constraints

If, in addition to survey actions, an area is determined to require the execution of some action that is performed after a survey (potentially by another UUV), there are several additional constraints necessary for implementing those post-survey efforts within the schedule. This applies to a variety of scenarios: when an object is found and must be further investigated, when something must be removed from the region, and when a vehicle with specialized sensing equipment must be brought in to a location that was pinpointed by the survey. In general, we consider all such situations to be actions that require effort that specifically occurs after a survey of an area. If additional UUVs are used for these actions, these UUVs are still to be transported by the same transport USVs as the survey UUVs, thus making the scheduling further complicated since these vehicles have a schedule dependence on previous activities of the survey UUVs. For notational convenience, we refer to the tasks associated with these additional actions as *post-survey actions*, as they require a specific action that is separate from the survey.

The first constraint for a post-survey action is that a post-survey action can only be executed if it is scheduled after a deployment action at a node $d \in \mathscr{D}_s$ or move action. Thus, we have the constraint

$$\forall m \in \mathscr{A}_u, \quad \forall n \in \mathscr{A}_s, \quad \forall p \in \mathscr{P}, \quad \forall d \in \mathscr{D}, \quad \forall s \in \mathscr{S} \tag{11.29}$$

$$\bar{\mathbf{I}}^{ps}_{m,p,s} = 1 \Rightarrow \sum_{n \in \mathscr{A}_s} \bar{\mathbf{I}}^{deploy}_{mn,p-1,d} + \sum_{d' \in D} \bar{\mathbf{I}}^{move}_{m,p-1,(d',d)} = 1 \tag{11.30}$$

where $(d', d) \in \mathscr{V}$. Next, we have the constraint that a post-survey action can only be performed if it is scheduled while the UUV is deployed

$$\forall m \in \mathscr{A}_u, \quad \forall p \in \mathscr{P} \qquad \sum_{s \in \mathscr{S}} \bar{\mathbf{I}}^{ps}_{m,p,s} = 1 \Rightarrow \bar{\mathbf{D}}_{m,p} = 0 \tag{11.31}$$

Additionally, every post-survey goal should be completed exactly once by the group of vehicles. Thus,

$$\forall s \in \mathscr{S} \qquad \sum_{m \in \mathscr{A}_u} \sum_{p \in \mathscr{P}} \bar{\mathbf{I}}^{ps}_{m,p,s} = 1 \tag{11.32}$$

Unlike the survey actions, we additionally add the constraint that a post-survey action on survey area s can only be performed after the required survey action. That is, we have

$$\forall m \in \mathscr{A}_u, \quad \forall s \in \mathscr{S}, \quad \forall p \in \mathscr{P} \qquad \bar{\mathbf{I}}^{ps}_{m,p,s} = 1 \Rightarrow \sum_{p' \in \mathscr{P}, \text{ s.t. } p' < p} \bar{\mathbf{I}}^{survey}_{m,p',s} \tag{11.33}$$

In addition to the above constraints, if post-survey actions are warranted for a given area, (11.7) must be augmented with an additional term $\sum_{s \in \mathscr{S}} f^{ps}_{m,s} \bar{\mathbf{I}}^{ps}_{m,p',s}$. Additionally, (11.21) and (11.23) must be augmented with an additional term $\sum_{s \in \mathscr{S}} \bar{\mathbf{I}}^{ps}_{m,p,s}$ in each addition.

We note that in many instances (such as MCM missions), the post-survey actions are probabilistic: not every search results in requiring a post-survey action after the fact. For example, if no objects of interest are found in an area, the follow-on post-survey action is unnecessary. In this case, constraints (11.29)–(11.33) must be represented probabilistically during plan time, and updated when new information is available. We now outline two methods of doing so, and will go into details of the second method in Section 11.4. The first method is to make the number of tasks probabilistic and create constraints that take the worst-case placement of those tasks into consideration for the initial plan. At specific intervals of mission execution, the plan then needs to be updated to take into account when tasks are then required. In the second method, we assume every search task would require a post-survey action; however, the time required for that post-survey action is probabilistic. In that situation, we tune the variation of the probabilistic task using the methods to be presented in detail in Section 11.4.

In the first method, we first choose a set fraction of the survey actions that are assumed to require post-survey actions, $N^{ps} \leq S$. The fraction may be a user-specified value or based on *a priori* situational information. Such *a priori* information may come from bathymetric knowledge of the area and the associated likelihood of objects of interest being found within that environment. Using N^{ps}, we then can impose the constraint

$$\forall \sum_{m \in \mathcal{A}_u} \sum_{p \in \mathcal{P}} \sum_{s \in \mathcal{S}} \bar{\mathbf{I}}^{ps}_{m,p,s} = N^{ps} \tag{11.34}$$

and relax constraint (11.32). We then iterate across every combination from the combinatorial set $C(S, N^{ps})$ and maximize across the selection as well in calculating the end time $\bar{\mathbf{T}}^{end}_{a,P}$ within the cost function to be discussed in Section 11.3.8. We leave this as an exercise for the reader.

In the second method, we assume that all survey actions must be associated with a post-survey action, but now we assume that the time cost of the post-survey action is probabilistic. In this case, we assume that c^{ps} is itself a nominal value with a distribution $p^{ps}(c^{ps*})$, where $c^{ps*} \in [0, c^{ps}]$. In this case, there are two scenarios in developing the schedule *a priori*: $c^{ps*} \leq c^{ps}$ and $c^{ps*} \geq c^{ps}$. If each scenario is associated with an external cost, we then determine the optimal time with which to pad the nominal time in order to minimize overall risk related to these costs. We develop this method formally in Section 11.4.

In both of these methods, it may be necessary to replan the schedule after more information is obtained. For it is only after this information is obtained when we determine what post-survey actions must be executed. This replanning is done by fixing the variables $\bar{\mathbf{I}}^{ps}_{m,p,s}$, $\bar{\mathbf{I}}^{ps}_{m,p,s}$, $\bar{\mathbf{T}}^{start}_{m,p}$, and $\bar{\mathbf{T}}^{start}_{m,p}$ for all $p \in \mathcal{P}$ such that phase p has already been executed, and then re-optimizing. This re-optimization problem is repeated to periodically replan as more information continues to be obtained.

11.3.8 Cost function

The cost function to be minimized in this MILP formulation is to minimize the end time of the last phase over all vehicles. That is, our objective is

$$\text{minimize} \max_a \bar{\mathbf{T}}^{end}_{a,P} \tag{11.35}$$

11.4 Robust scheduling extensions

We now extend our framework to incorporate uncertainty in the duration of various tasks being performed by the individual vehicles.[§] Our methodology is based on the general idea that there is a significant potential cost of schedule slips. Specifically, we define a *schedule slip* as when the end time of a task as scheduled exceeds the scheduled start time of the following task. Often, schedule slips can have catastrophic consequences in real-world multiagent systems. For example, a UUV finishing a survey task on time while a USV is significantly delayed could loiter in a dangerous location or potentially run out of fuel before docking could be performed. Similarly, a survey UUV being significantly delayed when expected to rendezvous with a USV could significantly delay the entire schedule for the remainder of the mission. In order to develop our robust scheduling scheme to mitigate these hazards with our robust survey UUV transport problem (R-SATP) extension, we first make the assumption that, for tasks taken by the individual vehicles, no tasks can be performed before their scheduled start time. Formally, this can be done by creating a constraint defining the true initial start time of a task as

$$\tilde{\mathbf{T}}_{a,p}^{start} = \max\left\{\tilde{\mathbf{T}}_{a,p-1}^{end}, \bar{\mathbf{T}}_{a,p}^{start}\right\} \tag{11.36}$$

where $\tilde{\mathbf{T}}_{a,p-1}^{end}$ is the true start time of the previous phase for vehicle a. We then define the true start time of the previous phase as

$$\tilde{\mathbf{T}}_{a,p-1}^{end} = \tilde{\mathbf{T}}_{a,p}^{start} + \sum_{k\in\mathcal{K}}\sum_{l\in\mathcal{L}_{\|}} \tilde{c}_{a,p,l}^{k}\bar{\mathbf{I}}_{a,p,l}^{k} \tag{11.37}$$

where $\tilde{c}_{a,p,l}^{k}$ is the true duration of task k during phase p performed by vehicle a on location set l. We now constrain $\bar{\mathbf{T}}_{a,p+1}^{start}$ from (11.14) to

$$\bar{\mathbf{T}}_{a,p+1}^{start} = \bar{\mathbf{T}}_{a,p}^{start} + \sum_{k\in\mathcal{K}}\sum_{l\in\mathcal{L}_k} c_{a,l}^{k}\bar{\mathbf{I}}_{a,p,l}^{k} + T_{a,p}^{buff} \tag{11.38}$$

where $T_{a,p}^{buff}$ is a notional buffer time to prevent schedule slips that we will define with optimization variables subsequently. Beforehand, we will characterize our general strategy of focusing on minimizing schedule slips during individual phases. We do so with the following proposition, which shows that if a phase is currently in a schedule slip situation, there must have been a phase in the past where the a schedule slip did not occur.

Proposition 1. *If*

$$\tilde{\mathbf{T}}_{a,p}^{start} = \tilde{\mathbf{T}}_{a,p-1}^{end} \tag{11.39}$$

[§]The robust scheduling formulation and Proposition 1 are reprinted from *Robotics and Autonomous Systems*, vol. 85, M. J. Bays and T. A. Wettergren, "Service agent-transport agent task planning incorporating robust scheduling techniques," pp. 15–26, Copyright 2015, with permission from Elsevier.

and

$$\tilde{\mathbf{T}}^{start}_{a,p} > \bar{\mathbf{T}}^{start}_{a,p-1} + \sum_{k\in\mathcal{K}}\sum_{l\in\mathcal{L}_k} \tilde{c}_{a,p,l}\bar{\mathbf{I}}^{k}_{a,p,l} \tag{11.40}$$

then there exists a $p' \leq p$ *for* $p \neq 0$ *such that*

$$\tilde{\mathbf{T}}^{start}_{a,p'} = \bar{\mathbf{T}}^{start}_{a,p'-1} + \sum_{k\in\mathcal{K}}\sum_{l\in\mathcal{L}_k} \tilde{c}_{a,p,l}\bar{\mathbf{I}}^{k}_{a,p,l} \tag{11.41}$$

Proof. Suppose (11.39) and (11.40) are true for some p. Since

$$\tilde{\mathbf{T}}^{start}_{a,p} = \tilde{\mathbf{T}}^{end}_{a,p-1} \tag{11.42}$$

$$= \tilde{\mathbf{T}}^{start}_{a,p-1} + \sum_{k\in\mathcal{K}}\sum_{l\in\mathcal{L}_k} \tilde{c}_{a,p,l}\bar{\mathbf{I}}^{k}_{a,p,l} \tag{11.43}$$

then

$$\tilde{\mathbf{T}}^{start}_{a,p-1} + \sum_{k\in\mathcal{K}}\sum_{l\in\mathcal{L}_k} \tilde{c}_{a,p,l}\bar{\mathbf{I}}^{k}_{a,p,l} > \bar{\mathbf{T}}^{start}_{a,p-1} + \sum_{k\in\mathcal{K}}\sum_{l\in\mathcal{L}_k} \tilde{c}_{a,p,l}\bar{\mathbf{I}}^{k}_{a,p,l} \tag{11.44}$$

$$\tilde{\mathbf{T}}^{start}_{a,p-1} > \bar{\mathbf{T}}^{start}_{a,p-1} \tag{11.45}$$

Due to the definition of $\tilde{\mathbf{T}}^{start}_{a,p}$ and (11.44),

$$\tilde{\mathbf{T}}^{start}_{a,p-1} = \tilde{\mathbf{T}}^{end}_{a,p-2} \tag{11.46}$$

$$= \tilde{\mathbf{T}}^{start}_{a,p-2} + \sum_{k\in\mathcal{K}}\sum_{l\in\mathcal{L}_k} \tilde{c}_{a,p-2,l}\bar{\mathbf{I}}^{k}_{a,p-2,l} \tag{11.47}$$

There are now three possibilities to the relation of $\tilde{\mathbf{T}}^{start}_{a,p-1}$ to $\bar{\mathbf{T}}^{start}_{a,p-1} + \sum_{k\in\mathcal{K}}\sum_{l\in\mathcal{L}_k} \tilde{c}_{a,p-2,l}\bar{\mathbf{I}}^{k}_{a,p-2,l}$:

$$\tilde{\mathbf{T}}^{start}_{a,p-1} > \bar{\mathbf{T}}^{start}_{a,p-2} + \sum_{k\in\mathcal{K}}\sum_{l\in\mathcal{L}_k} \tilde{c}_{a,p-2,l}\bar{\mathbf{I}}^{k}_{a,p-2,l} \tag{11.48}$$

$$\tilde{\mathbf{T}}^{start}_{a,p-1} < \bar{\mathbf{T}}^{start}_{a,p-2} + \sum_{k\in\mathcal{K}}\sum_{l\in\mathcal{L}_k} \tilde{c}_{a,p-2,l}\bar{\mathbf{I}}^{k}_{a,p,l} \tag{11.49}$$

$$\tilde{\mathbf{T}}^{start}_{a,p-1} = \bar{\mathbf{T}}^{start}_{a,p-2} + \sum_{k\in\mathcal{K}}\sum_{l\in\mathcal{L}_k} \tilde{c}_{a,p-2,l}\bar{\mathbf{I}}^{k}_{a,p-2,l} \tag{11.50}$$

Note that if (11.50) holds, then our proposition would be proven. Suppose (11.49) holds. Then

$$\tilde{\mathbf{T}}^{start}_{a,p-1} < \tilde{\mathbf{T}}^{start}_{a,p-2} + \sum_{k\in\mathcal{K}}\sum_{l\in\mathcal{L}_k} \tilde{c}_{a,p-2,l}\bar{\mathbf{I}}^{k}_{a,p-2,l} \tag{11.51}$$

$$\tilde{\mathbf{T}}^{start}_{a,p-2} + \sum_{k\in\mathscr{K}}\sum_{l\in\mathscr{L}_k} \tilde{c}_{a,p,l}\bar{\mathbf{I}}^{k}_{a,p-2,l} < \bar{\mathbf{T}}^{start}_{a,p-2} + \sum_{k\in\mathscr{K}}\sum_{l\in\mathscr{L}_k} \tilde{c}_{a,p-2,l}\bar{\mathbf{I}}^{k}_{a,p-2,l} \tag{11.52}$$

$$\tilde{\mathbf{T}}^{start}_{a,p-2} < \bar{\mathbf{T}}^{start}_{a,p-2} \tag{11.53}$$

which violates our definition of $\bar{\mathbf{T}}^{start}_{a,p}$ in (11.36). So we must assume that

$$\tilde{\mathbf{T}}^{start}_{a,p-1} > \tilde{\mathbf{T}}^{start}_{a,p-2} + \sum_{k\in\mathscr{K}}\sum_{l\in\mathscr{L}_k} \tilde{c}_{a,p-2,l}\bar{\mathbf{I}}^{k}_{a,p-2,l} \tag{11.54}$$

We now have the same assumptions as before for $p-1$. By induction, we can follow the same process to $p = 1$. From initial conditions,

$$\bar{\mathbf{T}}^{start}_{a,0} = \tilde{\mathbf{T}}^{start}_{a,0} = 0 \tag{11.55}$$

$$\tilde{\mathbf{T}}^{start}_{a,1} = 0 + \sum_{k\in\mathscr{K}}\sum_{l\in\mathscr{L}_k} \tilde{c}_{a,0,l}\bar{\mathbf{I}}^{k}_{a,0,l} \tag{11.56}$$

satisfying the proposition for $p = 1$. □

We use Proposition 1 as motivation to focus on treating the robustication of schedules as a Markov process. Namely, we note that the dominant factor in the probability of schedule slip in the current phase is the uncertainty in the action performed in the previous phase. In other words, we may mitigate the possibility of a schedule slip in the overall schedule by reducing the possibility of a schedule slip in every phase. Using (11.36)–(11.38), we now characterize our general strategy for reducing the risk of schedule slips.

We assume there is a numerical cost (measured in time) to the occurrence of either a schedule slip, or a delay beyond the buffer time

$$\sum_{k\in\mathscr{K}}\sum_{l\in\mathscr{L}_k} c^{k}_{a,l}\bar{\mathbf{I}}^{k}_{a,p,l} > T^{buff} \Rightarrow c^{slip} \tag{11.57}$$

$$\sum_{k\in\mathscr{K}}\sum_{l\in\mathscr{L}_k} c^{k}_{a,l}\bar{\mathbf{I}}^{k}_{a,p,l} < T^{buff} \Rightarrow c^{ineff}$$

Using the definition of Bayes risk, see [30], we may write the Bayes risk for this cost definition as

$$R = c^{ineff}P\left(\tilde{c}^{k}_{a,l} < T^{k,buff}_{a,l}\right)P^{ineff} + c^{slip}P\left(\tilde{c}^{k}_{a,l} > T^{k,buff}_{a,l}\right)P^{slip} \tag{11.58}$$

for a particular action k, where P^{ineff} and P^{slip} are prior probabilities of adverse consequences of either a delay, or a schedule slip, respectively. Figure 11.2 shows a number of different risk functions due to changing buffer time, as related to search actions, post-survey actions, and deployment actions and corresponding end time uncertainty distributions. For the specific risk functions shown in this figure, the end times follow $\mathscr{N} \propto (0, 30)$ for search actions, $\mathscr{N} \propto (0, 15)$ for post-survey actions, and $\mathscr{N} \propto (0, 10)$ for deploy actions. The cost functions were developed with $c^{ineff} = T^{k,buff}_{a,l}$ and $c^{slip} = 500$.

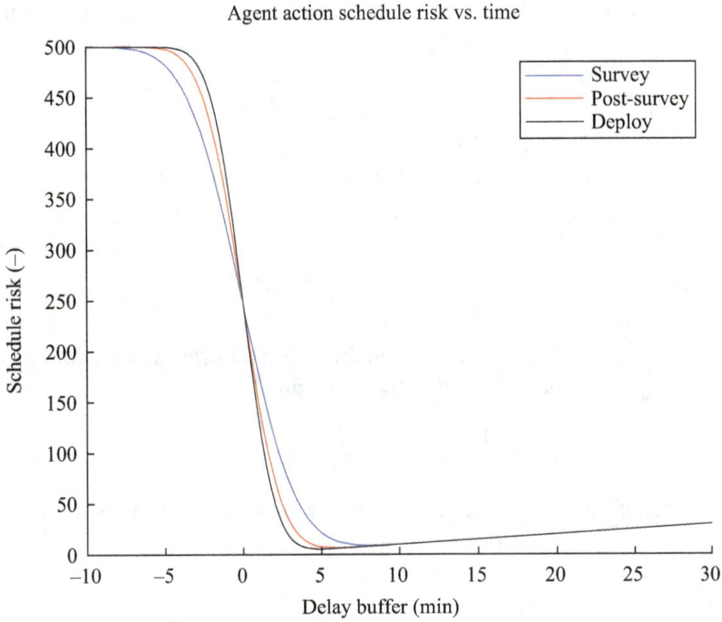

Figure 11.2 Risk function for schedule slips for search actions, post-survey actions, and deploy actions

Note that as written, the risk function is invariant in being performed during a particular phase p. As such, we may optimize the nonlinear optimization problem

$$T_{a,l}^{k,buff*} = \arg\min_{T_{a,l}^{k,buff}} \left\{ c^{ineff} P\left(\tilde{c}_{a,l}^{k} < T_{a,l}^{k,buff}\right) P^{ineff} + c^{slip} P\left(\tilde{c}_{a,l}^{k} > T_{a,l}^{k,buff}\right) P^{slip} \right\} \tag{11.59}$$

using numerical techniques prior to the schedule optimization. Once we develop an optimal buffer time for every action, location, and agent (type), we treat them as parameters in our robust scheduling extension. If P^{ineff} and P^{slip} are treated as uniform priors, (11.59) may be rewritten as

$$T_{a,l}^{k,buff*} = \arg\min_{T_{a,l}^{k,buff}} \left\{ c^{ineff} P\left(\tilde{c}_{a,l}^{k} < T_{a,l}^{k,buff}\right) + c^{slip} P\left(\tilde{c}_{a,l}^{k} > T_{a,l}^{k,buff}\right) \right\}. \tag{11.60}$$

Once the optimal buffer times for each individual task are found using (11.59), they may be factored into the overall optimization process. In order to incorporate the buffer time, we take advantage of the necessary delays between coordinated agent actions in the multiple UUV scheduling problem by using the wait times as part of the required buffer times. We define the synchronization time for agent a as the time difference between the end time of the current phase and the proceeding phase, written

$$\bar{\mathbf{T}}_{a,p}^{sync} = \bar{\mathbf{T}}_{a,p+1}^{start} - \bar{\mathbf{T}}_{a,p}^{end}. \tag{11.61}$$

We then ensure that the decision variable representing the *scheduled* buffer time $\bar{\mathbf{T}}_{a,p}^{sync}$ is at least the desired, optimal buffer time $T_{a,l}^{k,buff}$ with the constraint

$$\bar{\mathbf{I}}_{a,p,l}^{k} = 1 \wedge \bar{\mathbf{T}}_{a,p}^{sync} \leq T_{a,l}^{k,buff} \Rightarrow \bar{\mathbf{T}}_{a,p}^{buff} = T_{a,l}^{k,buff} - \bar{\mathbf{T}}_{a,p}^{sync}. \tag{11.62}$$

If the synchronization time is greater than the optimal buffer time parameter $T_{a,l}^{k,buff}$, then the buffer time is set to zero with the constraint

$$\bar{\mathbf{I}}_{a,p,l}^{k} = 1 \wedge \bar{\mathbf{T}}_{a,p}^{sync} \geq T_{a,l}^{k,buff} \Rightarrow \bar{\mathbf{T}}_{a,p}^{buff} = 0 \tag{11.63}$$

Finally, the robust start time is set as the deterministic start time plus the summation of all buffer times with the constraint

$$\begin{aligned} &\forall a \in \mathscr{A}, \quad \forall p \in \mathscr{P} \\ &\bar{\mathbf{T}}_{a,p+1}^{start,R} = \bar{\mathbf{T}}_{a,p+1}^{start} + \sum_{p' \in \mathscr{P} \text{ s.t. } p' \leq p} \bar{\mathbf{T}}_{a,p'}^{buff} \end{aligned} \tag{11.64}$$

In order to complete our robust scheduling extension, we modify (11.15) and (11.16) to use the new robust start times, creating the constraints

$$\begin{aligned} &\forall a \in \mathscr{A}, \quad \forall p \in \mathscr{P} \\ &\quad \bar{\mathbf{T}}_{a,p}^{end} = \bar{\mathbf{T}}_{a,p}^{start} + \sum_{k \in \mathscr{K}} \sum_{\mathscr{L} \in \mathscr{L}_k} c_{a,l}^{k} \bar{\mathbf{I}}_{a,p,l}^{k} \end{aligned} \tag{11.65}$$

$$\begin{aligned} &\forall m \in \mathscr{A}_{ur}, \quad \forall m \in \mathscr{A}_{s}, \quad \forall p \in \mathscr{P} \\ &\quad \sum_{d \in \mathscr{D}} \bar{\mathbf{I}}_{mn,p,d}^{dock} = 1 \Rightarrow \bar{\mathbf{T}}_{m,p}^{start} = \bar{\mathbf{T}}_{n,p}^{start} \end{aligned} \tag{11.66}$$

$$\begin{aligned} &\forall m \in \mathscr{A}_{ur}, \quad \forall m \in \mathscr{A}_{s}, \quad \forall p \in \mathscr{P} \\ &\quad \sum_{d \in \mathscr{D}} \bar{\mathbf{I}}_{mn,p,d}^{deploy} = 1 \Rightarrow \bar{\mathbf{T}}_{m,p}^{end} = \bar{\mathbf{T}}_{n,p}^{end} \end{aligned} \tag{11.67}$$

11.5 Simulation results

We next demonstrate the efficacy and computational performance of the proposed scheduling framework in a simulation setting. The IBM Optimization Studio was used as the MILP optimization solver [31]. A 2.8 GHz quad-core laptop containing 8 GB of RAM was used for all simulations.

11.5.1 End-to-end mission scheduling

We first demonstrate an end-to-end MCM-style mission optimization simulation, where the scheduler automatically determines a near-optimal schedule for performing the survey and reacquire/identify (RI) components of the MCM mission. In particular, we consider the RI component to be represented as a post-survey action that must be performed only after a survey. In this simulation, the post-survey action can be performed by the same UUV as survey actions. Figures 11.3 and 11.4 illustrate a simulation of $A_t = 1$ USVs, $A_s = 2$ UUVs, and $P = 18$ phases. In the simulations, we assume that a USV has a speed of 8 m/s, while UUVs have a speed of 1.2 m/s. We assume each of the $S = 4$ survey areas require $c^{survey} = 200$ min to survey, and $c^{ps} = 100$ min to perform a post-survey action. In the simulation, the algorithms first pare the decision space using heuristics to reduce the number of path-planning actions required in the schedule. We refer the reader to [32] for more information on complexity reduction heuristics involving UUV scheduling algorithms using MILP algorithms. As seen in the figures, the planner divides the tasks equally between the two UUVs, and determines that the best course of action is to perform post-survey actions immediately after survey actions.

11.5.2 Deterministic and robust scheduling results

We now present a comparison between the deterministic scheduling, the proposed robust scheduling approach, and a naive robust scheduling approach serving as a baseline with a mission consisting of only survey actions. In the following simulations,

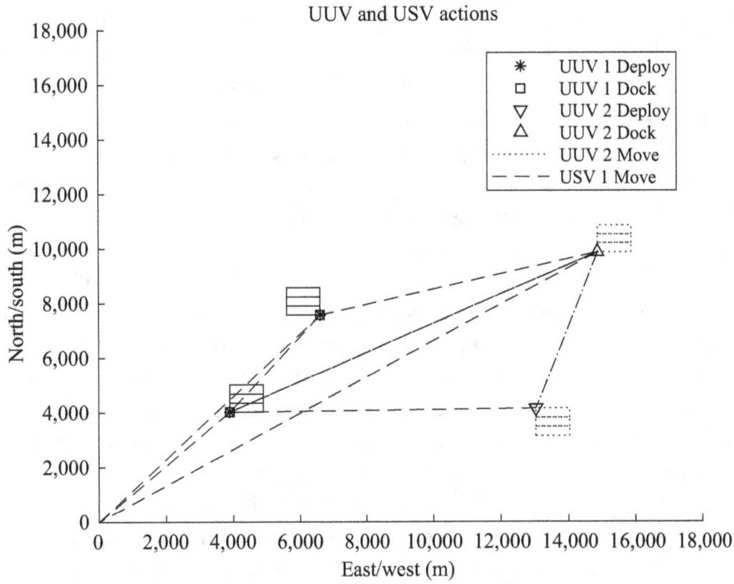

Figure 11.3 Plot of vehicle actions within an MCM area

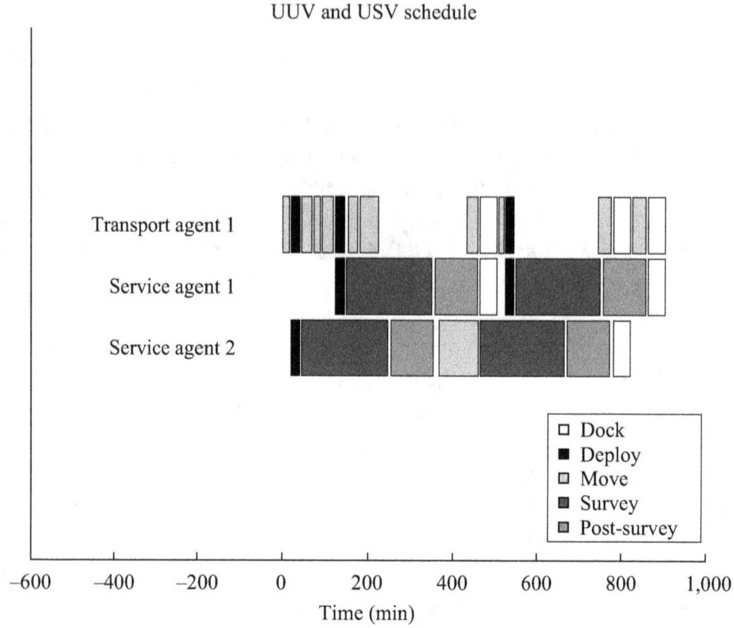

Figure 11.4 Simulated MCM schedule. Gantt chart represents the duration of each action performed by each vehicle at a point in time

we assume that the USV has a speed of 8 m/s, while UUVs have a speed of 1.5 m/s. We assume each of the $S = 4$ survey areas requires $c^{survey} = 100$ min. Furthermore, we assume that each docking action requires $c^{dock} = 20$ min, and $c^{deploy} = 10$ min. For the task uncertainty distributions, we used $\sigma^2_{dock} = 10$, $\sigma^2_{deploy} = 1$, $\sigma^2_{survey} = 30$, and $\sigma^2_{move} = \frac{\|(i,j)\|}{10}$ for nodes i and j, respectively. We present a direct comparison of simulations between the deterministic optimization problem, the naive robust optimization problem, and the minimum-risk optimization problem discussed in Section 11.4 for $N = 1$ USV and $M = 2$ survey agents within $P = 12$ phases. A naive approach in order to mitigate a schedule slip would be to add a fixed amount of buffer time, often as a function of an uncertainty distribution. In order to compare the deterministic method to our robust counterpart, we also implement this naive robust simulation involving three standard deviations of time being added to each action type. For example, a survey action could be planned for $c^{survey} = 116.425$ s. For the minimum-risk optimization problem, we set $c^{slip}_{a,l} = 500$ and $c^{ineff} = T^{buff}_{a,l}$ for all task types.

Table 11.1 shows the numerical results of a scenario using the three different optimization problems. As expected, the deterministic solution has the best completion time and the naive robust solution had the worst completion time. The minimum-risk robust scheduling solution had a completion time that is a significant improvement over the naive robust solution.

Table 11.1 Planned scheduling completion times for 1 USV and 2 UUVs performing 12 phases

Scheduling approach	Completion time (min)
Deterministic	310.90
Naive robust	364.90
Minimum-risk robust	345.45

11.5.3 Computational tractability and constraint violations

We now provide an empirical analysis of the computational tractability and reduction in constraint violations of the planning algorithms under the survey effort—post-survey effort mission paradigm discussed above. In order to gauge the computational tractability and constraint violations, we performed a Monte Carlo simulation of developing schedules for random sets of search areas. Each schedule planned to $P = 18$ phases in which task assignments may occur. The simulations assigned $N = 1$ transport USVs. The Monte Carlo simulations incrementally increased the number of search UUVs in the range $A_s = 1$ to $A_s = 4$ search UUVs, and assigned $A = 2A_s$ search areas for prosecution. For each combination of search UUVs and search areas, ten simulations randomly placing the service areas in a 1,000 m × 1,000 m field were performed. Increasing the number of search UUVs and search areas in this fashion dramatically increases the number of binary variables for the MILP from 3,967 binary variables in the one-search-UUV, nonrobust scheduling case to 48,462 binary variables in the four-search-UUV, robust scheduling case. Each simulation was set to return the best solution found within 600 s of optimization. We then perturb the resulting schedule using normal distributions as discussed in Section 11.4, and check for constraint violations where $\bar{\mathbf{T}}^{start}_{a,p+1} > \bar{\mathbf{T}}^{start}_{a,p}$. If so, we consider the scheduling constraint violated.

Figure 11.5 shows the results of the schedule that prosecutes all areas with a survey effort and a post-survey effort using both the standard and robust scheduling techniques. We note that for the robust case with $A_s = 3$ search UUVs, the robust scheduling technique was unable to find a solution that satisfied all constraints within the 600-s optimization time frame. As seen in the figure, the robust scheduling technique dramatically diverges in quality from the standard technique almost immediately. While one expects the robust technique to always have a higher completion time, as it is adding buffer times within the schedule, the magnitude increases due to the increase in variable complexity. Figure 11.6 shows the average number of constraint violations occurring for each simulation instance. As seen in the figure, using the robust scheduling techniques within R-SATP dramatically reduces the likelihood of schedule constraint violations. Additionally, the number of conflicts appears to linearly increase with the number of task assignments, and is not a function of optimization complexity.

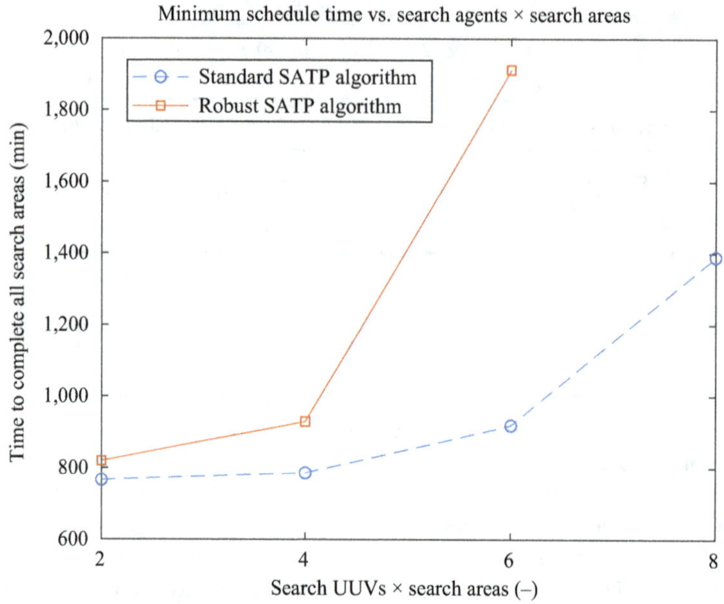

Figure 11.5 Plot of minimum schedule time vs. number of areas × number of service agents ($A_s \times A$)

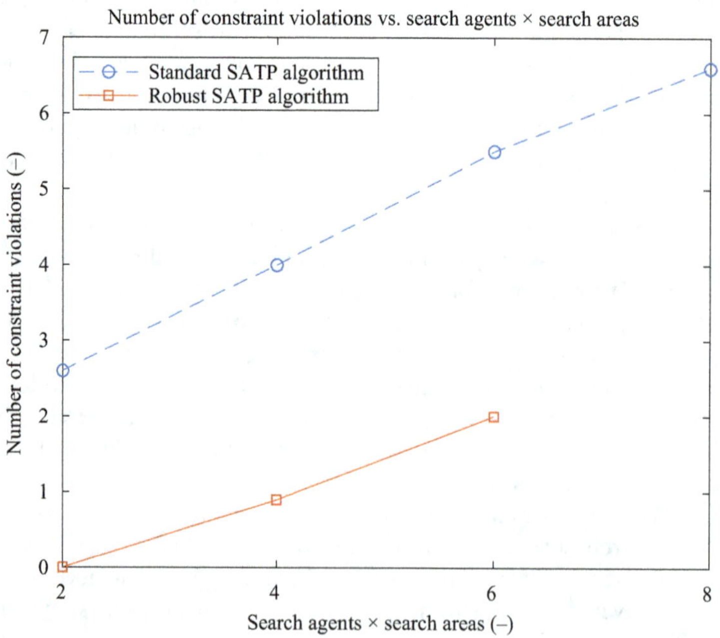

Figure 11.6 Plot of average number of constraint violations vs. number of areas × number of service agents ($A_s \times A$)

11.6 Conclusions

We have presented a framework for optimizing the scheduling of multiple heterogeneous unmanned vehicles in order to perform survey-type missions. The framework utilizes MILP in order to obtain optimal to near-optimal solutions, depending on the complexity of the problem and length of time the algorithm is allowed to search for a result. The framework specifically develops a tasking schedule applicable to multiple UUVs that includes docking, deployment, and movement actions performed throughout the area, and also accounts for scheduling of additional post-survey tasks that can only be done after the conclusion of a survey. A robust scheduling technique was introduced that extends the framework to include probabilistic constraints in order to account for uncertainty in the length of time required to perform tasks. Future work includes extending the framework to specific types of post-survey action tasks. Additionally, we will examine preprocessing heuristics to reduce the size and complexity of the MILP problem and lead to faster solve times for efficient solutions.

References

[1] Shafer AJ, Benjamin MR, Leonard JJ, *et al.* Autonomous Cooperation of Heterogeneous Platforms for Sea-Based Search Tasks. In: MTS/IEEE OCEANS; 2008.

[2] Jones EG, Browning B, Dias MB, *et al.* Dynamically Formed Heterogeneous Robot Teams Performing Tightly Coordinated Tasks. In: IEEE International Conference on Robotics and Automation; 2006.

[3] Estlin T, Gaines D, Fisher F, *et al.* Coordinating Multiple Rovers with Interdependent Science Objectives. In: Proceedings of the Fourth International Joint Conference on Autonomous Agents and Multiagent Systems; 2005.

[4] Sujit PB, Healey AJ, Sousa JB. AUV Docking on a Moving Submarine Using a K-R Navigation Function. In: International Conference on Intelligent Robots and Systems; 2011.

[5] Hobson BW, McEwen RS, Erickson J, *et al.* The Development and Ocean Testing of an AUV Docking Station for a 21″ AUV. In: IEEE/MTS OCEANS; 2007.

[6] Martins A, Almeida JM, Ferreira H, *et al.* Autonomous Surface Vehicle Docking Manoeuvre with Visual Information. In: International Conference on Robotics and Automation; 2007.

[7] Weaver JN, Frank DZ, Swartz EM, *et al.* UAV Performing Autonomous Landing on USV Utilizing the Robot Operating System. In: ASME Early Career Technical Symposium; 2013.

[8] Wenzel KE, Masselli A, Zell A. Automatic Takeoff, Tracking and Landing of a Miniature UAV on a Moving Carrier Vehicle. Journal of Intelligent Robot Systems. 2011;61:221–238.

[9] Floudas CA. Nonlinear and Mixed-Integer Optimization: Fundamentals and Applications. Oxford, UK: Oxford University Press; 1995.

[10] Bays MJ, Shende A, Stilwell DJ. An Approach to Multi-Agent Area Protection Using Bayes Risk. In: International Conference on Robotics and Automation; 2012.

[11] Booth KEC, Nejat G, Beck JC. A Constraint Programming Approach to Multi-Robot Task Allocation and Scheduling in Retirement Homes. In: Rueher M, editor. Principles and Practice of Constraint Programming. Switzerland: Springer International Publishing; 2016. pp. 539–555.

[12] Koes M, Nourbakhsh I, Sycara K. Constraint Optimization Coordination Architecture for Search and Rescue Robotics. In: International Conference on Robotics and Automation; 2006.

[13] Korsah GA, Stentz A, Dias MB, *et al.* Optimal Vehicle Routing and Scheduling with Precedence Constraints and Location Choice. In: International Conference on Robotics and Automation; 2010.

[14] Mandalandraki C, Daskin MS. Time Dependent Vehicle Routing Problems: Formulations, Properties, and Heuristic Algorithms. Transportation Science. 1992;26:185–200.

[15] Mathew N, Smith SL, Waslander SL. Optimal Path Planning in Cooperative Heterogeneous Multi-Robot Delivery Systems. In: Workshop on the Algorithmic Foundations of Robotics; 2014.

[16] Gombolay MC, Wilcox RJ, Shah JA. Fast Scheduling of Multi-Robot Teams with Temporospatial Constraints. In: Robotics: Science and Systems; 2013.

[17] Dechter R. Temporal Constraint Networks. Artificial Intelligence. 1991;49: 61–95.

[18] Sariel S, Balch T, Erdogan N. Naval Mine Countermeasure Missions. IEEE Robotics & Automation Magazine. 2008;15:45–52.

[19] Costa R, Wettergren TA. Managing the Areas of Responsibility for Collaborative Undersea Searchers. In: IEEE/MTS OCEANS; 2010.

[20] Hyland JC, Smith CM. Automated Area Segmentation for Ocean Bottom Surveys. In: Proc. SPIE 9454, Detection and Sensing of Mines, Explosive Objects, and Obscured Targets XX; 2015.

[21] Molineaux M, Auslander B, Moore PG, *et al.* Minimally Disruptive Schedule Repair for MCM Missions. In: Proc. SPIE 9454, Detection and Sensing of Mines, Explosive Objects, and Obscured Targets XX; 2015.

[22] Lin X, Janak SL, Floudas C. A New Robust Optimization Approach for Scheduling under Uncertainty: I. Bounded Uncertainty. Computers & Chemical Engineering. 2003;29:1069–1085.

[23] Janak SL, Lin X, Floudas C. A New Robust Optimization Approach for Scheduling under Uncertainty: II. Uncertainty with Known Probability Distribution. Computers & Chemical Engineering. 2006;31:171–195.

[24] Bertsimas D, Sim M. The Price of Robustness. Operations Research. 2004;52(1):35–53.

[25] Daniels RL, Kouvelis P. Robust Scheduling to Hedge against Processing Time Uncertainty in Single-Stage Production. Management Science. 1995;41(2):363–376.

[26] Lombardi M, Milano M, Benini L. Robust Scheduling of Task Graphs under Execution of Time Uncertainty. IEEE Transactions on Computers. 2013;62(1):98–111.

[27] Davenport AJ, Gefflot C, Beck JC. Slack-Based Techniques for Robust Schedules. In: Sixth European Conference on Planning; 2014.

[28] Wu SD, Byeon ES, Storer RH. A Graph-Theoretic Decomposition of the Job Shop Scheduling Problem to Achieve Scheduling Robustness. Operations Research. 1999;47(1):113–124.

[29] Bays MJ, Wettergren TA. A Solution to the Service Agent Transport Problem. In: IEEE International Conference on Intelligent Robots and Systems. vol. 89; 2015. pp. 15–26.

[30] Poor HV. An Introduction to Signal Detection and Estimation. Hiedelberg: Springer; 1994.

[31] IBM ILOG CPLEX User Manual; 2013.

[32] Bays MJ, Wettergren TA. Service Agent-Transport Agent Task Planning Incorporating Robust Scheduling Techniques. Robotics and Autonomous Systems. 2017;89:15–26.

Chapter 12
MONSUN: a swarm AUV for environmental monitoring and inspection

Erik Maehle[1], *Benjamin Meyer*[1], *Cedric Isokeit*[1] *and Ulrich Behrje*[1]

This chapter presents the development of the small swarm-capable autonomous underwater vehicle (*AUV*) MONSUN and its use for environmental monitoring and inspection tasks. A summarizing description of robot development in hardware and software is given, whereby efforts of the design process are explained in detail. After the focus on underwater communication techniques, swarm behaviours based on robust and scaleable localization principles are presented. Finally, the flexibility and functionality of the system are shown with the help of various experiments and field tests.

12.1 Introduction

The sense of responsibility for our own living balanced with the fragile nature is becoming a keystone in human activity nowadays. Life in all aspects is needed to be understood and saved for future and many research projects deal with different kinds of environments and their inhabitants. Particularly, diverse waterbodies are in focus of investigations, because of complex interaction of aquatic organisms and its vulnerability to climate changes and environmental hazards. The recently introduced so-called environmental monitoring attempts to gain an overview of water conditions of important waterbodies around the world. It is a costly undertaking because of the tremendous number of waters and with conventional methods hardly imaginable. The gained results serve as a basis for research of new environmental questions and for the development of goals and definitions for saving nature. Periodical and comprehensive observations not only help to check if required statutory guidelines are fulfilled, but give an answer if the guidelines have to be adapted to current circumstances.

An autonomous robot helping to measure the water qualities is introduced in this chapter. The presented MONSUN AUV is a small and inexpensive robot ('micro' AUV). It is characterized by its modular and versatile concept which allows for automatic environmental measurements in different kind of waterbodies.

[1]University of Lübeck, Institute of Computer Engineering, Lübeck, Germany

Being equipped with mission-specific sensors like cameras, sonars and lidars, MONSUN AUVs can also carry out various inspection tasks, e.g., in harbours, lakes or rivers. With the help of acoustic communication, the robot can be deployed in a swarm to enlarge the benefit of investigations increasing the covered area on the one hand and reducing the amount of time on the other hand. Furthermore, in case of a robot failure in the swarm, the mission does not need to be aborted, but the swarm can continue with decreased efficiency.

12.2 Motivation and related work

12.2.1 Micro AUV swarms for monitoring and inspection

Today's AUVs are mostly rather big torpedo-shaped vehicles, which are so heavy that they require a special infrastructure like ships with a crane to be operated. Besides the high price of the AUV itself, this makes their use complex and expensive. On the other side, recent technological progress with computing hardware and software, sensors and material has made it possible to build small inexpensive so-called micro AUVs. There is currently no clear definition of the term micro AUV. Here the assumption is made that micro AUVs are small, inexpensive and easy to develop AUVs of small length (usual less than 1 m) and low weight (usual less than 10 kg). Thus, they can be handled by one person without special infrastructure and be easily deployed and recovered from the shore or a small boat. Furthermore, due to their low cost, it is possible not only to use a single AUV, as it is common for the current large AUVs, but to use many of them cooperating in a swarm.

To employ such a swarm of small inexpensive micro AUVs has a lot of benefits. Due to several robots working in parallel, many missions like searching an area can be carried out much faster than with single AUVs. Swarms also offer fault tolerance: if a swarm member fails, the others can take over its tasks and the mission can be continued. Being small and agile, micro AUVs can operate in narrow and confined spaces which are out of reach for large AUVs.

There are many important applications of swarms of micro AUVs. In our view, their main application field is not the open and deep ocean, but local shallow waters like coastal waters, harbours or ports, lakes and rivers. Near the coast they can, e.g., be employed for monitoring the water quality or other environmental monitoring tasks. The vegetation on the seafloor in shallow water can, e.g., serve as an indicator of water quality. Instead of human divers, as done today, micro AUV swarms with cameras can autonomously investigate and by image processing evaluate the seafloor vegetation.

In harbours or ports, the water quality also plays an important role. It must not be polluted, e.g., by chemicals or oil. Micro AUV swarms can help to find such pollutions with suitable sensors and help to identify their sources. Furthermore, quay walls have to be inspected regularly for holes or other damages which can impair their stability. In big ports, there are kilometers of quay walls. Inspecting them by human divers, being necessary nowadays, is a tedious, dangerous and expensive task. Micro AUV swarms can help the divers to find the holes autonomously and then guide them there for a dedicated repair. Similar tasks can be carried out for other underwater constructions like bridge piers or walls of buildings. One more application in harbours is the inspection of ship hulls for fouling which increases the ship's fuel consumption due to its water resistance. A micro AUV swarm is fast enough to do this inspection during the short time when the ship is loaded and unloaded.

There are many other possible applications like the inspection of embankment type dams, the underwater part of offshore wind wheels or aquacultures. Also underwater cartography and archaeology or in general underwater searching and mapping can profit from micro AUV swarms. For scientific marine research, they are a versatile tool for various monitoring tasks. A concrete example will be presented later in this chapter.

As the application examples presented earlier show, swarms of micro AUVs are not so much an alternative to traditional large AUVs, but open many new application areas where the latter are either too expensive or not small and agile enough to be operated.

12.2.2 Related work on micro AUVs

Though for monitoring and inspection of larger areas, the deployment of a swarm of underwater robots has many benefits, there has not much work on swarms of small low-cost micro AUVs been done so far. A pioneering project was Serafina, where small swarm-capable AUVs were proposed and several important swarm algorithms were studied [1]. The company Hydromea from Lausanne, Switzerland, is currently working on a flexible AUV called VERTEX (Figure 12.1(a)) for environmental monitoring based on these concepts [2]. The CoCoRo Swarm project introduced an underwater robot swarm to study basic swarm behaviours (Figure 12.1(b)), but due to limited computational power and sensor payload, it is not suited for large-scale monitoring missions [3]. The SwarmDiver (Figure 12.1(c)) provides on-board sensors for environmental monitoring, but has no means to communicate underwater [4]. It relies on radio communication at the surface and then can dive only vertically down and up again for doing its underwater measurements. The swarm-capable SEMBIO micro AUV developed at our institute in a companion project (Figure 12.1(d)) uses similar concepts as MONSUN, but was mainly designed for energy efficiency and has not operated as a swarm yet [5].

12.3 The MONSUN micro AUV

The development of the AUV MONSUN started in 2009 at the Institute of Computer Engineering of the University of Lübeck [6]. The conceptional design is characterized by a robust and modular design, which allows for a high grade of adaptation to various tasks in the field of environmental monitoring and inspection. Furthermore, the robot concept is scalable to use the robot in a controlled swarm consisting of a number of independent cooperating AUVs. The following section will describe the vehicle in detail, beginning with a more precise description of the design concept. Afterwards, the robot's hardware and software are explained [7].

*Figure 12.1 Examples of small swarm-capable AUVs: (a) Hydromea VERTEX AUV, (b) CoCoRo Swarm, (c) SwarmDiver and (d) SEMBIO AUV**
**© Copyright Information: (a) Hydromea SA, Felix Schill; (b) Artificial Life Lab, Universität Graz, Thomas Schmickl; (c) APIUM, Inc., Joy Shapiro; (d) Universität zu Lübeck, Ammar Amory*

12.3.1 Design concept

The requirements for designing the Monsun underwater robot were to build a small low-cost micro AUV (see Figure 12.2), manageable by one person without mechanical help. The robot should be built in an attentive way simplifying the reproduction, which allows for using it in a swarm. The internal parts reflect the current state of the art of cost-effective hardware enabling a robot affordable by many institutions. Using an agile propulsion and an active diving system helps to dive easily in hard-to-reach areas on the one hand and increasing the fault tolerance on the other hand. Together with the open source Robot Operating System (*ROS*)[†] a platform is given, which is suitable for many tasks in the field of environmental monitoring and inspection. Finally, an expansion set which allows for specialization of the robot by using different docking points to include new sensors and devices in the system results in a vehicle differing from currently available platforms.

12.3.1.1 Robustness

The main focus of the design is a high robustness to external influences, e.g. water pressure in depths down to 100 m and mechanical stress while boarding the robot from a ship's side. Therefore, the robot's hull is made from glass-fibre-reinforced plastics

[†]Robot Operating System; Accessed: 2020-04-21. Available at http://www.ros.org/

Figure 12.2 MONSUN underwater robot designed for various tasks in the field of environmental monitoring. The basic version is small in size with a length of 60 cm, a diameter of 10 cm and has a weight of about 8 kg, which makes it easy to handle by a single person. The propulsion system consists of six thrusters, four horizontally and two vertically, which allows for precise and agile movements under water. With the help of different communication techniques, the robots can be applied in a swarm [8]

assembled with the help of aluminium profiles. Besides the gained impact-resistance of the material, it has an elasticity saving the hull from dangers caused by collisions with obstacles and other robots. Further, the material is corrosion-resistant and has a lower weight compared to conventional designs made of metallic hulls. The weight of the robot itself is balanced in a way to generate a positive buoyancy preventing a loss of a robot caused by possible damage to the electronics, the propulsion system or occurred software failures. This feature increases the robustness of the system further because of forcing the robot automatically to the surface in case there is no damage at the pressure chamber.

To access the interior electronics of the robot, the hull can be opened via an 100-mm aluminium bayonet closure in the centre. It is waterproof up to 100 m and manageable without using additional tools. Furthermore, at the robot's front a second bayonet closure with a diameter of 60 mm is used to attach a glass dome or various sensor types.

The propulsion system of MONSUN consists out of six independent brushless motors integrated in the robot's fins. Four of them are used for vertical movements, while the other two are placed in the middle of the robot, generating forces for horizontal movements. The propulsion system enables movements with five degrees of freedom (*DOF*) in the water. The use of these four vertical directed thruster increases the robustness and fault tolerance further, because of the ability of autonomous depth and posture control even if one thruster fails.

12.3.1.2 Modularity

Most commercial underwater robots are similar in construction and interior structure: a tubular waterproof hull with sensors and actuators at predefined positions along the robot. The size and propulsion force are often a multiple compared to a MONSUN robot. Therefore, the robots need large batteries and a reinforced hull, which makes it difficult to modify them, e.g. installing new sensor devices if these are needed for an upcoming mission.

Hence, the conceptional design of MONSUN was focused on a modular robot structure. Most components are independent from each other and interchangeable. The flexible hull provides different docking points to easily install mission-specific devices with low effort. Due to modern production processes, the parts of different MONSUN AUVs are identical, which makes it, e.g., possible to exchange the rear part with internal electronics between the robots in failure cases. This modular concept is utilized for internal electronics too. It is based on a bus-based communication architecture with uniform printed circuit board (pcb) sockets for sensors, actors and control units. Hence, defective components can be quickly exchanged and the behaviour of the robot can be flexibly adapted for new mission tasks.

12.3.1.3 Adaptability

The modular concept of the robot allows for flexible operation and a high degree of adaptability for various tasks of environmental monitoring and inspection. Besides a basic configuration consisting of the propulsion system and sensors for autonomous depth and posture control, uniform pcb sockets can be used to integrate a wide range of other sensors and devices. At the robot's front, devices for image processing suitable for inspection tasks and mutual member identification can be integrated. The bayonet closure in the middle of the robot can be used to extend it with independent modules changing the robot's elementary arrangement.

The adaptability of MONSUN is also applied in the software part. The use of the open source software framework ROS offers a high degree of abstraction and encapsulation of single software parts. Drivers and controllers can be independently implemented and use a common communication network based on TCP/IP packets. Hence, many parts of the software can be reused and the development can be done independent from the robot. If a new mission requires a special type of sensor, it will be connected to the robot with the help of a free pcb socket, and the software driving it will be simultaneously implemented and integrated in the ROS structure.

12.3.1.4 Scalability

The high-scale factors of nowadays underwater robots lead to the ability to carry more and more powerful sensor devices. With the help of sonars, Doppler velocity logs and inertial navigation systems, the challenging tasks of localization and navigation can be performed, but the cost factor of such devices is comparatively high. The aim of the MONSUN project is to compensate smaller sensor capacities by using the swarm capabilities associated with utilizing a number of robots to fulfil the tasks.

The realization of an AUV swarm depends on scalability of the system. While designing and manufacturing the MONSUN robot, special attention was paid to a simple shaping making a reproduction easily possible. Attachment parts can be 3D printed in short time and internal electronics need only small wiring efforts due to the bus-based architecture. The operating system and special software of the robot are stored on a memory card, which can be duplicated without much effort.

Furthermore, a reliable and robust communication of swarm members is needed to perform swarm algorithms and fulfil the mission together in a team. Thereupon the development of an own small and low-cost acoustic modem was in parallel started at the Institute of Computer Engineering of the University of Lübeck [9]. Today this work is continued at the Institute smartPort of Hamburg University of Technology [10].

12.3.2 Hardware

One of the big challenges while designing an underwater vehicle is the placement of all components and electronics in mostly cramped and rounded housings. In case of the rather small MONSUN micro AUV, a sophisticated design is needed for optimal utilization of the existing space on the one hand and to provide valuable connection options for periphery on the other hand. Figure 12.3 illustrates the main components needed to run the robot in its current configuration. As main control unit a Raspberry Pi 3 is used, which features a Quad Core 1.2GHz Broadcom BCM2837 64bit ARM CPU with 1GB RAM. It allows for using a Linux operating system capable of implying

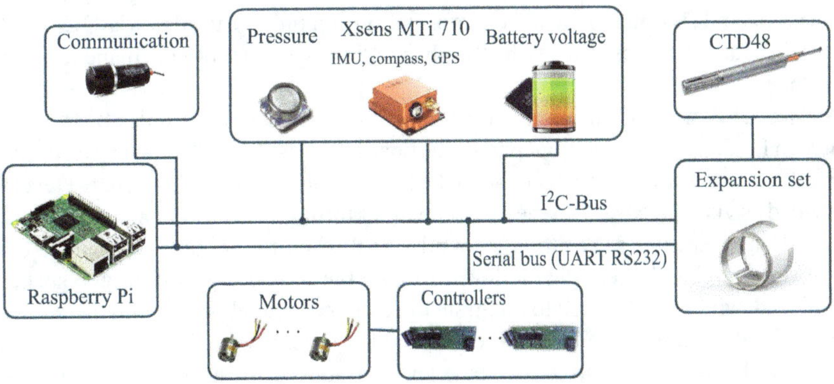

Figure 12.3 Main components of the MONSUN underwater robot, connected via a bus system using I2C and serial communication (UART RS232). As main controller a Raspberry Pi 3 is used, which features a Quad Core 1.2GHz Broadcom BCM2837 64bit ARM CPU. The essential sensor devices for an autonomous drive include a MS5803-14BA pressure sensor for depth control, a Xsens IMU positioning system with internal compass and GPS and a voltage monitoring system. The propulsion system uses six independent brushless motors, each driven by an own motor controller also connected via I2C-Bus. For communication with the user and other robots, a Wi-Fi connection at the surface and an acoustic modem under water can be used. The main system can be further expanded with modules of the expansion set explained in Section 12.3.4

the ROS framework while it features a comparable low-power consumption of 4.7 W at full loads. For communication with all sensors, actuators and other devices, the I2C-Bus with the Raspberry Pi acting as master node is provided. The propulsion system consists of six independent brushless motors, each driven by an own motor controller. The thrusters can be individually driven by the control system with an update rate of 10 Hz. For a useful operation of the robot, an essential number of sensors is needed to ensure an autonomous driving behaviour without external user commands. MONSUN features a temperature-compensated MS5803-14BA pressure sensor, which measures the depth with an accuracy of 2 cm down to a maximal depth of 100 m.

For position and posture control, an inertial measurement unit (IMU) with a set of gyroscopes and accelerometers and a compass are used, all together provided by a Xsens MTi-G-710 module. It delivers roll, pitch and yaw angles. Furthermore, the Xsens includes an additional GPS sensor with external antenna, which is used for localizing the robot at the surface. These mentioned sensors ensure an autonomous drive under water and offer a basis for intelligent high-level behaviours using them. A special feature of the sensory system of MONSUN is the use of various microcontrollers for each sensor. The pair consisting of sensor and microcontroller is called smart sensor in this context. The controllers associated to each sensor perform a pre-processing and filtering of measured information and provide the data via I2C-Bus to the main controller. This relieves the Raspberry Pi and causes a lower demand for serial interfaces.

The mentioned components are combined on the main electronics board illustrated in Figure 12.4. At the front part of the board the Raspberry Pi and the electronics for voltage monitoring are placed. With the help of coloured LEDs, the charge level is provided as visual feedback to the user if he looks through the glass dome. Associated, a small piezo speaker is emitting a sound signal when the robots gets switched on and in case of low battery charge level. Behind the Raspberry Pi, four pcb sockets are placed, which can be used to integrate different peripheral boards into the system. They provide access to the communication buses of the system, like the I2C-Bus and other serial data lines, and fixed voltage levels usable by the new device, which can choose between 3.3, 5, 7.4 and 11.1 V. This increases the modularity of the system on the one hand and minimizes the wiring effort for new devices on the other hand. Behind the socket area fuses and the main power switch are placed. It is a magnetic switch which enables the activation of the robot from outside the hull so it does not have to be opened in the operating area. At the end of the main board, a connector is placed to attach the electronics of the rear part of the robot or possible expansion modules via a flat ribbon cable.

The rear part of the robot is equipped also with an electronics board, including pcb boards for controlling the pressure sensor and the thrusters connected at the back. Furthermore, the place above this so-called backboard is used to house the main LiPo battery of the robot (11.1 V with 4,500 mAh). The battery system contains also a LiPo battery with 7.4 V and 2,700 mAh used for powering the controllers and sensors and one with 11.1 V and 1,300 mAh exclusive for the acoustic modem. Combined this makes an overall mission time of up to 4 h possible.

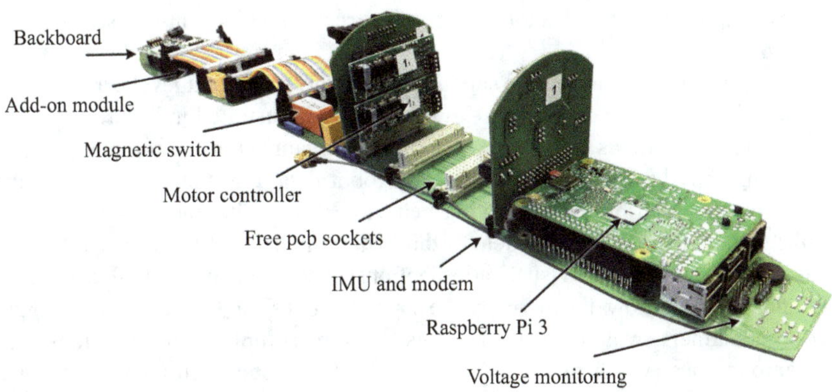

Figure 12.4 Bus-based electronics concept of the MONSUN robot. The main board can be subdivided in three parts: At the front of the board, the Raspberry Pi and the voltage monitoring electronics are placed. Behind the main controller board, four pcb sockets are placed. They provide access to the communication buses, various serial data lines and different usable voltage levels. The basic configuration uses two of these sockets for electronic boards driving the thrusters, the Xsens and the acoustic modem. At the rear part of the main board, a magnetic switch is placed which is used as main switch of the system operable without opening the robot's hull. Moreover, a connector to attach other electronics via flat ribbon cables is placed at the rear. Further, a board for an intermediate module and the backboard are illustrated, used to power the pressure sensor and thrusters placed at the rear part of the robot

12.3.3 Software

The software architecture of MONSUN is hierarchically structured and developed according to high grades of modularity and adaptability. The system is based on using the open source ROS, which was primarily developed for autonomous robots. It delivers various functionalities, e.g. hardware drivers and abstraction techniques, package management and communication between program implementations and parts of it. The code consists of so-called *Nodes*, *Packages* and *Stacks*, whereby, packages can contain nodes, libraries, configuration data or various helpful additions. Stacks contain a number of packages with a multitude of directly usable source code. Nodes are mainly implemented in C++, Python or Lisp and form the fundamental part of the software as including the main functionalities of the robot program. Finally, a so-called ROS-Core merges the nodes to a runnable software structure enabling the nodes to exchange messages according to the publisher and subscriber principle. Thereby, a node publishes a message assigned to a specified topic. All other nodes which have subscribed this topic receive the message then automatically. Additionally nodes can supply services as another message structure which can be used by other nodes too.

The MONSUN software can be subdivided into three different levels: a *Sense/Act Level*, a *Behaviour Level* and a *Task Level*, which are illustrated in Figure 12.5. The tasks define autonomous robot missions and are organized at the topmost level of abstraction. They have access to elementary robot data, e.g. the current time, and information provided by the underlying behaviour level. The missions are defined as finite state machines with the help of the Python library *SMACH*. This allows for an easy syntax and various visualization options of state machines, with associated transitions and data flow. Furthermore the state machines can be hierarchically organized in containers, which enables administration of complex robot missions such as an autonomous search of an underwater object in a specified area. These missions contain behaviours like driving to a waypoint, diving at this point and search in predefined depths with the help of different sensors. Depending on feedback of

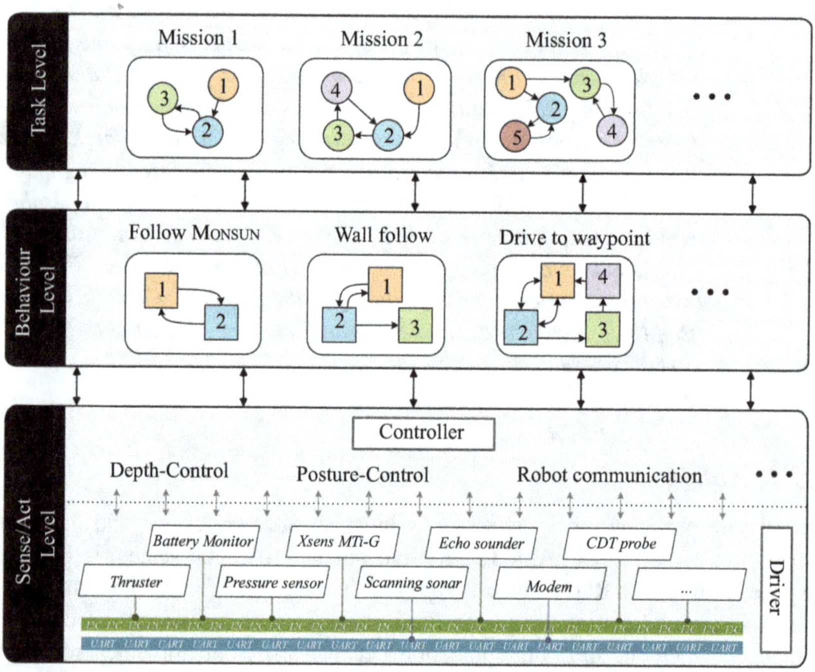

Figure 12.5 The ROS-based software architecture of the MONSUN underwater robot subdivided in three hierarchical levels: a Sense/Act Level, a Behaviour Level and a Task Level. The lowest level contains hardware drivers and controllers as well as basic functions like posture and depth control. These capabilities are used by the middle level for implementations of specific behaviours like driving to a given waypoint in a predefined depth. The top level uses finite state machines to organize full missions as sequences of different behaviours [7]

different behaviours, MONSUN can change its current state and continue its mission. Furthermore, all states of the currently performed mission contain a failure case, letting the robot get up to the surface in case of emergency.

The middle level is the so-called behaviour level, which consists on the one hand out of elementary behaviours like diving to a given depth and on the other hand rather complex driving behaviour like navigation to a specified location. This level of abstraction has a direct access to sensors and actors of the system, which allows for controlling of the robot. If the robot should drive to a specified waypoint, the behaviour will ask the Xsens sensor first, to get the current compass values and a GPS position. Then, the controllers will calculate a thruster speed to get to the destination point.

The lowest level is the so-called sense/act level, which generates the ROS-based interface to the hardware of the robot. It is further subdivided into four categories: the hardware drivers, the controllers, the navigation and localization and the image processing. The drivers interact with periphery containing sensors and actors via I2C-Bus. In case of sensors not supporting this communication standard, various ATmega microcontrollers are used as an intermediate station. They translate between different communication techniques and perform first data aggregations and analyses to transmit preprocessed data, which relieves workload of the main processor. Since many different sensors are connected via the bus system, a sort of regulation and scheduling is needed. Therefore, a so-called I2C-Core is implemented. It administrates the accesses to the bus via a semaphore-based system, which allows for an exclusive use of the bus at predefined time steps. It uses a FIFO (First In – First Out) queue to schedule the sensors one after another. If a sensor blocks the bus for a larger time period, it can be terminated by the system to prevent deadlocks.

The image processing is integrated in an own module and its software is considered as an independent part. Because of high computational load, it is outsourced to a separate microcontroller which publishes its data via a serial bus to the main processor. With the help of the OpenCV library, blob detection and member identification techniques are performed, which can be used to follow other robots in good visible water conditions [12]. The localization and navigation is outsourced in an own software stack as well. With the help of information delivered by acoustic modem, Xsens and possible sonars, a position determination is implemented in an ROS node.

The most important part of the sense/act level is the set of feedback controllers for autonomous drive in all five DOF. Normally, the behaviours generate target values for the position and depth of MONSUN and the controllers have to compute actuating values for the thrusters. The computation is done with the help of four independent feedback controllers, together implemented in an ROS node.

12.3.4 Expansion set

MONSUN AUVs can be flexibly modified in various ways to fulfil the requirements of many needed mission types. Therefore, the robot has three different docking points at the hull, which can be used to integrate stand-alone modules, sensors and devices to the system (see Figure 12.6). At the front of the robot, a small 60-mm bayonet

Figure 12.6 Modular expansion set of the MONSUN underwater robot classified in three different device types. With the help of a small bayonet closure at the front, an optical distance sensor (i), a sonar (ii) or an imaging device (iii) can be attached. Underneath the robot, an aluminium equipment carrier helps to attach external devices like a CTD probe (a) or an acoustic modem (b). Furthermore a handle with antenna mast (c) can be mounted. In the middle of the robot, self-contained modules (1–5) can be mounted at the main bayonet closure, which allows for fully different robot capabilities [11]

closure can be used to attach either an acrylic glass dome or different sensors with a small size factor to the robot. Besides imaging devices and distance sensors, an echo sounder sonar for frontal obstacle avoidance can be integrated in the system.

At the bottom of the robot, a flexible equipment carrier is mounted. It is made from aluminium and allows for attaching even heavier devices with the help of reliable 8-mm screws. Currently, the carrier is used to hold a stand-alone CTD (conductivity, temperature and depth) probe[‡] and the used acoustic modem[§]. Furthermore, more powerful T100 thrusters[‖] can be mounted at the carrier as horizontal thrusters to drive the robot in environments with strong currents and waves. Also other devices, like a multibeam sonar, can be attached to the carrier. If no devices are needed, an extra weight can be attached to tare the robot.

The most important docking point is placed in the middle of the robot. With the help of the 100 mm bayonet closure, self-contained expansion modules can be

[‡]Sea&Sun Technology CTD48; Accessed: 2020-04-21. Available from: https://www.sea-sun-tech.com/product/multiparameter-probe-ctd-48/

[§]Evologics S2C M 18/34; Accessed: 2020-04-21. Available from: https://evologics.de/acoustic-modem/18-34

[‖]Blue Robotics T100 Thruster; Accessed: 2020-04-21. Available from: https://bluerobotics.com/store/thrusters/t100-t200-thrusters/t100- thruster/

integrated, which allows for fully different robot capabilities while extending the robot only about the length of the module itself. Meanwhile, a full set of different modules has been developed. These include interface modules with sockets for many different sensors attached to the equipment carrier as well as sensor modules for localization and environmental monitoring or inspection. Probes for the most important environmental parameters can be used and even water samples can be taken for later analyses. Independent from different docking points, a handle bar can be attached to the robot. It includes a large antenna mast for external Wi-Fi and GPS antennas, which improve the radio range even at high waves surrounding the robot. The handle bar serves not only as transport help, it is also useful for deploying the robot from a ship with high sides.

12.4 Swarm behaviours

12.4.1 Localization and communication concept

The localization and communication concept of MONSUN swarms (Figure 12.7) assumes that some MONSUN AUVs stay at the surface (S-AUVs) and the others dive to a given depth (D-AUVs). The S-AUVs obtain their positions by GPS and communicate with each other and a control station at a boat or at shore by Wi-Fi. S-AUVs and D-AUVs communicate with each other by acoustic modems. The D-AUVs can easily measure their depth by pressure sensors and keep it stable by a depth control algorithm. Therefore, a 2D localization of the D-AUVs relative to the S-AUV in the corresponding plane is sufficient for them to compute their 3D positions. To achieve this, the distance to the S-AUVs at the surface is used.

This distance d can be obtained by the acoustic modems via the round trip time R of a sent message and its acknowledgement from the receiver according to

$$d = c_S \cdot \frac{R}{2} \tag{12.1}$$

where c_S is sound speed in water. Unfortunately, the speed of sound in water is not constant, but is a complex function of the temperature and pressure of the water, and in seawater also of the salinity. Typical values are about 1,482 m s^{-1} at 20°C in fresh water and about 1,501 m s^{-1} at 10 m depth, 15°C and 3% salinity in seawater. There exist various approximate equations for calculating the sound speed in water. One simple example for seawater is given by the Mackenzie equation [13]:

$$\begin{aligned} c_S = {} & 1{,}448.96 + 4.591T - 5.304 \cdot 10^{-2}T^2 + 2.374 \cdot 10^{-4}T^3 \\ & + 1.340(S - 35) + 1.630 \cdot 10^{-2}D + 1.675 \cdot 10^{-7}D^2 \\ & - 1.025 \cdot 10^{-2}T(S - 35) - 7.139 \cdot 10^{-13}TD^3 \end{aligned} \tag{12.2}$$

with T being the temperature of the water in degrees Celsius, S is salinity in parts per thousand and D the depth in meters of the modem. In MONSUN, the Evologics S2C M 18/34 acoustic modems are used, which support measurements of the round trip time R. T, S and D can be provided by MONSUN's sensors.

A problem with acoustic communication underwater is, however, its low bandwidth and poor reliability. Messages can easily get lost, e.g., due to reflections and

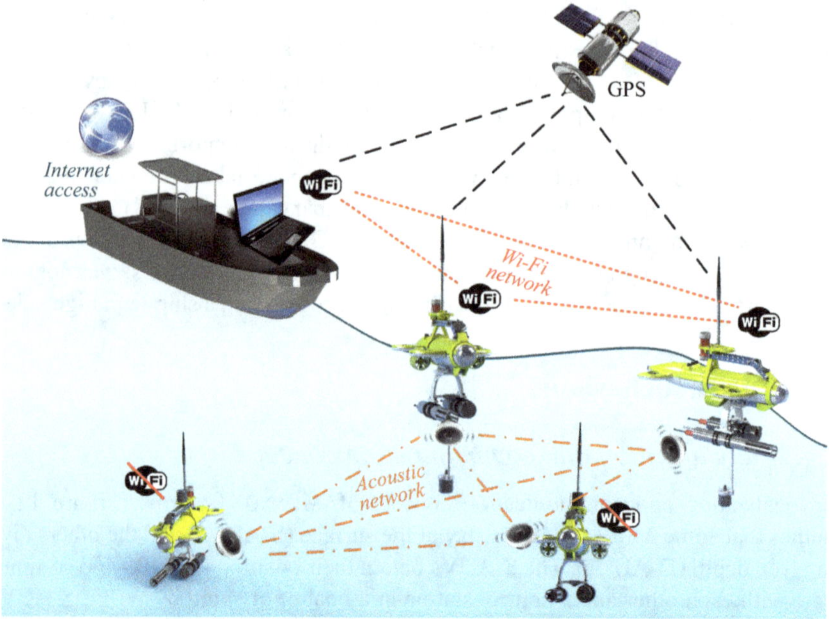

Figure 12.7 General localization and communication concept of M*ONSUN swarms. Some* M*ONSUNS stay at the surface, the other swarm members are underwater. The surfaced AUVs communicate among each other and with the user, e.g., on a boat by Wi-Fi and obtain their absolute positions by GPS. The submerged AUVs communicate with and measure the distances to other AUVs by acoustic modems. They use these distances then for their localization relative to the surfaced AUVs*

interference. To verify this, experiments with two AUVs were conducted in a local lake. The aim was to determine the reliability of the communication channel and the accuracy of the distance measurements. The two MONSUNS were placed fixed one after the other at three different positions. Thus, communication distances of 45, 25 and 10 m were achieved. For a set of 100 messages sent for each scenario, the average message receive rates were determined. To get a valid distance measurement, each communication has to be acknowledged. Therefore upon receiving a message, an acknowledgement is sent back immediately by the modem. In Figure 12.8, the results of all messages sent are depicted. The plot A illustrates the message reception from sender to receiver. For each received message, a message is sent back by the receiver to the sender. The corresponding message reception is shown in plot B. In the first case, the average receive rate was 87% and in the second case 83%. The second rate is always lower than the first one, because only received messages can result in a response message. The other experiments showed that different distances were not noticeably influencing the receive rates. Due to the fact that receive rates of above

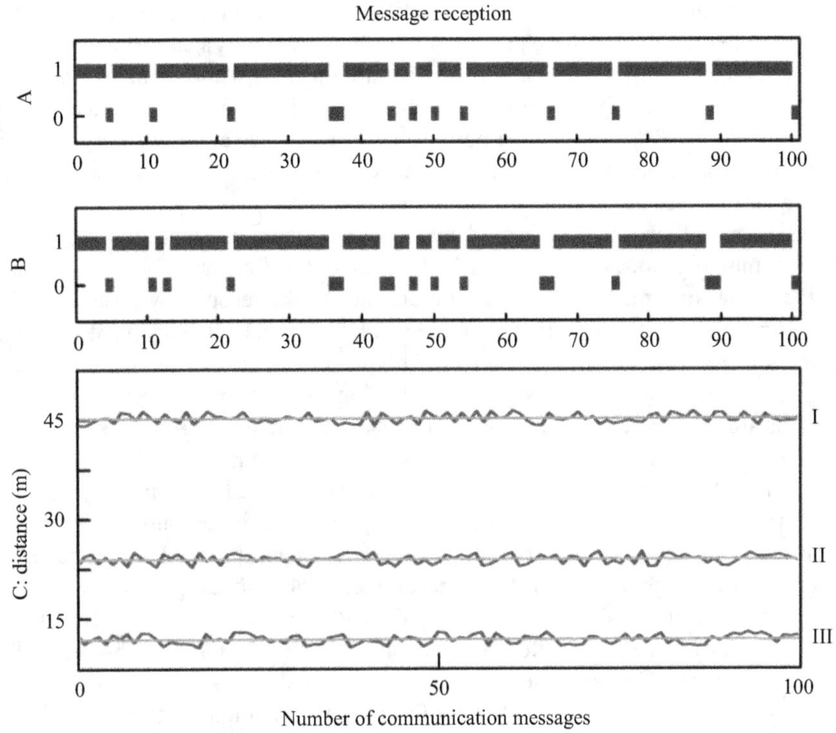

Figure 12.8 This experiment illustrates acoustic communication reliability between two AUVs at three fixed positions in a local lake. The distances between the robots were 10 m (C, III), 25 m (C, II) and 45 m (C, I). The mean square error of the measured distance is hereby 0.29 m with the biggest outlier being 0.49 m. Plots A and B display the message reception between sender and receiver and back again for messages in a distance of 25 m in a Gantt chart (1 = successful reception). Plot A shows a receive rate of 87% and plot B a rate of 83%

80% are high enough to warrant a periodic exchange of metadata between the AUVs, the experiment proved the principal appropriateness of using acoustic communication for a swarm of AUVs. The accuracy of the measured distances is displayed in plot C for all three scenarios. With a mean square error of 0.29 m and the biggest outlier being a difference of 0.49 m, the measurements are highly accurate when compared with the accuracy of basic GPS localization.

12.4.2 Communication principles for swarm operations

Based on the general localization and communication concept for MONSUN swarms described earlier (see Figure 12.7), in the following concrete communication, protocols for various scenarios will be described [14]. Since the acoustic communication is rather unreliable, these protocols must be very robust against lost messages and even longer outages. Due to the slow communication with low bandwidth, the protocols have to use only a minimum number of small messages. To avoid complex media access methods, only single point-to-point messages are used. Furthermore, the protocols shall be able to tolerate lost AUVs in order to be able to continue the mission also in case of failures.

As described earlier, acoustic modems can allow accurate distance measurements to their communication partners which can be used for localization. Here, it is assumed that the sender of a message provides the distance to the receiver by measuring the round trip time of the transmitted message itself and an acknowledgement message automatically generated by the receiver. This is, e.g., the case for the Evologics S2C modems used for MONSUN. In the following localization and communication protocols are first presented for the important special case of V-formations including meander-shaped paths and then for general swarm operations.

For monitoring and inspection tasks, it is often required to thoroughly cover a whole area of a waterbody. To achieve this, a V-formation is a suitable approach which can, e.g., traverse the area along meander-shaped paths. The members of the V-formation each take individual measurement tracks at their respective depth. According to MONSUNS' general localization and communication concept, some AUVs drive at the surface (called S-AUVs in the following) taking absolute GPS positions an some AUVs are dived (called D-AUVs) making use of relative localization by acoustic modems. A simple case for such V-formations is shown in Figure 12.9. Two S-AUVs S_1 and S_2 are used on the surface communicating with each other via Wi-Fi. Up to n D-AUVs, D_1 to D_n follow them in a V-formation dived to the depth required by the mission. S-AUV S_1 plays the role of a Master controlling the formation and communicating by Wi-Fi with a user station on a boat or at the shore.

The V-formation is first built up on the surface using GPS. Then the D-AUVs submerge to their given depth. Also the desired distances d_i between S- and D-AUVs as well as the distance between the two S-AUVs are assumed to be given as parameters of the formation. For maintaining this formation underwater and thus localization of the D-AUVs, the Master S_1 initiates the following simple protocol using acoustic communication. The protocol consists of rounds of only three messages. First S-AUV S_1 sends message 1 to D-AUV D_1. This message contains among others the desired course and speed as well as the required distances d_i between S- and D-AUVs of the formation. Having received this message, D_1 first sends message 2 to the second S-AUV S_2 and subsequently message 3 back to Master S_1. Since the sender of a message also gets the distance to the receiver as described earlier, D_1 can measure the distances d_1 to S_1 and d_2 to S_2, respectively. Knowing these two distances D_1 can compute its relative position and control its speed and course such that the V-formation is maintained.

For speed control, a comparison between measured and required distances for d_1 and d_2 is made. In case the measured distances are smaller than required, D_1 is too close to the S-AUVs and has to reduce its speed. If they are greater, D_1 is too far away and has to increase its speed. The course control makes also use of the distances d_i.

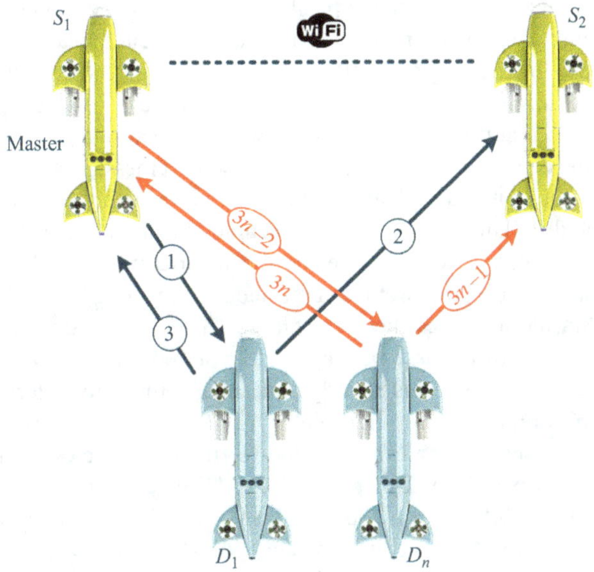

Figure 12.9 Communication principle of the message exchange sequence of a robot swarm [14]. The yellow coloured AUVs drive at the surface with the ability to communicate via Wi-Fi whereas the blue robots are submerged and have to use their acoustic modems. The robots at the surface build a V-shaped formation together with each submerged robot. The Master S_1 communicates in message triples with the dived MONSUN robots. For n dived AUVs, the message complexity adds up to 3n. Furthermore, the localization for dived AUVs is based on surface reference values. In order to reduce the needed amount of surface vehicles, only the two surface AUVs S_1 and S_2 are used. The depth can be determined by the pressure sensor of each AUV. The distances to the surfaced robots can be computed using the round trip time of the messages. By using this information, the position uncertainty is reduced to two points: one in front and one behind the surface vehicles. By providing a known start position for the formation, one point is hence eliminated and the position can be determined unambiguously relative to S_1 and S_2

For simplicity, it is first assumed that D_1 shall stay on a middle line between S_1 and S_2. In this case, the course is controlled such that both measured distances d_1 and d_2 stay equal – i.e. if d_1 gets smaller than d_2, the AUV turns to the right, and if d_1 gets greater than d_2, the AUV turns to the left. In case D_1 shall not stay in the middle but on the left or right side behind S_1 and S_2, the same principles can be applied, but only the required distances are different from each other.

Speed and course control are not typical continuous feedback control loops, since the acoustic communication is unreliable and thus the distance measurements may be unavailable due to lost messages. So, the D-AUV must be able to drive autonomously to bridge these outages – i.e. the distance measurements can only be used to correct the course and speed at the points in time when they are available. Note that the D-AUV receives the course and speed always with the first message and has a compass and an IMU for autonomous navigation which are, however, rather inaccurate due to sensor noise and drift. Unfortunately, this inaccuracy accumulates the longer the path gets. But for short paths between course and speed corrections, it is low enough.

The generalization of the protocol presented earlier to n D-AUVs is straightforward. The Master addresses the dived slaves D_1 to D_n round robin (see Figure 12.9). Since only rounds of three messages are required for each D-AUV, the total complexity for n slaves is $3n$. This linearly increasing complexity in terms of number of participants n naturally reduces the scalability of the method for larger formations. Assuming 0.5 s per message, a typical value for the used Evologics modems, the protocol takes 1.5 s per slave. Up to $n = 5$ D-AUVs seems to be realistic, i.e. a position update every 7.5 s. For higher numbers of AUVs, clusters of V-formations could be formed whose Masters communicate by Wi-Fi.

The proposed protocol also shows high robustness against lost messages and provides fault tolerance in case of failed AUVs. The Master initiates a new round in a given time interval independent from getting back a message from a slave. By this, the protocol continues also in case of lost messages, only a course and speed correction may be lost for this round which is bridged by autonomous driving. Since all slaves are addressed independently from each other by the Master, also a failed slave is easily tolerated. The rest of the formation simply continues the mission, while the failed AUV will usually surface due to its positive buoyancy. If one of the S-AUVs fails, one of the D-AUVs can surface and take over its role, including that of the Master.

For many applications not only an accurate position of the AUVs is required for geo-referencing the measurement data (e.g. from payload sensors), but also the data itself has to be communicated to the user, e.g. for online visualization on a Graphical User Interface (GUI). The proposed protocol supports this by using the last two messages of each round for allowing to send payload data from the slaves to the two surfaced AUVs which can in turn forward them by Wi-Fi to a user station.

So far, it was assumed that the submerged D-AUVs follow the two surfaced S-AUVs in the V-formation. It is also possible that the D-AUVs drive in front of the S-AUVs using the same protocol as described earlier for keeping this kind of V-formation. However, it must be known, if the D-AUVs are in front or behind the S-AUVs, because the protocol cannot distinguish between these two cases. Therefore, the formation should be formed in the beginning at the surface to yield a unique positioning. Mixed formations with some D-AUVs behind and some in front are also possible, but they must not change their relative positions underwater.

Often not only a line but a whole area has to be scanned by the AUV swarm for monitoring purposes. An obvious strategy to achieve this is a meander-shaped path (Figure 12.10). Since MONSUN can turn on the spot by 90 degrees, a single MONSUN can easily drive such a path. However, for a V-formation, turning becomes more

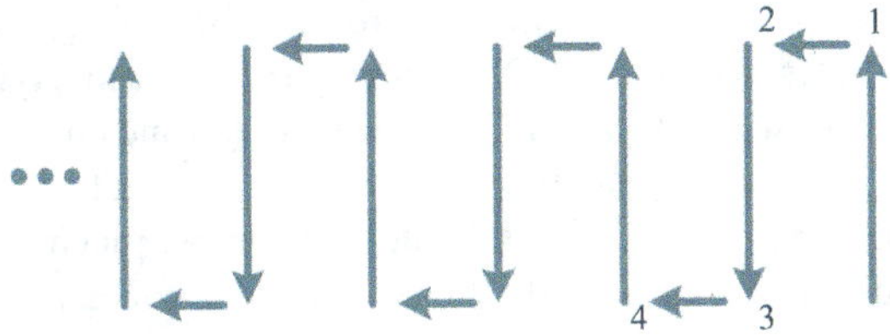

Figure 12.10 Meander-shaped path for area monitoring. The AUVs drive upward along a long straight line and then turn on the spot by 90 degrees to the left (1). Then after driving a short distance determining the spacing between the long scan lines, they turn by another 90 degrees on the spot (2) and follow a long line downward, turn to the left again (3) and after short sideward path turn upward again (4), etc. In this way, a large area can be covered

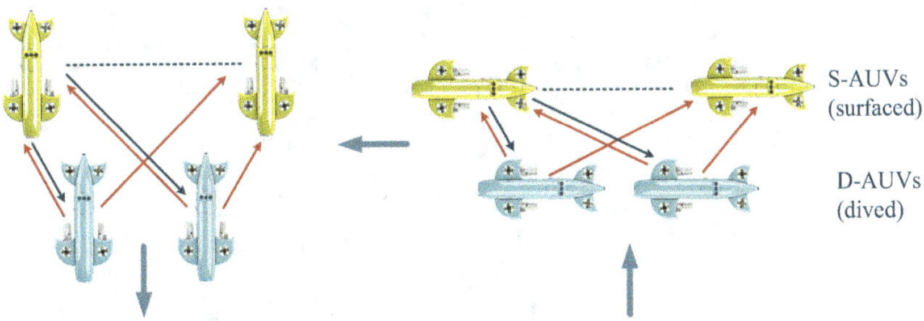

Figure 12.11 Turning of a V-formation on top corners of meander-shaped paths. After the long upward path the surfaced S-AUVs and the dived down D-AUVs following behind them in the V-formation turn on the spot by 90 degrees to the left at the same time in order to drive the short sideward path. Then they all turn left again by 90 degrees simultaneously to follow the long downward path. The submerged D-AUVs are now driving in a V-formation in front of the surfaced S-AUVs

challenging, since the formation has to be preserved. A possible solution which does not require complicated manoeuvres is shown in Figure 12.11.

Having reached the upper right turning point (1) of the meander after the long upward path, both S- and D-AUVs turn on the spot to the left by 90 degrees and drive in the new direction along the short sideward path until they reach the upper left turning point (2). There, they all turn again by 90 degrees for following the long meander downward path. Thus, the V-formation is kept, but now with the D-AUVs driving in front of the S-AUVs which can be achieved with the same protocol as described earlier. During the short sideward path, the D-AUVs use as usual the distance to the S-AUVs for their localization while the latter use GPS. The two meander turning points (3) and (4) at the meander's low side are handled analogously. This strategy ensures that S-AUVs as well as D-AUVs follow parallel meander-shaped paths for covering the area under scan. The fact that on the short top and bottom meander paths S- and D-AUVs drive behind each other, respectively, does not negatively affect the overall area coverage and can easily be tolerated. Otherwise much more complicated manoeuvres would be required since the AUVs would have to change their positions in the formation.

Not for all applications V-formations are an adequate solution, but an arbitrary position of D-AUVs may be desirable. In this case, 3 S-AUVs S_0, S_1 and S_2 can be used forming a triangle on the surface (Figure 12.12). They know their absolute

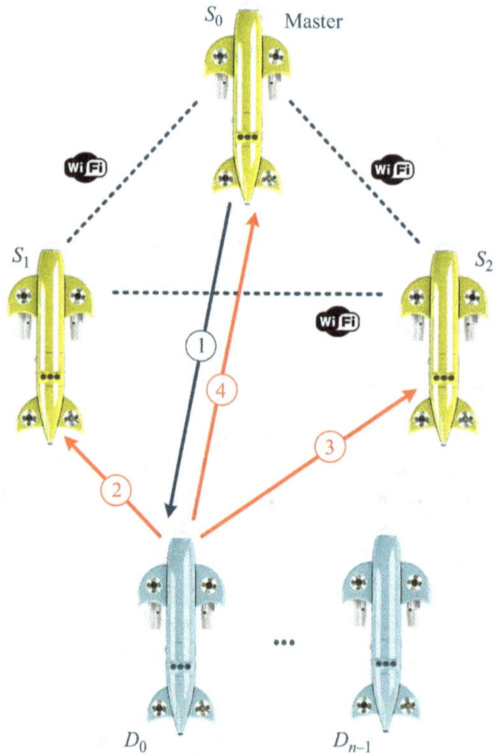

Figure 12.12 Communication protocol for an AUV swarm with arbitrary positions of submerged swarm members. Due to the fact of three so-called anchor stations at the surface, the position of a submerged robot can be computed unambiguously. The Master S-AUV S_0 sends a message 1 to the D-AUV D_0 which then sends sequentially messages 2,3 and 4 to S-AUVs S_1, S_2 and S_0 yielding the respective distances to them. Having the three distances D_0 uses trilateration for computing its location. The same protocol is used for the other D-AUVs D_1 to D_{n-1}

position by GPS and communicate via Wi-Fi. The submerged D-AUVs D_i can then use triangulation to locate themselves independent of their position if they know the distances to S_0, S_1 and S_2, respectively. This localization can be obtained by the following simple protocol. First Master S_0 sends a message 1 to D_0, which then sends messages 2 and 3 sequentially to S_1, S_2 and finally message 4 back to S_0. Thus, D_0 gets the required three distances for the trilateration. The same message protocol is carried out by S_0 to the other D-AUVs D_i. Since four messages are required for each D-AUV, the total complexity is $4n$ for n D-AUVs. As the Master initiates each round anew, the protocol is again very robust against message losses and lost D-AUVs. Failed S-AUVs can be replaced by emerging D-AUVs or by falling back to the V-formation protocol which only requires two S-AUVs.

S-AUVs have lower energy consumption than D-AUVs, since the latter have to use their vertical motors for active depth control due to MONSUNS' positive buoyancy. Furthermore, they usually employ additional payload sensors like sonars or cameras consuming energy. As a consequence the mission time of the whole swarm is dependent on the shorter endurance of the D-AUVs. In order to increase the overall mission time, a load balancing strategy can be employed. If the remaining energy of the accumulators of the D-AUVs is under a certain threshold, they exchange their position and role with one of the S-AUVs which still have a higher remaining energy. Simulation studies with our Marine Simulator MARS [15] in [16] have shown that an extension of the mission time up to about 30% can be achieved with this method.

12.5 Experiments

Verifying the potential capabilities and performance by using the AUV MONSUN is one of the key challenges while developing such kind of robot. It has to perform not only in controlled environments but also in natural waterbodies like lakes or rivers up to coastal areas and the ocean itself. Therefore, this section shows various types of performed experiments and measurement campaigns, showing the capability of the system under real conditions.

12.5.1 Clockwork Ocean

The Expedition Clockwork Ocean was carried out under the direction of Helmholtz Zentrum Geesthacht in the Baltic Sea between the islands of Bornholm and Usedom in June 2016 [17]. The aim of the measurement campaign was to better understand the importance of sub-mesoscale eddies and fronts for the ocean's energy scale, not only in global scale, but down to small turbulences and their likely significant role in phytoplankton production. The data were gathered by a combination of high-resolution remote sensing data from an airplane and an airship together with in-situ observations done by four vessels carrying various sensors including MONSUN AUVs (Figure 12.13). The combined measurements delivered data with a spatial and temporal resolution in order of a few meters and minutes, respectively, which allows for observing eddies and turbulences from generation to decay.

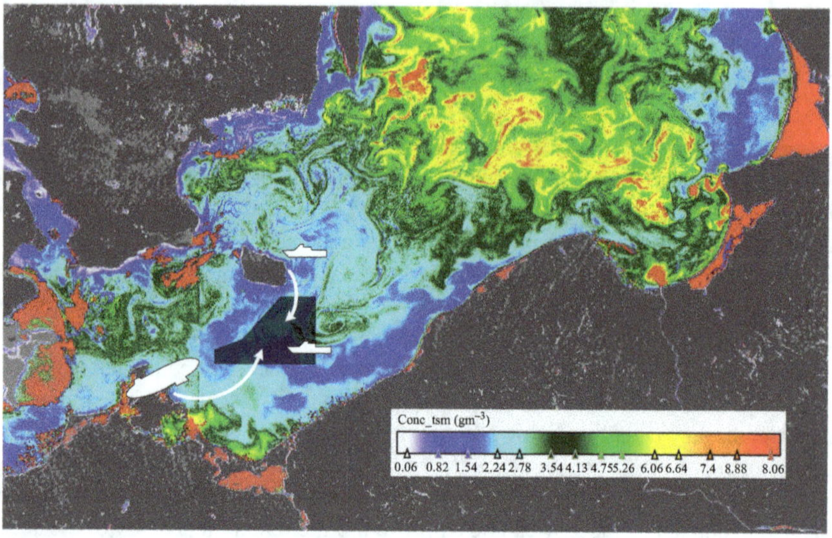

Figure 12.13 Satellite image taken of the Baltic Sea in June 2016. Depicted is the number of suspended solids to illustrate the turbulences and currents. Clearly visible is an eddy at the south of the island of Bornholm, there the main experiment field was chosen

The measurement strategy was subdivided into different stages. First, a motor glider equipped with infra-red cameras localized interesting structures at the sea surface. If promising data were found, the airship, equipped with hyper-spectral and infra-red cameras was sent out for stationary measurements above the eddy or sub-mesoscale structure with the ability to measure temperature fluctuations within 0.03°C on the surface. The research vessels carrying devices for in-situ measurements were then guided to the most promising place in real time. Besides so-called GPS-drifters for current measurements, the vessels made use of gliders, MONSUN robots and mainly towed instrument chains to measure environmental data underneath the water surface. The used sensors had probes for temperature, salinity, oxygen concentrations and chlorophyll attached. The aim was to measure the value of potential energy in the eddy and how micro algae are influenced by it.

The MONSUN AUVs were placed in the centre of an eddy from a boat and operated as an experimental robot team measuring conductivity and temperature values while driving trajectories through the inner structure of the turbulence. For the campaign only two robots could be used that were available at that time. One robot was driving at the surface navigating with its GPS sensor and the other robot was following the first one in submerged condition down to 5 m. To measure the full water column, the following robot was diving in a periodical saw-tooth pattern to measure the environmental data at each point on its way. Hence, this measurement strategy can be seen as a modular extension of the towed instrument chains attached to the research vessels but with changing measurement depths.

The submerged robot is not able to rely on GPS localization for the behaviour and the geo-referencing of the gathered data. Instead an acoustic modem is used to follow the robot at the surface, communicating current compass values and the distance between each other. The trajectory of the robot at the surface was programmed by the user on the vessel and transmitted to it via Wi-Fi connection [18].

During the expedition, MONSUN robots were used on three different measurement days collecting data of various structures. Figure 12.14 shows data from the second measuring day, the 23rd of June; 37 km south-east of Bornholm, an oceanographic front together with a cyanobacteria bloom was discovered. The robots travelled a combined distance of 1,896 m with a velocity of 1.5 kn, whereby the second robot followed the first one in a saw-tooth like diving pattern down to 5 m in distance of 10 m. The infra-red image at the top shows the oceanographic front and the transect travelled by the robots through it. Clearly visible is the diving pattern in the measured temperature values of the second robot.

The expedition proved on the one hand that MONSUN robots are able to collect usable environmental data over a longer time period, and on the other hand the possibility of using the robots under real conditions. The techniques of following with the help of acoustic communication and the ability of autonomous movements commanded by user interrupts from a nearby vessel over Wi-Fi communication showed the flexibility of the system and its capability to adapt to various environmental conditions. Using a swarm of more MONSUNS will allow to gather more measurement data at a time and makes this instrument even more useful and versatile.

12.5.2 V-formation

The communication principle presented in Section 12.4.2 is the foundation for tests of a V-formation. The experiment was conducted in a shallow lake with a size of 14 km^2 called Ratzeburger See and located near Lübeck, Germany. At a pier, three MONSUN AUVs were launched next to each other. The objective was to move south-east in V-formation along a straight line. For further auxiliary conditions, the two outer robots were expected to move in parallel with an inter-AUV distance of 20 m, while the third AUV at the tip of the V was supposed to follow them in a distance of 10 m. Considering a point of view located on the pier, the left MONSUN was given a set of GPS coordinates south-east of the starting position as a navigation target, while the right robot drove on a parallel course and by using compensation movements tried to reach the desired distance of 20 m. In order to test the reliability of the acoustic communication link as well as the described communication principle, a straight line was chosen as a course. A straight line as a test scenario shifts the focus on the communication, as the requirements for the motion controller are relatively moderate compared to more complex manoeuvres. Furthermore, the speed correction of the third MONSUN relative to the other AUVs was not yet used for this test. The following robot in the middle submerged with a time delay of 10 s to a depth of 1 m after the start of the mission. Afterwards, it started to follow the surfaced AUVs while attempting to remain equidistant, thus building a V-formation.

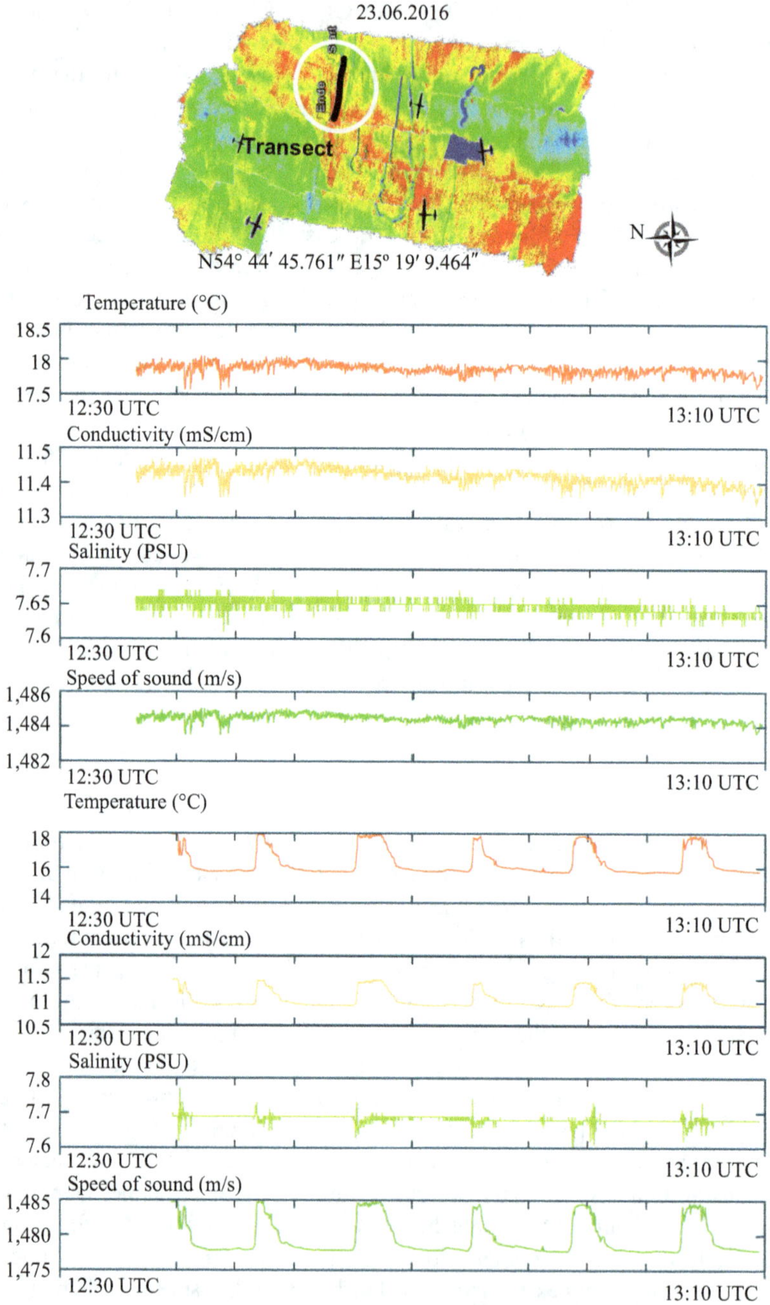

Figure 12.14 *Results of the second measurement day 37 km south-east of Bornholm. On top of the figure a patched together infra-red image taken from air is shown with a transect of both AUVs highlighted. Below both quartets of measurement data are displayed. The saw-tooth pattern of the diving robot is visible in the four lower plots*

The two MONSUN robots at the surface used continuous position measurements of their position with the built-in GPS and communication via Wi-Fi to reach the inter-AUV distance of 20 m in spite of being influenced by water currents. Utilizing the acoustic modems for repeated distance measurements to the surfaced MONSUNS, the submerged AUV followed them in a V-shaped formation. Figure 12.15 illustrates the place of experiment with its 100 m test track. The courses of the AUVs at the surface are depicted in yellow and the Master robot is labelled with S_1. The test started at a pier and the inter-AUV distance between S_1 and S_2 was 8 m. While the Master AUV took a straight course to the target, the other surface AUV needed to compensate while maintaining a parallel course to reach the target distance of 20 m to the Master S_1. The desired formation of the surfaced robots was reached after two thirds of the mission time. The track of the diving AUV D_1 is shown as a blue area. Without being able to receive GPS coordinates in a submerged state, a ground truth could only be taken at the start and the end of the track, where the robot surfaced. The width of the corridor was calculated using the GPS deviation of the surface vehicles to get a statement of position information compared to the other robots. It has an average width of 4.6 m, which is comparable to most widely used GPS systems.

Figure 12.16 shows the acoustically measured distance of the submerged AUV D_1 to the AUV S_2 at the surface over time and the message reception in a Gantt chart.

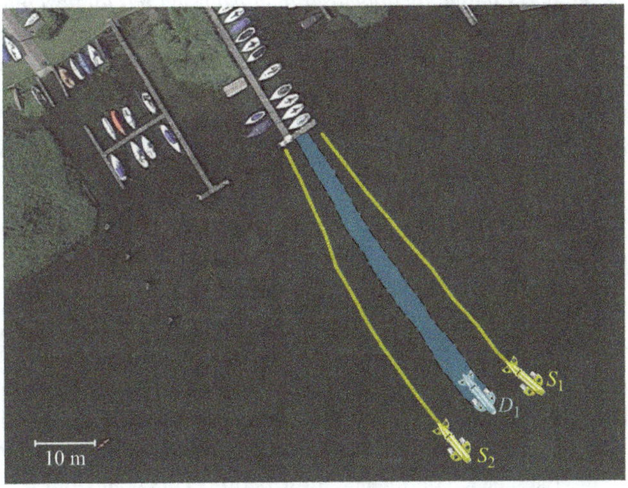

Figure 12.15 The formation building behaviour was tested with three MONSUN robots in a shallow waterbody. Two robots drive at the surface next to each other to a desired GPS position and a third robot follows them in a distance of 10 m while diving in a depth of 1 m. The yellow lines depict the transects of the Master AUV S_1 and the second AUV S_2. The blue corridor between them shows the estimated positions of the following robot based on the distance measurements to the surfaced robots. In average, the corridor has a width of 4.6 m, which is dependent on GPS deviations of the other AUVs

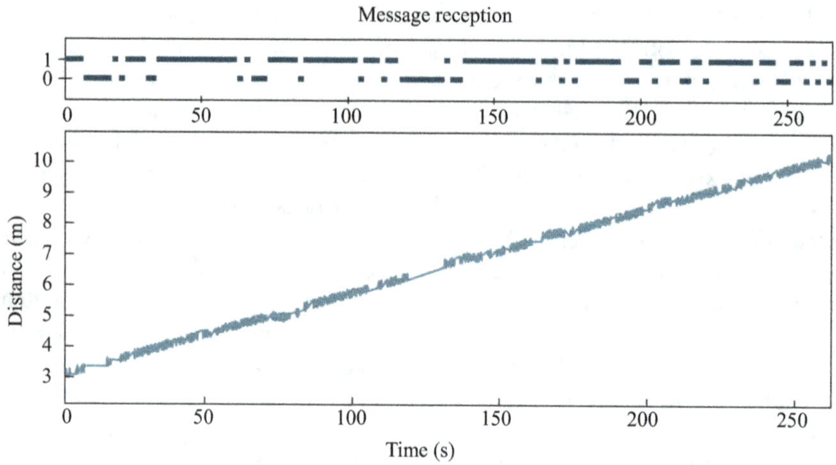

Figure 12.16 The Gantt chart at the top of the figure illustrates the message reception rate of the communication between the submerged Monsun *and the Master at the surface. During the test scenario depicted in previous figure, 78% of acoustic communication messages reached their destination. It could be observed that the communication outages commonly appeared in clusters with a longest outage of 13 s. Below the distance measurement is seen. The robots reached their desired distance of 10 m after 4 min. The noise of the measurements was in average not higher than 20 cm*

After 4 min, the desired distance of 10 m was reached. The measurements show a mean variation of only 20 cm. Considering this accuracy, the GPS uncertainty has a far higher influence on the localization of the submerged AUVs. Furthermore, a total of 78% of acoustic communication messages reached their destination. This validates the results previously collected in the experiment depicted in Figure 12.8. In addition, it could be observed that outages appeared in clusters. The longest acoustic outage lasted 13 s. This could mean that the highest influence on the channel reliability are changing physical conditions. In times with no communication, the robots use their own sensors to navigate based on the last information. This behaviour is sufficient as long as the times without communication are not getting longer.

12.6 Conclusions and future work

This chapter has presented the low-cost swarm micro AUV Monsun for environmental monitoring and inspection tasks. Monsun's hardware and software design is robust, modular, scalable and easily adaptable to a wide variety of applications. The communication among swarm members is based on Wi-Fi on the surface and acoustic modems underwater. The localization concept assumes that some AUVs stay at the surface using GPS while the other swarm members submerge and localize with the help of acoustic communication and distance measurements provided by the modems. Since no further infrastructure is required, the swarm can be easily deployed from a small boat or the shore.

Depending on the application, various swarm behaviours are possible. A robust and fault-tolerant protocol for a V-formation with two surfaced MONSUNS and *n* submerged ones has been described in detail. It can, e.g., be used for scanning an area following meander-shaped paths. Also a protocol for general swarms with three surface and *n* underwater MONSUNS at arbitrary positions is given which is based on trilateration.

First field experiments have been successfully been made to prove the MONSUN concept. In the expedition Clockwork Ocean of Helmholtz Zentrum Geesthacht, two MONSUNS were successfully used for measuring sub-mesoscale eddies in the Baltic Sea. A V-formation with three MONSUNS using our robust protocol designed especially to work with unreliable acoustic communication has been tested under realistic conditions in a lake.

Future work will, e.g., consist of using MONSUN micro AUV swarms for monitoring tasks in ports and harbours in case of water pollution by dangerous chemicals (BMBF Project MoSAIk) [19] or inspection of quay walls and ship hulls (BMVI Project SecurePort). For this, the swarm behaviours are further developed to be tailored to these tasks and additional sensors like oil sensors or imaging sonars will be integrated. Additionally, a GUI for easy handling of the swarm also by inexperienced users is under development. Further field experiments are planned with a higher number of MONSUNS in different realistic scenarios.

Acknowledgements

This work has been partially supported by the Ministry of Economic Affairs, Employment, Transport and Technology of Land Schleswig-Holstein, Germany, in the project Sea Our Future under grant number 122-13-005, the German Federal Ministry of Education and Research (BMBF) in the project MoSAIk under grant number 13N14153 and the German Federal Ministry of Transport and Digital Infrastructures (BMVI) in the project SecurePort under grant number 19H17008B. Our special thanks go to Prof. Baschek from Helmholtz Zentrum Geesthacht, Germany, who supported us in testing our MONSUN AUVs as part of the Expedition Clockwork Ocean.

References

[1] Kottege N, Zimmer UR. Acoustical methods for azimuth, range and heading estimation in underwater swarms. In: Acoustics '08. Paris, France; 2008.

[2] Quraishi A, Bahr A, Schill F, *et al.* Autonomous feature tracing and adaptive sampling in real-world underwater environments. In: 2018 IEEE International Conference on Robotics and Automation (ICRA). IEEE; 2018. pp. 1–6.
[3] Schmickl T, Thenius R, Möslinger C, *et al.* CoCoRo – the self-aware underwater swarm. In: Fifth IEEE Conference on Self-Adaptive and Self-Organizing Systems Workshops; 2011. pp. 120–126.
[4] SwarmDiver; Accessed: 2020-03-24. Available from: https://www.aquabotix.com/
[5] Amory A, Maehle E. SEMBIO – a small energy-efficient swarm AUV. In: OCEANS 2016, MTS/IEEE, Monterey; 2016. pp. 1–7.
[6] Osterloh C, Litza M, Maehle E. Hard- and software architecture of a small autonomous underwater vehicle for environmental monitoring tasks. In: German Workshop on Robotics; 2009. pp. 347–356.
[7] Meyer B, Ehlers K, Isokeit C, *et al.* The development of the modular hard- and software architecture of the autonomous underwater vehicle MONSUN. In: Proceedings of the 41st International Symposium on Robotics; ISR/Robotik; 2014. pp. 1–6.
[8] Meyer B, Maehle E. MONSUN – a modular testbed for swarms of autonomous underwater vehicles. In: ICRA2013 – Workshop on Many Robot Systems: Crossing the Reality Gap – From Single to Many Robot Systems; 2013. pp. 1–2.
[9] Osterloh C, Maehle E. Low-power microcontroller-based acoustic modem for underwater robot communication. In: ISR 2010 (41st International Symposium on Robotics) and ROBOTIK 2010 (6th German Conference on Robotics); 2010. pp. 1–6.
[10] Renner C, Gabrecht A, Meyer B, *et al.* In: Zerr B, Jaulin L, Creuze V, *et al.*, editors. Low-power Low-cost Acoustic Underwater Modem. Cham: Springer International Publishing; 2016. pp. 59–65.
[11] Meyer B, Renner C, Maehle E. Versatile sensor and communication expansion set for the autonomous underwater vehicle MONSUN. In: Advances in Cooperative Robotics: Proceedings of the 19th International Conference on Clawar. World Scientific; 2016. pp. 250–257.
[12] Osterloh C, Meyer B, Amory A, *et al.* MONSUN II – towards autonomous underwater swarms for environmental monitoring. In: IROS 2012 - Workshop on Robotics for Environmental Monitoring; 2012. pp. 1–6.
[13] Mackenzie KV. Nine-term equation for sound speed in the oceans. The Journal of the Acoustical Society of America. 1981;70(3):807–812.
[14] Behrje U, Isokeit C, Meyer B, *et al.* A robust acoustic-based communication principle for the navigation of an underwater robot swarm. In: OCEANS 2018 – Kobe; 2018. pp. 1–5.
[15] Tosik T, Schwinghammer J, Feldvoss MJ *et al.* MARS: A simulation environment for marine swarm robotics and environmental monitoring. In OCEANS 2016, Shanghai; 2016. pp. 1–6.
[16] Amory A, Tosik T, Maehle E. A load balancing behavior for underwater robot swarms to increase mission time and fault tolerance. In: 2014 IEEE

International Parallel Distributed Processing Symposium Workshops; 2014. pp. 1306–1313.
[17] Helmholtz-Zentrum Geesthacht. Clockwork Ocean; Accessed: 2020-04-21. Available from: http://clockwork-ocean.com.
[18] Meyer B, Isokeit C, Maehle E, *et al.* Using small swarm-capable AUVs for submesoscale eddy measurements in the Baltic Sea. In: OCEANS 2017 - Anchorage; 2017. pp. 1–5.
[19] Renner C, Heitmann J, Maehle E, *et al.* Automatische Ortung von Gefahrenstoffen im Hafenbecken. In: 29th Hamburger Kolloquium zur Abwasserwirtschaft, Hamburg; 2017. pp. 1–10.

Chapter 13

Reconfigurable, adaptable, multi-modality, mobile, wireless, energy-efficient, underwater sensor network of hover-capable AUVs

Vladimir Djapic[1]

Autonomous underwater vehicles (AUVs) have routinely been used for both commercial and military ocean exploration missions. In particular, there is an interest in low-cost, rapidly deployable, environmentally friendly, unmanned systems that can work for months at the time and detect, track, and identify air, surface, and subsurface targets. Recent AUV research is focused on long-endurance missions in unknown areas with minimal operator supervision or intervention in persistent ISR (Intelligence, Surveillance, Reconnaissance) operations.

Moreover, AUVs are an important tool for underwater research and commercial use as they are applied to carefully study the seafloor before hydrocarbon offshore production and laying underwater pipelines, to lay underwater cables, to inspect underwater structures, to conduct large-scale underwater research in different areas (hydrophysics, marine biology, chemistry, geology, climate science, underwater archeology, etc.), and to search for the sunken ships and airplanes [1].

The three primary methods for navigation of AUVs are (1) dead-reckoning and inertial navigation systems, (2) acoustic navigation, and (3) geophysical navigation techniques [2]. While dead-reckoning and inertial navigation systems can only be used for short-duration missions due to the navigation drift, acoustic navigation and geophysical navigation, in conjunction with inertial navigation, provide drift-free navigation and in combination with Global Navigation Satellite System (GNSS) can provide absolute positional accuracy [3–8].

Traditional approaches in underwater passive sonars are based on the utilization of hydrophone arrays, usually towed by ships. In more recent applications, traditional and lightweight AUV towed arrays are being utilized in the development of littoral autonomous sensing networks for applications such anti-submarine warfare, marine mammals, or ambient noise measurements [9]. Although robots are being put to work to tow such systems, the ideas how such arrays are employed are similar as in cases when arrays are towed by ships or submarines. Fully autonomous launch and recovery

[1]American Haval Motor Technology, Farmington Hills, MI 48335, USA

of such systems is not feasible. Therefore, ship would still be needed to support AUV missions.

Multiple low-cost AUVs carrying the low-cost sensors (acoustic, magnetic, etc.) can be autonomously deployed and recovered with other unmanned systems, for example, autonomous surface vehicles (ASVs); therefore, array bending is prevented (sensors fixed to the hull of AUVs) and left-right (port/starboard) ambiguity [10] can be overcome as multiple line arrays are utilized in the detection process.

In the last couple of years, there has occurred a "storm" of small (micro), size, weight, and power (SWaP) of AUVs coming to the market. The decrease in SWaP has come in great reduction in price as some common high-performance AUVs used to cost $250k while now base configuration of micro AUV can be purchased for $10–15k. The goal is to utilize only low-power, low-cost, and small-size sensors, computing and advanced manufacturing techniques. This can lead to massive deployments of low-cost, rapidly deployable, environmentally friendly, AUV systems that can perform ocean exploration for months at the time. Recent acquisitions of smaller and start-up companies that produce such micro AUV systems by larger, well-established marine/defense technology companies justify that we are heading in such direction as General Dynamics Mission Systems acquired Bluefin Robotics in 2016 and BAE Systems acquired Riptide Autonomous Solutions in 2019.

In contrast to a traditional wired acoustic array (stationary or towed by surface or underwater platforms), a "virtual" array can consist of multiple, untethered, detection sensor-equipped AUVs maintaining desired distance (based on a target signature desired to detect) using the acoustic communication links. Because of the adaptivity of its physical configuration, a virtual array can be both steered in angle and tuned via element spacing—to optimize response. Micro AUVs can be used to create reconfigurable, adaptable, multi-modality, mobile, wireless, underwater sensor network [11].

Multi-robot coordination and cooperation have a potential to significantly improve ocean exploration missions. For multiple AUV operations, coordination refers to AUVs keeping their formation and cooperation association to achieve specified mission objectives. Teams of coordinated AUVs can then cooperatively work with other teams of coordinated AUVs.

This vision for such ocean exploration dates back two decades ago with Autonomous Ocean Sampling Network (AOSN) [12], but now it is a reality with several technological advancements. Cost of small AUVs has significantly decreased but more importantly some expensive sensors needed for precise AUV navigation are not needed if coordinated teams are employed. For example, costly fiber optic or laser ring gyros can be replaced with microelectromechanical systems (MEMS) counterparts and a Doppler velocity log (DVL), often considered as a main navigation sensor for AUV navigation, needs be installed only on one AUV that serves as the main navigation node for the team of AUVs.

Coordination can be accomplished with less sensor sophisticated AUVs with the aid of low-cost underwater communication modems, both acoustic [28] and optical [13], and utilization of inexpensive and low-power low-drift clocks, for example, chip-scale atomic clock (CSAC) [14] and decentralized distributed control and localization algorithms. Cooperation can be accomplished with advances in underwater networking schemes. Some related work can be found in [15–22].

This chapter details the combination of cooperative AUVs, accurate timing and clock synchronization, underwater sensors, and acoustic communication for the purposes of detection of various underwater targets. The overall theme is the novel use of the aforementioned systems to create a coordinated network of accurately positioned and navigated AUVs that can be used for seismic surveys, surveillance, marine mammals detection applications, and many other ocean exploration applications. This network can cooperate with another group of coordinated AUVs in order to maximize coverage rate while reducing the detection time.

The following technologies contribute to AOSN made with heterogeneous AUVs:

- small (micro) and SWaP AUVs;
- Accurate Position, Navigation, and Timing (PNT) sensors;
- wireless communication modems (acoustic and optical) and spiral beacon technology.

13.1 Micro AUVs

AUVs have routinely been used for both commercial and military ocean exploration missions. The decrease in SWaP has come in great reduction in price as some common high-performance AUVs used to cost $250k while now base configuration of micro AUV can be purchased for $10–15k. The goal is to utilize only low-power, low-cost, and small-size sensors, commuting and advanced manufacturing techniques. This can lead to massive deployments of low-cost, rapidly deployable, environmentally friendly AUV systems that can perform ocean exploration for months at the time.

Micro AUVs can be used complementary with other larger AUVs. Micro AUVs equipped with navigation suite that enables accurate underwater positioning can then work together with AUVs, ASVs, or manned surface/subsurface vessels equipped with much higher-grade navigation sensors. It is this cooperation that will allow for fast coverage rates and reduced time in spatial and temporal ocean sampling and explorations. Multi-robot cooperative systems have the potential for completing tasks more efficiently. In order to cooperate, these systems must communicate using low-bandwidth, slow, and unreliable communication channels, making cooperative operations very challenging. Only some of the micro AUV examples are presented in this chapter—there exist many more on the market.

The REMUS M3V [23], built on a proven Hydroid and Kongsberg technology, can achieve speeds of 10 knots and is equipped with an articulated tail control system and a positioning and navigation system as standard. It can dive up to 300 m in depth. The vehicle is true A-size (36 × 4.875) vehicle with no fins or appendages outside the A-size envelope and can be used for multi-domain deployments.

Riptide Autonomous Solutions [24] is offering a new, highly flexible, open-source software autonomous undersea vehicle that provides users a state-of-the-art, low-cost development solution ideally suited for developers of autonomy and behaviors, power systems, subsea sensors, and new payloads. Endurance of these compact (4.875-inch diameter) AUVs is projected to be 30 h with alkaline, 48 h with lithium ion (Rechargeable), 144 h (6 days) with lithium primary, and 400 h (over 2 weeks) with aluminum seawater battery batteries. Thus, at 3-knot speed, they can travel 90, 144, 432, and 1,200 nautical miles, respectively.

The VERTEX AUV [13] developed by Hydromea is the smallest, most integrated commercial AUV in the world. Designed to be portable, it is easy to deploy in large numbers, and in places that are otherwise difficult to reach. The AUVs are only 70-cm long, and at 7 kg weight can easily be carried in one hand. The transport case can travel as standard check-in luggage.

The Bluefin SandShark [25] is a one-person-portable, low-cost AUV designed to leverage today's miniaturized sensors and also be the development platform for the next generation. The standard tail section houses all core vehicle systems and connects in seconds to any number of fully customizable and modular wet or dry payload sections. Small enough to be carried in a backpack, the Bluefin SandShark is a low logistics and rapidly deployable solution for virtually any underwater need.

Many of these micro AUVs have been already used and improved by different research groups; for example, modular design with hovering capability and ultra-short baseline (USBL) navigation was added to COTS Riptide micro AUV by Naval Information Warfare Center Pacific [11], COTS SandShark micro AUV was equipped with one-way travel-time (OWTT) ranging and inverted USBL (iUSBL)—full navigation fix via triangulation of range and angle—by Massachusetts Institute of Technology (MIT) [26], and low-cost nano-acoustic modem was integrated onto ecoSUB micro AUV by National Oceanography Centre and ecoSUB Robotics, Ltd. [27].

The compact design is crucial when deploying multiple AUVs in a group—it can easily allow up to ten vehicles be deployed from the shore, or a small boat, without any special equipment. By using up to ten AUVs in parallel, it is feasible to obtain detailed spatial measurements within hours with a level of detail—real data of the full water column, comparable only to simulation models or surface satellite images. The aforementioned micro AUV systems are shown in Figure 13.1.

Thus, there is a growing interest in deploying low-cost, small-size AUV as swarms in coordinated and distributed fashion. However, traditional acoustic navigation methods utilizing LBL (long baseline) and USBL based on two-way acoustic transmission do not scale with number of AUVs. They might work well for a couple of AUVs but not for a swarm of AUVs. New acoustic communication and clock synchronization methods have to be employed.

13.2 AUV communication and PNT

Most commonly used technology for underwater communication is acoustic modem; however, low-cost optical modems can be employed if ranges among AUV operating in swarm is limited to less than 50 m or if a data muling (optical communications

Figure 13.1 On the top left, VERTEX AUV; left bottom, RIPTIDE AUV. On the top right, SandShark AUV; right bottom, REMUS M3V AUV

can be enabled only when the sensors and the mobile mules are in close proximity) is desired. Low-cost Hydromea LUMA 500ER optical modem [27] offers data rate of 500 Kbit/s with fairly low power consumption (receiving 1 W, transmitting 2–5 W) and advertised range is 50 m. It can be used as modems or as access points, arranged in pairs or in a mesh network. Optical communication is capable of much higher data transmission rates compared to acoustic communication.

13.2.1 Nano modem, sea modem, and CSAC

Numerous capable acoustic modems used with larger AUVs exist. Their cost is proportional to the cost of the AUV system. This section focuses on the miniature, low-cost, low-power acoustic modems. One such modem is being developed at Newcastle University; see (top) Figure 13.2. The work in [27] describes the development of a network of AUVs for ocean monitoring and surveillance in dynamic environments. The key component of the system is the ecoSUB vehicle, a small and low-cost AUV that acts as a mobile node of the network and the results toward the implementation of such a network, describing the communication and localization are described. To make the ecoSUB a mobile node of an underwater acoustic network, one of the key objectives of the project is the development and use of a low-cost, low-power nano-underwater modem [28]. Nano acoustic modem is only 42 mm in diameter with the length of 60 mm. It operates in the acoustic frequency band 24–28 kHz and utilizes a simple but

Figure 13.2 (Top) Nano-modem developed at Newcastle University. (Bottom) The Hydromea LUMA 500ER optical modem

highly robust spread spectrum modulation scheme based on binary orthogonal keying using chirp waveforms which has proven highly effective in commercially available modem products [28]. It can achieve data transmission at 40 bps over ranges up to 2 km transmitting a source level of 168 dB re 1 Pa @ 1 m (0.5 W) while the receiver power consumption is <10 mW. Range between AUVs can also be measured to a resolution of 10 cm. An enhanced version of these modems is in the final stages of testing at the time of writing, with near identical dimensions, power and cost but with wider acoustic bandwidth and increased data rate up to 640 bps enabled by binary orthogonal code keying. Its very low standby power consumption makes it particularly well suited for long-term deployments along with battery-powered instruments.

Time synchronization and localization are key requirements for distributed underwater systems consisting of numerous low-cost AUVs. In [22], the authors investigate the problem of time synchronization and tracking for submersibles that only have the capability to receive acoustic signals. Eustice *et al.* have reported integrating a

Figure 13.3 SeaModem and CSAC

temperature clock into an acoustic modem, with a skew of only 0.02 ppm [29]. Similar levels of accuracy have also been achieved through a novel and inexpensive approach using two crystals. While the effect of clock drift is negligible when an accurate clock is used, the effect of the motion can still be significant and the authors in [22] propose to compensate for motion as much as possible by adding of an inertial measurement unit (IMU).

The Control Engineering Group of the Autonomous Systems Laboratory at the DIMES Department of the University of Calabria has developed, in collaboration with Applicon s.r.l., a compact low-power underwater acoustic modem for AUV navigation and communication applications [30]. SeaModem is the commercial name of this modem actually produced by Applicon s.r.l. It is an frequency-shift keying (FSK) underwater acoustic modem well suited for shallow water communications. The working frequency band ranges between 25 and 35 kHz, with selectable data rates of 750 bits/s, 1,500 bits/s, and 2,250 bits/s. An UART (Universal Asynchronous Receiver/Transmitter) interface is used to host the modem. SeaModem hardware is being extended with a CSAC. CSACs with a very low drift will enable the SeaModem clock to remain synchronized for many hours without remarkable error in the range calculation. The SeaModem and CSAC are shown in Figure 13.3. The authors in [16] and [31] demonstrated the utility of using synchronized atomic clocks with SeaModem for one-way ranging which can be used for navigation updates to team of AUVs broadcasting only one acoustic message. Several experiments have been performed to estimate the accuracy of both two-way and one-way ranging estimation to localize underwater nodes. In the experiments (ship based in open waters) south of Cartagena, Spain, it was shown that both the ranging techniques achieve very good results, with a minimum standard deviation of 0.11 and 0.28 m when using one-way and two-way ranging, respectively [31]. During the experiments in Amantea, Italy, further enhancements were made and the performance was improved. The one-way distance estimated by a SeaModem node equipped with CSAC is very stable and very precise for the whole duration of the experiment. In particular, the node estimated an average distance of 17.00 and 16.98 m with a standard deviation of 0.05 and 0.04 m, respectively, resulting in an average error of 0.02 m.

13.2.2 Acoustic communications with JANUS

Although the status of the technology and the user maturity has rapidly improved, so far the use of swarm of vehicles that are able to communicate and cooperate to accomplish more challenging tasks has been only explored in research but current trends that industry is starting to take a lead. To cooperate, however, one has to communicate. It has therefore become clear that underwater communications and networking play a critical enabling role for the marine technology community. Acoustic is currently the main technology used for communications in water. It however introduces several challenges, such as long propagation delays, low data rates, shadow zones, and high variable link qualities over time. Luckily, recent decisions for standardization of underwater communications will help in deployments of multi-AUV cooperative systems.

JANUS [32] is a physical layer standard for public use pioneered by NATO Centre for Maritime Research and Experimentation (CMRE) with Academia and Industry—a physical layer coding scheme allowing assets to transmit information in a common format that can be decoded by compliant devices. JANUS became the first international standard for underwater communications in 2017 (STANAG 4748). The two primary purposes for JANUS are to announce the presence of a node and to establish the initial contact between dissimilar nodes. These purposes are similar to the Channel 16 radio usage at sea and on land. JANUS has been demonstrated as a robust method to transmit information underwater. The mathematical tools necessary to create a JANUS signal encode it in a desired frequency band and decode a received signal are publicly available. JANUS, which is named after the Roman god of gateways, has been tested at 900 Hz to 60 kHz frequencies at distances of up to 28 km. However, it's optimized for sending data underwater at up to 10 km. JANUS assigns the 11.5 kHz band for initial discovery and defines a procedure for handshaking, synchronization, and 80 bps data transmission using 56-bit packets. JANUS includes a standard modulation encoding scheme called Frequency-Hopped Binary Frequency Shift Keying (FH-BFSK) that describes how acoustic waveforms are encoded. It also provides redundancy checking for reducing errors caused by interference from varying water temperatures, currents, and Doppler effects caused by motion. Integration of JANUS to the AUV platforms is being reviewed to maximize communication and awareness where applicable, and to potentially allow long-range information about AUV location, as well as sharing of real-time information.

In [33], the authors presented the implementation and in-field testing of a "bilingual" underwater acoustic modem supporting the JANUS standard and the proprietary SeaModem modulation scheme. The bilingual modem has been interfaced with the SUNSET networking framework [21] and integrated with static and mobile nodes (ASV) available at the Naval Information Warfare Center Pacific. JANUS has been used to support first contact, discovery, and parameter's negotiation among the nodes in the network and as a reliable robust channel to exchange short control messages. After the negotiation phase, the nodes were then switching to a faster communication language to exchange larger data messages.

13.2.3 Ambient field clock synchronization

One-way communication for updating position requires synchronized clocks. Synchronization requirements are dictated by position accuracy needs and the speed of sound in water: assuming that the speed of sound in water is approximately 1,500 m/s, synchronizing to within 1 ms corresponds a position estimate accurate to 1.5 m. For short-duration missions, a lack of synchronization may not be an issue. However, for long-duration missions it can be (over a month, a CSAC can lose up to 350 μs [34]). The work in [22] shows that by using two simple hardware enhancements (an accurate clock-crystal and an onboard IMU), submersibles that can only receive acoustic transmissions can be accurately tracked and synchronized using Time of Arrival based Tracked Synchronization (ToA-TS) technique. Further, if submersibles have acoustic communication capability, the proposed approach can be applied for virtually no communication overhead on the part of submersibles. A potential means to synchronize clocks could be based on the detection of ambient fields. In close proximity to 60 Hz power lines, one could sense the emitted magnetic field [35]. For subsurface environments, low-frequency signals could be used. With sensitive magnetometers, one could exploit the diurnal variations of the Earth's magnetic field [36]. Besides requiring a total field measurement, a noted limitation of this method is the need for sensors with sufficiently high enough sampling rates: a magnetometer with a sampling frequency of 1 kHz can at best provide synchronization to 1 ms. Another issue to consider is the information embedded in the ambient field. This can be given by a Cramer Rao lower bound: in general, high bandwidth signals and high signal-to-noise ratios improve synchronization capabilities [37]. A simulated demonstration of the method described in [36] is shown in Figure 13.4. Instead of using total field magnetometers, two, single-axis fluxgate magnetometers were aligned next to each other such that they sensed the same magnetic field. A delay corresponding to 5,000 samples was introduced to one of the signals. The cross-correlation of the two signals has a maximum at the time corresponding to the delay at 1,037 s.

13.2.4 Acoustic synchronization and ranging estimation

Clock synchronization can be also performed acoustically to avoid the surfacing of the underwater vehicles while performing a long-term mission for GNSS clock fixing. In particular, a reference node, such as an ASV, can be equipped with a stable GNSS clock and can serve as a moving beacon that broadcasts one-way messages to acoustically synchronize the clocks of the underwater platforms. At the same time, the underwater nodes are able to estimate their distances to the reference node measuring the OWTT with a very high accuracy. In order to perform acoustic synchronization and one-way ranging estimation, as mentioned above, the CSAC was integrated with the SeaModem acoustic modem [30]. The modem firmware has been improved to provide transmission scheduling and transmission and reception timing information to the upper layers of the protocol stack. The SUNSET framework [21] has been used to provide networking capabilities. In particular, new protocols have been designed and developed to acoustically synchronize the clocks of multiple node to that of a reference

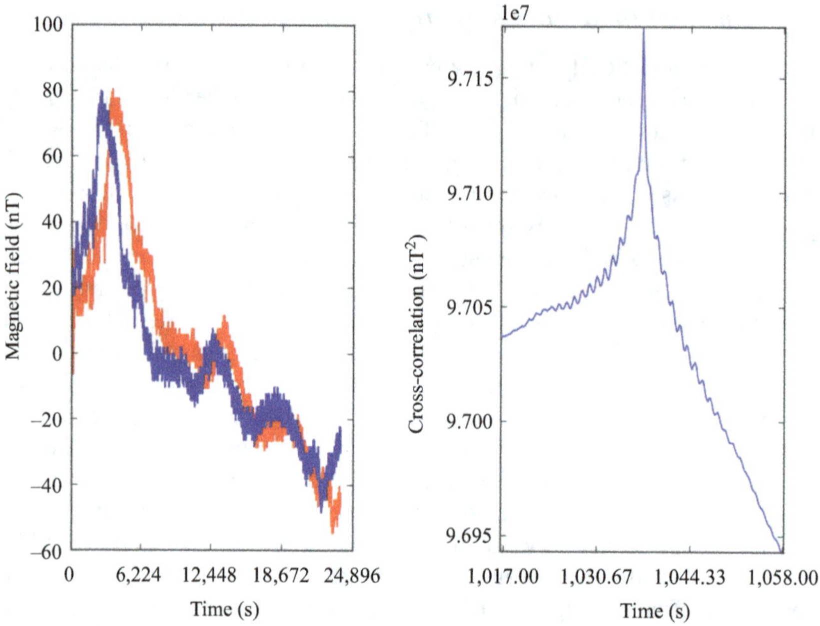

Figure 13.4 A simulated demonstration of clock synchronization using ambient magnetic fields. The skew introduced between the two signals creates a maximum in the cross-correlation that corresponds to this time

node and at the same time to estimate their distance through one-way messages. These new solutions can be used with any commercial acoustic modems that implement firmware features described above. Different at-sea trials were performed, [31] and [16], to demonstrate that acoustic synchronization and one-way ranging estimation are highly accurate, low overhead and energy consuming, fast and scalable and thus suitable for supporting cooperation and coordination of multiple AUVs where high accuracy and reduced overhead are required. Synchronized sensor data and accurate inter-AUV distances are the key requirements for the concept of wireless sensor line arrays to work.

13.2.5 Spiral beacon technology

Similar to the VOR system used in aerial traffic as a beacon whose mix of two signals informs the recipients of their bearing toward the beacon, an underwater acoustic analogy was developed by Naval Research Laboratory, Washington, USA [38], as a system using a coaxial set of a spiral wave-front beacon and a circular wave-front beacon. The circular wave-front beacon can be a classic beacon or even an acoustic modem, while the spiral wave front can be generated using a physical spiral beacon or a circular array of beacons fired in a rotational manner in order to synthetically create a spiral wave front. The difference in phase (or time) between the spiral wave-front signal and the circular wave-front signal informs the recipient of its bearing toward the beacon (see Figure 13.5). The separation between those two signals can be done in time, frequency or a mix of the two methods. Furthermore, in the synthetic spiral method, a changing chirp signal in relation to bearing can be used (as is normally used in scanning sonar and radar systems) to increase angular resolution and robustness.

One of the goals is the development of an underwater acoustic localization system for multiple robots using a spiral wave-front beacon in conjunction with a standard (circular) acoustic modem. Although, an inverted USBL (iUSBL) can be used on each AUV for full navigation fix via triangulation of range and angle as demonstrated by MIT [26], using the difference in time of arrival between the spiral wave front and the modem circular wave front, receivers will be able to determine the bearing to the source using only one hydrophone.

Synchronizing vehicles' clocks with the beacon in the beginning of the mission and during the longer mission (clock synchronization using compact magnetometers and e-field sensors) will ensure the vehicles ability to also calculate its distance from the beacon upon every message reception. The update loop of the system we are proposing does not grow with the number of vehicles being informed, and the latency is minimized to the travel signal time. It can be said that with this method a form of localization field is produced for any participant to use as needed. Use of underwater vehicle teams/swarms for various types of missions raises a need for such a system to ensure high-quality localization with a short position update loop, and minimum latency, for all vehicles without a need for costly components and even acoustically active components, thus saving energy or remaining silent, depending on the mission requirements.

Standard localization methods (LBL, SBL, and USBL) use a return signal to determine range and bearing from a target and then inform their targets of their positions. It takes a period of time (minimally 2 s) to inform one vehicle of its position. Adding more vehicles that period gets prolonged due to iterative interrogation of each one of them. Incorporating the range and bearing from the beacon with a known beacon position into the properties of the acoustic signal being transmitted from it allows any number of vehicles listening to know their position at any given moment. The update loop of the system we are proposing does not grow with the number of vehicles being informed, and the latency is minimized to the travel signal time. It can be said that with this method a form of localization field is produced for any participant to use as needed. Use of underwater vehicle teams/swarms for various types of missions raises a need for such a system to ensure high-quality localization with a short position update loop, and minimum latency, for all vehicles without a need for costly components and even acoustically active components, thus saving energy or remaining silent, depending on the mission requirements.

In Figure 13.5, red is the spiral wave front while blue is the modem circular wave front. The time scale represents the received signal in relation to the start of each second ($t = 0$). The figure depicts the composition of the received signal by the

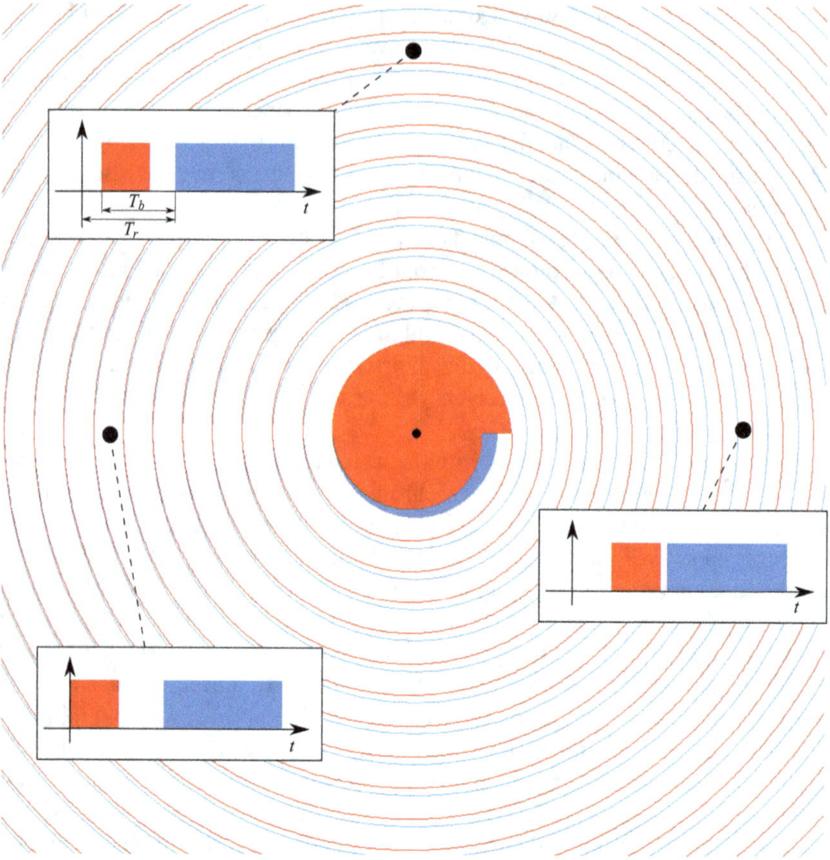

Figure 13.5 *Red is the spiral wave front while blue is the modem circular wave front. The time scale represents the received signal in relation to the start of each second (t = 0). The figure depicts the composition of the received signal by the vehicle dependent on its bearing to the spiral wave-front acoustic modem beacon. Representation of the acoustic localization field generated by the spiral wave-front acoustic modem beacon (T_b carries bearing information and T_r carries range information)*

vehicle dependent on its bearing to the spiral wave-front acoustic modem beacon. Difference in time of arrival between the spiral wave-front beacon and the circular wave-front modem message carries bearing information, while the elapsed time from the start of the current second carries range information. A system like this, which informs the vehicle of its position rather than informing a central station, can be easily adapted to be used by an AUV as an underwater navigation aid, and the passive nature of the recipient can guarantee longer autonomy of such a receiver and stealth of the AUV using it.

13.3 AUV navigation

One of the main requirements for successful missions of AUVs is their high-precision navigation. The three primary methods for navigation of AUVs are (1) dead-reckoning and inertial navigation systems, (2) acoustic navigation, and (3) geophysical navigation techniques [2].

Inertial/dead-reckoning navigation uses accelerometers and gyroscopes for increased accuracy to propagate the current state; however, all of the methods in this category have position error growth that is unbounded.

Techniques used in acoustic navigation are based on measuring the time of flight (TOF), time difference of arrival (TDoA) of signals from acoustic beacons, or modems to perform navigation. Traditional LBL and USBL and more recent single underwater transponder positioning acoustic positioning navigation systems require at least two if not three acoustic messages to be exchanged. The difference between the two methods is that the two-way (or three-way) approaches do not scale as well as compared to one-way methods. Traditional acoustic navigation methods make the key assumption that beacon transmissions occur nearly concurrently in time. While this assumption is ensured in small LBL deployments it does not hold as the size of the system scales up [22].

Geophysical techniques use external environmental information as references for navigation. This is done with sensors and processing that are capable of detecting, identifying, and classifying some environmental features most often with sonars (side and forward-looking and video cameras).

13.3.1 AUV navigation sensors

The AUV navigation is a term to measure vehicle geographic coordinates (latitude and longitude), and to measure vehicle motion parameters (heading, velocity, and depth) and attitude angles (yaw, pitch, and roll).

The usual AUV navigation suite includes the following sensors:

- DVL is a primary navigation sensor. Usually it is a phased array 600 kHz acoustic sensor which generates four beams at 30-degree angles relative to vertical and spaced 90 degrees around a circle. Beam orientation for DVL is at 45 degrees, 135 degrees, 225 degrees, and 315 degrees compared to the forward-backward centerline of the AUV.
- An attitude and heading reference system (AHRS): This sensor measures changes in roll and pitch up to 200 degrees/s, measures acceleration in three axes, and measures heading (compass). The AHRS automatically considers magnetic declination if it is fed the current GNSS position. This eliminates unintended large heading errors when relocating to a different area on Earth than the "home" area. The AHRS is factory calibrated for hard iron magnetic signature and can be further

two-dimensional (2D) or three-dimensional (3D) calibrated locally if desired. As periodic calibration is required, we designed an onboard calibration routine that helps with compensating for hard iron signature of the device, which can be performed in air or in water. It should be noted that the gyros of an AHRS, which assist in balancing heading, will always drift over time at a relatively slow rate. This can result in additional accumulated position error while swimming. If the AUV stays submerged for many hours, this drift can result in increased position error.

- Depth sensor: AUVs utilizes a depth sensor that provides a 0.1-m precision and an accuracy of ±0.5 m.
- GNSS and GNSS antenna: The GNSS antenna is a multi-frequency, multi-constellation, miniature ceramic antenna. The GNSS receiver has multiple options from single-frequency, single-constellation receivers to multi-frequency and multi-constellation receivers. Additional Military Selective Availability Anti-spoofing Module (SAASM) or M-code receivers may be chosen, but special considerations and cost are associated. AUVs rely on industry standard National Marine Electronics Association (NMEA) strings for communication with the GNSS receiver.

13.4 Acoustic navigation

Acoustic positioning is traditionally used for AUV navigation as inertial navigation fails due to accumulation of AUV position and heading errors during long autonomous navigation, and the usage of GNSS is often impractical because of significant loss of time during AUV surfacing for position updates.

LBL method consists in measuring the range to several (three or more) spaced beacons with known coordinates and solving the relevant system of equations, where AUV coordinates are unknown [1,39]. The method can be based on measuring the travel time (TT) of signals transmitted from the beacons to AUV (search for sphere crossing point) or TDoA (search for hyperbola crossing point). In the first case, if AUV and beacon clocks can work synchronously, signal OWTT or TOF is measured. If synchronous operation of AUV and beacon clocks is not guaranteed, two-way messaging time is measured. In the second case, TDoA is measured. Then only beacon clocks should be synchronized; AUV clocks do not require synchronization.

There are two types of LBL methods: (1) using stationary (generally bottom) beacons with known coordinates determined during installation as in Figure 13.6 (top) and (2) using moving (generally drifting) buoys as in Figure 13.6 (bottom). This figure is courtesy of [1]. As moving buoys, usually special drifting GNSS Intelligent Buoys (GIBs) are used, whose coordinates are determined by GNSS. LBL version for the ice-covered regions can be referred to the second type. GIBs are frosted into the ice with GNSS antenna being under the ice, and acoustic emitter, above the ice. LBL beacons and GIB beacons are often other autonomous vehicles either underwater or surface.

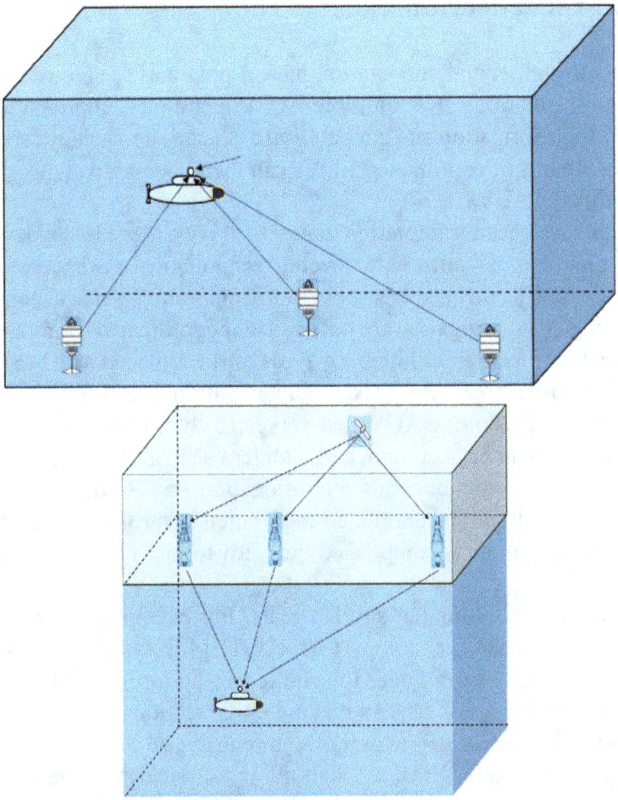

Figure 13.6 (Top) LBL method using bottom acoustic emitters with known coordinates as beacons. AUV position is determined by measuring the TT (or TDoA) of signals emitted simultaneously by all the beacons. (Bottom) LBL method using GIBs as beacons. AUV position is determined by measuring TT (or TDoA) of signals emitted simultaneously by all the buoys and carrying information on buoy current coordinates obtained using GNSS

The traditional acoustic navigation methods are acceptable for single-vehicle deployments but less desirable for multi-vehicle deployments because the interrogation-response navigation update period increases linearly with the number of vehicles (thereby proportionally decreasing each vehicle's overall navigation update rate). In practice, this limits multi-vehicle LBL navigation to networks of a few vehicles. The goal of the work of Eustice *et al.* was to enable the task of navigating multiple AUVs over length scales of O (100 km), while maintaining error tolerances commensurate with conventional LBL transponder-based navigation systems (i.e., O (1 m)), but without the requisite need for deploying, calibrating, and recovering seafloor anchored acoustic transponders [29].

13.5 Multi-robot coordination

Multi-robot coordination and cooperation have a potential to significantly improve ocean exploration missions. For multiple AUV operations, coordination refers to AUVs keeping their formation and cooperation association to achieve specified mission objectives. Teams of coordinated AUVs can then cooperatively work with other teams of coordinated AUVs.

This vision for ocean exploration dates back two decades ago with AOSN, see [12], but now it is a reality with several technological advancements. Cost of small AUVs has significantly decreased but more importantly some expensive sensors needed for precise AUV navigation are not needed if coordinated teams are employed. For example, costly fiber optic or laser ring gyros can be replaced with MEMS counterparts, and a DVL, often considered as a main navigation sensor for AUV navigation, needs be installed only on one AUV that serves as the main node for the team of AUVs. Coordination can be accomplished with less sensor sophisticated AUVs with the aid of underwater communication and utilization of nowadays inexpensive and low-power CSAC [14] and decentralized distributed control and localization algorithms. Cooperation can be accomplished with advances in underwater networking schemes.

Ongoing efforts are aiming at similar goals. For example, the Widely scalable Mobile Underwater Sonar Technology (WiMUST) [40] system will be composed of many AUVs carrying hydrophones to acquire sub-bottom profiling acoustic data. Instead of traditional arrays of sensors that are either stationary or towed by the ship forming the fixed geometry of the acoustic antenna, new and exciting technology advances to be integrated in WiMUST will allow to quickly, on-the-fly, change the geometry of the acoustic antenna during which it holds potential to drastically improve ocean surveying.

The team of researchers at Naval Information Warfare Center Pacific, the University of Rome and the University of Calabria, has recently demonstrated [16,31], the utility of using synchronized atomic clocks for one-way ranging which can be used for navigation updates to team of AUVs broadcasting only one acoustic message. This message can be sent by an AUV equipped with more advanced navigation sensors (FOG gyros and DVL) and can contain its x and y positions. By virtue of receiving a message from the "master" AUV, other inexpensive AUVs can update their position estimate.

Another technology that can have utmost in AUV swarms is Spiral Beacon Technology. Similar to the VOR (very high frequency (VHF) Omni Directional Radio Range) system used in aerial traffic as a beacon whose mix of two signals informs the recipients of their bearing toward the beacon, an underwater acoustic analogy was developed by Naval Research Laboratory, Washington, USA [38], as a system using a coaxial set of a spiral and a circular wave-front beacons. Using this technology together with synchronized clocks enables both bearing and range from the "master" to be known at each AUV.

These technologies not only have great potential for precise navigation of inexpensive AUVs but also significantly affect the endurance (mission time) of the AUVs as they do not need to transmit acoustic messages for navigation. Transmission of an acoustic message is much more costly compared to the reception of the message, about 40 W for transmission vs. about 40 mW for reception. Low-power onboard navigation sensors and the elimination of acoustic transmissions can enable AUVs to navigate underwater for weeks at the time and their mission time can come closer to currently employed underwater gliders so AUVs can be utilized for glider type of missions with much greater precision.

13.6 Underwater SLAM and geophysical navigation

Advanced in underwater simultaneous localization and mapping (SLAM) contribute in geophysical AUV navigation. Core component of underwater SLAM system is a dead-reckoning algorithm which includes high-frequency depth estimates, altimeter altitudes, GNSS fixes, and IMU estimates of roll, pitch, and heading are fused with actuation values (orientation of the servoed propeller and the estimated propeller revolutions per minute (RPM)) using a typical extended Kalman filter to produce an estimate of the vehicle position and uncertainty. The following papers present recent AUV navigation reviews: [41], [42], and [43]. In benign current-free conditions, with careful tuning and excellent compass calibration this procedure produced a dead-reckoning estimate with about 1–2% error per distance traveled. However as we transitioned to more realistic test locations, a more typical error was of the order of 3%. By comparison, Fiber Optic Gyroscope-based (FOG) navigation systems can achieve drift rates below 0.1% in the open ocean. SLAM-related work can be found in [3–8].

Many similarities exist between the requirements for autonomous land vehicles and those for autonomous marine systems. Many recent advances in the field of driverless cars may be applicable to autonomous surface vessels, underwater vehicles, and even aerial vehicles. Most driverless cars heavily rely on lidar (light detection and ranging) sensors, by many considered as the key self-driving car sensor, which bounces laser beams of nearby objects to create 3D maps of their surroundings. Differences in laser return times and wavelengths can then be used to make digital 3D representations of the target. It is both used for mapping and perception of the obstacles.

Similarly imaging sonars transmit sound pulses and convert the returning echoes into digital images in underwater applications of object detection and sea bottom mapping. Both lidars and sonars are superior to optical cameras as they can "see" in poor visibility conditions because they are unaffected by variations in ambient light. The resolution of these sensor correlates to the beam separation as smaller angles between beams equal higher resolution. Best lidars have 64 or 128 beams separated by an angle of 0.4° while best imaging sonars have beams separated by an angle of 0.18°. This makes them have a range resolution of 1.3 cm.

A common denominator for both driverless cars and AUVs is the need for accurate positioning. Both autonomous systems exhibit great challenges as GNSS is unavailable (underwater) or severely degraded as in urban canyon area with high buildings.

Thus, both driverless cars and underwater mapping systems must rely on onboard sensors. IMUs are common to both while DVLs (underwater) and wheel encoders (driverless cars) are used for dead-reckoning. These sensors together with GNSS fixes when available and, in some cases, map matching when previously acquired sonar/lidar data are available are used for producing valuable data products.

13.7 Detection potential of multiple AUV sensor network

Lightweight AUV towed arrays are being utilized in the development of littoral autonomous sensing networks for applications such anti-submarine warfare, marine mammals, or ambient noise measurements [9]. Although robots are being put to work to tow such systems, the ideas how such arrays are employed are similar as in cases when arrays are towed by ships or submarines. Fully autonomous launch and recovery of such systems is not feasible. Multiple low-cost AUVs carrying the low-cost sensors (acoustic, magnetic, etc.) can be autonomously deployed and recovered with other unmanned systems, for example, ASVs; therefore, array bending is prevented (sensors fixed to the hull of AUVs) and left-right ambiguity can overcome as multiple line arrays are utilized in the detection process.

Nonmilitary- or industry-driven (oil and gas, for example) research (such as marine mammal, environmental) is often conducted on restricted financial budgets by academic institutions from boats or ships of opportunity towing sensor arrays often comprising only limited number of elements. Cooperation of multiple low-cost AUVs carrying the low-cost sensors (acoustic, magnetic, etc.) can change this as the systems with comparable detection performance to long wired arrays but much lower cost and manning requirement can be constructed. The WiMUST [40] system will be composed of many AUVs carrying hydrophones to acquire sub-bottom profiling acoustic data. Instead of traditional arrays of sensors that are either stationary or towed by the ship forming the fixed geometry of the acoustic antenna, new and exciting technology advances that will be integrated into WiMUST will allow for rapid, on-the-fly, changes to the geometry of the acoustic antenna. This has the potential to drastically improve ocean surveying.

In [44], passive sonar technology was applied to detect and to track vocalizing animals when they were submerged and therefore not visible. Sperm whales are known to emit intense clicks at regular intervals for vocalizations which include directional clicks, from less than 100 Hz to 30 kHz. The majority of energy in these clicks is in the 5–25 kHz range. Sperm whales can cruise at 2 m/s; thus, they can be not only detected but also tracked by the group of micro AUVs (such as Riptide AUVs) that typically travel from 0.5 to 4 m/s.

Cooperation of multiple vehicles has received more attention in recent years as group performances can possibly lead to higher efficiency, robustness, and greater aperture. As opposed to centralized formation tracking control, distributed formation tracking control utilizes information of vehicle's own states and that from its neighbors. We will consider delayed communication for the proposed distributed controllers, estimate parameter uncertainties by adaptive control methods, and design a dynamics-based controller with the aid of kinematic-based controller and backstepping methods [31]. For our AUVs presented in Section 13.8.2, the new distributed estimation algorithms are proposed.

The overall goal is to have a group of AUVs, named the "followers," that can be commanded by an ASV or another AUV, named the "master." This allows for a team of underwater vehicles to remain in place while messages, data, and GNSS signals are gathered by the one master vehicle and transmitted to the others. The master vehicle can also serve as the decision-making hub for the team of robots. With localization techniques applied across several vehicles, the known position of each system becomes more accurate, allowing for improved path planning for the desired missions.

Upon receiving a command, the master vehicle decodes the message and determines the desired mission to execute. Many missions can be loaded onto the vehicle, so the received command could specify a position change, deployment to scout an area, or request for data. If necessary, the master vehicle then relays instructions to the followers so they can drive or reposition as desired.

The main advantage of this setup is that the followers can remain in a resting position while waiting for a command and save battery power. If the master vehicle is an ASV, it can either track the UUVs from the surface or periodically pass by the AUVs' vicinity and relay information via acoustic modems. This allows the ASV to patrol an area and identify a target location for AUVs' missions, and adds the capability for the ASV to be deployed as necessary from a dock or larger ship while the network of UUVs remains hidden in place. If the master is an AUV, it could also pre-scout a desired area while remaining more inconspicuous than an ASV. As the master AUV still faces the power restraints of the followers, the role of "master" could be passed between the vehicles to optimize for battery lifespan and not jeopardize losing the mothership controller. A wireless, "virtual" array consisting of mobile AUVs could utilize the sensor detection or acoustic command to change their inter-distances to optimize the detection performance against a specific parameter (frequency, target signal strength, environment, bottom type, direction of target travel), to switch from a horizontal to a vertical array (both 2D) or mix of the two (3D), or to adapt the orientation of the array to maximize resolution. This allows the wireless sensor network to automatically control the sensor detection parameters. Moreover, failures in single sensors that can be replaced by new ones, or substituted by existing ones without compromise the general mission objective. As opposed to the static array design where the number of sensing elements is determined a priori, the employment of the network of AUVs can allow for new sensors to joint and multiple arrays to be formed on the fly.

Every sensor able to communicate with the array can contribute to the overall system performance through the fusion of data and information from the others, increasing both the system and individual node situational awareness. For instance, some sensors can perform explorative missions and then rejoin the coordinated array to share the acquired information opening up the concept of an autonomous, active perception, reconfigurable networked monitoring system. Distributed arrays are crucial when the signal of interest is characterized by longer wavelengths where a single cabled array would have unrealistic length, would require a large platform for deployment, or would be prohibitively expensive. Large wavelength sensing is particularly appropriate for the concept as, at lower frequencies, performance is less sensitive to errors introduced by timing and positioning uncertainties. For applications where TDoA techniques (cross-correlation) are instead envisioned, the array sparsity allows a significantly increase in the resolution of the direction of arrival even when the wavelength is much smaller than the sensor spacing, given sufficient signal bandwidth.

13.8 Example: modular micro AUVs and communication

The methods presented in this example utilize advances in CSACs for synchronizing AUV and sensor timing. For long duration missions where even the drift of CSACs can affect cooperative sensor network performance or when the use of CSACs is prohibitive, alternative forms of clock synchronization based on ambient field detection could be used with a low-drift clock crystal.

Each AUV is assumed to be equipped with a MEMS IMU and navigation software. The goal is to utilize only low-power, low-cost, and small-size sensors together with innovative software solutions to replace power-hungry, higher-cost, larger- size, active traditional sensor networks. Next generation AUVs equipped with these components can then work together with larger more advanced AUVs or ASVs or manned surface and subsurface vessels equipped with much higher-grade navigation sensors. It is this cooperation that will allow for fast coverage rates and reduced time in spatial and temporal ocean sampling and explorations. In addition, recent advances with multi-AUV navigation seem promising. These methods are discussed herein.

Stationary bottom-deployed sensor arrays and ship or AUV towed arrays have been investigated for many decades. Significant advancements to improve sensitivity, low self-noise, and response across a wide range of frequencies have been made to stationary and mobile (ship or AUV) sensor arrays. Although sensor arrays towed by AUVs offer the requirement of deployment infrastructure (less manning, smaller ships), the fully autonomous deployment for such systems is a challenge because the sensor network is decoupled from the AUV. It cannot be guaranteed that the sensor array will not be damaged during autonomous launch and recovery. In contrast to a traditional wired acoustic array (stationary or towed by surface or underwater platforms), a "virtual" array can consist of multiple, untethered, detection sensor-equipped AUVs maintaining desired distance (based on a target signature desired to detect) using the acoustic communication links. Because of the adaptivity of its physical configuration, a virtual array can be both steered in angle and tuned via element spacing—to optimize response. Acoustic communication does not affect the sensor performance as [45] and [46] show that the acoustic modem-induced noise is outside the detection band of a target (ship) passed by the stationary sensor and does not inhibit the magnetometer's detection capabilities.

Our first prototype of multiple low-cost AUVs carrying the low-cost sensors that can be autonomously deployed and recovered with other unmanned systems, for example, ASVs, uses Riptide Autonomous Solutions micro AUVs. The first sensors being integrated are two types of fluxgate magnetometers, commercial from the shelf Bartington surveillance (MAG649-100, MAG634, and MAG651) and Naval Information Warfare Center Pacific single and couple core magnetometers [47]. In addition, a 16-mm KraitArray miniaturized acoustic array suitable for low-speed towed or static applications will be placed on the top of each AUV. All these are designed for power, low-cost, compact sensors for passive surveillance and battery-powered applications and will represent multi-modality sensor network.

13.8.1 Three Riptide micro AUVs

Naval Information Warfare Center Pacific has three Riptide Autonomous Solutions micro AUVs. Two of them are shown in Figure 13.7. Micro AUV is a new, highly flexible, open-source autonomous undersea vehicle that provides users a state-of-the-art, low-cost development solution ideally suited for developers of autonomy and behaviors, power systems, subsea sensors, and new payloads. The micro AUV features open hardware and software interfaces to provide users a reliable and robust platform to advance technology development. The vehicle design is optimized for high efficiency with the best hydrodynamic signature in its class. A SeaModem with CSAC [30] and a MEMS IMU will be added to the micro AUVs for the development of Distributed Localization and Cooperative Control and underwater networking algorithms. Endurance of these compact (4.875-inch diameter) AUVs is projected to be 30 h with alkaline, 48 h with lithium ion (Rechargeable), 144 h (6 days) with lithium primary, and 400 h (over 2 weeks) with aluminum seawater battery (available in 2016) batteries. Thus, at 3-knot speed, they can travel 90, 144, 432, and 1,200 nautical miles, respectively.

13.8.2 Modular design for customizable AUVs

Work is being done to transform the Riptide micro AUVs into a modular design, where sensor and thruster sections can be added or removed to change the desired functionality. The current AUV consists of three watertight sections: a nose section holding the onboard Linux computer, a central battery holder, and a tail that controls velocity with fins and a forward thruster. Sealing caps have been designed for these sections, allowing flooded payload pieces to be added on either side of the battery section. New sections can be 3D printed and coated with resin to create unusual waterproof geometries that accommodate any additional hardware. These sections can have normalized weight ratios in order to preserve balance around the center of the AUV. In the render below, one can see that two thrusters, an acoustic modem transducer, two magnetometers, hydrophone array, and landing skids have been added to the original design, and can be rearranged as desired.

The AUV runs on Mission Oriented Operating Suite (MOOS) [48], a common marine vehicle operating system developed by the MIT comparable to the Robot Operating System (ROS). A central mission is loaded onto the AUV, which activates individual MOOS programs designed for each module. This allows for modularity within the software; when adding a new physical module, the user only needs to program the mission to call the corresponding pre-written software package.

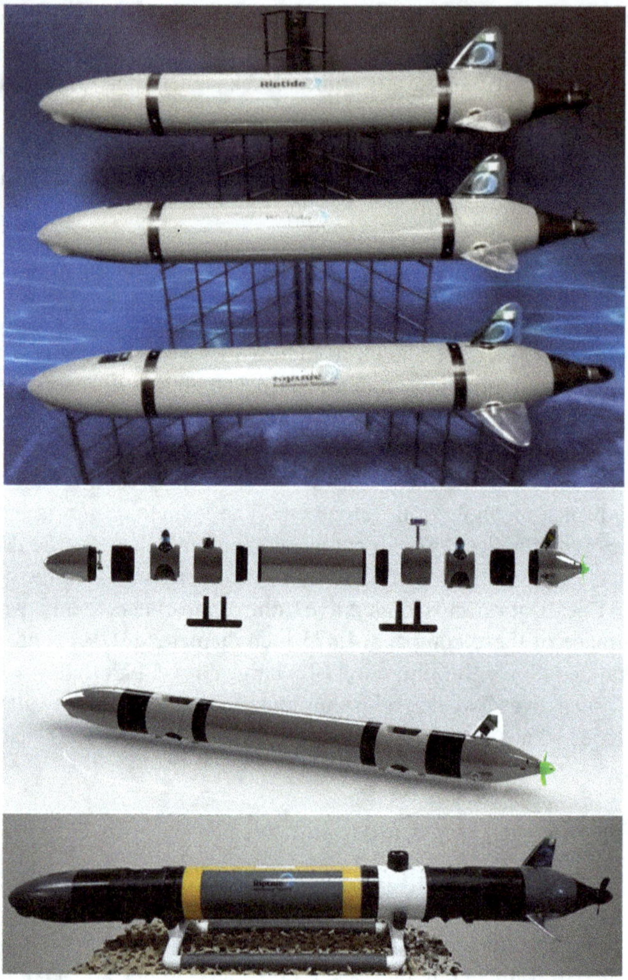

Figure 13.7 Three Riptide micro AUVs (top), modular components broken up (second from the top), thruster modules only (second from the bottom), modified-modular-more capable micro AUV (bottom)

The addition of the thrusters is a key step for enhancing long-term AUV operation. In order to be stationed at the bottom of the ocean, the vehicle needs a reliable method of settling on and ascending from the ocean floor. The current design can only alter vertical position by angling the fins and driving forward, thus making quick depth changes within a small area unachievable. With thruster modules that add force in the vertical plane, the AUV is given greater position precision and reduces wasted movement. Additional skid modules can be mounted to the thruster, lowering the probability of the AUV becoming buried in sediment or being rotated.

A second immediate application with the modularity is to add a section with a magnetometer, with the goal of improving detection performance by transmitting data between multiple AUVs through acoustic modems. Three AUVs positioned in a 2D or 3D space are key enablers for reconfigurable, adaptable, multi-modality, mobile, wireless, underwater sensor network. This modular design further allows users to create their own sections to incorporate hardware not originally intended for the Riptide AUV. In the future, an open-source catalog of low-cost module designs could share knowledge between users, allowing people to customize their AUV to use one vehicle for a variety of applications.

13.8.3 Simulation

The vertical thrusters are controlled from a BeagleBone Black (BBB) Linux computer and an Arduino microprocessor. The BBB runs Mission Oriented Operating Suite—Interval Programming (MOOS-IvP), which is a common operating system for marine vehicles developed by Oxford University and the MIT. MOOS runs a central database, from which individual applications can interact with the stored variables. To control the vertical thrusters, a MOOS application was written that watches for a variable assigned to vertical thrust percentage. When that variable is updated by the helm, the application sends a command through an I2C communications bus to an Arduino, from where the message is decoded and converted to a pulse width modulation microseconds command. The Arduino interfaces directly with a BlueRobotics T200 thruster in a module on the AUV. The MOOS application itself can be easily expanded to control several modules by replicating a few lines of code and adding an identifier corresponding to each thruster.

Screenshots shown in Figure 13.8 illustrate a simulated mission of the described behavior utilizing the existing structure that lacks the vertical structures. Without independent vertical controllers in the vertical plane, the vehicle must travel forward to rise and settle. The simulation was written for MOOS-IvP's pMarineViewer simulator. Figure 13.8 (top left) demonstrates the first stage of the AUV behavior, where the vehicle has just begun the mission and is descending to the ocean floor. The vehicle is traveling in a straight line along the positive x-axis, and has been programmed to enter the PARK mode upon reaching the precoded depth of 20 m. When in deployment, the AUV would ideally measure its height from the bottom by using sensor data and then enter PARK, but 20 m was selected as an arbitrary point for this simulation. Figure 13.8 (top right) shows the next stage of the behavior, where the AUV has entered PARK mode and is waiting for a command. At this stage, the thruster and fins are turned off and the machine enters a low-power mode, but the computer remains listening for an external stimulus to begin the next phase. In Figure 13.8 (bottom left), the AUV is rising from the ocean floor and traveling to its defined mission to begin surveying. This simulation awakened the AUV by poking the database with a specific variable, which is the equivalent of a sensor reporting that a new command has been received. The octagon shape has been pre-defined for this variable, and other variables can correspond to different sets of paths. The last stage of the behavior is shown in Figure 13.8 (bottom right), where the AUV is traversing the defined

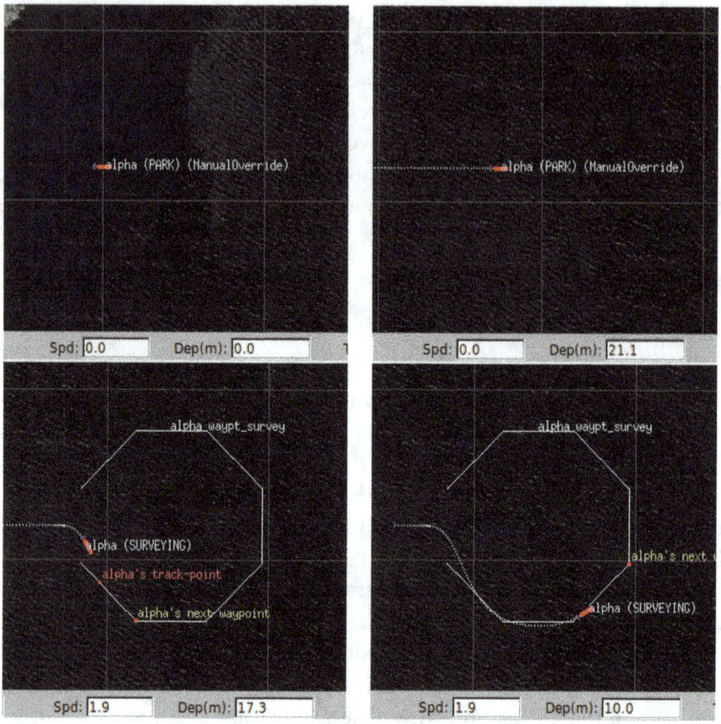

Figure 13.8 AUV descending to the ocean floor (top left); the AUV has entered PARK mode and is waiting for a command (top right); the AUV is rising from the ocean floor and traveling to its defined mission to begin surveying (bottom left); and the AUV is traversing the defined mission upon receiving a command or target detection input (bottom right)

mission upon receiving a command or target detection input (to be implemented in future). Upon completion, the machine can be programmed to restart the cycle and return to the ocean floor, or may be given new other instructions based on the mission requirements.

13.9 Conclusion

This work details the combination of cooperative AUVs, accurate timing and clock synchronization, underwater sensors, and acoustic communication for the purposes of detection of various underwater targets for seismic surveys, surveillance, and marine mammals detection applications. Future work will include sensor fusion of passive sensor array of sensors of same kind or passive sensor array of sensors of different kind

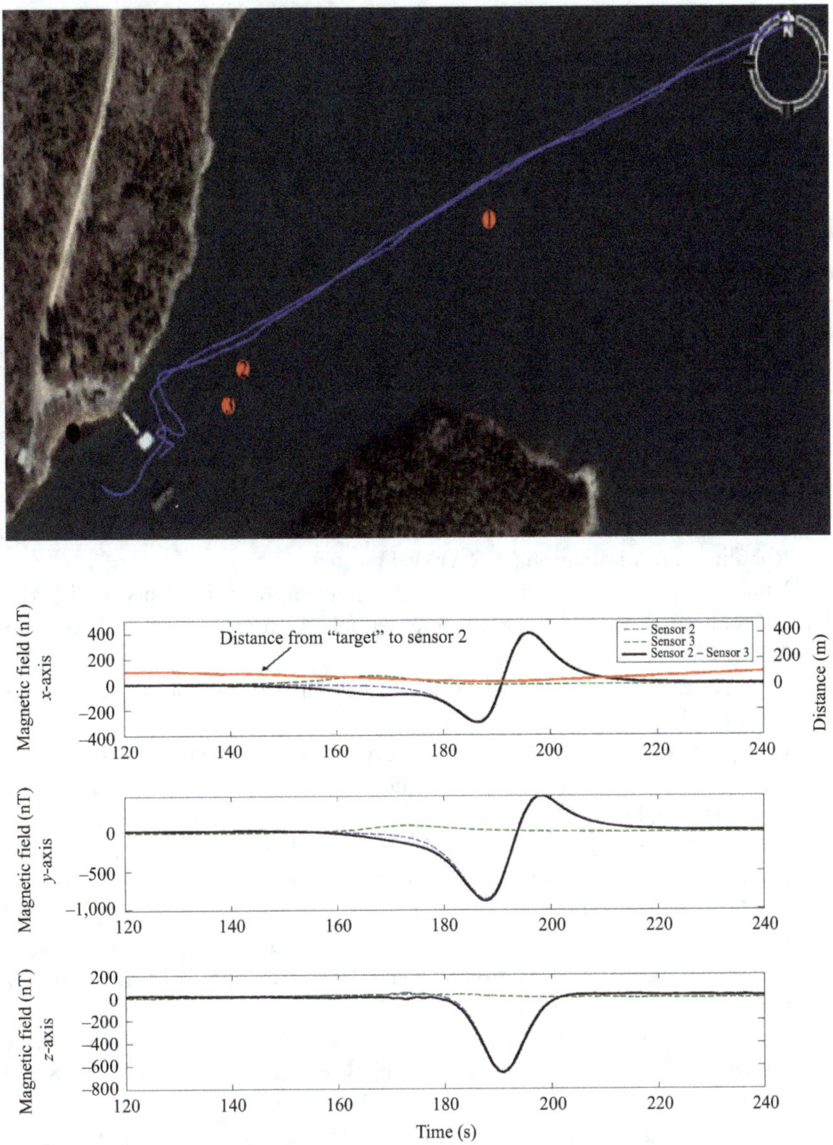

Figure 13.9 Experimental setup with three stationary magnetometer sensors (in red) shown in the top figure. Detection result in fusing the detection from two sensors (nodes 2 and 3)

(magnetic and acoustic) similar to an underwater monitoring systems by introducing autonomous triggering based on acoustic presence indicators of marine species [49]. Data in Figure 13.9 (bottom) show that the fusion of two sensors can improve the detection performance.

Acknowledgments

This work has been partially supported by Office of Naval Research program's Center for Innovative Naval Design (CINT) and Naval Research Enterprise Internship Program (NREIP). We would like to thank Mr. Seth Wijnhamer and Dr. Anthony Jones for the design and manufacturing of AUV modular components.

References

[1] Kebkal KG, Mashoshin AI. AUV acoustic positioning methods. Gyroscopy and Navigation. 2017;8(1):80–89.
[2] Leonard JJ, Bennett AA, Smith CM, *et al.* Autonomous underwater vehicle navigation. In: MIT Marine Robotics Laboratory Technical Memorandum; 1998.
[3] Fallon MF, Kaess M, Johnnsson H, *et al.* Efficient AUV navigation fusing acoustic ranging and side-scan sonar. In: Proceedings of IEEE Intl. Conf. on Robotics and Automation (ICRA); 2011.
[4] Paull L, Huang G, Seto ML, *et al.* Communication-constrained multi-AUV cooperative SLAM. In: IEEE International Conference on Robotics and Automation (ICRA); 2015. pp. 509–516.
[5] Bahr A, Leonard JJ. Cooperative localization for autonomous underwater vehicles; 2008. pp. 387–395.
[6] Paull L, Seto ML, Leonard JJ, *et al.* Probabilistic cooperative mobile robot area coverage and its application to autonomous seabed mapping. I J Robotics Res. 2018;37(1):21–45.
[7] Fallon MF, Folkesson J, Mcclell H, *et al.* Relocating underwater features autonomously using sonar-based SLAM. IEEE Journal of Oceanic Engineering. 2013;38:500–513.
[8] Fallon MF, Papadopoulos G, Leonard JJ. A measurement distribution framework for cooperative navigation using multiple AUVs. In: IEEE Intl. Conf. on Robotics and Automation (ICRA); 2010. pp. 4803–4808.
[9] Maguer A, Dymond R, Mazzi M, *et al.* SLITA: A new slim towed array for AUV applications. Acoustical Society of America Journal. 2008;123: 3005.
[10] Canepa G, Tesei A, Troiano L, *et al.* Comparison of computation time and accuracy of the real-time implementation of two beamforming algorithms. In: 4th Underwater Acoustics Conference and Exhibition on Underwater Acoustics (UACE 2017); 2017.
[11] Djapic V, Fiscko C, Bulsara A, *et al.* Reconfigurable, adaptable, multimodality, mobile, wireless, underwater sensor network. In: Proceedings of IEEE Sensors Applications Symposium (SAS). Glassboro, USA; 2017.
[12] Curtin TB, Bellingham JG, Catipovic J. Autonomous oceanographic sampling networks. Oceanography. 1993;6(3):86–94.

[13] VERTEX. VERTEX AUV; 2018. http://www.hydromea.com/.
[14] Microsemi. SA.45s CSAC—Chip Scale Atomic Clock; 2015. http://www.microsemi.com/document-portal/docdownload/133305-quantum-sa-45s-csac.
[15] Olson E, Leonard J, Teller S. Robust range-only beacon localization. In: Proceedings of Autonomous Underwater Vehicles; 2004. pp. 66–75.
[16] Cario G, Casavola A, Djapic V, *et al.* Clock synchronization and ranging estimation for control and cooperation of multiple UUVs. In: Proceedings of MTS/IEEE OCEANS 2016. Shanghai, China; 2016. pp. 1–9.
[17] Djapic V, Nad D, Ferri G, *et al.* Novel method for underwater navigation aiding using a companion underwater robot as a guiding platforms. In: Proceedings of MTS/IEEE OCEANS 2013. Bergen, Norway; 2013. pp. 1–10.
[18] Djapic V, Dong W, Jones A, *et al.* Advanced underwater acoustic networking and cooperation of multiple marine robots. In: Proceedings of MTS/IEEE OCEANS 2015. Genova, Italy; 2015. pp. 1–9.
[19] Djapic V, Dong W, Stipanov M. Decentralized cooperative localization and control with non-GPS clock synchronization and one-way messaging. In: 3rd Underwater Acoustics Conference and Exhibition on Underwater Acoustics (UACE 2015). IEEE; 2015.
[20] Djapic V, Gravelle E, Ouimet M, *et al.* Cooperative navigation of low-cost autonomous underwater vehicles. In: 4th Underwater Acoustics Conference and Exhibition on Underwater Acoustics (UACE 2017); 2017.
[21] Petrioli C, Petroccia R, Potter JR, *et al.* The SUNSET framework for simulation, emulation and at-sea testing of underwater wireless sensor networks. Ad Hoc Networks. 2015;34:224–238.
[22] Yi J, Mirza D, Schurgers C, *et al.* Joint time synchronization and tracking for mobile underwater systems. In: WUWNet. ACM; 2013. pp. 38:1–38:8.
[23] REMUSM3V. REMUS M3V (Micro 300 Meter Rated Vehicle); 2018. https://www.hydroid.com/REMUS-M3V.
[24] RIPTIDE. RIPTIDE Micro-UUV; 2018. https://riptideas.com/micro-uuv/.
[25] SandShark. SandShark AUV; 2018. https://gdmissionsystems.com/products/underwatervehicles/bluefin-sandshark-autonomous-underwater-vehicle.
[26] Rypkema NR, Fischell EM, Schmidt HR. Closed-loop single-beacon passive acoustic navigation for low-cost autonomous underwater vehicles. In: IROS. IEEE; 2018. pp. 641–648.
[27] Fenucci D, Munaf A, Phillips AB, *et al.* Development of smart networks for navigation in dynamic underwater environments. In: 2018 IEEE/OES Autonomous Underwater Vehicle Workshop (AUV). IEEE/OES; 2018.
[28] Neasham JA, Goodfellow G, Sharphouse R. Development of the Seatrac miniature acoustic modem and USBL positioning units for subsea robotics and diver applications. In: OCEANS 2015 – Genova. IEEE/OES; 2015.
[29] Eustice RM, Whitcomb LL, Singh H, *et al.* Experimental results in synchronous-clock one-way-travel-time acoustic navigation for autonomous underwater vehicles. In: 2007 IEEE International Conference on Robotics and Automation; 2007.

[30] Cario G, Casavola A, Lupia M, *et al.* SeaModem: A low-cost underwater acoustic modem for shallow water communication. In: OCEANS 2015 - Genova. IEEE/OES; 2015.

[31] Djapic V, Dong W, Spaccini D, *et al.* Cooperation of coordinated teams of autonomous underwater vehicles. In: Proceedings of the 9th IFAC Symposium on Intelligent Autonomous Vehicles 2016. Messe Leipzig, Germany; 2016. pp. 1–6.

[32] McCoy K. JANUS: From primitive signal to orthodox networks. In: Underwater Acoustic Measurements: Technologies and Results. Nafplion, Greece; 2005.

[33] Petroccia R, Cario G, Lupia M, *et al.* First in-field experiments with a "bilingual" underwater acoustic modem supporting the JANUS standard. In: Proceedings of MTS/IEEE OCEANS 2015. Genova, Italy; 2015.

[34] Lutwak R, Emmons D, English T, *et al.* The chip-scale atomic clock—Recent development progress. In: Proc. 34th Annual Precise Time and Time Interval Systems and Applications Meeting, San Diego, CA. IEEE; 2003.

[35] Rowe A, Gupta V, Rajkumar R. Low-power clock synchronization using electromagnetic energy radiating from AC power lines. In: SenSys. ACM; 2009. pp. 211–224.

[36] Tomic M, Hughes MR. Network clock synchronization using total-field magnetometers and the spatial coherence of the Earth's geomagnetic field. Active; 2015. US Patent No. 8954608.

[37] Weinstein E, Weiss A. Fundamental limitations in passive time-delay estimation—Part II: Wide-band systems. IEEE Transactions on Acoustics, Speech, and Signal Processing. 1984;32(5).

[38] Dzikowicz BR, Hefner BT, Leasko RA. Underwater acoustic navigation using a beacon with a spiral wave front. IEEE Journal of Oceanic Engineering. 2015;40(1):177–186.

[39] Miller PA, Farrell JA, Zhao Y, *et al.* Autonomous underwater vehicle navigation. IEEE Journal of Oceanic Engineering. 2010;35(3):663–678.

[40] Al-Khatib H, Antonelli G, Caffaz A, *et al.* The widely scalable Mobile Underwater Sonar Technology (WiMUST) project: An overview. In: OCEANS 2015—Genova. IEEE; 2015.

[41] Kinsey JC, Eustice RM, Whitcomb LL. Underwater vehicle navigation: Recent advances and new challenges. In: IFAC Conf. on Manoeuvring and Control of Marine Craft. Lisbon, Portugal; 2006.

[42] Paull L, Saeedi S, Seto M, *et al.* AUV navigation and localization: A review. IEEE Journal of Oceanic Engineering. 2014;39(1):131–149. Available from: https://doi.org/10.1109%2Fjoe.2013.2278891.

[43] Stutters L, Liu H, Tiltman C, *et al.* Navigation technologies for autonomous underwater vehicles. IEEE Transactions on Systems, Man, and Cybernetics, Part C (Applications and Reviews). 2008;38:581–589.

[44] Zimmer WMX. Three-dimensional beam pattern of regular sperm whale clicks confirms bent-horn hypothesis. The Journal of the Acoustical Society of America. 2005;117(3):1473–1485.

[45] Tomic M, Sullivan PT, McDonald VK. Wireless, acoustically linked, undersea, magnetometer sensor network. In: OCEANS 2009—Biloxi. IEEE; 2009.

[46] Hughes M, Tomic M. UXO survey using vector magnetic gradiometer on autonomous underwater vehicle. In: OCEANS 2010 MTS/IEEE SEATTLE. IEEE; 2010.

[47] Ando B, Baglio S, Bulsara AR, *et al.* Residence times difference fluxgate magnetometers. IEEE Sensors Journal. 2005;5:895904.

[48] Benjamin MR, Schmidt H, Newman PM, *et al.* Nested autonomy for unmanned marine vehicles with MOOS-IvP. J Field Robotics. 2010;27(6):834–875.

[49] Wilby A, Slattery E, Hostler A, *et al.* Autonomous acoustic trigger for distributed underwater visual monitoring systems. In: WUWNet. ACM; 2016. p. 10.

Part V

Methodology

Chapter 14

Task specification and behavior verification for UUV behavior design

Signe A. Redfield[1], Elizabeth I. Leonard[1] and Jamie Lennon[1]

14.1 Introduction

The robotics community is deeply interested in both platform design and behavior design, but we lack tools to connect the two. The platform, the behavior design, and the environment work together to determine the robot's actions, but our tools visualize the design of the hardware and the design of the behaviors separately. We lack tools that allow us to visualize the relationship between the platform and the behavior. To address this gap, we introduce a new design method based on a tabular representation called Capability Analysis Tables.

The Capability Analysis Table enables the designer to define the constraints on a behavior design based on the platform it will be used on, and to define the constraints on the platform based on the behavior design. It gives the customer and the designer an opportunity to more clearly specify the desired behavior. Environmental factors are implicit in the platform interface definitions—sensory perception filters environmental inputs (colored objects can only be seen by sensors that produce color information) and actuators filter environmental outputs (wheeled robots typically do not provide depth control). Communications are handled explicitly as outputs (e.g. "send message") and as either operator inputs (e.g. "new task") or remotely gathered sensory data (e.g. "other vehicle pose").

Note that this design method can be used for autonomous and unmanned vehicles—the principles are the same.

There are two use cases for Capability Analysis Tables. In the first case, the designer can use the platform capabilities and knowledge of the job to be done as inputs to the capability analysis. The output will be a specification for the behaviors needed to accomplish that job using that platform. This is shown in Figure 14.1(a).

Alternatively, the designer may use the behavior design to specify a platform. In this case, the result of the analysis is a specification for the information the platform

[1]Naval Research Laboratory, Washington, DC, USA

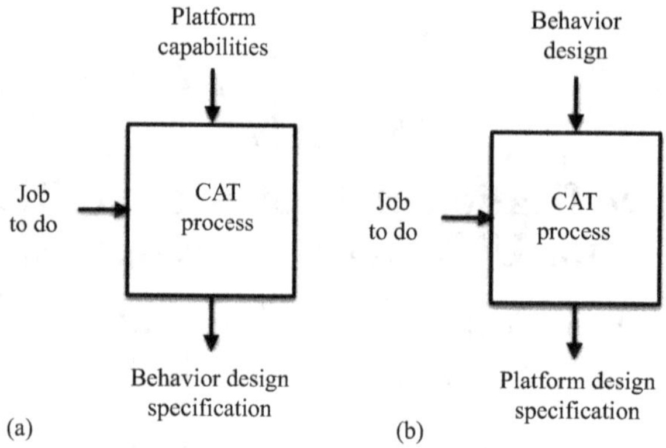

Figure 14.1 Uses of the Capability Analysis Table process: (a) behavior specification development process; (b) platform specification development process

must be capable of providing to the behaviors and a specification for the control options it must offer. Figure 14.1(b) shows this use of the Capability Analysis Table process.

In both cases, designers are already doing this work and are generating these specifications, but generally in an *ad hoc* manner. This table is a new representation that provides the ability to formalize the process and to better document the resulting specification, as well as to support later verification efforts.

14.2 Background

14.2.1 Notional platform

Throughout this chapter, we use an autonomous underwater vehicle as our notional platform. We assume that the vehicle in question has a sensor that enables it to locate items of interest to the operator and provides both the item's position registered on a shared map and the vehicle's confidence in its identification of the item. The vehicle is capable of moving through its environment and knows both approximately how accurately it is maneuvering and how accurately it is able to represent the position of the sensed item of interest relative to itself. It can communicate this information within a shared reference frame, and it can receive similar information from other team members and represent it on the shared map. It can keep track of time and receive task-specific messages relating to the region to be surveyed, the amount of time available to perform the survey, and which subregions within the larger region have or have not been surveyed. It can keep track of its battery status.

14.2.2 Platform specification

Platform specification is typically performed in terms of hardware, information flow, and default functions—if the platform is specified *a priori*, this information will be available at the outset of the process. If the platform is intended to be selected once the behavior design is mature, the important aspects of the platform will be defined as the behavior design matures. Generally, a designer will at least define some assumptions about the platform capabilities to guide the design process.

Platform design tools are tightly focused on hardware interactions (the physical and electronic designs, typically CAD and circuit design tools) and on software functions and interactions (including dependency graphs and process diagrams) which provide functionality at the platform level rather than at the behavioral or capability level. In general, behaviors in the robotics sense of "physically embodied actions taken in the world in response to stimulus" are not represented.

14.2.3 Behavior representation

Behavior design tools are generally tightly focused on the interactions between behaviors and the sequencing of behaviors. Their primary focus is to enable the designer to clearly express what causes a behavior to start and stop, and what behaviors need to be precursors for what other behaviors, and when. They can support hierarchical levels of abstraction if necessary, but capture only a subset of the interactions of the system. The behaviors are typically represented as abstractions within a theoretical structure, independent of platform and with the environment abstracted to consider only salient features relevant to the specific behaviors. The supporting variables and platform capabilities are abstracted away in these representations. For the purposes of this chapter, we will use the capability analysis to design the behaviors resident on our example platform.

14.2.3.1 Instantaneous and temporal visualization tools

Instantaneous representations capture the triggers that cause behaviors to start and stop and organize the mechanisms that drive the robot's actions. Precursors and successors are defined by their connections rather than by properties inherent to the components themselves. These representations include state-based visualization tools [1,2] such as topological task graphs [3] and UML diagrams [4]. State-based tools are also used for representation of Markov decision processes [5]. Skill trees [6] and behavior trees [7] are used in the learning community to keep track of the various capabilities possessed by the robot and their relationships to each other.

In these tools, behaviors are represented as atomic structures while their precursors and successors are defined by the connections between the behaviors and the situations and triggers that cause the transitions.

Autonomy architectures use similar tools to visualize the relationships between the components of the system [8–11].

Planner and scheduler visualization tools provide more insight into the sequential and temporal aspects of behavior interaction, but less insight into the relationships between individual behaviors themselves. These tools typically provide a notation for representing the actors in a given process along one axis and time along the other, with the behaviors represented as blocks or notes along the time axis as they are activated.

In T-REX [12], the *y*-axis of the display enables visualization of hierarchical relationships between plans and subplans, with higher-level plans represented by bars in higher-level rows at the top and lower-level actions represented by smaller bars at the bottom. Information transfer and replanning stages appear as arrows between the bars and as increases in the processor burden over time.

Both the instantaneous and the temporal representations are focused on helping the designer visualize the way the behaviors interact as a task is performed—how the capabilities combine to produce the desired actions. They can range from very abstract [13] to extremely detailed [4]. Although platform assumptions are often implicit in the behaviors and their interactions, the details of how a given behavior interacts with a given robot is abstracted away from these representations.

14.2.3.2 Formal behavior representations

Formal methods techniques and tools rely on precise, mathematical representations to specify system behavior. Common approaches include representing the system behavior in a state-based formalism such as Statecharts [14] or Input/Output Automata [15], in a tabular representation of a state machine such as SCR [16] or SpecTRM [17], and as a set of formulas in a logic such as Linear Temporal Logic [18] or Computation Tree Logic [19]. Because these methods have underlying sound, mathematical foundations, they can be used as the basis for formal verification using model checking [20] or theorem proving [21]. In addition, some formal methods have associated tool support for analyzing specifications for problems such as inconsistencies and incompleteness [16,22], as well code synthesis [22,23] and test case generation [22,24]. Formal methods have been applied in a wide range of safety-critical domains including avionics [25], robotics [26,27], and unmanned systems [28].

While these methods are capable of addressing design questions at the intersection of the behavior and platform properties, developing the precise mathematical representations that enable specification of the system behavior requires significant effort and expertise, as well as deep *a priori* understanding of the system as a whole. Behavior developers often lack both the expertise and the deep understanding of either the platform or the desired behavior to use these methods effectively. The process of developing a Capability Analysis Table drives the behavior designer to refine their understanding of the detailed system properties and desired behavior until it reaches a level of precision suitable for analysis with formal methods.

14.2.3.3 Simultaneous development

The real-time control system (RCS) [29] approach to system design is the most rigorously defined in the robotics community. It uses competency-based task representation with a hierarchical deconfliction mechanism to generate planned behaviors. These planned behaviors enable the identification of the information that must be represented in the world model, which in turn determines the necessary sensors and the necessary accuracy of the information provided by the sensors, which can then be turned into system specifications [5]. Once the system has been designed to these specifications, it should be capable of implementing the behaviors using the hierarchical planning mechanism to implement the desired task. While a competency-based task representation may be used with the RCS framework, it is fundamentally a sense-plan-act framework, resting on Sensory Processing, World Modeling, and Behavior Generation modules. The sensory processing and world modeling modules provide the sense portion, while the behavior generation module encapsulates both the plan and act portions. This limits the behavior designs that RCS can be applied to. RCS assumes specific system design choices and constrains the designer to a world model-based design paradigm. It cannot support reactive or subsumption-based designs and assumes that the autonomy designer has the authority and ability to drive the platform design.

14.2.4 Capability definitions

The robotics community does not currently have an accepted definition of "capability." For this chapter, we define capability as an abstraction of robot behavior, where behavior is defined in the robotics sense as "the algorithms, software modules, and/or actions of a robot in a given context (designed and observed)" [30]. However, the ontology development communities are beginning to develop frameworks to represent tasks and capabilities.

14.2.4.1 Ontologies

Ontologies are used to define less ambiguous terminology with associated semantic meaning to describe properties or functions of the system. Several ontologies already exist or are in work to describe planning [31], map information [32], and hardware [33,34]. An autonomous robot ontology [35] is also under development, but its focus is on the expression of terms related to autonomy and the ability to define the properties of an existing autonomous robot and its behaviors rather than on the design of new behaviors.

14.2.4.2 Capability taxonomy

Taxonomies are useful guides to categorize types of capabilities available to or required by a given design. Existing taxonomies can be used to categorize the capabilities that enable grasping and manipulation operations [36,37], but outside of the grasping and manipulation communities, attempts to define taxonomies often become mired in excessive detail or weighed down by too broad a focus. As ontologies mature, development of capability taxonomies in more domains should follow.

14.3 Capability Analysis Table

The Capability Analysis Table is a structure that maps available or required information to control and data outputs through specific behaviors. This can be a very complex mapping, so the table uses a structured form to track the various categories of information and outputs. The structure can be replicated at various levels of abstraction to further simplify the visualization and documentation of the design. The basic structure is shown in Figure 14.2. It consists of three basic elements:

- the columns, which identify specific information available as inputs to behaviors;
- the rows, which identify specific commands provided to the robot's control system and data generated by behaviors;
- the cells, which contain the specific behaviors used to define the desired capability and their triggers.

14.3.1 Information available

Information available to the robot (the columns of the table in Figure 14.2) is broken into four main categories. The *None* category enables representation of physical constraints and drivers of control outputs, the *Internal* category holds information generated on board the robot and information provided *a priori* by the operator. The *External* category holds all the information about the environment and the robot that is available to the behaviors in the table, including environmental and sensory information as well as communications sources. The *Sensory Input* section covers information about the world and the robot itself, including internally sensed information like proprioception, locally sensed information about the world and robot's contact with it, and nonlocally sensed information about the world. It encompasses both internal information and external information.

14.3.1.1 None

This column is reserved for the use of behaviors that rely on physical properties of the platform and its environment rather than numeric quantities. Purely automatic operations that do not rely on internal or external inputs for triggering or operation are represented in this column.

14.3.1.2 Internal information

This section holds information provided *a priori* by the operator, like thresholds and parameters, information generated by the robot for internal use, and information about the internal status of the platform.

Fixed a priori information

This section holds information that is provided *a priori* by the operator before the task starts and that does not change over the course of the task. It includes fixed assumptions about the environment or the robot's configuration. For example, this might include robot maneuverability or sensor configuration information and map structure and limits. If the robot is told that it has a minimum turning radius defined by its propulsion system, this is fixed *a priori* information that could be used in a path planning algorithm (and therefore needs to be represented in the table), but there is no information that the vehicle can modify while it is running.

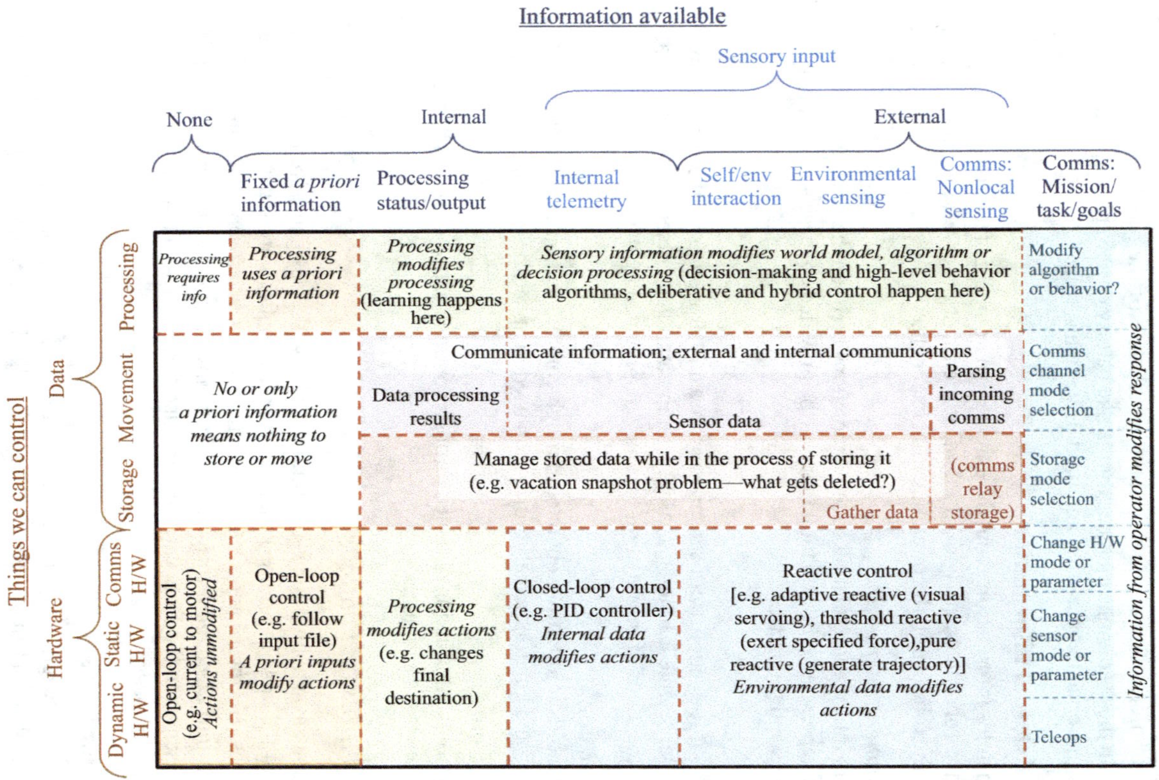

Figure 14.2 Capability Analysis Table structure overview: columns contain inputs, rows contain output and cells contain behaviors and behavior triggers

Processing status/output

This section holds information that is generated onboard via internal processing. Every column in this section must be produced in a row of the table.

Internal telemetry

This is where information relating to the internal state of the robot is represented. Information in these columns includes things like battery or power level, time, motor current draw, internal temperature, and hard drive status.

14.3.1.3 External information

This section holds information about the external environment of the robot, including things that the robot interacts with, things that the robot observes, and information provided to the robot through communication channels.

Self/environment interaction

This is where information relating to the interactions between the robot and its environment are represented. Information that may be provided by these sensors includes contact events and properties.

External sensing

This is where information from sensors that measure properties of the robot's environment are represented. Imagery, salinity, temperature, and object locations are all examples of the kinds of information that may be provided by these sensors.

Nonlocal sensing

This section holds information obtained through communication channels related to other vehicles and environmental information used to update a world model.

Mission/task/goals

This section holds information obtained through communication channels that define new or updated tasks and goals.

14.3.2 Controls and outputs

Control options available to the robot (and, by proxy, the designer) are represented in the rows of the table in Figure 14.2..These control options are separated into two core categories, *Hardware* and *Data*. The outputs of the *Hardware* rows are further broken down into *Dynamic hardware*, *Static hardware*, and *Communications*. The outputs represented in the *Data* category are also further broken down, into *Storage*, *Movement*, and *Processing*.

14.3.2.1 Hardware

This section of the diagram captures the commands and control mechanisms available to the designer that control and configure specific hardware components.

Dynamic hardware

Commands in this category control both the robot's motion through the environment and motion of components relative to the robot's body. It also includes active sensors that have moving parts, like cameras with adjustable focal lengths and active sonar.

Static hardware
Commands in this category enable control of components like heaters that can be controlled and/or configured, but do not move. It also includes passive sensors that do not have actively controlled moving parts, like microphones.

Communications
Commands in this category enable control and configuration of communications hardware. This could include mechanical signaling devices, wireless modems, and microphone/speaker combinations. If the same hardware is used for both environmental sensing and communications, the designer may choose where to represent it. These rows capture hardware configuration rather than processing of incoming or outgoing messages.

14.3.2.2 Data

This section of the diagram captures the information generated by the robot as it operates. This is where incoming messages, stored information, and intermediate variables are generated and tracked.

Data storage
Information and commands generated in this category relate to the destination of specific information, enabling the designer to capture specific data management mechanisms. If a given behavior causes specific data to be deleted or stored in a specific location, the commands that make that happen appear on these rows.

Data movement
This category captures commands and configuration options related to both externally and internally communicated information. When messages are generated to external parties like other vehicles or the operator, the commands that send those messages are in this section.

Data processing outputs
This category holds all the internally generated variables and information required to implement the capabilities tracked in the table. Information generated on these rows are replicated in the *Processing Status/Output* columns and used as inputs to other functions. The processes used to generate the data are held in the cells of the table.

14.3.3 Behaviors

The columns of Figure 14.3 capture the information provided by the example vehicle from Section 14.2.1. The rows of Figure 14.3 capture the controls provided by the example vehicle from Section 14.2.1. Now that we have defined the structure of the table, we can begin to populate the behaviors, including their triggers, their interfaces, and their constraints.

Individual behaviors fill cells in the table. A given behavior fills the cells at the intersection of the information required as input and the controls it produces as output. If a given piece of information is required to trigger the behavior to start, the behavior appears in that column with the condition that is used to trigger it. If a

Example platform	Object Info (pose, pose accuracy, ID, & confidence)		Self Info (battery status, pose, & accuracy)		Task Info (region, time limit, & Progress)		Time		Position Sensor		Object Sensor		Comms In	
	Cond	Beh	Cond	Beh	Cond	Beh	Cond	Beh	Cond	Beh	Cond	Beh	Cond	Beh
Object Info														
Self Info														
Task Info														
Message Out														
Vehicle Motion Control														

Figure 14.3 Empty Capability Analysis Table for example platform

given piece of information is required as input on an ongoing basis while the behavior runs, its condition column is left blank or marked "available." Within the column, the behavior appears in every row for which it produces output. If the behavior has multiple modes, where it requires different configurations of incoming information, each configuration is given its own row, even if the outputs are all the same. AND is represented within a single row for each output (illustrated in Figure 14.4), while OR is represented with each instance on its own row (illustrated in Figure 14.5).

Figure 14.4 shows a table with a single behavior—in this case, a perceptual algorithm that detects objects in the Object Sensor data and generates their pose and pose accuracy relative to the vehicles current position based on the Position Sensor data.

The perception algorithm is useful, but unless the vehicle is moving, there will be no objects to look at, so we add a Survey behavior to drive the vehicle. Figure 14.5 shows the Survey behavior added to the example table. In this case, the Survey behavior produces three outputs. The Survey behavior runs as long as the time limit is not reached. While it is running, it continuously generates progress updates and generates outgoing messages on an event-driven basis. The messages containing the pose and pose accuracy associated with a found target. Outgoing commands generated by the Survey behavior drive the vehicle along the desired survey path (the Vehicle Motion Control output). Each of these outputs is represented on a single row. The Object Info input column is the output of the Object Info row. We can read the behavior sequence from the table as follows: Position Sensor and Object Sensor feed the Object Detection behavior, which outputs Object Info. This output becomes an input (column header) to the Survey behavior. The Survey behavior controls vehicle motion whenever it is active, so there are no conditions on the self- and task information needed in order for the behavior to be active. However, the Survey behavior is only active before the time limit expires, so there is a condition imposed in the Time column.

As the table becomes more complex, it becomes helpful to add a column to capture notes to the designer addressing details of the various behaviors. Figure 14.6 illustrates this for our example and finishes populating the table with appropriate behaviors.

14.4 Complexity and levels of abstraction

If the goal is to provide the behavior designer with a complete picture of the tools available, the table must be complete and all available information must be represented. For complex and capable systems, this table will be very large. In practice, the designer will only work with subsets or abstractions of the table at any given time. It is useful to abstract the vehicle properties into categories of information and controls, with smaller tables capturing the details of each section. In this way, the designer can select only the relevant information for display in the main design table. The level of abstraction to use depends on the desired level of abstraction of the behavior specification or design.

Example platform	Object Info (pose, pose accuracy, ID, & confidence)		Self Info (battery status, pose, & accuracy)		Task Info (region, time limit, & progress)		Time		Position Sensor		Object Sensor		Comms In	
	Cond	Beh	Cond	Beh	Cond	Beh	Cond	Beh	Cond	Beh	Cond	Beh	Cond	Beh
Object Info										Object Detection		Object Detection		
Self Info														
Task Info														
Message Out														
Vehicle Motion Control														

Figure 14.4 Object Detection perception algorithm behavior represented in Capability Analysis Table for example platform

Example platform	Object Info (pose, pose accuracy, ID, & confidence)		Self Info (battery status, pose, & accuracy)		Task Info (region, time limit, & progress)		Time		Position Sensor		Object Sensor		Comms In	
	Cond	Beh	Cond	Beh	Cond	Beh	Cond	Beh	Cond	Beh	Cond	Beh	Cond	Beh
Object Info										Object Detection		Object Detection		
Self Info														
Task Info														
Message Out	New object pose & pose accuracy	Survey												
Vehicle Motion Control				Survey		Survey	Time < t	Survey						

Figure 14.5 Survey behavior added to Capability Analysis Table, for example, platform

Example platform	Object Info (pose, pose accuracy, ID, & confidence)		Self Info (battery status, pose, & accuracy)		Task Info (region, time limit, & progress)		Time		Position Sensor		Object Sensor		Comms In		Notes
	Cond	Beh	Cond	Beh	Cond	Beh	Cond	Beh	Cond	Beh	Cond	Beh	Cond	Beh	
Object Info										Object Detection		Object Detection			Always on
Self Info										Pose Estimator					Always on
Task Info														Task Spec	Defines Task Info from received message
						Survey									Progress updates generated by Survey behavior
Message Out	New object pose & pose accuracy	Survey													Generates message when new object detected
Vehicle Motion Control				Survey		Survey	Time< t	Survey							Generates survey trajectory within time limit

Figure 14.6 Addition of a Notes column to the Capability Analysis Table

Braitenberg vehicle 2b	Left illumination sensor (LS)		Right illumination sensor (RS)		Notes
	Cond	Beh	Cond	Beh	
Right motor (RM)		RM = a * LS + b			Right motor turns faster when left sensor value is higher
Left motor (LM)				LM = a * RS + b	Left motor turns faster when right sensor value is higher

Figure 14.7 Low-level Capability Analysis Table for Braitenberg vehicle 2b [38]

At a low level of abstraction, the vehicle properties involve the output of specific hardware: raw sensor data that need to be run through perceptual processing, commands that directly control motors, and low-level information available through whatever software is running on the robot. A small table assembled at a low level of abstraction is shown in Figure 14.7, capturing a simple Braitenberg vehicle 2b [38] that approaches a light source. In this figure, the equations that relate the sensory input to the motor output are used in place of behavior names and the Notes column contains a human-readable explanation of the effect of the equations on the vehicle's motion.

In contrast, Figure 14.8 shows a table defining desired functionality at a high level of abstraction. In this example, the shoreside is accepting abstract information from the vehicles performing the survey and is further abstracting that information. Instead of defining the details of the information required to support the Survey Replan activity, abstractions called "Subtask Progress" and "Map" and "Region Allocation" are used.

Similarly, if a vehicle provides behaviors as part of its typical complement of capabilities available to the behavior designer, those capabilities may be represented in two ways. Either the designer can represent the provided behaviors as control outputs or the designer can explicitly include them in the behavior design as behaviors in the cells of the table. If the vehicle already has a suite of maneuvering and manipulation commands, the behavior outputs might be vehicle-provided behaviors like "go to waypoint" or "pick up cup" instead of thruster or motor commands.

14.5 Representing a behavior design

In Section 14.3.3, we built up a Capability Analysis Table from a vehicle, using the vehicle properties to constrain and specify a behavior design that could produce the desired effect. However, it is also possible to build a table from a desired behavior and use it to specify the required vehicle capabilities. Instead of starting with a platform specification, we can start with an initial behavior specification.

Survey: shoreside	Map		Region Allocations		Object List		Time		Comms In: Subtask Progress		Notes
	Cond	Beh	Cond	Beh	Cond	Beh	Cond	Beh	Cond	Beh	
Map						Map Update					Update map when Object List updated
Target List									New target	Update Target List	Update Object List when new object found
Region Allocations		Survey Replan		Survey Replan					Progress= subtask done	Survey Replan	Replanning requires the current Map and Region Allocation; triggered by news that vehicle has completed a region
Comms Out: status update to operator						Status Update Generation				Status Update Generation	Status update includes progress and object list status
Comms Out: allocate survey subtask to vehicle			New allocation	New Allocation Message Generation							Once Replan has generated a new region allocation, the allocation is transmitted to the vehicle
Comms Out: home							Time> t	Send homing command			Finish when time limit expires
									All subtasks complete	Send homing command	Finish when survey complete

Figure 14.8 High-level Capability Analysis Table for shoreside component of multi-vehicle Survey task

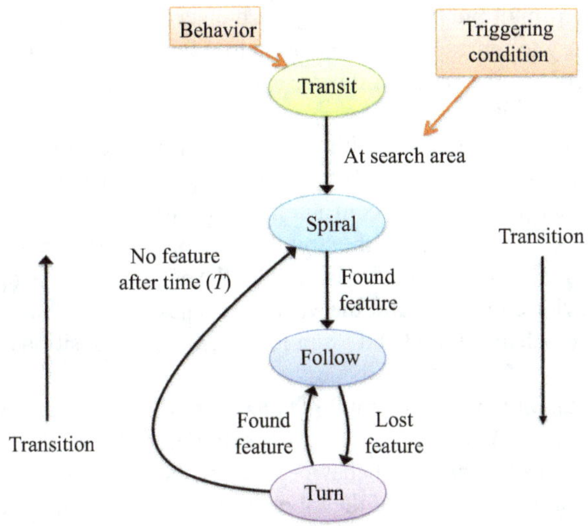

Figure 14.9 State machine diagram for feature-following behavior, adapted from [39]

Adaptive Feature Detection	Position on Map		Time		Feature Status		Task Status		Notes
	Cond	Beh	Cond	Beh	Cond	Beh	Previous State	New State	
Vehicle Position Control	Not yet at search area	Transit					Start	Transit	Start in Transit behavior
	At search area	Spiral					Transit	Spiral	Start Spiral from Transit
			No feature after time (T)	Spiral			Turn	Spiral	Start Spiral from Turn
					Found feature	Follow	Spiral	Follow	Start Follow from Spiral
					Found feature	Follow	Turn	Follow	Start Follow from Turn
					Lost feature	Turn	Follow	Turn	Start Turn from Follow

Figure 14.10 Capability Analysis Table for the state machine in Figure 14.9

Given the behavior described in the state machine diagram in Figure 14.9, the table can be generated by filling in the cells and using the definitions of the behaviors to drive the information and commands and intermediate variables needed. In order to maintain the sequential nature of the relationships represented in the original diagram, an additional set of columns called "Task Status" are added to the Capability Analysis Table. The first new column holds the behavior the system transitions from and the second holds the behavior the robot is transitioning to.

The resulting table is shown in Figure 14.10. As before, a specific behavior is represented in each row. A given row may only represent the information required by one behavior, so if multiple behaviors may each control the same element, multiple rows are necessary to represent them. Both go to waypoint and trajectory following behaviors may control the same hardware, but as different behaviors they would be represented on different rows.

All the conditions associated with a behavior on a given row must be true in order for the behavior to be active. If there are multiple configurations of the conditions that are acceptable, each configuration is represented on its own row. In Figure 14.10, this is shown where the Spiral behavior is called under two different sets of conditions. The Spiral behavior is called at the beginning of the operation when the Transit behavior completes and the vehicle reaches the search area. It is called again after the vehicle has found and lost a feature, when it transitions from the Turn behavior.

A single behavior may rely on multiple pieces of information, so within a row, a given behavior may appear in multiple columns. In Figure 14.11, the SLAM (simultaneous localization and mapping) behavior relies on the current position of the robot in the Map, the estimated motion of the robot as provided by the Inertial Nav Sensor, and the Feature Status to generate a new robot position on the Map.

The information represented in the Capability Analysis Table is the information available to the behaviors and the control options and information generated by the platform. If a behavior requires certain information or commands to be available, and the vehicle does not provide them, the table forces the designer to specify how the information will be acquired or how the command will be implemented using the available tools.

14.6 Improving the design

Once the initial design is in the table, the table can be used to improve and refine the design. In cases where the table was built from a vehicle specification and the behavior was designed around it, the various unused and undeveloped capabilities latent in the platform can be used to refine the design.

14.6.1 Identifying unused capability

For a capability table derived from a platform rather than a behavior, the highest level global capability table contains all the information and control mechanisms that the robot provides in the baseline design, along with information and control mechanisms that may help with future missions that do not require alterations to the hardware and are theoretically possible to implement given sufficient software resources. One purpose of this table is to identify both unused capability and potentially useful but currently undeveloped capability.

This section assumes that the designer has already generated a table that contains all the baselined behaviors that the system is expected to perform. To identify sensor data or information that is superfluous to the current operation, look for empty columns. If a column has no behaviors, there is either (1) a gap in the baselined

Adaptive Feature Detection	Position on Map		Feature Status		Time		Inertial Nav Sensor		Altimeter Sensor		Task Status		Notes
	Cond	Beh	Cond	Beh	Cond	Beh	Cond	Beh	Cond	Beh	Previous State	New State	
Position on Map		SLAM		SLAM				SLAM			Start	SLAM (ongoing)	Keep track of position on map
Feature Status										Feature Detector	Transit	Feature Detector (ongoing)	Keep track of feature once you reach the search area
Vehicle Position Control	Not yet at search area	Transit									Start	Transit	Start in Transit behavior
	At search area	Spiral									Transit	Spiral	Start Spiral from Transit
					No feature after time (T)	Spiral					Turn	Spiral	Start Spiral from Turn
			Found feature	Follow							Spiral	Follow	Start Follow from Spiral
			Found feature	Follow							Turn	Follow	Start Follow from Turn
			Lost feature	Turn							Follow	Turn	Start Turn from Follow

Figure 14.11 Source for Position on Map and Feature Status input in Figure 14.10; addition of Task Status columns

capability, (2) a problem with the baselined capability interface definition, or (3) information that is unused and superfluous.

If we start with the Survey behavior and the example platform in Figure 14.6, we can refine the table to generate a more complete representation of the vehicle. For example, we have abstracted the communications infrastructure into the "Message Out" and "Comms In" items. In practice, the vehicle would actually have an acoustic modem for underwater communications and additional communications capability (using Wi-Fi, Ethernet, or satellite) for surface communications. If we represent both these capabilities in the table, we find ourselves with the table in Figure 14.12.

We now have an empty "Surfaced Message Out" control output row and an empty "Acoustic Comms In" information input column. This should trigger three kinds of design activity:

- Identification of a gap in the behavior specification: Does the vehicle actually need to receive and use the information from an acoustic message? Does the vehicle actually need to send the operator status updates or upload raw data post-mission via a surfaced message in order for the job to be considered complete?
- Consistency checking: Is the content of the outgoing acoustic message related in any way to the incoming surfaced message? For example, does the outgoing message trigger a surfaced channel response? If there's a gateway buoy providing translation between acoustic and surfaced communications channels, does it need to be represented in the table or does it imply that the choice of communications channel is irrelevant and we *should* be abstracting the communications channel out of the table?
- Identification of superfluous capability: If the Task Spec information comes in over the acoustic channel rather than the surfaced channel, can we eliminate the surfaced channel and use a less capable vehicle to perform the task?

To identify superfluous information items, look for empty columns, and to identify superfluous control elements, look for empty rows. If a row or column has no behaviors, then either (1) it is unused and that control mechanism or information item is presumably unnecessary or (2) the robot needs to use that control mechanism or information item, but the baseline behavior specification is wrong or the design is incomplete.

14.6.2 Identifying undeveloped capability

To identify potential undeveloped capability, look at both occupied and unoccupied cells of the table in turn and ask whether there is some controller or behavior, not yet developed, that could use that information to produce that control output. If a potentially useful controller or behavior is identified, it may be added to the table as a theoretical capability of the system, currently undeveloped. In general, undeveloped capabilities should be represented with different formatting (e.g. different colored text, no fill).

For example, the cell at the intersection of the Object Sensor column and the Object Info row in Figures 14.6 and 14.12 is currently occupied by a perceptual

Example platform	Object Info (pose, pose accuracy, ID, & confidence)		Self Info (battery status, pose, & accuracy)		Task Info (region, time limit, & progress)		Time		Position Sensor		Object Sensor		Acoustic Comms In		Surfaced Comms In		Notes
	Cond	Beh	Cond	Beh	Cond	Beh	Cond	Beh	Cond	Beh	Cond	Beh	Cond	Beh	Cond	Beh	
Object Info										Object Detection		Object Detection					Always on
Self Info										Pose Estimator							Always on
Task Info																Task Spec	Defines Task Info from received message
						Survey											Progress updates generated by Survey behavior
Acoustic Message Out	New object pose & pose accuracy	Survey															Generates message when new object detected
Surfaced Message Out																	
Vehicle Motion Control				Survey		Survey	Time< t	Survey									Generates survey trajectory within time limit

Figure 14.12 Example platform Survey behavior table with less abstract Communications control and information representation

algorithm called "Object Detection." There are several computations that could fill that role. If we wanted to capture potential capability, we could document that the vehicle's object sensor data could be used by a deep learning identification algorithm or a Bayesian classifier or a data fusion algorithm to generate object identifications and associated confidences, depending on the nature of the raw data, the type of object, and the availability of training data. These different capabilities are represented as new rows, with the different behaviors populating each one, as shown in Figure 14.13. A sensor registration algorithm using environmental data like water temperature or salinity could be used to generate an improved object pose estimate.

Similarly, additional behaviors can be tracked on the Vehicle Motion Control row. In addition to a high-level Survey behavior, we could note that the vehicle is also capable of both GoToWaypoint and TrajectoryFollowing behaviors.

In the original Figure 14.6, the intersection of the Self Info column and the Task Info row is unoccupied. However, we could add a behavior at that point that generates a progress value that includes the vehicle's current position as well as the task status. This potential but undeveloped capability is shown in the table in Figure 14.13.

14.6.3 Behavior analysis

Regardless of whether the table was built from a vehicle specification or a behavior specification, specifying the behavior in a tabular form with a well-defined semantics makes it possible to perform additional analysis of the specification, including evaluation of the consistency and completeness of the specification. This part of the process enables the designer to elicit improved requirements by asking concrete questions about the desired behavior of the system as they work to refine the design. For example, if the desired behavior is ongoing rather than time- or space-limited, it may be appropriate for there to be no transition to an end state.

Checking a table for consistency will identify nondeterminism in the specification. Inconsistency in a table is identified by examining the set of rows defining the behavior of a single controlled output. If more than one row can hold at the same time, but result in different behavior for the controlled output, then the specification is inconsistent.

A table can also be evaluated with respect to completeness to identify missing cases not addressed in the specification. In the feature following behavior from Figure 14.9, there are several ways in which the behavior is not well specified. What is the condition that causes the behavior to end? A search area is defined—is the vehicle constrained to remain within the search area? What happens if it leaves? Does it need to transit back to a retrieval area? Is there a time limit associated with retrieval? Is the progress reported back to the operator while the vehicle is running? Can the operator send new commands and interrupt the current operation?

In addition to incomplete specification of the desired behavior, the design initially captured in the state machine diagram in Figure 14.9 was obviously incomplete, since the algorithms necessary to support the required perceptual inputs (SLAM and Feature Detection) were not identified. However, even the version in Figure 14.11 can be improved.

Example platform	Object Info (pose, pose accuracy ID, & confidence)		Self Info (battery status, pose, & accuracy)		Task Info (region, time limit, & progress)		Time		Position Sensor		Object Sensor		Comms In		Notes
	Cond	Beh	Cond	Beh	Cond	Beh	Cond	Beh	Cond	Beh	Cond	Beh	Cond	Beh	
Object Info (general)										Object Detection		Object Detection			Always on
Object ID, Confidence												Deep Learning ID			Optional object classification algorithm
												Bayesian Classifier			Optional object classification algorithm
												Data Fusion Detector			Implements several classification algorithms on the same input data and fuses the results
Object Pose, Accuracy										Sensor Registration		Sensor Registration			Generates accuracy pose estimates from sensor properties and environmental conditions tracked by position sensor
Self Info										Pose Estimator					Always on
Task Info														Task Spec	Defines Task Info from received message
				Progress Update		Progress Update									Progress updated based on self *pose and Task Info
						Survey									Progress updates generated by Survey behavior
Message Out	New object pose & pose accuracy	Survey													Generates message when new object detected
Vehicle Motion Control				GoTo Waypoint											Support for waypoint-based navigation
				Trajectory Following											Support for path following-based navigation
				Survey		Survey	Time< t	Survey							Generates survey trajectory within time limit

Figure 14.13 Example platform with potential capabilities identified

Adaptive Feature Detection	Map		Position on Map		Feature Status		Time		Inertial Nav Sensor		Altimeter Sensor		Task Status		Notes
	Cond	Beh	Cond	Beh	Cond	Beh	Cond	Beh	Cond	Beh	Cond	Beh	Previous State	New State	
Map		SLAM		SLAM		SLAM				SLAM			Start	SLAM (ongoing)	Generates updated map
Position on Map		SLAM		SLAM		SLAM				SLAM			Start	SLAM (ongoing)	Keep track of position on map
Feature Status												Feature Detector	Transit	Feature Detector (ongoing)	Keep track of feature once you reach the search area
Vehicle Position Control			Not yet at search area	Transit									Start	Transit	Start in Transit behavior
			At search area	Spiral									Transit	Spiral	Start Spiral from Transit
							No feature after time (*T*)	Spiral					Turn	Spiral	Start Spiral from Turn
					Found feature	Follow							Spiral	Follow	Start Follow from Spiral
					Found feature	Follow							Turn	Follow	Start Follow from Turn
					Lost feature	Turn							Follow	Turn	Start Turn from Follow

Figure 14.14   Adding a Map variable to the feature-following behavior in Figure 14.11

Because we used a SLAM algorithm to provide the vehicle's position on the map, and because we track the vehicle and feature positions on the map, the existence of a map is implicit in the variable structure. The SLAM algorithm inherently generates a map as well as vehicle localization within it. We can improve the documentation of the design by making the generation of the map explicit.

Figure 14.14 shows the SLAM behavior producing both the Position on Map output (localization) and an update to the Map itself (mapping). When a single behavior produces more than one control output, the behavior may appear on more than one row. For example, a behavior that produces an object pose within a sonar image, a confidence value associated with that pose, and a registered pose in the physical world would appear on three Data Processing rows. The SLAM algorithm demonstrates this.

14.7 Conclusions

We have introduced Capability Analysis Tables and shown how they can be used to connect behaviors and platforms to create a detailed behavior specification. Making this connection explicit helps to identify missing or unused capabilities, incompleteness, and inconsistencies in the design. This design method is agnostic to platform and application area, and has been applied to behavior specification in space robotics and underwater operations.

We also see a further extension of this method: providing a framework for robotic systems that makes them precise enough for formal methods. Formal methods are a powerful tool, but are not widely accessible to roboticists. The Capability Analysis Table represents a connection between the robotics community and the formal methods community. In the future, we expect to be able to link the Capability Analysis Table representation to one or more formal methods representations, enabling formal analysis of arbitrary behavior designs. Today, it allows roboticists to improve the level of precision in their design without committing to developing the expertise necessary to apply formal methods.

Capability analysis does not replace state machine diagrams or similar behavior-centric representation, just as it does not replace platform design representations. It is a complementary design method that represents information related to both the behavior design and the platform design to illuminate the facet of the behavior design problem that focuses on its interaction with the platform.

References

[1] Cleeremans A., Servan-Schreiber D., McClelland J.L. "Finite State Automata and Simple Recurrent Networks." Journal of Neural Computation. 1989; **1**(3):372–381.

[2] Harel D. "Statecharts: A Visual Formalism for Complex Systems." Science of Computer Programming. 1987;**8**:231–274.

[3] Abbas T., MacDonald B.A. "Generalizing Topological Task Graphs from Multiple Symbolic Demonstrations in Programming by Demonstration (PbD) Processes." *Proceedings of the 2011 IEEE International Conference on Robotics and Automation*; Shanghai, China; 2011. pp. 3816–3821.

[4] Kong J., Zhang K., Dong J., *et al.* "Specifying Behavioral Semantics of UML Diagrams through Graph Transformations." The Journal of Systems and Software. 2009;**82**(2):292–306.

[5] Dibangoye J.S., Amato C., Buffet O., *et al.* "Optimally Solving Dec-POMDPs as Continuous-State MDPs." Journal of Artificial Intelligence Research. 2016;**55**:443–497.

[6] Konidaris G., Kuindersma S., Grupen R., *et al.* "Robot Learning from Demonstration by Constructing Skill Trees." International Journal of Robotics Research. 2012;**31**(3):360–375.

[7] Colledanchise M., Marzinotto A., Ogren P. "Performance Analysis of Stochastic Behavior Trees." *Proceedings of the 2014 IEEE International Conference on Robotics and Automation (ICRA)*; vol. 26; 2014. pp. 3265–3272.

[8] Ingrand F., Lacrois S., Lemai-Chenevier S., *et al.* "Decisional Autonomy of Planetary Rovers." Journal of Field Robotics. 2007;**24**(7):559–580.

[9] Benjamin M.R., Schmidt H., Newman P.M., *et al.* "Nested Autonomy for Unmanned Marine Vehicles with MOOS-IvP." Journal of Field Robotics. 2010;**27**(6):834–875.

[10] Pirjanian P., Huntsberger T.L., Trebi-Ollennu A., *et al.* "CAMPOUT: A Control Architecture for Multirobot Planetary Outposts." McKee G.T., Schenker P.S. (eds.). *Proceedings of the SPIE Symposium on Sensor Fusion and Decentralized Control in Robotic Systems III*; vol. 4196. Boston, MA: SPIE; 2000. pp. 221–230.

[11] Tang F., Parker L.E. "A Complete Methodology for Generating Multi-Robot Task Solutions Using ASyMTRe-D and Market-Based Task Allocation." *Proceedings of the 2007 IEEE International Conference on Robotics and Automation (ICRA)*; Roma, Italy; 2007. pp. 3351–3358.

[12] McGann C., Py F., Rajan K., *et al.* "T-REX: A Model-Based Architecture for AUV Control." *Third Workshop on Planning and Plan Execution for Real-World Systems*; Providence, RI; 2007. pp. 1–8.

[13] Arkin R.C. "Robot Behavior." *Behavior-Based Robotics*. 3rd edn. Intelligent Robotics and Autonomous Agents. Cambridge, MA: The MIT Press; 1998. p. 85.

[14] Harel D. "Statecharts: A Visual Formalism for Complex systems." Science of Computer Programming. 1987;**8**(3):231–274. Available from: http://www.sciencedirect.com/science/article/pii/0167642387900359

[15] Lynch N. *Distributed Algorithms*. San Mateo, CA: Morgan Kaufman; 1996.

[16] Heitmeyer C., Archer M., Bharadwai R., *et al.* "Tools for Constructing Requirements Specifications: The SCR Toolset at the Age of Ten." Computer Systems: Science and Engineering. 2005;**20**(1):19–35.

[17] Leveson N.G., Reese J.D., Heimdahl M.P.E. "SpecTRM: A CAD System for Digital Automation." *Proc. 17th Digital Avionics Systems Conf.*; Bellevue, WA: IEEE/AIAA/SAE; 1998.

[18] Pneuli A. "The Temporal Logic of Programs." *Proceedings of the 18th Annual Symposium on Foundations of Computer Science*; SFCS '77. Washington, DC, USA: IEEE Computer Society; 1977. pp. 46–57.

[19] Clarke E.M., Emerson A.E. "Design and Synthesis of Synchronization Skeletons Using Branching-Time Temporal Logic." *Proceedings of the Workshop on Logics of Programs*; Berlin, Heidelberg: Springer-Verlag; 1981. pp. 52–71. Available from: http://dl.acm.org/citations.cfm?id=648063.747438

[20] Jhala R., Majumdar R. "Software Model Checking." ACM Computing Surveys (CSUR). 2009;**41**(4):21:1–21:54. Available from: http://doi.acm.org/10.1145/1592434.1592438

[21] Marić F. "A Survey of Interactive Theorem Proving." Zbornik Radova. 2015 July.

[22] Heimdahl M.P.E., Whalen M.W., Thompson J.M. "NIMBUS: A Tool for Specification Centered Development." *Proceedings of the 11th IEEE International Conference on Requirements Engineering*; RE '03. Washington, DC, USA: IEEE Computer Society; 2003. p. 349.

[23] Rothamel T., Heitmeyer C., Leonard E., *et al.* "Generating Optimized Code from SCR Specifications." *ACM Conf. on Languages, Compilers, and Tools for Embedded Systems (LCTES 2006)*; 2006.

[24] Gargantini A., Heitmeyer C. "Automatic Generation of Tests from Requirements Specifications." *Proc. ACM 7th ACM SIGSOFT Symp. on Foundations of Software Eng. (ESEC/FSE99)*; 1999.

[25] Faulk S.R., Finnerman L., Kirby Jr. J., *et al.* "Experience Applying the CpRE Method to the Lockheed C-130J." *Proc. 9th Annual Conf. on Computer Assurance (COMPASS '94)*; 1994.

[26] Mitsch S., Ghorbal K., Vogelbacher D., *et al.* "Formal Verification of Obstacle Avoidance and Navigation of Ground Robots." International Journal of Robotics Research. 2017;**36**(12):1312–1340.

[27] Sarid S., Xu B., Kress-Gazit H. "Guaranteeing High-Level Behaviors While Exploring Partially Known Maps." *Proceedings of Robotics: Science and Systems 2012*; 2012.

[28] Webster M., Fisher M., Cameron N., *et al.* "Formal Methods for the Certification of Autonomous Unmanned Aircraft Systems." Flammini F., Bologna S., Vittorini V. (eds.). *Computer Safety, Reliability, and Security*; Berlin, Heidelberg: Springer; 2011. pp. 228–242.

[29] Albus J.S., Huang H.M., Messina E.R., *et al.* "4D/RCS Version 2.0: A Reference Model Architecture for Unmanned Vehicle Systems." NIST; 2002. 6910.

[30] Redfield S.A., Seto M.L. "Verification Challenges for Autonomous Systems." Lawless W.F., Mittu R., Sofge D., *et al.* (eds.). *Autonomy and Artificial Intelligence: A Threat or Savior?* Springer, Cham; 2017. pp. 103–128.

[31] Bermejo-Alonso J., Salvador J., Sanz R. "Towards an Ontology for Task and Planning in Autonomous Systems: An Emergency Scenario." Ollero A., Sanfeliu A., Montano L., *et al.* (eds.). *ROBOT 2017: Third Iberian Robotics Conference*; vol. 1 of Advances in Intelligent Systems and Computing. Springer; 2018. pp. 429–440.

[32] "IEEE Standard for Robot Map Data Representation for Navigation." 1873-2015 IEEE Standard for Robot Map Data Representation for Navigation. 2015 October; pp. 1–54.

[33] "IEEE 1872-2015, IEEE Standard Ontologies for Robotics and Automation." IEEE Std 1872-2015. 2015 April; pp. 1–60.

[34] ISO/TC 299 Robotics. ISO/TC 299 Robotics (ed.). "ISO 8373:2012, Robots and Robotic Devices—Vocabulary." Organisation Internationale de Normalisation (ISO); 2012. ISO Standard. Available from: https://www.iso.org/standard/55890.html [cited June 27, 2018].

[35] IEEE. "IEEE 1872-2015.2, IEEE Standard Ontology for Autonomous Robots'; Standard Under Development.

[36] Cutkosky M.R. "On Grasp Choice, Grasp Models, and the Design of Hands for Manufacturing Tasks." IEEE Transactions on Robotics and Automation. 1989;**5**(3):269–279.

[37] Bullock I.M., Dollar A.M. "Classifying Human Manipulation Behavior." *2011 IEEE International Conference on Rehabilitation Robotics*; ETH Zurich Science City, Switzerland: IEEE; 2011. pp. 1–6.

[38] Braitenberg V. *Vehicles: Experiments in Synthetic Psychology*. Cambridge, MA: MIT Press; 1984.

[39] Bennett A.A., Leonard J.J. "A Behavior-Based Approach to Adaptive Feature Detection and Following with Autonomous Underwater Vehicles." IEEE Journal of Oceanic Engineering. 2000;**25**(2):213–226.

Chapter 15
Advanced AUV fault management

Mae Seto[1] and Kathleen Svendsen[2]

15.1 Introduction

The application of autonomous robots, like autonomous underwater vehicles (AUVs; Figure 15.1), for commercial, scientific and military purposes has escalated over the last decade. Commercial examples include monitoring and maintenance of offshore structures to localizing and recovering submerged targets; scientific exploration of the oceans to track carbon content and algal blooms; and military missions like mine counter-measures, intelligence surveillance and reconnaissance and anti-submarine warfare – to name a few.

The increase AUV use is due to several reasons: miniaturization of sensors and sub-systems that use less power; higher-energy density power sources; more powerful embedded processors that are also less power hungry (and now parallelizable); modularity (plug-and-play) of sub-systems, ease with which to network sub-systems; access to satellites for remote AUV updates and control, reduction in overall cost, etc. With these near-transformative developments, AUVs are valued for their portability, stealth and scalability (distribute as needed to sense a phenomenon at the required spatial and temporal resolutions). The role of the AUV is evolving and they are tasked to ever more complex missions in challenging environments.

Fault management enables AUVs to provide persistent ocean observation – i.e. able to deploy for extended periods on complex missions in unstructured environments at great ranges from operators. It refers to an on-board response to an AUV fault that could be reactive or deliberative with the intent to recover from, or mitigate, the fault. Otherwise, the AUV is unavailable for the entire duration of the mission. On long deployments, there will inevitably be hardware failures and data corruption. Cascading failures could jeopardize a mission or the vehicle [1]. The objective is to provide some ability for the AUV to manage the faults triggered by the AUV, the environment around it or its interaction with the environment. The advanced fault management within the on-board autonomy is an enabler to realize the potential for the complex roles expected of AUVs.

[1]Faculty of Engineering and Faculty of Computer Science (cross-appointed), Dalhousie University, Halifax, Canada
[2]Faculty of Engineering, Dalhousie University, Halifax, Canada

Figure 15.1 The Arctic Explorer AUV is among the most advanced in the world

On-board autonomy [2] facilitates the AUV fulfilling complex goals with minimal operator intervention as it has some abilities to (1) make decisions given a set of automated planned options and if these plans are inadequate, it makes new ones; (2) analyse and interpret in-situ sensor data; (3) diagnose and recommend solutions for robot problems detected through on-board sensing and monitoring; (4) learn and adapt; and (5) collaborate with other systems. Fault management is a component of the robot autonomy under (3).

Advanced fault management is a means to bridge the gap for routine operation of long endurance, complex missions with AUVs in unstructured environments. The proposed advanced fault management would be more deliberative approaches in the recovery. However, this is at an early stage with AUVs compared to other areas and the requirement for operations in complex environments for extended durations is here [3]. Autonomous spacecrafts have similar requirements to long-endurance AUVs and are used as an example for what AUVs might aspire to.

The objective of this chapter is to review fault tolerance in general for AUVs. This includes an update on an earlier review that addresses reactive AUV fault-tolerance strategies [4]. The chapter concludes with motivation and suggestions on research areas that could have high impact.

The rest of this chapter is as follows. First, the fault management terms used are defined as there is variability in the literature and amongst users. Then, the state-of-the-art AUV and its operating environment is described to provide perspective on the fault management challenges. This is followed by a description of fault management requirements and the motivation for deliberative approaches in light of the vehicle and its environment. Then, traditional fault tolerance for AUVs is described followed by deliberative fault tolerance with some applications to AUVs. The application of artificial intelligence tools that support deliberative fault tolerance is briefly reviewed followed by metrics that assess the efficacy of fault-tolerance measures. The chapter concludes with recommendations on timely research areas with high impact.

15.2 Definition of terms

Fault management terms can vary depending on the application. This subsection endeavours to define the ones used here.

An *anomaly* or *anomalous event* is unexpected and could trigger faults and failures. It is a motivation for a fault management system.

A *fault* is an unintended or unanticipated change, i.e. an anomaly, or defect in the AUV's system function or state which interferes with its nominal operation by causing performance deterioration. The fault can be an incorrect step, process or damage to the AUV.

A fault can lead to a *failure* which is an inability of a system to perform its required functions within specified performance requirements.

Fault *detection* is the process of monitoring for faults. This involves instrumenting the AUV with health sensors for vehicle and sub-system states and performance. With AUVs, this includes sensors to detect leaks, over-depth, over-pitch, failure of sub-components like a compass or Doppler velocity log (DVL; see next), battery power left, performance of propulsion motor, ability of control surfaces to deflect over the range and rates they were designed for, health of the payload sensor, etc.

Fault *identification*, *classification* or *diagnosis* is the ability to localize the sub-system that the fault occurred in. This is a prerequisite for fault *recovery*.

Fault recovery is a response to a detected fault with the intent to recover from, accommodate, or mitigate, the fault. It comprises any steps necessary to continue safe operation albeit with potentially degraded performance. Fault detection, identification and recovery (FDIR) are different individual mechanisms in an FDIR system. FDIR can also be referred to as health monitoring and management in some applications [5].

Fault *isolation* is used inconsistently. Sometimes it refers to classifying the severity and location of the fault [5]. Other times it refers to isolation of the fault to protect the rest of the system and is thus more a recovery action [6,7]. Identification will be the term used here for actions that functionally involves identification, classification and diagnosis. Recovery, here, will implicitly include isolation.

A *fault-tolerant* system is one capable of continued operation, possibly with degraded performance, in the event of faults in its hardware or software. Fault tolerance is desired in autonomous systems for economic, survivability and robustness reasons given ever more complex mission expectations.

A system is *robust* if it can operate reasonably well even under a load that is higher than expected. For fault tolerance, the load includes inherently poor designs, faults, failures or environmental disturbances.

An *automatic* system follows a script and does not make choices for itself. If it encounters the unexpected, it cannot adapt, and waits for operator help.

An *autonomous* system has *autonomy* and thus the ability to make decisions, enforce decisions by applying actions, evaluate results and adapt.

The operability of a vehicle depends on its *survivability* and *reliability*. Survivability is concerned with faults and failures from external interactions or damage (e.g. collisions) and is not discussed here. Fault tolerance enhances the overall AUV reliability and pertains to the hardware and software.

Reliability is the ability of a system to function under specified conditions for a stated period. The sub-systems collectively determine whether the system functions. Their configurations can affect reliability. In a series system, all sub-systems must work; for example, the battery, motor and propeller on an AUV must individually function to propel the AUV. A parallel system fails if every system fails. As an example, AUVs can have parallel navigation systems. A compass and DVL which aids an inertial navigation system (INS) would be one system and a parallel system could be the INS aided only by a compass. The AUV would be unable to navigate if both failed. Parallel systems afford some redundancy and graceful degradation in the event of faults/failures. The reliability of a system or sub-system can be quantified through the mean-time-between-failure (MTBF) metric. Reliability analysis builds a system model from the sub-system and component MTBFs and their dependencies to assess risk.

Risk management provides analysis on the system reliability and all sub-systems that depend on it. It may achieve this through an analysis of the system performance, probability of anomalous (unexpected) events, etc., to identify issues related to the system and its sub-systems. At the simplest, the relationship between equipment risk and reliability [8] is the following:

$$\text{risk (\$/year)} = \text{consequences (\$)} \times \frac{\#\text{ failure opportunities}}{\text{year}} \times (1 - \text{reliability})$$

where reliability is the probability of not failing, so (1 − reliability) is the probability of failure. This type of analysis could be applied to critical AUV systems as it is to autonomous spacecraft [9]. The objective with risk analysis is to spot the potential for cascading failures that could threaten a mission or AUV.

15.3 The state-of-the-art AUV

AUVs are used for military, scientific and commercial applications. Examples include surveying the continental shelf (under ice) for territory gain [10], inspecting underwater pipelines [11], hazardous chemical spill monitoring [12] and ocean health monitoring [13]. The most developed military mission for AUVs is naval mine-countermeasures (NMCMs). AUVs have been used in underwater mine-warfare since the Gulf War [14]. This mission will be described next as it has driven AUV development.

An underwater area that could be potentially mined (e.g. harbour approaches) would be surveyed with the side-scan sonar on the AUV, when there are no mines, to obtain a baseline image. This is the route survey. When the area is suspected to be

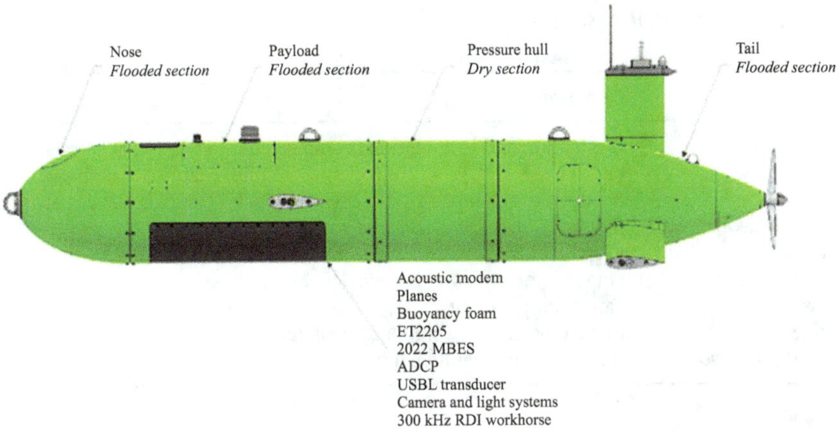

Figure 15.2 The ISE Ltd. Arctic Explorer Class AUV general arrangement

mined, a minehunting survey is performed and differenced against the route survey through change detection [15] methods.

The state-of-the-art AUV performs this mission with its side-scan sonar and vehicle telemetry data streams processed in-situ on-board the AUV to detect, classify and identify targets that are mine-like objects (MLOs). Of those with a high likelihood of being a mine, the AUV (or another asset) can plan a mission to re-acquire the targets with a higher frequency (higher resolution) sonar and perhaps lower vehicle altitude. This target re-acquisition confirms its location and identity and determines whether subsequent action like disposal is necessary. The use of AUVs for NMCM has the advantage of increasing the ship's reach, more persistence than divers, and it keeps the ship out of a potential minefield.

The modern AUV is a collection of networked hardware and software systems. AUV fault tolerance is concerned with hardware and software as both impact the overall operability and availability of the vehicle. These AUV components will be discussed next.

15.3.1 AUV hardware

AUVs (e.g. Figures 15.1 and 15.2) are battery-powered and free-swimming autonomous underwater vehicles (robots). Some sections are free-flooded and others are pressure vessels. They propel themselves with a drive train and propeller, maintain attitude and altitude/depth in the water column and navigate and localize. From the payload sensor's perspective, the AUV is a stable platform to house, power, control, transport the payload sensor to its measurement locations, process the sensor data into information, communicate the sensor information, receive and interpret commands for the sensor mission and provide an on-board framework to reason on, among other things, what the AUV might do with the in-situ sensor information.

Table 15.1 Typical AUV subsystems

Subsystem	Function
Vehicle sensors	Navigation and localization (compass, INS, DVL, GPS, ultra-short base line (USBL)), attitude, depth/altitude, energy distribution, conductivity–temperature–density, acoustic Doppler current profiler
Payload sensors	Side-scan sonar, bathymetric sonar
Thrusters	Propulsion
Communications	Underwater acoustic modem, Wi-Fi, radio
Ballasting	Tanks, drop weights
Vehicle processor	Basic closed-loop vehicle controller
Payload processor	Implement deliberative algorithms (autonomy)

The AUV has sub-systems that run the vehicle and the payload sensors. An example is shown in Table 15.1. The AUV's endurance and the amount of computations (CPU activity) it can perform is limited by the available energy carries. Fault management that is computationally intensive can affect the vehicle endurance. Larger working class AUVs have greater endurance since they have the capacity to carry more batteries. Therefore, it is the larger AUVs that are deployed for long mission durations (e.g. Autosub 6000, Arctic Explorer, HUGIN 1000, REMUS 6000, Bluefin-21) and they are also the ones that require the advanced fault tolerance. As a class, they can dive to depths of 3–6 km with in-water endurance for missions that vary from 20 to 70 hours depending on the payload sensors. The energy density of the batteries carried is among the highest possible.

Long endurance AUVs, by necessity, will have high-performance INS and DVL as their missions can be up to 1,000 km or more in length. The AUV's navigation and positioning is fundamental to the AUV operation and faults in this severely compromise the mission. This is briefly introduced as an example of integrated hardware and software.

Navigation and positioning – integrated hardware and software

The AUV's position must be known to some tolerance for its sensor measurements to be meaningful. Underwater navigation methods can be inertial, acoustic or geophysical. Inertial methods are common. An on-board INS uses motion measurements and state estimation filters (described later) [16] to determine the dead-reckoned AUV position, orientation and velocity. This is subject to continuous cumulative errors which grow unbounded. An on-board DVL acoustically measures the AUV speed-over-ground to correct the inertially derived speed and increases the AUV operational range.

To obtain the absolute AUV position, this dead-reckoned localization is fused with external position references from ship-based acoustic baseline methods, deployed transponders or beacons [17]. This extends the AUV operational range still more to the range of acoustic communications (<10 km). AUVs can also surface for a GNSS position reference if they are not under ice. However, in deep (>1 km) water, the accuracy of a surface GNSS calibration is reduced after spiral diving to depth mostly without the DVL (designed to operate <200 m from the seabed). Methods that do not require surfacing like simultaneous localization and mapping and the geophysical terrain-aided navigation [18] are active research areas. The DVL losing bottom-lock and a non-responsive INS are mission critical faults that the fault tolerance must address.

The on-board software is divided into (i) the basic closed-loop vehicle controller hosted on the vehicle processor and (ii) the deliberative autonomy-driven controller hosted on a payload processor.

15.3.2 AUV software

15.3.2.1 Vehicle processor

The vehicle processor hosts the vehicle controller which implements the closed-loop control for the AUV yaw, pitch, roll, depth/altitude, heading, speed and waypoint following, at a minimum. The AUV is instrumented to measure the feedback values for these quantities with an INS or inclinometer. At the beginning of the mission, the operator inputs setpoints for these quantities and the closed-loop control, which can be a basic proportional-integral-derivative (PID) controller, would attempt to zero the error between feedback and setpoint to control the AUV.

Mostly, mission plans are an ordered set of waypoints that the AUV travels through at a prescribed speed while logging its payload sensor measurements. The AUV reaches a waypoint when it is within a pre-set radius from it. Upon acquiring a waypoint, there may be changes made to the payload sensor settings (e.g. side-scan sonar frequency) and vehicle setpoints (e.g. altitude). The vehicle controller can also implement the vehicle navigation solution and positioning (discussed in next subsection), the logging of vehicle sensors and the fusion of vehicle sensors. On some AUVs, the off-board communications with the acoustic modem may be controlled through the vehicle processor. The vehicle processor also has responses for basic vehicle faults. These built-in responses tend to be basic and mostly reactive.

With basic NMCM survey missions, the concept of operations is to deploy the AUV from a ship or jetty, let it perform its mission logging side-scan sonar data then, the AUV is recovered, its data downloaded and its data analysed to detect geo-referenced MLOs. This contrasts with the more advanced concept of operation at the beginning of the section. The basic survey mission does not use on-board deliberation. If the AUV was unable to adapt to the evolving vehicle systems, environment, or mission, it would perceive it as a fault and the usual reactive response is to surface.

With expectations for long-duration missions, the AUV, environment and mission will evolve. This increases the probability of triggering faults or failures in the hardware and software. There may be damage or an event that requires a reconfiguration of the AUV as a remedy to recover. This may be due to anomalous events (lower energy, compromised control of the AUV, inability to communicate with the support vessel, off course, sensor failure, etc.). The coveted persistence desired in deploying a robot requires that its fault tolerance be addressed. These are deliberative (autonomy) capabilities hosted on the payload processor.

15.3.2.2 Payload processor

The work at the Massachusetts Institute of Technology [19,20] was among the earliest to introduce the concept of a payload processor or back-seat driver where goal-based planning and deliberative processing (autonomy) could take place. This concept left the vehicle processor untouched and it could function stand-alone if there was a failure in the payload processor. The autonomy on the payload processor abstracts away the vehicle specifics and could be mostly ported, as is, to another similar AUV to be interfaced to its vehicle processor.

The payload processor is networked with the vehicle processor for sustained (Ethernet) communications between the two as the vehicle processor monitors the AUV's status and communicates it to the payload processor. The autonomy on the payload processor requires this for its deliberation. In the event of a vehicle fault, the payload processor may defer to the vehicle processor's reactive response to say a leak. As mentioned earlier, the vehicle processor has mostly reactive responses to faults. It does not process vehicle health measurements or prescribe remedies to complex faults. This deliberation is performed on the payload processor.

Given the AUV capabilities and limitations, the AUV operating environment is discussed next as it also drives the fault-tolerance requirement.

15.3.3 State-of-the-art AUVs

As mentioned earlier, the commercially available AUVs with more advanced fault tolerance include Autosub 6000, Arctic Explorer, HUGIN 1000, REMUS 6000 and Bluefin-21. Two of the AUVs are introduced here as examples.

15.3.3.1 Autosub 6000

These AUVs are developed by the National Oceanography Centre for research in geophysical, marine science and oceanographic surveying as well as missions under ice in the arctic [21]. This AUV can dive to 6 km depth and has a range of 400 km at 1.6 m/s. The Autosubs are primarily a research vehicle and have employed different fault-tolerance methods. One such solution was to use dynamic Bayesian networks (BNs) where expert knowledge is used to produce conditional probability tables that are applied to the AUV for diagnosing a fault [22]. Alternatively, Markov chains were used to model risk in different phases and scenarios given expert knowledge to calculate the probabilities of transition.

Additionally, their fault diagnosis system, Livingstone 2 (discussed later), was tested on the vehicle [23]. This fault system was originally developed for spacecraft and deployed on-board Goddard's Earth Observing One (EO-1) remote sensing satellite [24]. This used a model-based diagnosis system which compares the ideal model of the vehicle to the estimated model of the vehicle and determines if a fault has occurred.

15.3.3.2 Arctic Explorer

The Arctic Explorer was developed by ISE Ltd for under-ice surveying in the Canadian Arctic. One of its missions was to collect critical bathymetric data for Canada's

UNCLOS (United Nations Convention on the Law of the Sea) claim. This was performed in 2010 and 2011 over multiple Arctic deployments. The Arctic Explorer has a sophisticated variable ballast system that allows it to park on the ocean floor or up against the ice cap. It has a maximum depth of 5 km. For the UCLOS missions, it dove to 3.6 km depth. Its in-water endurance is 3 days with a total distance travelled of 310 km (one-way) during the UNCLOS missions. Over a continuous 10-day operating period, it travelled over 1,000 km.

The Arctic Explorer had an innovative hydrophone-based homing system which homed to a drifting ice camp that moves up to 10 km/day (30 km drift for the 3-day mission), from a range of 50 km away. For the 2010/2011 UNCLOS deployments, the Arctic Explorer used a sophisticated fault response table for its operations [25]. The fault response tables drew on an extensive expert knowledge base in the operation of these vehicles. As with most fault-tolerance methods that use fault response tables, a detected fault is mapped to a response action for the AUV.

15.4 The AUV operating environment

The underwater environment is different from above-water and space in that well-developed communication technologies and methodologies using radio frequency and optical means cannot be used at ranges beyond a kilometre as they are rapidly attenuated. This limitation impacts the AUV fault-tolerance measures possible. Another difference from the above-water environment is the lack of universal positioning systems like the Global Navigation Satellite System (GNSS) or Global Positioning System (GPS) satellites. The impact of underwater acoustic communications is described next.

15.4.1 Underwater acoustic communications

Electromagnetic and optical waves quickly attenuate underwater, leaving acoustic waves as the best for underwater communications using acoustic modems. However, acoustic communications are limited to low-frequency carrier bands due to frequency-dependent absorption [2]. This attenuation varies as frequency or frequency squared (depends on frequency regime) so acoustic propagation is best supported at lower frequencies. The carrier frequencies tend to be lower (e.g. 20–30 kHz). As the bandwidth of a modulated signal is a fraction of the carrier frequency, the bandwidths are now quite low.

Network bandwidths are further reduced by scattering, natural and anthropogenic ambient noise sources and time-varying sound speed profiles. These factors limit typical mean bandwidths to bytes per second (B/s) versus the kB/s and MB/s of in-air radio networks. Bandwidth improvement via multiplexing is challenging as one frequency is typically shared. The fixed-cycle, fixed data rate time-division multiple access (TDMA) access scheme is common for survey-type collaborative AUV missions; however, code-division multiple access (CDMA) is gaining in popularity [26].

Acoustic modems provide a variety of data rates, allowing operators to trade bandwidth for reliability through transmission rate selection [27].

The speed of sound in water is dependent on temperature, depth and salinity and thus varies temporally and spatially in the water column. Consequently, the propagation direction is determined by the sound speed gradients. Variable sound speeds contribute to multi-path as sound rays from a source arrive at a point in space based on the sound speed of the path taken. (Multi-path is also created by physical discontinuities like the water surface, sea bed and density stratification.) The low sound speed causes latency in the received acoustic signals and a message may be obsolete when it is received. Additionally, the low sound speed makes the underwater acoustic channel susceptible to Doppler shifts from moving boundaries like the water surface or even the low speeds that AUVs transit at (<6 knots). The low sound speed further undermines the quality of the received acoustic signal.

The operator cannot communicate well, if at all, with the AUV beyond short ranges. This drives the AUV requirement for on-board fault recovery. Like other robots deployed in remote and harsh environments, the responses to complex faults cannot be driven by an operator at extended ranges. Fault responses must mostly come from on-board the AUV.

15.4.2 Other environmental considerations on AUV faults/failures

Other AUV environmental challenges include corrosion, ice, high sea states and high pressures at depth which can trigger faults if the AUV is not robustly ruggedized to tolerate them. High sea states and the noise from breaking ice can increase the acoustic ambient which makes underwater communications difficult. The AUV could have a fault or fail due to

1. a physical fault in a critical component, whether internal (manufacturing defect or wear that causes a leak) or external (collision with ice);
2. a design or specification error that leads to healthy components being used incorrectly, e.g. software error;
3. an anomalous environmental event like an unexpected thick freshwater layer for an AUV that was ballasted to work in salt water; and
4. operator error, e.g. in the mission script [28,29].

The underwater environment and the AUV can exhibit varying degrees of uncertain behaviour. The operational status of the AUV is dependent on internal sub-system and component reliability, external environment factors affecting system reliability and safety (e.g. high sea states and pressures) and on AUV–environment interactions (e.g. stress factors). These can cause run-time anomalous events, mission degradation and system failures.

Given the AUV strengths and limitations and the environment they work in, the AUV requirements for fault management are summarized next.

15.5 AUV fault management requirements

The environment aspects that impact on-board fault management are as follows:

1. trade-off between range and bandwidth for communications with the operator;
2. poor communications' signal quality due to various underwater acoustic propagation mechanisms;
3. a dynamic environment which is not easily amenable to predictions and is characterized by noisy or unobservable information;
4. operators at more than arms length; and
5. lack of universal localization or navigation references.

In the case of anomalous events, the AUV should assess the risks associated with possible actions to degrade gracefully. Constraints that impact this include limited on-board computational resources and energy.

Autonomous spacecrafts suffer the same difficulties and limitations. The development of spacecraft autonomy has been driven by large programs. Given the similarities between space and underwater, there are lessons learned from space that could be applied. The argument for deliberative fault management is made next.

15.5.1 Motivation for deliberative AUV fault management

Though dated, the US Navy Unmanned Undersea Vehicle Master Plan [14] notes that the least sophisticated sub-system on-board AUVs was the autonomy with technology readiness levels [30] rarely exceeding 5 out of 9.

Prior to on-board autonomy, AUVs executed scripted missions on the vehicle processor. The AUV could not adapt to anomalous events in itself, the environment or mission. In this case, a successful mission was achieved if there was no unexpected circumstances that caused a critical AUV fault or failure and it performed as its mission scripted it to. For long-duration complex missions, this is unrealistic given the difficult environment. Beyond supporting the sensor and performing the measurements adaptively to better achieve the mission, the AUV's persistence is determined by another autonomy requirement – fault tolerance.

When active and passive recovery methods no longer address a fault or failure, another solution is for the AUV to reconfigure itself and make use of what capabilities it has left considering the fault/failure [31]. That is the potential of autonomy. The types of autonomy are discussed next.

With a remote or teleoperated system, the operator controls it with all the bandwidth necessary (e.g. remotely operated vehicle). With supervisory control, the operator has overall control but can transfer limited control over a specific function (e.g. change altitude). With semi-autonomous control, the system can accomplish a subset of its defined tasks without operator interaction (e.g. waypoint following) and the operator completes the rest (after waypoint-following, operator analyses data). In true autonomous control, the system accomplishes all its defined tasks without operator interaction. Automatic control, which is a scripted response, can have elements of supervisory or semi-autonomous control. All AUVs have this. Every AUV possesses

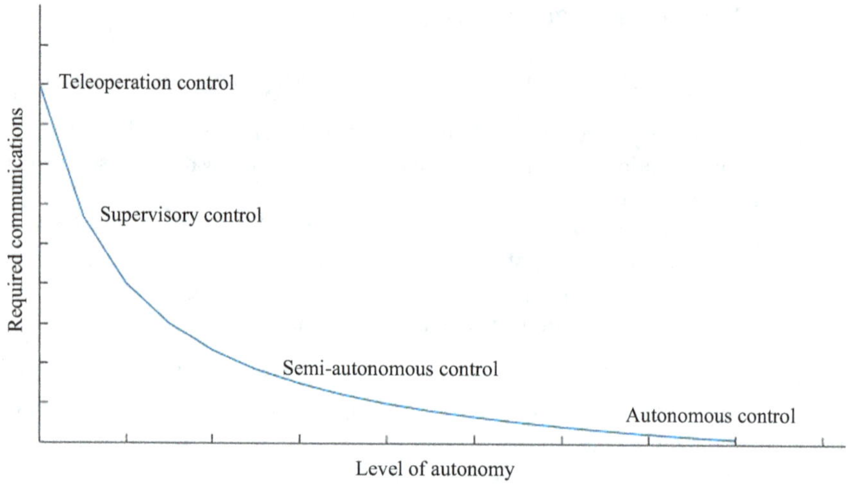

Figure 15.3 Spectrum of autonomy – level is inversely correlated to the communications bandwidth

some degree of autonomy to address losing contact with its operators or after serious failures. AUVs possess modes of operation which they can autonomously switch between depending on circumstances.

The required communication frequency and bandwidth with the operator decreases with increase autonomy (Figure 15.3). With the case of autonomous control, which is of interest here, the operator would not interact with the AUV much. The value of autonomous control is clear when the AUV has poor or intermittent communications with its operators. From this point on, 'autonomy' refers to autonomous control.

With insight into the AUV's capabilities, environment and mission, the autonomy can reason about an anomalous event even if the event is dynamic, only partially observable AUV and environment variables are available, and there is uncertainty in the sensor observations and their adequacy. The objective is to put the AUV in a safe mode and autonomously respond to anomalies to mitigate or suppress their impact on the ongoing mission.

The next section provides a review on AUV fault management work to date.

15.6 Traditional AUV fault management

15.6.1 AUV system hardware failures – reactive responses

Table 15.1 lists AUV sub-systems whose faults and failures are studied. A fair bit of attention has focussed on thrusters, actuators and sensors. These studies address a fault or failure in an individual sub-system. The fault recovery targets that sub-system in one of two ways. First, to change the vehicle model to reflect its debilitated state, or second, to activate behaviours to rectify the fault or failure. A popular behaviour is to abort the mission and surface or to use an alternative method to complete the mission. This can be the result from a fault response table. These static tables are developed by expert operators and coded onto the vehicle computer.

Fault responses can be reactive or deliberative. Generally, reactive responses map pre-set actions to a sensor measurement. For example, a response to a detected large change in pitch may be to abort the mission and surface. This action does not require deliberation. A deliberative response may reason that the large pitch was caused by an abrupt salinity change from correlating past and present salinity measurements with depth-keeping effort. Therefore, the autonomy may choose to recover by adjusting the vehicle ballast and continuing with the mission.

Reactive on-board 'safety rules' screen for, and respond to, basic faults like over-pitch, over-depth, over-roll, unsuccessful dive (after some number of attempts), DVL does not have bottom-lock (speed-over-ground is not available), sensor failure (altimeter, INS, side scan, DVL, etc.), waypoint not reached, battery voltage low or minimum power, leak detected, no forward progress and no upward progress. In many cases, the response is a mission abort and the vehicle surfaces and if able, travels along a safe return path to 'park' or wait for the operator. A survey of reactive fault-detection and fault-tolerance methods has been provided in [4] for AUVs and underwater remotely operated vehicles. The focus was on model-based fault detection for hardware systems.

These measures do not so much constitute fault tolerance as they are responses or vehicle safety rules. To be fault tolerant, there must be some degree of self-recovery possible. With these reactive fault responses, the assumption is that the AUV would be recovered and an operator would fix the problem.

15.6.2 Limitations of traditional fault management

The current state-of-the-art FDIR is based on design-time analysis of faults and failures [32], fault tree analysis [33] and run-time observations of the system's operational status (health monitoring). The objective is timely fault detection and application of recovery actions which are mostly fault response tables. This is usually with the objective of putting the AUV in a safe configuration (e.g. surfaced). Classic FDIR has shortcomings. As mentioned earlier, the system and its environment are only partially observable. Consequently, there is uncertainty in the interpretation of observations for the system status. This prevents on-board procedures from being devised since on-board knowledge and models should take this into account. The AUV is usually surfaced and recovered so that knowledgeable and experienced operators can resolve the fault/failure. The knowledge and experience of the operators is not represented on board.

Classic FDIR provides acceptable results for statically captured system configurations but performs poorly for dynamic aspects such as recovery actions and reconfigurations. Standard on-board FDIR processes do not reflect probabilistic causal dependencies between the faults and general system capability or between the system environment interaction evolution and system dependability characteristics. This is where artificial intelligence methods can contribute to FDIR on AUVs [3]. Needed is a capability to reason about anomalous events based on knowledge of the AUV capabilities, environment and environment interactions in the presence of uncertainty, dynamic evolution and partial observability.

AUV fault tolerance realized through payload deliberative capabilities is discussed next.

15.7 Deliberative AUV fault management

AUV missions involve operating a vehicle in an inherently uncertain environment of which prior knowledge is often sparse or low-resolution. The lack of an accurate prior, coupled with poor situational awareness and potentially significant sensor noise, presents substantial engineering challenges in navigation, localization, state estimation and control. Software responses can be on-line or off-line. On-line software responses, where possible, are of interest due to their timeliness and in-situ ability to reconfigure or re-optimize the hardware and/or software to adapt to a changed condition.

AUV fault tolerance can apply to faults in the vehicle and its sub-systems; fusion of sub-systems for an adaptive solution (e.g. navigation system); the vehicle interaction with its environment (e.g. fighting currents) and the mission. Deliberative fault management is an enabler for long AUV endurance missions. In our experience, the performance of large working class AUVs are limited by their ability to manage faults as much as they are by their endurance. Operationally, the objective of fault management is to increase the AUV's readiness, system availability and performance while at the same time, reducing false alarms and life-cycle costs. There are generally three categories of fault recovery: 100% functional recovery using a redundant component, functional recovery using an alternate solution or a degraded recovery.

Deliberative fault tolerance can use the following tools to achieve its objectives: particle filters, evolutionary algorithms, dynamic BNs, dynamic decision networks, hidden Markov models, fuzzy (fault trees), digital signal processing, ontology-based semantic knowledge representation, design and performance optimization, risk assessment and formal methods (model checking) – to name a few. Some of these will be discussed. The work to date on adaptive mission-planning is discussed first.

15.7.1 Single AUV mission fault tolerance

Mission-planning has a pivotal role in the ability for the AUV to have a persistent presence in an area. The interest here is global deliberative mission planning and not so much local reactive mission planning to avoid obstacles. AUV mission fault tolerance has had less attention than hardware fault tolerance. Mission fault tolerance has focussed on multi-vehicle methods where other vehicle(s) replace those which experienced a fault [34]. Other work has addressed this as an adaptive in-situ path-planning problem [35–37]. The work of [36] underscored the value of adaptive mission-planning for AUV tasks that require long-duration missions like following underwater features like thermoclines and not so much fault response.

Alternatively, when an AUV mission fault occurs, one can consider if portions of the mission are salvageable. This may require the AUV minimize its survey area or modify the mission goals to be more achievable given the fault [38]. Other mission fault tolerance considers graceful performance degradation to work around the fault and achieve partial mission completion. This prevents cascade failures for the AUV that may jeopardize both the vehicle and the mission. This can include isolating the fault to a single sub-system, modifying the mission requirements/re-planning [39], or reconfiguring the system to work around the fault to achieve objectives. Patron *et al.*'s work [39] is a unique approach where a semantic world model is used towards a hierarchical distributed representation of knowledge. Their work was validated through in-water AUV trials where faults were injected.

The work of [40] presents a declarative goal-based approach for adaptive mission-planning that adapts and recovers from failures. This is a development on the static and procedural mission plans still used for maritime missions. The focus addresses embedded service-oriented agents that require higher levels of data representation for contextual, temporal and behavioural awareness. This type of information resides with a human expert or is inferred from processed sensor measurements. The authors validated a semantic-based framework that provided the architecture for knowledge representation, model-based diagnosis and adaptive mission plan in at-sea trials.

The work of [41] builds on this and implements an on-board mission planning fault-tolerance capability with diagnostic (identification) and mitigation (recovery) agents. Their method can address faults triggered by the AUV or the environment. It is implemented through multiple agents for a service-oriented architecture. Mission plan faults are mitigated through re-planning or plan repair. Re-planning develops a new plan based on the current system state and the original goals. Repair uses salvageable portions of the original plan with the intent to save time. However, if the salvageable portions conflict with the original goals their agents revert to re-planning. They tested in simulation on failures in the agent, operator control unit and robot failure during its positioning, docking or survey mission. Their mitigation responses are part of standard operations (e.g. abort mission, come back along safe return path, flag the operator).

15.7.2 Multi-vehicle mission fault tolerance

Deploying multiple AUVs makes it possible to capture the high spatial and temporal variations of a physical, biochemical or geophysical ocean process better especially where the process spans ranges and time intervals that are greater than one AUV could observe. The benefit of multiple vehicles is the scalability of the AUV. Adaptive sampling, however, better exploits the full value of deploying multiple vehicles. The sampling is adapted so that subsequent observations are informed on where and how to measure. For example, to determine the location for a hydrothermal vent the AUV could track the highest concentration of sulphides based on a running gradient descent. The fault-tolerance challenge that is distinct from the single AUV case would be in the common mission. Multi-AUV fault tolerance addresses faults and failures of a team member when it is unable to perform its role the mission. To be fault tolerant is to have the AUVs autonomously modify their common mission or reconfigure individual physical arrangements to manage the fault(s).

There is the obvious redundancy with deploying multiple vehicles where an AUV that can no longer perform its part of the mission can have it re-distributed amongst the others that are still able. This is especially so for missions that are concerned with sensor coverage [42]. Multi-vehicle missions can also minimize fault probability by effectively coordinating mission tasks to optimally perform the mission (such as a survey) with little overlap [43]. A more efficient mission minimizes the probability of a fault occurring.

A weights-based algorithm [44] is used for decentralized coordination of robot swarms to find targets and avoid obstacles. The approach provides an efficient way to cover an area and find all the targets. One potential application is search and rescue. The fault tolerance addresses gaps in communications between the robots to avoid collisions.

AUV swarms could increase the efficiency of environmental monitoring through redundancy and fault tolerance. A load-balancing behaviour [34] divides half the swarm to be on the surface (and limits its activities to conserve energy) and the other half submerges to perform the monitoring. Then, the roles are switched when an AUV becomes low in energy or suffers faults. This means the mission time can be increased, but more vehicles are also deployed.

15.7.3 Fault detection

To be useful, fault detection should be achieved early by detecting incipient effects associated with the fault before it becomes serious. Classic fault detection methods use limit monitoring, trending, transient filtering, etc. More advanced FDIR will also use Kalman filtering, weighted sum-squared residual test, generalized likelihood test, random sample consensus method and simulations. Kalman filtering computes the least-squares estimate of the real state from noisy or drifting sensors. Weighted sum-squared residual (or chi-squared) test accumulates Kalman filter residuals (inconsistencies between the nominal and faulty operating state) over a time window and is more sensitive to small errors than a single-sample test.

15.7.4 Fault identification/diagnosis

Classical fault diagnosis relies on the compilation of knowledge to create an expert system from which rule-based approaches can be used. This is contingent on the adequacy and completeness of the expert system. Given the uncountable permutations of anomalies that could be generated by AUVs in their environments, this is not the best answer.

Model-based diagnosis creates nominal and fault behaviour models of the system. At runtime, they are used by the diagnosis engine to estimate the current system state. An example is Livingstone 2, which is a model-based fault diagnosis and recovery engine. Livingstone 2 can arbitrate and diagnose multiple faults and detect unforeseen ones. It generates actions for the system to recover from the diagnosed faults. It was used on the Earth Observation Satellite 1 [24]. When a fault is detected, Livingstone 2 performs a diagnosis by searching for combinations of faults that are consistent with the observations.

Most notably for AUVs, Livingstone 2 is used on-board the Autosub 6000 AUV as its fault diagnosis and recovery engine. A challenge in applying it to AUVs was that unlike autonomous spacecraft, Autosub 6000 changes its configuration every mission so that diagnostic model had to change constantly. Software behaviours are not modelled as they are assumed correct and fully captured in the commands issued to the hardware (not necessarily true). Another challenge is that the missions change and the Livingstone 2 architecture was not designed for the diagnosis engine to access the mission scripts. Their solution was to automatically generate a diagnosis model of the mission script [23,28] to integrate with the hardware models. They found that the inclusion of the mission script in the diagnosis model allowed them to detect unforeseen faults and, consequently, resulted in a more complete diagnosis capability.

Probabilistic methods like BNs could be applied to AUV fault diagnosis [3] implemented as probabilistic graphical models [45]. This is discussed in the next section.

15.7.5 Fault isolation/recovery

Fault recovery takes the identified fault or failure and gives, as output, a possible path to recover the system functionality. When a failure has been detected and diagnosed, the action for isolation and recovery must be chosen. 'Isolation' refers to isolating the failure component so that is cannot harm the vehicle or its operation. This can be achieved by changing the configuration of the vehicle, for example, by switching to a redundant, spare part or by switching to an operational mode that compensates for and does not use the failed component [31]. Traditionally, the usual recovery actions are chosen to be a safe one like 'surface' which ensures the AUV survival but is not functional. The mission may be interrupted until the operators analyse the problem and put the AUV into a functional configuration. Recovery should place the AUV in a configuration that isolates the failed part but is still as functional as possible.

A recovery method must deal with decisions and be able to assess the risks associated with multiple potential solutions. Probabilistic approaches provide a decision theoretical framework to model the effects of possible actions to estimate the risks or utilities associated with them [45]. This is discussed in the next section.

15.8 Artificial intelligence to support deliberative FDIR

The artificial intelligence techniques to support FDIR are summarized in Table 15.8. Their use is not common place for AUVs, but it is increasing. Their application to mostly space vehicles will be described with the hope that others will consider it for AUVs.

Table 15.2 Tools that can be applied to deliberative FDIR

FDIR mechanism	Intelligence tool
Detection	Neural networks
	State estimation
Identification/diagnosis	State estimation
	Neural networks
	BNs
	MDP
Isolation/recovery	Dynamic BNs
	Partially observable MDP
	Decentralized partially observable MDP

15.8.1 Deliberative fault management tools

This section describes methods and tools to implement advanced FDIR that draw on state estimation, machine learning and probabilistic methods.

15.8.1.1 State estimation

The state (e.g. location) for the AUV as well as its environment is only partially observable and has uncertainty (system evolution, perception, resulting autonomous action). There may be noisy measurements of the state. State estimation techniques provide refined estimates of the AUV state. Only the commonly used Kalman and filters are presented here.

The Kalman filter iteratively uses joint probability distributions to predict state variables that are not directly observable. It fuses measurements and predictions based on the prior state with a weighted average to update the predicted vehicle state. An example was described in section. There are several variants of the Kalman filter such as the extended Kalman filter and the unscented Kalman filter to name a few. These filters are applicable to linear systems and can be robust against noise and uncertainty in the input data. They are widely used in AUVs for navigation [17]. Fault control Kalman filters can be applied to detect state changes [46] or fault isolation [47].

The particle filter is a non-parametric method that can capture multiple hypotheses on the AUV state unlike the Kalman filters which are unimodal and assume Gaussian distributions for the sensor models and state transitions. The particle filter estimates the posterior probability density of state variables given the observed variables. It can be used on non-linear systems and is often used with hidden Markov chains [48]. Particle filters can be applied towards fault detection and identification. They do suffer from large computation cost which can limit their use with AUVs, however; modified particle filters, such as risk-sensitive particle filter or variable-resolution particle filter, could be used to minimize computational load [49].

15.8.2 Learning

A neural network is a computing system which consists of many interconnected processing elements. It is a framework for machine learning algorithms to work together and was originally intended to emulate the human brain. Unlike conventional anomaly detection and diagnosis, neural networks do not need externally supplied parameters (e.g. limit values, rules, system behaviour models), as it can learn them, or their equivalent, given feedback. These systems can learn to perform tasks like fault detection from examples or a training set rather than a human expert [50]. Once a network has been 'trained', it can be applied to perform fault detection (or identification) on data that is different from the training set. They however still need training sets and the results may not be optimal.

Neural networks can be applied to fault detection and identification for AUVs since they can compare the current vehicle state with the desired model and can extract characteristics which can help with identification [51].

Once the neural network is trained then possibly distilled [52], it can be deployed as a single model on potentially an autonomous system. Distilling the knowledge from a neural network is a newly emerging way to transfer the knowledge from machine learning into a smaller form factor. It has not received much attention for fault detection, identification or recovery but has considerable untapped potential.

15.8.3 Probabilistic methods

A partially observable system is one where the entire state of the system is not fully measurable (observed) by the FDIR process/sensor, e.g. the absolute position of a submerged unmanned underwater vehicle. This introduces uncertainty in the interpretation of measurements (observations). Additional challenges for the AUV include the uncertainty and dynamic evolution of the extended duration AUV and its environment. Given this complexity and uncertainty, probabilistic methods can capture fault system models well. Probabilistic reasoning in the form of BNs and model-based approaches (causal networks) are appropriate fault diagnosis tools to consider.

This section presents modelling probabilistic causal dependencies between an AUV state and its faults/failures using BNs and then their dynamic evolution through Markov decision processes (MDPs).

Bayesian networks

Within a complex system, a fault or failure can manifest itself through different observable variables and behaviours. The diagnosis or identification part of FDIR considers all this evidence jointly and infers the most likely common cause – the component that failed. A Bayesian (belief) network (BN) models a system through the joint probability distribution of all the relevant observable (e.g. sensor values) and unobservable/hidden (e.g. whether the sensor works) variables to a system or sub-system. The probabilistic calculus used in BNs lends itself to the use of decision theory [45] to estimate the risks associated with possible future action. Consequently, models based on probability theory are important for safety-critical problems like AUV FDIR. For fault management, BNs make it possible to build and reason about models that relate faults to observations (sensor measurements) and potential recovery actions.

Dynamic BNs
BNs that model sequences of variables are dynamic BNs. Dynamic BNs can naturally address the partial observability, uncertain system evolution and system-environment interaction, as well as the prediction and mitigation of imminent faults/failures. With fault tolerance, they can be used to model and understand cascade failure propagation [53] which is useful towards determining actions to mitigate them.

Decision networks
A decision network is an extension of a standard BN with decision and utility nodes for using the decision-theoretic framework. Decision nodes correspond to controllable variables related to possible actions and utility nodes correspond to the quantities (risks or utilities) be optimized. Solving the decision network amounts to finding values for the decision nodes so that the expected utility is maximized (or risks minimized) given the observed beliefs.

Dynamic decision networks
Dynamic decision networks are networks that map actions and their effect on the state. These networks are dynamic since they can be updated with new information on how actions are affecting the state. These networks are useful for fault resolution since they can model components having more than one state such as faulty or non-faulty [54].

The temporal evolution of the system variables modelled through BNs can be captured through the Markov assumption and a discrete time model. With the Markov assumption, future states only depend on the current state with hidden/unobservable states. This is discussed next.

Hidden Markov model
A hidden Markov model is a simple form of a dynamic BN. These models cannot directly know the current state but only the outcome. Each state has a probability distribution given the observed outcomes. This methodology can be applied to fault diagnosis/identification to determine fault states given the observable malfunctions of the vehicle [55].

Markov decision processes
MDPs are mathematical methods that map actions to their probable effect on their state. These networks apply the Markov assumption where the probability of an event depends only on the state from the previous event and the state is fully observable. They can however become complex depending upon their scope [56].

Partially observable MDPs
Partially observable MDPs (POMDPs) map actions to their probable effect on the vehicle state while maintaining a probability distribution on what the vehicle state is. These networks use Markov's assumption where the probability of an event depends only on the state in the previous time step. Additionally, they account for the lack of

certainty in knowing the actual state. The probability distribution allows the system to approximate its state given the likelihood of being in any state and the probability of the state change given previous actions. POMDPs are applied in fault recovery to choose the best recovery action given the most probable state of the system [57].

Decentralized partially observable MDPs
Decentralized partially observable MDPs (Dec-POMDPs) account for multiple systems/vehicles that can only estimate their state and an effects. This means that the systems had a probability distribution for what their current state is and what their actions will do in changing the state. This method can be used to make decisions for multiple AUVs [58]. Dec-POMDPs are a generalization of the MDPs and POMDPs.

Model checkers
Model checking is a formal method to automatically verify the correctness of a system against a finite-state system model. The AUV as an autonomy-enabled cyber-physical system is not easily amenable to traditional model checking due to its large state space which inevitably causes state explosion. However, the work [59] was able to control the state explosion using model checking by decomposing a system with a large intractable state space into multiple concurrent sub-systems with tractable state spaces. This makes the strengths of model checking accessible for fault tolerance of systems with autonomy. The earlier work of [60] used model checking to assess the functionality of the Autosub 6000 AUV under degraded conditions. Their key contribution was to integrate a fault injection approach into the model to study fault diagnosis and recovery. This proved to be a useful tool for engineers to evaluate their designs.

The metrics to assess the efficacy of FDIR measures is discussed next.

15.9 Metrics for fault-tolerance effectiveness

It is difficult to assess the efficacy of fault-tolerance and survivability methods. First, fault-tolerance measures are custom-built for a system. These systems are not yet modules or commodities that can be purchased commercial-off-the-shelf. Second, fault-tolerance techniques target a sub-system. Finally, fault-tolerance measures have an overhead in terms of extra time required (latency) and resources (memory, bandwidth, processing). This overhead can outweigh the benefits of the fault-tolerance measure. The work of [32] suggested metrics in the form of cost functions that should be optimized for a fault-tolerant system based on requirements for reliability, availability, performance probabilities, maintenance and repair, spare parts, integration and test, development cost and life-cycle cost. These are good broad engineering metrics to start with.

For test and development, Cleveland *et al.* [61] built a framework that assesses the effectiveness of fault-tolerance measures by injecting faults into the system. The faults are injected using aspect-oriented programming which also naturally lends itself to metric collection and logging. The system software itself is not altered to accommodate the fault injection – additional test software is inserted into the system software.

This allows the fault tolerance to be evaluated during development without impacting the system's run-time performance. Not all software architectures are amenable to this type of insertion. They realized their framework for a publish–subscribe architecture which is used in quite a few AUVs. Their approach automatically injects the fault, analyses the results and builds a library of faults for further testing. At the macroscopic level, the system responses to the injected faults are assessed through the volume of message (published and subscribed) traffic and other metrics like message throughput and latency.

The work of Mohsin *et al.* [62] applied to web-centric mobile applications is another framework to quantify the fault-tolerance metrics. Their efforts are motivated by the lack of fault-tolerance metrics for measurement and benchmarking for quality purposes. They identify attributes (e.g. down time, responsiveness, time-to-failure) which could be used towards defining fault-tolerance metrics for hardware and software and they target web-centric mobile application which include cyber-physical systems like AUVs.

The work of [63] also acknowledges the lack of fault-tolerance metrics in robotic systems. They define metrics to assess the impact of fault tolerance on system performance for multi-robot teams. These metrics account for intelligence, reasoning and learning on the system's fault tolerance – beyond basic redundancy of robots or components. This is a notable development beyond traditional fault-tolerance metrics up to that point. They validated their metrics by applying them to a team of heterogeneous robots. Their work is even more relevant as the intelligence on board the AUV has grown to the point where reasoning and learning, as exemplified by deliberative fault-tolerance measures, are emerging as on-line capabilities for AUVs.

A coarser (but no less important) metric of fault tolerance is the AUV survivability, i.e. that the vehicle is recovered. The work of Brito and Griffiths [64] presents data that acknowledge the AUV reliability during a mission depended on whether it was operated by developers or non-developers and there was a marked difference between the two. What made the difference was the on-board fault and failure prevention and mitigation measures. As well, that the efficacy of these measures should be accounted for in the risk profile of an AUV. Towards that, they develop a Bayesian inference framework that is based on the insight into the probability of failure prevention and mitigation and the number of subsequent deployments where the failure does not occur. This was applied to the ISE Explorer (Figure 15.1).

15.10 Concluding remarks

Deliberative AUV fault tolerance has not had as much attention compared to autonomous land or space-based vehicles. This is due to the smaller industry and challenges in communication and control of the vehicles on extended long-range missions. Reactive fault methods for vehicle hardware sub-systems has had some attention but there is far less for deliberative fault tolerance.

15.11 Recommendations

The AUV FDIR currently employed are basic and not adaptable for complex faults. Fault tolerance is a growing field for AUVs which can exploit previous gains in space and other types of unmanned vehicles. There is research into vehicle sub-system fault detection and a growing trend in fault recovery but there is a lack of combined methods for multiple sub-systems with in-water testing of more complex methods. Such integrated AUV solutions do not exist yet in the same way as they do for space systems.

Promising areas to further AUV FDIR development include learning and probabilistic methods.

References

[1] Ernits J, Deardon R, Pebody M. Fault diagnosis challenge in a flight-class autonomous underwater vehicle. In: Proc. 22nd Int. Workshop on Principles of Diagnosis; 2014. pp. 1–4.

[2] Seto M, Paull L, Saeedi S. Introduction to autonomy for marine robots. In: Seto M, editor. Marine Robot Autonomy. New York: Springer; 2012. pp. 1–40.

[3] Seto M, Bashir A. Fault tolerance considerations for long endurance AUVs. In: Proc. Annual Reliability and Maintainability Symposium; 2017. pp. 1–6.

[4] Antonelli G. Fault detection/tolerance strategies for AUVs and ROVs. In: Underwater Robots. Switzerland: Springer International; 2014. pp. 101–116.

[5] Tipaldi M, Bruenjes B. Spacecraft health monitoring and management systems. In: Proc. IEEE Metrology for Aerospace; 2014. pp. 68–73.

[6] Guiotto A, Martelli A, Paccagnini C. SMART-FDIR: Use of artificial intelligence in the implementation of a satellite FDIR. In: Proc. DASIA 2003 (ESA SP-532); 2003.

[7] Holsti N, Paakko M. Towards advanced FDIR components. In: Proc. DASIA 2001: Data Systems in Aerospace; 2003.

[8] What is the connection between equipment risk and equipment reliability and how does it affect your maintenance strategy selection? [homepage on the Internet]. Lifetime Reliability Solutions; 2005-2019 [cited 2018 September]. Available from: https://www.lifetime-reliability.com/cms/free-articles/reliability-improvement/only-path-to-reliability-excellence/link-between-risk-and-reliability/.

[9] NASA's exploration systems architecture study – Final report, Part 8. Risk and reliability. National Aeronautics and Space Administration; 2005. NASA-TM-2005-214062. Sponsored by the National Aeronautics and Space Administration.

[10] Crees T, Kaminski C, Ferguson J, *et al.* UNCLOS under ice survey: An historic AUV deployment in the Canadian High Arctic. In: Proc. MTS/IEEE Oceans Conf. IEEE; 2010. pp. 1–8.

[11] Hansen R, Sb T, Callow H, *et al.* Interferometric synthetic aperture sonar in pipeline inspection. In: Proc. MTS/IEEE Oceans Conf. IEEE; 2010. pp. 1–10.

[12] Camilli R, Reddy C, Yoerger D, *et al.* Tracking hydrocarbon plume transport and biodegradation at deepwater horizon. Science. 2010;330(6001):201–204.

[13] Manley J. NOAA's AUV vision: Status and opportunities. In: Proc. MTS/IEEE Oceans Conf.; 2006. pp. 1–5.

[14] The Navy Unmanned Undersea Vehicle (UUV) Master Plan. United States Navy; 2004.

[15] Wei S, Leung H, Myers V. An automated change detection approach for mine recognition using sidescan sonar data. In: Proc. IEEE Int. Conf. Systems, Man, and Cybernetics. IEEE; 2009. pp. 553–558.

[16] Dinc M, Hajiyev C. Integration of navigation systems for autonomous underwater vehicles. J Marine Engineering and Technology. 2015;14(1):32–43.

[17] Paull L, Saeedi S, Seto M, *et al.* AUV navigation and localization: A review. IEEE J Oceanic Engineering. 2014;39(1):131–149.

[18] Melo J, Matos A. Survey on advances on terrain-based navigation for autonomous underwater vehicles. J Ocean Engineering. 2017;139:250–264.

[19] Eickstedt DP, Sideleau SR. The backseat control architecture for autonomous robotic vehicles: A case study with the Iver2 AUV. J Mar Tech Soc. 2010;44(4):42–54.

[20] Benjamin M, Schmidt H, Newman PM, *et al.* Nested autonomy for unmanned marine vehicles with MOOs-IvP. J Field Robotics. 2010;27(6):834–875.

[21] McPhail S. Autosub 6000 – A leap forward in capability: A long range and deep diving AUV. National Oceanography Centre; 2007.

[22] Griffiths G, Brito M. Predicting risk in missions under sea ice with autonomous underwater vehicles. In: Proc. IEEE/OES Autonomous Underwater Vehicles; 2008. pp. 1–7.

[23] Dearden R, Ernits J. Automated fault diagnosis for an autonomous underwater vehicle. J Oceanic Engineering. 2013;38(3):484–499.

[24] Hayden S, Sweet A, Shulman S. Lessons learned in the Livingstone 2 on Earth observing one flight experiment. NASA Ames Research Center, US; 2005. Document ID: 20060006365.

[25] Kaminski C, Crees T, Ferguson J, *et al.* 12 Days under ice: An historic AUV deployment in the Canadian High Arctic. In: Proc. 2010 IEEE/OES Autonomous Underwater Vehicles. IEEE; 2010. pp. 1–11.

[26] Kumar P, Kumar P. CI/DS-CDMA Scheme for autonomous underwater vehicle communication. In: Proc. IEEE ICC – Workshop on Radar and Sonar Network; 2015. pp. 2501–2506.

[27] Micro-Modem Software Interface Guide. Woods Hole Oceanographic Institution, WHOI Acoustic Communications Group; 2014. Document ID: 401040 Version 3.25.

[28] Ernits J, Deardon R, Pebody M, *et al.* Diagnosis of Autosub 6000 using automatically generated software models. In: Proc. 21st Int. Workshop on Principles of Diagnosis; 2010. pp. 1–8.

[29] Ernits J, Deardon R, Pebody M. Automatic fault detection and execution monitoring for AUV missions. In: Proc. IEEE/OES Autonomous Underwater Vehicles. IEEE; 2010. pp. 1–10.

[30] Technology readiness level [page on the Internet]. National Aeronautics and Space Administration [updated Aug 7, 2017; cited September 2018]. Available from: https://www.nasa.gov/directorates/heo/scan/engineering/technology/txt_accordion1.html.

[31] Seto ML. An agent to optimally re-distribute control in an underactuated AUV. Int J Intelligence Defence Support Systems. 2010;4(1):3–19.

[32] Babcock IV P, Zinchuk J. Fault-tolerant design optimization: Application to an autonomous underwater vehicle navigation system. In: Proc. Symposium on Autonomous Underwater Vehicle Technology; 1990. pp. 34–43.

[33] Aslansefat K, Latif-Shabgahi G, Kamarlouei M. A Strategy for reliability evaluation and fault diagnosis of autonomous underwater gliding robot based on its fault tree. Int J Advances in Science Engineering and Technology. 2014;2(4):83–89.

[34] Amory A, Meyer B, Osterloh C, *et al.* Towards fault-tolerant and energy-efficient swarms of underwater robots. In: Proc. Int. Symposium on Parallel and Distributed Processing Workshops and PhD Forum; 2013. pp. 1550–1553.

[35] Rajan K, McGann C, Py F, *et al.* Robust mission-planning using deliberative autonomy for autonomous underwater vehicles. In: Proc. Int. Conf. Robotics and Automation – Workshop on robotics in challenging and hazardous environments; 2007. pp. 1–6.

[36] Brito MP, Bose N, Lewis R, *et al.* The role of adaptive mission planning and control in persistent autonomous underwater vehicles presence. In: Proc. IEEE/OES Autonomous Underwater Vehicles; 2012. pp. 1–9.

[37] Paull L, Saeedi S, Seto M, *et al.* Sensor-driven online coverage planning for autonomous underwater vehicles. IEEE/ASME Trans Mechatronics. 2013;18(6):1827–1838.

[38] Seto ML. On-line learning with evolutionary algorithms towards adaptation of underwater vehicle missions to dynamic ocean environments. In: Proc. Int. Conf. Machine Learning Applications. IEEE; 2011. pp. 236–240.

[39] Patron P, Miguelanez E, Petillot Y, *et al.* Fault-tolerant adaptive mission planning with semantic knowledge representation for autonomous underwater vehicles. In: Proc. Intelligent Robots and Systems Conf. IEEE; 2008. pp. 2593–2598.

[40] Lane D, Brown K, Petillot Y, *et al.* An ontology-based approach to fault tolerant mission execution for autonomous platforms. In: Seto M, editor. Marine Robot Autonomy. New York: Springer; 2013. pp. 225–255.

[41] Insaurralde C, Petillot Y. Dependability in autonomous maritime vehicles: Building resilience into service-oriented agent robots. In: Proc. Annual IEEE Systems Conference. IEEE; 2016. pp. 1–6.

[42] Sujit P, Zhang Y, Healey A. Multi-AUV fault tolerant adaptive sampling. Presented at 2013 Robotics: Science and Systems Workshop; 2016. p. 8 pages.

[43] Amory A, Meyer B, Osterloh C, *et al.* Towards fault-tolerant and energy-efficient swarms of underwater robots. In: Proc. Int. Symposium on Parallel and Distributed Processing Workshops and PhD Forum. IEEE; 2013. pp. 1530–1553.

[44] Aniketh R, Manohar E, Yazwa G, *et al.* A decentralized fault-tolerant weights based algorithm for coordination of swarm robots for a disaster scenario. In: Proc. IEEE Annual India Conf., INDICON 2016. IEEE; 2016. pp. 1–5.

[45] Codetta-Raiteri D, Portinale L. Dynamic Bayesian networks for fault detection, identification, and recovery in autonomous spacecraft. IEEE Trans Systems, Man, and Cybernetics: Systems. 2015;45(1):13–24.

[46] Nabil E, Sobaih A, Abou-Zalam B. Active fault-tolerant control system. In: Proc. Int. Conf. on Computer Engineering & Systems. IEEE; 2010. pp. 274–279.

[47] Zhu Y, Cheng X, Wang L, *et al.* An intelligent fault-tolerant strategy for AUV integration navigation systems. In: Proc. Int. Conf. Advanced Intelligent Mechatronics. IEEE; 2015. pp. 269–274.

[48] Zhao, B., Blanke, M., Skjetne. Particle filter for fault diagnosis and robust navigation of underwater robot. IEEE Transactions on Control Systems Technology. October 2014; 22(6):2399–2407.

[49] Verma V, Gordon G, Simmons R, *et al.* Real-time fault diagnosis [robot fault diagnosis]. IEEE Robotics & Automation Magazine. 2004;11(2):56–66.

[50] Yairi T, Kawahara Y, Fujimaki R, *et al.* Telemetry-mining: A machine learning approach to anomaly detection and fault diagnosis for space systems. In: Proc. 2nd IEEE Int. Conf. Space Mission Challenges for Information Technology. IEEE; 2006. pp. 1–8.

[51] Liu Q, Zhu D. Fault-tolerant control of unmanned underwater vehicles with continuous faults: Simulations and experiments. Int J Advanced Robotic Systems. 2009;6(4):301–308.

[52] Hinton G, Vinyals O, Dean J. Distilling the knowledge in a neural network; 2015. arXiv:1503.02531 [stat.ML].

[53] Bobbio A, Codetta-Raiteri D, Montani S, *et al.* Modeling cascading failure propagation through dynamic Bayesian networks. In: Proc. Workshop on Dependable Control of Discrete Systems. IFAC; 2009. pp. 209–214.

[54] Portinal L, Codetta-Raiteri D. Using dynamic decision networks and extended fault trees for autonomous FDIR. In: Proc. Int. Conf. Tools with Artificial Intelligence. IEEE; 2011. pp. 480–484.

[55] Alvarez MDL, Lane DM. A hidden Markov model application with Gaussian mixture emissions for fault detection and diagnosis on a simulated AUV platform. In: Proc. MTS/IEEE Oceans Conf. IEEE; 2016. pp. 1–4.

[56] Beccuti M, Codetta-Raiteri D, Franceschinis G, *et al.* A framework to design and solve Markov decision well-formed net models. In: Proc. 4th Int. Conf. Quantitative Evaluation of Systems. IEEE; 2007. pp. 165–166.

[57] Fernández J, Sanz R, Diéguez A. Probabilistic models for monitoring and fault diagnosis: Application and evaluation in a mobile robot. J Applied Artificial Intelligence. 2004;18(1):43–67.

[58] Liu M, Sivakumar K, Omidshafiei S, *et al.* Learning for multi-robot cooperation in partially observable stochastic environments with macro-actions. In: Proc. 2017 IEEE/RSJ Int. Conf. Intelligent Robots and Systems; 2017. pp. 1853–1860.

[59] Liu H, Yang T, Wang J. Model checking for the fault tolerance of collaborative AUVs. In: Proc. IEEE 17th Int. Symposium on High Assurance Systems Engineering. IEEE; 2016. pp. 244–245.

[60] Ezekiel J, Lomuscio A, Molnar L, *et al.* Verifying fault tolerance and self-diagnosability of an autonomous underwater vehicle. In: Proc. 22nd Int. Joint Conf. on Artificial Intelligence; 2011. pp. 1659–1664.

[61] Cleveland J, Loyall J, Hanna J. An aspect-oriented approach to assessing fault tolerance. In: Proc. IEEE Military Communications Conference. IEEE; 2014. pp. 1374–1381.

[62] Mohsin A, Naqvi S, Khan A, *et al.* A comprehensive framework to quantify fault tolerance metrics of web centric mobile applications. In: Proc. Int. Conf. on Communication Technologies. IEEE; 2017. pp. 65–71.

[63] Kannan B, Parker L. Fault-tolerance based metrics for evaluating system performance in multi-robot teams. In: Proc. Performance Metrics for Intelligent Systems Workshop; 2006. pp. 1–8.

[64] Brito MP, Griffiths G. Updating autonomous underwater vehicle risk based on the effectiveness of failure prevention and correction. J Atmospheric and Oceanic Technology. 2018;35:797–808.

Chapter 16

Designing, building and testing a fleet of novel AUV under the elevated requirements of an XPRIZE competition

Gunnar Brink[1]

The chapter describes the application of agile management methods in large hardware projects using the example of the development of autonomous underwater vehicles (AUVs). Our case study concerns the development, construction and testing of a fleet of five new AUVs with associated unmanned surface vehicles (USVs) over a period of less than 2 years. First, however, the most important terms of classical project management and agile methods are compared. The *Agile Manifesto* and agile management methods including SCRUM and Kanban, in comparison with SCRUM alone, as well as eXtreme programming (XP) from software development, are presented. The possibilities of agile methods in the development of physically based systems such as automobiles are shown by the example of eXtreme manufacturing (i.e. Wikispeed). The experiences of the project itself, along with its participation in the Shell Ocean Discovery XPRIZE, are presented and discussed, deriving from it a development methodology.

16.1 Introduction: speed versus cost

Many different requirements play a role in the development of AUVs. Vehicles need to be easy to transport, easy to deploy and recover, able to withstand underwater pressure, to be able to navigate without access to Global Positioning System (GPS), to communicate between vehicles and many other requirements. The development of such a project of course means setting up an interdisciplinary team that is well-coordinated. Yet it is not only the team that plays a decisive role in the development process: When managing larger projects, it is always essential to find a balance between different dimensions of project management and at the same time be able to guarantee proper project results. Most readers are probably familiar with the 'iron triangle' model. It is formed by the three dimensions of time, cost and performance and can serve as part of the planning basis, but it also plays a role as a basis for risk management and as a reference point for project changes.

[1]Fraunhofer Institute of Optronics, System Technologies and Image Exploitation IOSB, Karlsruhe, Germany

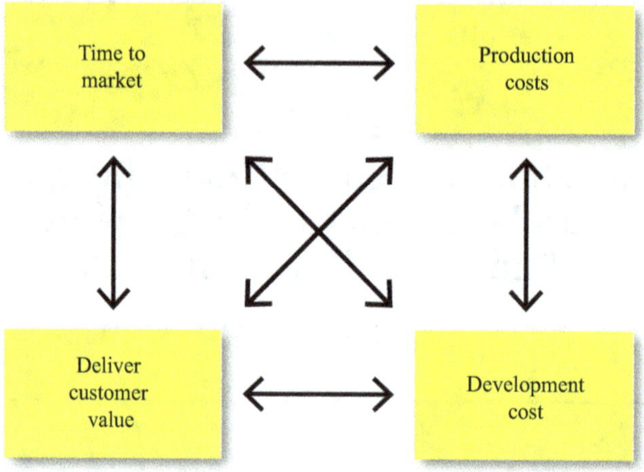

Figure 16.1 The four dimensions of product development according to Smith and Reinertsen

Because 'time to market' is becoming increasingly important as a dimension, Smith and Reinertsen postulated the model of 'four trade-offs' in their much-quoted book [1]. In contrast to the iron triangle, it takes into greater account the increasing time pressure in project management. The costs are also considered to be more differentiated, and the two dimensions of product and development costs are separated from each other. Smith and Reinertsen worked out a total of four dimensions to make a project as efficient as possible (see Figure 16.1).

The first dimension, development speed, is defined as the time between the first moment in which the development programme is dealt with and the finished product, wherein a large quantity, is available for customers to purchase.

The second dimension is product cost, which includes production, sales and marketing costs. It is the total cost of the product delivered to the customer. Product performance is defined as the third dimension. It describes how the product is evaluated, in the eyes of either customers or users. The last dimension consists of the development costs, including all costs that are incurred during the development of a specific project.

Since it is hardly possible to optimise all four of these dimensions simultaneously, it is necessary to make compromises between these individual dimensions. This arrangement applies to the three dimensions time, cost and performance of the iron triangle, too.

In classical project management, it is assumed that the optimisation of a single dimension always leads to a deterioration in relation to another dimension, i.e. one cannot accelerate a project without sacrificing either cost or performance. In this view, a reduction in costs delays product completion, whereas accelerating the project must be bought at a higher cost.

Industrial psychology offers a completely different perspective on the speed of development. The so-called flow is the mental state first proposed in the 1970s by the positive psychologist Mihaly Csíkszentmihályi. In flow, people are so absorbed in a sense of energising focus, full of participation and joy in the process of activity that they lose the sense of space and time. Or as Steven Kotler describes it in *Rise of the Superman*,

> In the state [of flow], every action, each decision, leads effortlessly, fluidly, seamlessly to the next. It's high-speed problem solving; it's being swept away by the river of ultimate performance. Over 100 years of research shows that flow sits at the heart of almost every athletic championship; underpins major scientific breakthroughs; and accounts for significant progress in the arts […] In recent years, flow has also become exceptionally critical to business.

According to the so-called Yerkes–Dodson Law, performance increases to a certain degree of stress when physiological or mental excitement (stress) increases [2].

If one takes into account the findings of Csíkszentmihályi and the industrial psychology work that has been built upon it, three aspects need to be given special attention:

1. formulating clear objectives in relation to tasks and activities;
2. a good balance between the challenges of the activities and the understanding of one's own abilities; and
3. good intermediate feedback and early adaptation of requirements to increasing competencies.

We have taken further information from critical chain project management, which was developed and published by Eliyahu Goldratt in 1997. It is based on the basic assumption that before the start of a project every task is subject to a certain degree of uncertainty. The duration of each task is often overestimated by the team members or task owners. This is typically done to add a safety margin to the task to ensure that it is completed within the specified time. In most cases, it would, therefore, be possible to complete the tasks earlier than the estimated time that the margin allows for. Even if the assumed margin of safety is not needed, it is often wasted in practice: If the task can be completed earlier, the successor task may still not start earlier because the resources required are often not available until the scheduled time. In other words, the time saved cannot be passed on so that the project is completed ahead of schedule. However, if there are delays that go beyond the estimated timelines, they are always passed on and, in most cases, delay the overall project. On the basis of the above assumptions, the critical path methodology of project management recommends bundling the task buffers of each task and only scheduling them at the end of the critical path.

Because of our tight schedule, we had been looking for a way in which an interdisciplinary team that from the outset would not include an unnecessary buffer in

Figure 16.2 This image which was taken at the Shell Ocean Discovery XPRIZE final in Kalamata, Greece, shows one crew member and three of the five USVs

the individual tasks to be completed. Hence, current workload would always adapt to the increasing self-confidence of the team, providing a way for the team itself to be maximally motivated to do the right thing, wherein each team independently seeks and finds new solutions.

In our search for a suitable project management approach, we doubted that traditional approaches were suitable for us. First, the project challenge was highly ambitious. It was a project in a competition organised by the XPRIZE foundation. Their motto in their own words is 'to create the breakthroughs that pull the future forward, make the impossible possible by incentivising great minds to make a difference and bring us closer to a world of abundance [_sic_]'.

When we spoke to a CEO of an existing company, which is developing AUV technology, before we started our project, we heard that their internal estimate was that it would take 25 million dollars and 5 years at normal development speed to meet the requirements of Shell Ocean Discovery XPRIZE (see Figures 16.2 and 16.3).

However, our situation required us to manage with about one-tenth the budget estimated by those professionals and less than one-half the time.

We needed to build on new, creative ideas that would only emerge during the course of the project. We bought components, such as our Inertial Navigation and Doppler velocity log (DVL) systems and the real atomic clocks that we used, which are built in small quantities. Neither item was full featured with all functions being documented, nor was customer-oriented service available.

Most of our team members did not come from shipbuilding or the construction of submersible vehicles, but rather from outside, such as from the fields of embedded

Figure 16.3 *After about 1 year after the project start, five autonomous deep sea robots were built and ready for the competition*

programming, mechanical engineering or sensor technology because everyone we talked to, who came from 'the trade', considered the tasks unsolvable.

It was impossible for us to estimate beforehand how quickly they would become familiar with the diverse topics, what creative solutions they would come across and how quickly their own technical expertise would develop. In this situation, how could it have been possible, at the start of the project, to develop a meaningful project plan that could estimate and plan the appropriate periods and the right financial requirements for the respective task up to a year in advance?

Instead, a completely different approach to classical project management methods was followed by agile methods, which are regarded to be particularly suitable for the implementation of a complex project. This agile approach to project management and, in particular, the decision not to plan the project from the beginning through some large-planning exercise, steered the project iteratively, in individual time slices. In doing so, focus was particularly placed on the current self-assessment of the team's efficiency.

16.2 Agile project management

Within software development, agile principles were formulated in the *Agile Manifesto*, a common set of overarching values and principles. It describes 4 core values for enabling powerful teams (Box 16.1) and a set of 12 principles (www.agilemanifesto.org).

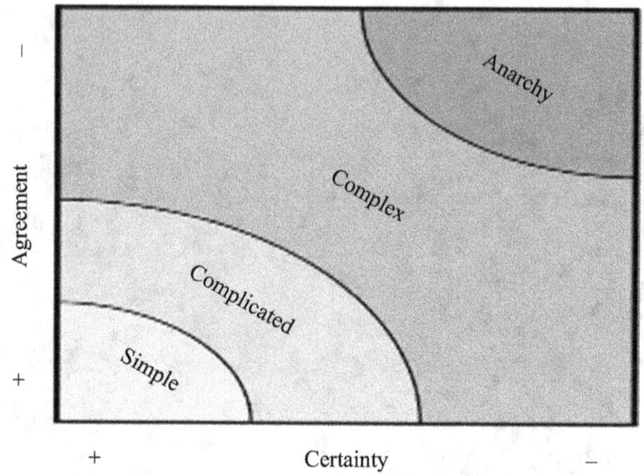

Figure 16.4 The Stacey graph

Box 16.1
The Agile Manifesto
We are uncovering better ways of developing software by doing it ourselves and keeping others doing the same. Through this work, we have come to value

- individuals and interactions over processes and tools,
- working software over comprehensive documentation,
- customer collaboration over contract negotiation and
- responding to changes over rigidly following a plan.

That is, although there is value in the items on the right, we assign greater value to the items on the left.

(Agilemanifesto.org)

When should agile methods be used? The so-called Stacey complexity model is often used to solve this question [3]. It divides projects into four different categories, simple, complicated, complex and anarchistic, and then suggests the best approach for each category.

In this model, there are two dimensions: first, the uncertainty of the requirements, which indicates how secure (or uncertain) the project requirements are and how likely it is that the requirements will change in the future.

The first dimension says something about the degree of uncertainty about what to do or build. The second dimension is the uncertainty of the technology, which shows how uncertain the chosen approach or technology is to accomplish a task or create an application. It indicates the degree of uncertainty involved with *how* to build a project.

Simple tasks are well defined and easy to solve. Classic project management is very easy to use and perhaps even dispensable for those tasks where it is clear what needs to be done and how to approach it. However, with complicated tasks, it is not quite clear what to do and how to approach them. Tasks of this type can easily be solved by classical project management. If things do not work out, a task can be tracked and corrected using a simple Deming cycle approach.

In the anarchy realm, one cannot predict anything and nobody knows how to do things. This is often the area for those who want to develop completely new technical foundations for a new technological paradigm. Unfortunately, such challenges are not that rare and can even occur several times during a project cycle with new technologies if unforeseen technical problems arise that could not have been expected before the start of the project [4]. Such tasks require particularly creative scientists or inventors and a high degree of freedom to find the right solution.

Complex tasks show a rather high degree of uncertainty about what to do and how to do it. In our opinion, the development of new autonomous diving robots usually falls under this category. The requirements and technical challenges are not clear from the beginning and change a lot during the development of its mechanics, electronics, sensors and software. And also, the technology is not so clear upfront. Typically, nobody knows from the beginning how to tackle certain problems, as will be explained shortly; one often does not even know what problems will be encountered.

Since one cannot plan for everything perfectly from the start anyway, it is best to spend relatively little time planning all the details at the beginning of the project and simply start it as early as possible. Further planning is shifted to the iterations in the project. This shift is, in a nutshell, the core of the agile approach. Experience has shown that doing so allows even complex tasks to be solved within a relatively short timeframe.

The *Agile Manifesto* was drawn up and signed in 2001 by software developers representing the various agile methods. The *Agile Manifesto* provides a guide for those who want to use agile methods in their projects. Agile here means that one can react quickly to changing conditions.

The best-known agile methods include SCRUM, Lean Kanban, XP and Crystal [5]. They are supposed to help to do the double work in one-half of the time [6].

Hardly any project uses an agile method in its purest form, but elements of several agile methods can usually be found in combination of actual project management.

SCRUM was developed in the early 1990s by Ken Schwaber and Jeff Sutherland and is used especially in the development of complex products. The term *SCRUM* is based on rugby and means utilising as much of a crowd as possible. While traditional project management is compared with a relay race where each runner has to cover his distance alone and then pass on his result, in rugby, several players cover the distance together by playing the ball flexibly back and forth.

The SCRUM team consists of an interdisciplinary and self-organised development team. There are no subdivisions within the team; the whole team is accountable for its work. It is assumed that there is never a finished version of a product, but only versions that can be delivered.

Table 16.1 Organisation of the SCRUM method

Box/level	Project	Epic	Sprint	Daily SCRUM
Period	2–3 years	2–3 months	1–4 weeks	Daily
Documentation	Backlog	Backlog	Task board	Daily task

SCRUM assumes that the project is planned in small steps and in detail. It works in small teams with a maximum of nine members, based on the assumption that larger teams become too unproductive.

The work process consists of prescribed activities that have a time limit (a 'time box'). The organisation of the tasks takes place in the SCRUM approach on three levels, i.e. the three different 'time boxes', which envisage different time periods (see Table 16.1). The first level represents the project and covers a period from 1 to 3 years. The second level consists of the 'sprint', which can last from 1 to 4 weeks. Among other things, the team defines the tasks until the next sprint meeting. The smallest level is the 'daily SCRUM', in which all team members update themselves daily on their work and discuss what they have planned for the next daily SCRUM and the difficulties they are facing. Through these regular meetings, it is possible to get regular feedback and adapt the development as quickly as possible.

From time to time, another level is used, which is indispensable for us in a complex project like AUV development: 'epics' or 'epic stories' are defined as a summary of backlogged items that would be difficult to estimate in this form or to complete within a given iteration.

Epics are often used as placeholders for new ideas that have not yet been fully elaborated. Epic stories may be used to create an initial backlog. Later, however, they should be broken down into smaller items that are easier to manage and where the requirements of each item are much more precisely defined. Epics serve as placeholders for extensive requirements, better overview of the entirety of all stories.

All known product requirements are recorded in a product backlog. All items that are called for the next sprint are recorded in the sprint task board. During the sprint, the team can work on all items from the sprint task board without being disturbed or influenced from inside or outside. At the end of each sprint, a sprint review takes place in which the team presents the new functionality, or the new product increment.

The SCRUM method also includes three personnel levels: the product owner, the SCRUM master and the team (see Table 16.2). The product owner is responsible for the work of the entire development team and for maximising the value of the product. The product owner is the only person responsible for managing the product backlog, and the entire organisation must accept his or her decisions. A SCRUM master is responsible for understanding and implementing SCRUM. He or she helps to collaborate with outsiders and removes obstacles to the project.

The team carries out the tasks defined in the daily and sprint meetings.

Table 16.2 Roles and related tasks in the SCRUM approach

Role team	Product owner	Scrum master	Team
Task	Defines the requirements and prioritises backlog sequence	Regulates the timings and eliminates obstacles	Determines the scope of tasks to be processed in parallel and carries them out

A SCRUM sprint is the result of various work steps. First, all items for the product are recorded in the backlog, with exact details of what each task contains.

Since a realistic estimation of the effort required for the individual tasks is often difficult and time-consuming, the approach of using numbers from the Fibonacci sequence has proved its worth. The listed tasks are evaluated with integers from the Fibonacci sequence, relatively assigned according to difficulty, starting with the most difficult and longest period items. After this step, the work sprint is planned. The team estimates which tasks they can complete within the given time period, based on previous sprints.

The work within the current sprint is made visible by creating a task board. This consists of the three columns: do, doing and done. In this table, the tasks are sorted according to their respective stage of development.

During the sprint, stand-up meetings of about 5–15 min are held every day to explain yesterday's and today's work and the difficulties encountered. These meetings are led by the SCRUM master. At the end of each sprint, there will be a live demonstration of the existing product increment. The team will reflect on the past sprint together before starting another sprint. Important points in the use of the SCRUM concept are transparency, review and customisation.

Like SCRUM, Lean Kanban is also a pull system, which means that both ensure that the product comes from the product backlog to the customer as quickly as possible. Both frameworks are very similar, but there are some differences. The SCRUM master role at Kanban is called the *agile coach*. Unlike SCRUM, Kanban does not consist of sprints, but is a continuous process. The pull system also works differently. The tasks are divided into the areas build, test and done. As soon as there are no more tasks in the '*to do* area', the team itself adds new tasks from the backlogged projects list.

One of the essential aspects of Lean Kanban is that it limits the amount of work that can be done at the same time, i.e. work in progress (WIP). The limitation of WIP depends on the development capacity of the team. In general, a WIP limit is set for each column and sprint. However, it is also acceptable to set limits per person or per team. Setting a WIP limit for a particular column does not mean that the team cannot add another task to that column. Rather, it means that if the limit is exceeded, the entire team should take responsibility for it. Members should try to understand why such a situation has occurred so to improve it and prevent a WIP overshoot from recurring.

Another well-known agile method is XP. It implies that all levels are cyclical. The smallest unit is pair programming. Two people always code together, so that there is constant mutual control and errors – or a cumbersome design – can be pointed out as quickly as possible. Constantly, nearly in minute cycles, permanent checks of the code take place, as this is where test-driven development occurs.

Another important feature of XP is that the unit tests are written before the code. In this focus on test-driven development, the tests that are constantly carried out play a special role. As the name suggests, the tests then drive the development of the product. So, it is not built first and then the test is designed, but the development is subject to constant testing, so that difficulties can be dealt with flexibly. Code integration still takes place, which means that the newly developed components are integrated into the executable overall system at least daily. This means that errors can be detected very quickly.

As with SCRUM and Kanban, daily stand-up meetings take place in which the development team briefly reflects on the progress of the project. The regular exchange prevents undesirable developments. A weekly iteration plan ensures that the customer always receives executable system versions and can then test them. Communication can help to keep technical requirements up to date.

Very short release cycles, which are usually limited to 1 month, generate developments into the business world at a very early stage. Thus, projects developed with XP should be self-financing much earlier than classically developed projects.

The following order in test-driven software development is based on the Beck scheme [7].

According to this scheme, software development consists of five consecutive steps that are repeated cyclically:

1. **Test generation**
   In test-driven development, each new feature begins with the writing of a test. You write a software test that defines a function or improves one, which should be concise. To write the test, the developer must clearly understand the specifications and requirements of the feature. He or she can achieve this through use cases and user stories to cover the requirements and exceptional situations, thus coding in the appropriate test framework for the software environment.

   This step is a key differentiator of test-driven development from writing unit tests after writing the code. It makes the developer aware of all requirements, along with indicating what is *not* required, before writing the code.
2. **Testing the test execution**
   At this point, all tests are performed, including the new test. This test should fail, because the new code has not been implemented yet. This fact confirms that the automated test framework works correctly, shows that the new test will not pass without new code because the required capabilities are already present and excludes the possibility that the new test is buggy and will always pass.
3. **Code development**
   The next step is to write code that ensures that the new test is passed. The new code written at that time is not perfect, and it is not a problem if it passes the test

in an inelegant way. The programmer must not write code that goes beyond the functionality the test checks.

4. **Test execution**
   If all tests are now passed, the programmer can be sure that the new code meets the test requirements and that no existing functions have been damaged or degraded. Otherwise, the new code must be corrected.
5. **Refactoring the code**
   The growing software must be regularly adjusted and optimised during test-driven development.

16.2.1 Agile project management using the example of a hardware development project

One of the best-known examples of the successful application of agile methods is found in a hardware project called Wikispeed. Under the direction of Joe Justice in Seattle, the Wikispeed project succeeded in developing a roadworthy car within 3 months. The Wikispeed claims that in some respects, it even exceeded the standards of normally produced cars. With the development of this car, the 44-member team took part in the Progressive Automotive XPRIZE competition from 2008 to 2010 [8,9].

Due to the very short development cycles and the high quality, Wikispeed now sees itself as a competitor with other car manufacturers. Their development is oriented towards customer demand. Regular tests avoid undesirable developments. These tests are defined before the work begins, so that each team member is precisely informed about the goals to be achieved. Contrary to the principles of SCRUM, the Wikispeed team consists of considerably more than nine people, and the resulting difficulties are to be resolved by better communication. As with XP, Wikispeed always develops in pairs, as this principle reduces costs and accelerates innovation. Also, the training time and the necessary documentation can thereby be reduced. In order to facilitate necessary changes, the entire development is kept as uncomplicated as possible. Communication within the team is always peer-to-peer, so that there is no classic hierarchy that could restrict the team in development.

The agile methods are all based on the assumption that there is a stable team with a lively exchange. Frequently changing team members must not be part of the plan, at least in theory.

16.2.2 Description of our case study project

The Arggonauts are one of 32 teams from around the world that participated in the Shell Ocean Discovery XPRIZE. The participating teams will develop AUVs capable of mapping the seabed and taking high-resolution images. In the competition, people were not allowed to stay at sea to steer or control UAVs or unmanned boats.

The timing of the competition was also extremely ambitious. By the time of the semi-finals in November and December 2016, the Arggonauts had succeeded in meeting 11 criteria. Less than 1 year after the semi-final, the final had to be carried out with a real sea test at up to 4,000 m of ocean depth. However, the project was financed only 9 months before the semi-final, in February of the same year, putting the team under enormous time pressure. In contrast to the normal development times in AUV development, these periods were extremely short.

The Arggonauts passed the semi-finals and advanced to the final, which took place in December 2018, as one of five final teams.

A special feature of our case study is that the Arggonauts are not a permanent team, since many employees permanently employed at Fraunhofer are mostly involved in projects on a long-term basis and could not be released at short notice for the duration of the competition. In order to ensure the necessary development capacity with such short notice, the team consisted of a mixture of permanent employees, students, volunteers and freelancers.

16.2.2.1 Selection of software platform

We decided to use the middleware Robot Operating System (ROS) as the framework for software development. Most parts of our software were realised in C++. We had sold our previously developed proprietary software exclusively to another company, so there was no way around a new development. The criteria for the selection were how well the middleware was suited to support unmanned maritime vehicle development, how easy it would be to find and integrate new, experienced developers and whether the use of the middleware could cause licensing problems if the technology were to be sold.

In addition to ROS, we considered the use of MOOS (Mission Oriented Operating System), which is even more tailored for use in AUV, and LCM (Lightweight Communications and Marshalling) when selecting the optimal middleware platform for our project. These and many other middleware variants are compared by Li *et al.* [10].

One argument for the most widely used software platform for robots, ROS, was that ROS was developed from the beginning with the goal of promoting collaborative robotics software development. Due to communication via TCP/IP, ROS can be programmed in different languages: C++, Python and Java are supported. Ultimately, however, the selection of ROS for use in our AUVs and USVs was based on the skills and experience available in our team and in the accessible freelancer community.

16.2.3 Introduction of the team to an agile development methodology

The team of the first hour was trained in a 2-day intensive SCRUM training in this agile development method. In the first half year of the project, experienced SCRUM masters supported us as externals at the daily SCRUMs. Afterwards, members of the team took over this role.

The daily SCRUM meetings were well rated by our team members, especially by those who were familiar with how each scientist stands and works alone at a laboratory bench in classical research. Moreover, only technical results were communicated among members during the 10 min of daily exchange. The team members highly appreciated that this ritual strengthened the feeling of togetherness and enabled them to think and cooperate with others.

Another observation relates to the selection of backlog items by the team. The literature on SCRUM suggests that product owners tend to burden teams with too many tasks. Therefore, SCRUM assumes that the team itself plays a role in deciding how many items from the backlog it selects at the beginning of each sprint and plans for the next few weeks.

In our project, we had the opposite experience. In view of the high ambitions and the external pressure of the XPRIZE competition, the team took on more tasks in almost every sprint than they could handle. As a result, the feeling of success was missing. Individual tasks were often not moved from the 'to-do' category to the in-progress and done columns every day. Even at the end of the sprint, the task board left column was often not left empty.

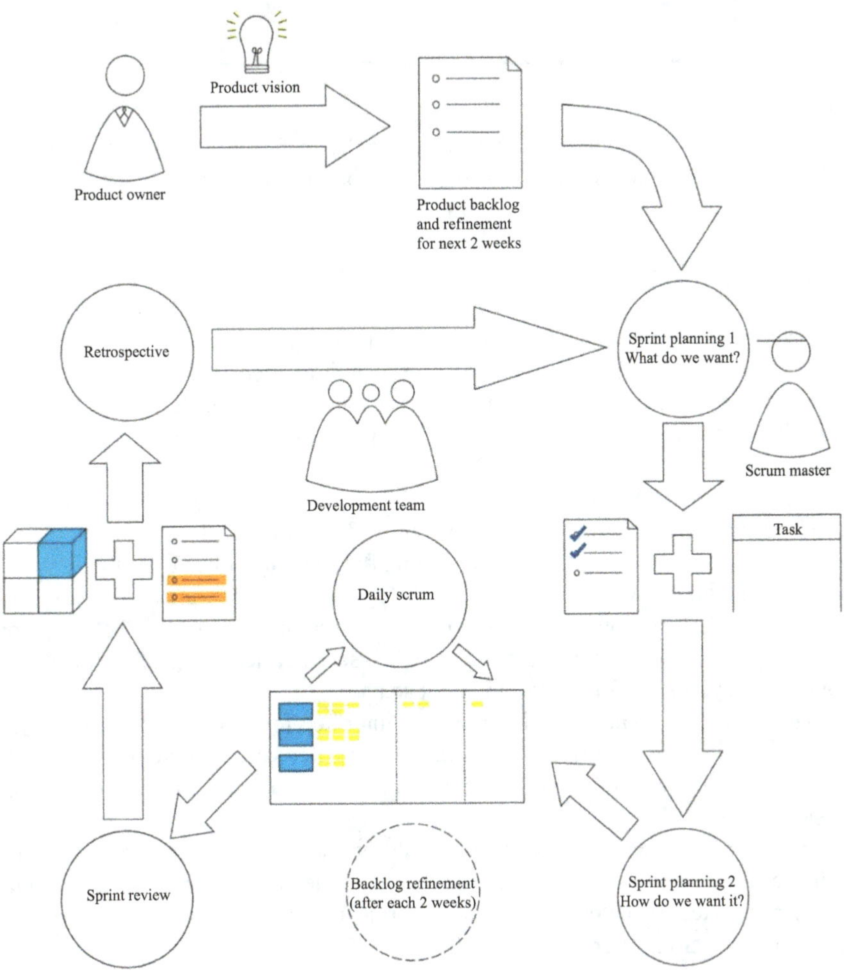

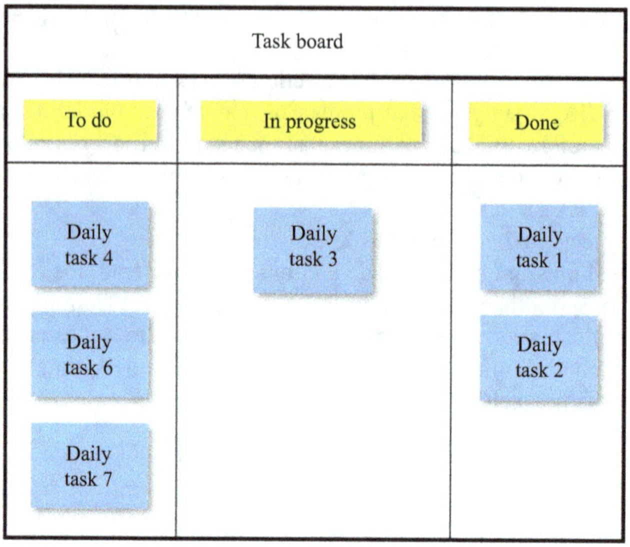

To improve the situation, we decided during the course of the project to apply the WIP concept and allow the teams to add tasks from the backlog, but never to accept more than three or four tasks in a column.

16.2.4 Test-driven development

Test-driven development plays a special role in the two agile methods: Crystal and XP. But how can the above five steps be sensibly applied to the development of a mechatronics system with mechanics, electronics and software and, even more difficultly, to the development of a fleet of ten vehicles and their control centre?

We researched and found no suitable answers to this question in the literature. The literature on the Wikispeed project did not help us either. Thus, we experimented ourselves and gleaned some insights that can be found in the following sections. We only identified many of them upon reflection after the project had ended.

In a comparable situation, in software development in the future, we would place the building of a user interface for programming and monitoring a vehicle mission relatively early in the development and take the user interface and the functionality it provides as a yardstick for all development steps.

The next step would be to build a user interface for testing and experimental parameterisation and determine how good the parameterisation should be. What scale should be chosen to evaluate the inevitable waviness in the vehicle path during the mission?

Instead of the control interval, we would probably take the length of the path of the actual mission compared with the length of the desired ideal path as a single quality parameter and determine before the project which lengths we will accept for the planned mission patterns.

Such an early creation of a user interface has the additional advantage that one does not have to go to the test area with all team members for every optimisation.

Third, next time we will design the user interfaces for hardware acceptance tests relatively early. The pressure vessels of our AUVs contain purchased electronic components, single board computers and home-made electronic components that require hundreds of electrical connections.

A constant concern was the crosstalk of weak sensor signals in the electronics and their being interfered with by high voltage from the sonar, the acoustic modem or the high currents at start of the vehicle's deployment.

Investing in an appropriate user interface at the beginning would have accelerated the acceptance tests of purchased components and externally outsourced wiring work. It would have ensured that our computer scientists would have had their heads free for other tasks in the later phases of the project, such as when the big hour of software development would actually was, and that other team members would have been able to relieve them of more routine test tasks.

Immediately after defining the hardware acceptance tests, we would have had to define electrical and electronic interfaces. The high pace of prototype development brings with it the temptation to use simple plugs to connect different boards and components. But the complexity leads to the fact that – at least in our context with beginners, all-rounders, external assemblers, service providers and students – one has a lot of effort to look for connection errors, to rule out the possibility that new connection errors will be generated during a correction.

The next time we would put much more thought into suitable and error-avoiding connections than the impatient common sense would suggest.

After about 4 months of development, when the very first hardware prototype was ready, our weekly tests began in the quarry lake. We used a quarry lake that was about 800 m wide and up to 70 m deep for routine tests. During the following year and a half, we also tested twice in the Baltic Sea, once in the Bay of Biscay in the Atlantic and twice off Greece in the Mediterranean Sea.

It is still very difficult to obtain permits for tests abroad, even within the so-called 'European Research Area' of the EU Commission which should be 'a unified area open to the world, in which scientific knowledge, technology and researchers circulate freely'. Despite this European Research Area, developers and scientist still must involve the highest governmental bodies and apply up to 6 months prior to testing in order to obtain such approvals abroad.

In our experience, however, despite the high effort to get a permit, these mostly 2-week test missions abroad were worth 'their weight in gold'. They not forged the team's togetherness; they also resulted in many new insights being gained each time, which have flowed into the further development and improvement of the system.

Some tests, such as the buoyancy of the vehicle at different salt contents of the water, could be carried out in the pool. Furthermore, a separate pressure chamber was available to test individual components of the vehicle for sufficient compressive strength.

Externally, hydrodynamic tests could be carried out in a flow channel. Weekly tests could be carried out in the mentioned quarry lake, a branch of the Rhine river.

16.2.5 Design freeze

We had heard from a few colleagues at automotive suppliers and from the industrial plant construction industry that it had proved best with their efforts to deviate from the agile method during the final phase of development projects and to return to classical project management.

Even before we received this information, we had tried to arrange a design freeze 4 weeks before each major field test.

After this design freeze, the goal was to complete the vehicles in normal project management mode with individually responsible persons and PDCA (Plan, Do, Check, Act) checklists ready for testing and competition. In the logic of the Stacey graph (see Figure 16.4), we were in a project phase in which the '*what*' and '*how*' were much clearer than they were during other project phases.

However, it turned out that the design freeze was difficult to enforce, given the pressure of the competition and of time. It was not always easy to accept known weaknesses of the system and to concentrate on getting the overall system ready as far as the next test required. This approach was therefore not always successful, not only with newer team members, but especially with our particularly creative minds in the team. The author has no doubt that our performance in the competition would have been even better if we had stuck more consistently to periods of design freeze before the corresponding test missions.

In this context, a study that our colleagues from another Fraunhofer Institute already published in 2008 came to mind [11]. The scheme found at that time was somewhat different from the Stacey graph. Spath and Warschat [11] divided the technology development process into five different levels of innovation (page 11): Market – Product/Performance – Function – Technology – Competence. In an ensuing chapter, Heubach *et al.* described the technology development process [4]. The path of a technology is generally derived from the technology development process phase's endeavour to recognising technology potential – developing technology – thus integrating technology into product and establishing technology pertaining to market progress and, initially, the level of competencies, then the technology, then the function and finally the product.

Yet, in the case of new technologies, from time to time one has to go backwards again through the levels, sometimes up to the competence level. Completely new and unexpected problems arise again and again, which can only be mastered by moving away from concrete technical solutions and resolving fundamental questions, creating new competencies.

We believe that this view can also be applied to the Stacey graph. At least that was the case in our case study. As we progressed from general system concepts to concrete technologies and the selected solution, new problems arose that required completely new technical approaches.

For example, our DVL sensor, which measures the vehicle's height above ground level and its speed above ground, was specified to be suitable for heights up to 160 m above ground. Tests in the open ocean showed that during a moderate diving process, in which the vehicle was to dive down to an 800-m-deep rocky seabed, the DVL only provided a usable altitude signal 8 m before impacting the sea floor. Because of the rapidly changing values and the uneven topography, the software had suppressed the output of the altitude values in the purchased component.

How should we resolve the problem? No usable hints came from the manufacturer of the DVL. Probably it would have made sense to install a second sensor, a point echo sounder (i.e. altimeter), for the altitude signal. However, this addition was not possible at that time, because there was neither space in the vehicle body nor a bore in the pressure vessels made of titanium, through which a plug could be passed. Last but not least, the delivery time for the corresponding plugs was very long. So we looked for a way to use the height data from the sonar for all situations in which the DVL did not provide useful data.

In the beginning, we intended to record only raw mission data, which would not be processed in real time. In cooperation with the manufacturer of the sonar, it was possible to extract the height data from the data stream and make it available in real time. So we had to go from the lower left corner of the Stacey graph to the upper right corner to achieve a good and fast solution.

In other cases, it was also the case that individual team members had switched to an anarchistic solution mode for problem-solving, but later it turned out that systematic problem analysis in the team would have led to better and more stable solutions.

16.3 Our approach: eXtreme innovation

On the basis of these experiences, we then developed our own method for project management, related to the XP process which, among other things, consists of first writing the test routines, on top of which the actual software is then developed.

This process is not possible with the development of an AUV, because it is not a pure software product.

Instead, a test and verification plan is set up, upon which one develops. The process is somehow based on the assembly, integration and verification (AIV) procedure, which is used in aerospace engineering. AIV is a standardised process by which the individual components are first checked before they are integrated into the overall system. It is also based on the ECSS-E-HB-10-02A of the *European Cooperation for Space Standardisation*, a handbook that defines the guidelines for the standards. Early system tests are carried out along a verification strategy, from which a verification plan then followed. During the development of the AUVs in the Arggonauts team, system tests were carried out at regular intervals.

In the test plan, only one epic is tested and developed at a time. Pre-prototyping prevents one from investing a lot of time and money in a prototype that will not work afterwards. Therefore, a pre-prototype is developed in a short time and without great effort [12]. This ensures that the possibilities of failure are already apparent in advance and that it is possible to check whether the product in this form would even be accepted by the customer.

It is important that the backlogged items always be designed only with the verification strategy rather than other goals in mind. Doing so can minimise losses and avoid errors. Our verification strategy is analogous to the user stories at SCRUM.

Especially in mechanics and electronics, a design freeze is very important at the end of the development. Through versioning, even after the release of a first version, further improvements can be made to the product, which then appear in a later version. There is always the danger that too much work is done on the development of the hardware, leaving too little time for testing, integration and software.

That is why we also worked with the *Toyota Five Why* in our project. This method means questioning everything five times with a 'why'? (i.e. Why not faster? Why not earlier? Where do you have to start to save one-third of the time?). These are the questions we ask when we set up the verification plan and backlog refinements. This implementation automatically applied the critical chain: the theory of constraints applied to project management by Eli Goldratt and Graham Rand, so that the buffer is removed from the individual backlog items.

Our main elements are the epics and daily tasks. It has proved successful to align the effort estimation and the 'burndown chart' with the epics and not with the individual backlog items because the team was too heterogeneous. The product owner is the one who determines what the test plan will look like, which epic is being worked on, and what the points on the burndown chart are.

In the test plan, only one epic is tested and developed at a time. Pre-prototyping prevents one from investing a lot of time and money in a prototype that will not work afterwards. Therefore, a pre-prototype is developed in a short time and without great effort. Doing so ensures that the possibilities of failure are already apparent in advance and that it is possible to check whether the product in this form would even be accepted by the customer.

It is important that the backlog items are always designed only with regard to the verification strategy and not with regard to other goals. This can minimise losses and avoid errors. Our verification strategy is analogous to the user stories at SCRUM.

16.4 Discussion

The use of agile methods makes sense mainly when a project is complex, and a good organisation is possible. If the teams are too large, it is hardly possible to guarantee a communication that is conducive to the progress of the project. In this case, difficulties are more likely to arise because the requirements of the individual tasks are significantly lower than in classic management.

Agile project management was applied to the Arggonauts project with the SCRUM method and a test-driven development strategy as far as possible. Due to the general conditions caused by the participation in the XPRIZE competition, we could not influence on the duration of the project; the dates for the semi-final and the final were fixed from the beginning of the project. The team was put together quickly at the beginning of the project and was able to learn many new things during the duration of the project. Through the application of the SCRUM method and the associated daily and sprint meetings, there was close cooperation between the team members and new members were able to integrate themselves quickly. However, we could see that the team members, due to their different personalities, were able to deal with this methodology and the resulting proximity differences well.

In retrospect, it would have been important to implement simple user interfaces for mission planning very early with regard to the test-driven development approach. The early development of automated hardware acceptance tests would also have been advantageous in this context.

In an academic approach, a model is designed at the beginning of a project and validated by means of simulation. The next step is to build this model until an end product is created. Only at this stage are tests for functional capability carried out under real conditions. This approach involves the risk that the real conditions differ from the conditions assumed in the simulation, so that the developed end product is not functional.

In a test-driven approach, there is no scientific model at the beginning of the project, but instead develops simultaneously with the construction of the end product. Through permanent testing, this model or preliminary product is constantly evolving. The experience with the development of the AUV in the Arggonauts project shows that this test-driven approach worked well, even if a few details could have been solved better in retrospect.

References

[1] Smith, P. G., and Reinertsen, D. G. (1997). Developing Products in Half the Time. Wiley. ISBN-13: 978-0471292524.

[2] Gino, F. (2016). Are you too stressed to be productive? Or not stressed enough? Harvard Business Review. Retrieved from https://store.hbr.org/product/are-you-too-stressed-to-be-productive-or-not-stressed-enough/H02SBU. Last accesed on 20.03.2020.

[3] Ralph D. Stacey (2011). Strategic Management and Organisational Dynamics: The Challenge of Complexity. Pearson Laureate Education Inc. ISBN-13: 978-1256990826.

[4] Heubach, D., Slama, A., and Rüger, M. (2008). Der Technologieentwicklungsprozess. In Bullinger, H. J. (Ed.), Fokus Technologie: Chancen erkennen-Leistungen entwickeln. Hanser, Munich, pp. 13–43.

[5] Ashmore, S., and Runyan, K. (2014). Introduction to Agile Methods Addison-Wesley Professional

[6] Sutherland, J. (2014). The Art of Doing Twice the Work in Half the Time. New York, NY: Crown Publishing Group.

[7] Beck, K. (2003). Test-driven development by example. Upper Saddle River, NJ: Addison-Wesley.

[8] Denning, S. (2012). How agile can transform manufacturing: the case of Wikispeed. Strategy and Leadership, 40(6), 22–28.

[9] Socha, D., Folsom, T. C., and Justice, J. (2013). Applying agile software principles and practices for fast automotive development. In Proceedings of the

FISITA 2012 World Automotive Congress (pp. 1033–1045). Springer, Berlin, Heidelberg.
[10] Li, X., Martínez, J. F., Rodríguez-Molina, J., and Martínez, N. (2016). A survey on intermediation architectures for underwater robotics. Sensors, 16(2), 190.
[11] Spath, D., and Warschat, J. Innovation durch neue Technologien. (2008). In Bullinger, H. J. (Ed.), Fokus Technologie: Chancen erkennen-Leistungen entwickeln. Hanser, Munich, pp. 13–43.
[12] Savoia, A. (2011). Pretotype It: Make Sure You Are Building the Right It before You Build It Right. Harper Collins Publishers.

Further reading

[1] Goldratt, E. M. (2017). Critical Chain: A Business Novel. UK: Abingdon-on-Thames.
[2] Kotler, S. (2014). The Rise of superman: Decoding the Science of Ultimate Human Performance. Orlando, FL: Houghton Mifflin Harcourt.
[3] Csíkszentmihályi, M. (2004). Good Business: Leadership, Flow, and the Making of Meaning. New York, NY: Penguin.
[4] https://www.xprize.org/ retrieved 22 May 2019

Part VI

Application

Chapter 17
AUV navigation, guidance, and control for geoseismic data acquisition

Lorenzo Pollini[1], Gianluca Antonelli[2], Filippo Arrichiello[2], Andrea Caiti[1], Giuseppe Casalino[3], Daniela De Palma[4], Giovanni Indiveri[3], Matteo Razzanelli[1] and Enrico Simetti[3]

This chapter describes the main challenges, and the corresponding solutions, encountered during the development of the guidance, navigation, and control (GNC) systems for the autonomous underwater vehicles (AUVs) and the autonomous surface vehicles (ASVs) involved in the Widely scalable Mobile Underwater Sonar Technology (WiMUST) project [1].

17.1 Introduction

WiMUST is a project funded by the European Community within the H2020 framework, which started in February 2015 and ended in January 2018. The ultimate goal of the project was to design and test a system of cooperating AUVs able to perform innovative geotechnical surveying operations. In particular, the WiMUST system is composed of a small fleet of AUVs carrying hydrophones to acquire sub-bottom profiling acoustic data. Contrary to the classical technology based on ship-towed streamers, the WiMUST solution allowed a greater survey flexibility since the geometry, position, and depth of the acoustic antenna could be changed during the mission: something that had not been achieved in practice and represented a potential to significantly improve ocean surveying.

The vehicles involved in WiMUST were built by three different project partners: the Medusa AUV and the Delfim catamaran were built by Istituto Superior Tecnico [2,3]; the Folaga AUV was built by GraalTech [4]; and the Ulisse catamaran was built by the University of Genova node of the inter-university center on Integrated System for the Marine Environment (ISME) [5].

[1]Università di Pisa, Dipartimento di Ingegneria dell'Informazione, Pisa, Italy
[2]Università degli Studi di Cassino e del Lazio Meridionale, Dipartimento di Ingegneria Elettrica e dell'Informazione "Maurizio Scarano", Cassino (FR), Italy
[3]Università di Genova, Dipartimento di Informatica, Bioingegneria, Robotica e Ingegneria dei Sistemi, Genova, Italy
[4]Università del Salento, Dipartimento di Ingegneria dell'Innovazione, Lecce – Monteroni – LECCE (LE), Italy

17.2 Geoseismic acquisition with AUV's main challenges

The WiMUST project aims at conceiving, designing, and engineering an intelligent, manageable, distributed, and reconfigurable underwater acoustic array that could drastically improve the efficiency of the methodologies used to perform geophysical and geotechnical acoustic surveys at sea. The deployment of the WiMUST system will be beneficial in a vast number of applications in the fields of civil engineering and oil and gas industry, where seabed mapping, seafloor characterization, and seismic exploration are fundamental operations. The novel key feature of the WiMUST system consists in the use of a team of cooperative autonomous marine robots, acting as intelligent sensing and communicating nodes of a reconfigurable moving seismic recording network.

The vehicles are equipped with hydrophone streamers of small aperture, such that the overall system behaves as a large distributed seismic recording array, capable of acquiring seismic data obtained by illuminating the seabed and the ocean sub-bottom with strong seismic waves sent by a seismic source installed on-board a support ship/boat. By actively controlling the geometry of the robot formation, it becomes possible to change the shape of the seismic recording array, according to the needs of the considered application. The resulting operational flexibility holds tremendous potential advantages, as it allows improving the seabed and sub-bottom resolution and obtaining side lobe rejection at almost any frequency and for any plane.

The availability of the proposed system, other than improving the quality of the acquired data, greatly facilitates the operations at sea, thanks to the lack of physical ties between a surface ship and the acquisition equipment. All the mission phases, from deployment to recovery, are therefore simplified with consequent substantial savings of mission time and operating costs. Moreover the above capability paves the way for the design and implementation of optimal geometric formation control systems, aimed, for example, at maximizing the information content extracted from the acquired seismic signals, by selecting the most suitable shape for the situation at hand.

The reference scenarios of the WiMUST project cover both the 2D and 3D marine seismic reflection exploration and geotechnical seafloor characterization scenarios. These are henceforth referred to as "geotechnical surveys". Due to resource limitations, it has been decided to limit the scope of the demonstration to 3D geotechnical requirements.

The main challenges, with respect to GNC, that were faced in this project can be summarized in trajectory and formation control at the surface and underwater, underwater localization with minimal acoustic communications, and scalability of the communication and control architecture to fleets of many vehicles (tens or hundreds).

17.3 AUV navigation system for surface operations

The navigation system used when the vehicles are at the surface fuses vehicle odometry and Global Positioning System (GPS) measurements using a Kalman filter (KF).

A different navigation system, developed by IST-ID, is used when underwater. The underwater navigation system is composed of a KF and a maximum likelihood estimator (MLE), developed by Instituto Superior Técnico for Research and Development (IST-ID). The estimator is a simplified version of the one presented in [6]. These navigation systems are mainly based on a kinematic model of the AUVs and were implemented unmodified on the Folaga and the Medusa vehicles. Further details can be found in [7].

This section describes an alternative navigation filter based on a dynamic vehicle model, rather than on a pure kinematic one; a validation analysis is included exploiting experimental data collected during the WiMUST sea trials. Preliminary results relative to this dynamic navigation filter have been presented in [8].

Following the standard marine vehicle model described in [9], the vehicle kinematics and dynamics equations are the following:

$$\dot{\boldsymbol{\eta}} = J(\boldsymbol{\eta})\boldsymbol{\nu} \tag{17.1}$$

$$\boldsymbol{\nu}_r = \boldsymbol{\nu} - \boldsymbol{\nu}_c \tag{17.2}$$

$$M\dot{\boldsymbol{\nu}}_r + C(\boldsymbol{\nu}_r)\boldsymbol{\nu}_r + D(\boldsymbol{\nu}_r)\boldsymbol{\nu}_r + \mathbf{g}(\boldsymbol{\eta}) = B\mathbf{u} \tag{17.3}$$

where $\boldsymbol{\eta} = (x, y, z, \phi, \theta, \psi)^\top = (\mathbf{x}^\top, \boldsymbol{\Phi}^\top)^\top$ is the 6D pose in *<NED>* (North, East, Down) frame being $\boldsymbol{\Phi} = (\phi, \theta, \psi)^\top$ the attitude modeled with the roll, pitch, and yaw angles. Vector $\boldsymbol{\nu} \in \mathbb{R}^6$ is the generalized velocity in body frame, namely $\boldsymbol{\nu} = (\boldsymbol{\nu}_1^\top, \boldsymbol{\nu}_2^\top)^\top$. The term $\boldsymbol{\nu}_1 = (u, v, w)^\top$ is the inertial linear velocity vector projected in body frame (u is surge, v is sway, and w is heave). The term $\boldsymbol{\nu}_2 = (p, q, r)^\top$ is the vehicles angular velocity vector in body frame, i.e. $\boldsymbol{\nu}_2 = {}^b\boldsymbol{\omega}_{b/ned}$, where

$${}^b\dot{R}_{ned}\,{}^bR_{ned}^\top = S({}^b\boldsymbol{\omega}_{ned/b}) = -S(\boldsymbol{\nu}_2) \tag{17.4}$$

being $S(\cdot) \in \mathbb{R}^{3\times 3}$ the skew symmetric matrix associated to the cross product $\mathbf{a} \times \mathbf{b} = S(\mathbf{a})\mathbf{b}$ for any $\mathbf{a}, \mathbf{b} \in \mathbb{R}^3$. Moreover, $\boldsymbol{\nu}_c = \begin{bmatrix} {}^bR_{ned}\,\mathbf{v}_c \\ 0_{3\times 1} \end{bmatrix}$ is the irrotational ocean current velocity in body frame. Indeed the linear current velocity term $\mathbf{v}_c$ is assumed to be constant in *<NED>* frame. Vector $\boldsymbol{\nu}_r = (\boldsymbol{\nu}_{r1}^\top, \boldsymbol{\nu}_{r2}^\top)^\top$ is the AUVs *relative* velocity in body frame.

We are interested in the estimation of the position of the vehicle, the linear relative velocity of the vehicle, and the linear velocity of the ocean current. Therefore, we will consider only the translational parts of the vectors in (17.1)–(17.3). The inertia matrix, the Coriolis and centripetal matrix, the hydrodynamic damping matrix, the restoring forces and moments [9], and the allocation matrix in (17.3) can be written in a compact form as

$$M = \begin{bmatrix} M_{11} & M_{12} \\ M_{21} & M_{22} \end{bmatrix}; \qquad C(\boldsymbol{\nu}_r) = \begin{bmatrix} C_{11}(\boldsymbol{\nu}_{r2}) & C_{12}(\boldsymbol{\nu}_{r2}) \\ C_{21}(\boldsymbol{\nu}_r) & C_{22}(\boldsymbol{\nu}_r) \end{bmatrix}; \tag{17.5}$$

$$D = \begin{bmatrix} D_{11} & D_{12} \\ D_{21} & D_{22} \end{bmatrix}; \qquad g(\boldsymbol{\eta}) = \begin{bmatrix} g_1(\boldsymbol{\Phi}) \\ g_2(\boldsymbol{\Phi}) \end{bmatrix}; \qquad B = \begin{bmatrix} B_1 \\ B_2 \end{bmatrix} \tag{17.6}$$

where each term $M_{ij}, C_{ij}, D_{ij}, B_i \in \mathbb{R}^{3\times 3}$ and g_i is a (3×1) vector, $\boldsymbol{\eta}$ and $\boldsymbol{\Phi}$ are defined as in (17.1)–(17.3), and $\boldsymbol{\nu}_{r2}$ is the vehicle relative angular velocity vector in body frame. With regards to the knowledge of the added mass and hydrodynamic parameters in (17.5), we resort to the values used in the Folaga simulator described in [10]. Note that for the sake of simplicity, hydrodynamic drag has been modeled with a linear term only. Assuming the attitude and the angular velocity of the vehicle to be measured, the attitude dynamics is completely known. In this hypothesis, the model in (17.1)–(17.3) for the linear degrees of freedom results in

$$\dot{\mathbf{x}} = {}^{ned}R_b\,\boldsymbol{\nu}_{r1} + \mathbf{v}_c \tag{17.7}$$

$$\begin{aligned} &M_{11}\dot{\boldsymbol{\nu}}_{r1} + M_{12}\dot{\boldsymbol{\nu}}_{r2} + C_{11}(\boldsymbol{\nu}_{r2})\boldsymbol{\nu}_{r1} + C_{12}(\boldsymbol{\nu}_{r2})\boldsymbol{\nu}_{r2} \\ &\quad + D_{11}\boldsymbol{\nu}_{r1} + D_{12}\boldsymbol{\nu}_{r2} + \mathbf{g}_1(\boldsymbol{\eta}) = B_1\mathbf{u} \end{aligned} \tag{17.8}$$

$$\dot{\mathbf{v}}_c = \mathbf{0} \tag{17.9}$$

Hence, the model in (17.7)–(17.9) can be rewritten as

$$\dot{\mathbf{x}} = {}^{ned}R_b\,\boldsymbol{\nu}_{r1} + \mathbf{v}_c \tag{17.10}$$

$$\dot{\boldsymbol{\nu}}_{r1} = A^*\boldsymbol{\nu}_{r1} + B^*\mathbf{u} + D^* \tag{17.11}$$

$$\dot{\mathbf{v}}_c = \mathbf{0} \tag{17.12}$$

where

$$A^* = -M_{11}^{-1}(C_{11}(\boldsymbol{\nu}_{r2}) + D_{11}) \tag{17.13}$$

$$B^* = -M_{11}^{-1}B_1 \tag{17.14}$$

$$\begin{aligned} D^* &= -M_{11}^{-1}((C_{12}(\boldsymbol{\nu}_{r2}) + D_{11})\,\boldsymbol{\nu}_{r2} - \mathbf{g}_1(\boldsymbol{\eta})) \\ &\quad - M_{11}^{-1}M_{12}\,\dot{\boldsymbol{\nu}}_{r2} \end{aligned} \tag{17.15}$$

Note that, even if $\dot{\boldsymbol{\nu}}_{r2}$ in (17.15) should not be known, in most cases it can be neglected. Indeed, often the inertia matrix M has a diagonal form, so that the term M_{12} in (17.15) that multiplies $\dot{\boldsymbol{\nu}}_{r2}$ is a null matrix ($M_{12} = 0_{3\times3} \Rightarrow M_{12}\dot{\boldsymbol{\nu}}_{r2} = \mathbf{0}$). Moreover, the majority of motions of practical interest are straight line trajectories, characterized by a null angular velocity ($\boldsymbol{\nu}_{r2} = \mathbf{0}$, $\dot{\boldsymbol{\nu}}_{r2} = \mathbf{0}$) or trajectories with constant angular velocity ($\boldsymbol{\nu}_{r2} = \text{constant} \Rightarrow \dot{\boldsymbol{\nu}}_{r2} = \mathbf{0}$), leading both to a null value of $\dot{\boldsymbol{\nu}}_{r2}$.

Defining a state vector $\mathbf{z} = [\mathbf{x}^\top, \boldsymbol{\nu}_{r1}^\top, \boldsymbol{\nu}_{c1}^\top]^\top \in \mathbb{R}^9$, (17.10)–(17.12) can be reformulated in a linear time variant state space setting:

$$\begin{aligned} \dot{\mathbf{z}} = \begin{bmatrix} \dot{\mathbf{x}} \\ \dot{\boldsymbol{\nu}}_{r1} \\ \dot{\boldsymbol{\nu}}_{c1} \end{bmatrix} &= \underbrace{\begin{pmatrix} 0_{3\times3} & {}^{ned}R_b & I_{3\times3} \\ 0_{3\times3} & A^* & 0_{3\times3} \\ 0_{3\times3} & 0_{3\times3} & 0_{3\times3} \end{pmatrix}}_{\bar{A}} \begin{bmatrix} \mathbf{x} \\ \boldsymbol{\nu}_{r1} \\ \mathbf{v}_c \end{bmatrix} \\ &\quad + \begin{bmatrix} 0_{3\times9} \\ B^* \\ 0_{3\times9} \end{bmatrix} \mathbf{u} + \begin{bmatrix} 0_{3\times1} \\ D^* \\ 0_{3\times1} \end{bmatrix} \end{aligned} \tag{17.16}$$

The measurements available for the navigation are the following:

- position in <*NED*> frame acquired through GPS while the vehicle is on the surface;

- surge velocity in body frame estimated by an experimentally identified thruster-based model linking thruster commands to steady-state surge values (details about the identification of such model are not reported in this book chapter for the sake of brevity);
- range measurements from one or more anchor vehicles (ASVs) having known positions, acquired mainly when the vehicle is underwater and has access only to anchors by means of acoustic modems.

Thus, the output equation is given by

$$\mathbf{y} = \begin{bmatrix} \mathbf{x} \\ u \\ \|\mathbf{x} - \mathbf{x}_{a_1}\| \\ \vdots \\ \|\mathbf{x} - \mathbf{x}_{a_n}\| \end{bmatrix} \tag{17.17}$$

with $\mathbf{x}_{a_i}$ the positions of the ith anchor vehicle ($i = 1, \ldots, n$) assumed known (they could come, by example, from the GPS onboard the ASVs that transmit the data to the AUVs). Given the nonlinearity of the output equation (17.17), due to range measurements, an extended KF (EKF) or unscented KF (UKF) [11] could be used for the estimation of the state variables. Given the time-varying and nonlinear nature of the model, its observability properties are nontrivial. The observability analysis is here omitted for the sake of brevity. Specific results relative to the observability of range-based navigation filters has been already studied in [12–14]. Of course, in the absence of range measurements, the output equation becomes linear and a simpler observer like a standard KF can be adopted.

The effectiveness of the proposed dynamic navigation filter has been validated on the experimental data collected during the last WiMUST trials in Sines. Such data include

- measures of roll, pitch, and yaw angles acquired through inertial on board sensors (compass/inclinometer);
- GPS readings when the vehicle is on the surface;
- range measurements from two anchors ASV ($n = 2$) acquired through acoustic modems when the vehicle is underwater;
- anchors ASV position acquired through on-board GPS and transmitted to the AUVs;
- surge velocity from a model based on thrusters commands; and
- the command imparted to the thrusters, eight pumps and one propeller, as depicted in Figure 17.2.

These experimental data refer to 2 h of lawn mowing motion both on surface and underwater. Such data have been post-processed with a KF/EKF applied on the dynamic AUV model (17.16)–(17.17).

During the surface navigation, a standard KF can be applied on the dynamic AUV model (17.16)–(17.17) taking into account the surge velocity and GPS readings

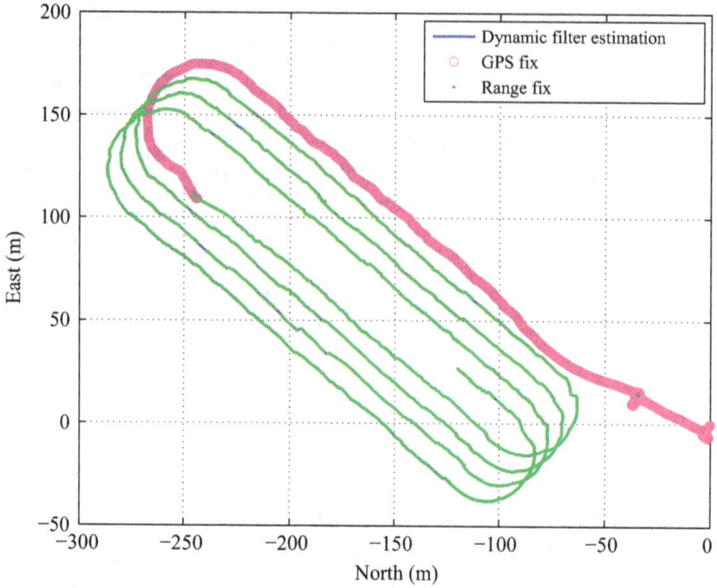

Figure 17.1 Vehicle trajectory estimated using the dynamic navigation filter with GPS corrections and range corrections depicted as magenta circles and green points, respectively

only. Indeed, when the AUV is on the surface, the range measurements are very few or not available at all due to the geometry of acoustic paths from the AUV to the anchors. It should be noted that, in order to implement the dynamic navigation filter here described, the rotation matrix ${}^{ned}R_b$ is computed by exploiting the measures of roll, pitch, and yaw angles, while the angular velocity $\boldsymbol{\nu}_{r2}$ has been derived using the relation in (17.4), and the $\dot{\boldsymbol{\nu}}_{r2}$ has been assumed to be null. Note also that no ground truth is available; thus, the navigation filter is assessed with reference to the available GPS data.

During underwater navigation, range measurements to anchors play an important role since the AUV has no longer access to the GPS readings. In this case, the dynamic model-based EKF was used. It should be noted that when processing such data, range outliers should be rejected in order not to jeopardize the estimation. Indeed, acoustic measurements are often contaminated by outliers. Here, a few outlier candidate data values have been removed prior to processing the data. Alternative online outlier robust filtering techniques are available in the literature [15,16] and could be possibly exploited. The position estimated with the dynamic navigation filter along the whole experiment is shown in Figure 17.1. In particular, Figure 17.1 highlights the estimations obtained including in the measurement channel the GPS readings when the AUV is on the surface, and those exploiting the range measurements when the AUV is underwater. Indeed, the resulting estimation appears to be reasonable also during underwater navigation, confirming the effectiveness of the proposed dynamic navigation filter.

17.4 AUV control system

The motion control system of the AUV has three main goals: regulation of surge speed, heading, and depth; in addition, output of the controllers, namely the AUV control input, must be allocated to the thruster that are located in different position with respect to the center of gravity. This section describes the AUV control system: first, the control allocation technique is presented with reference to the actuators mounted on the vehicle, then a controller composed by three decoupled control loops is presented.

17.4.1 Control allocation

The Folaga AUV is actuated by eight pumps, four mounted horizontally and four mounted vertically. Pumps generate forces and moments on the vehicle by pumping water from the inlet to the outlet only and their direction cannot be reversed. The vehicle eight pumps are mounted as four couples of opposite pushing pumps, two at the bow and two at the stern. This allows an abstraction layer in which an actuation system is controlled in order to generate forces and moments, along the Y- and Z-axes. Thus, the command pumps can assume positive or negative values. Forces along the X-axis can be generated using a propeller mounted at the stern. Moments around the X-axis cannot be generated, due to the pumps mounting positions; however, roll stability of the vehicle is guaranteed by a built-in mechanical system. Pitch equilibrium is obtained by using both the movable vehicle battery pack mounted on a slip inside the vehicle, and a flooding chamber that is able to change the vehicle buoyancy and make it neutral or floating on software command. Figure 17.2 shows the pumps position on the vehicle.

17.4.2 Surge control

For a typical torpedo-shaped AUV, the surge speed controller can be implemented as a Proportional-Integral-Derivative (PID) controller. Given the surge speed tracking error $e_s = u_{ref} - u$ (with u_{ref} the desired surge velocity and u the current surge velocity of the vehicle), the control law is (in continuous time) as follows:

$$\frac{u_s(s)}{e_s(s)} = K_{Ps} + \frac{K_{Is}}{s} + K_{Ds}\frac{s}{1 + s\tau_{hs}} \tag{17.18}$$

where the pole in $-\frac{1}{\tau_{hs}}$ is needed both to make the derivative system causal and to damp the effect of noise. The current velocity u must be estimated by the AUV navigation system as detailed in Section 17.3. The continuous time-surge controller transfer function is then discretized by using the forward Euler methods; the sampling frequency was set to 10 Hz due to communications constraints with the low-level layer of interface with sensors and actuators. An anti-windup technique is also designed

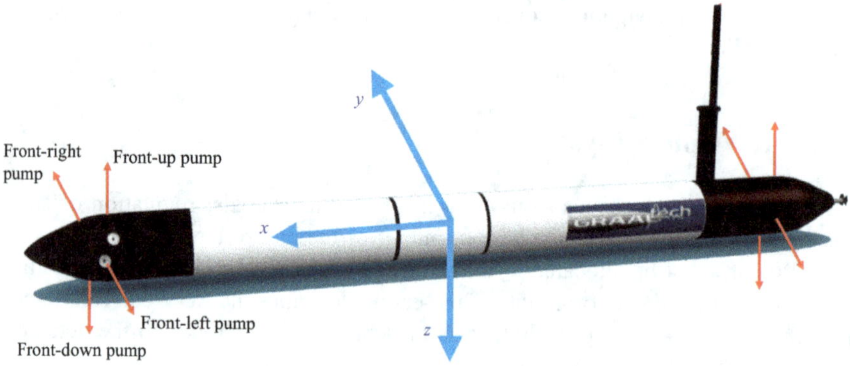

Figure 17.2 AUV actuation system in its body frame

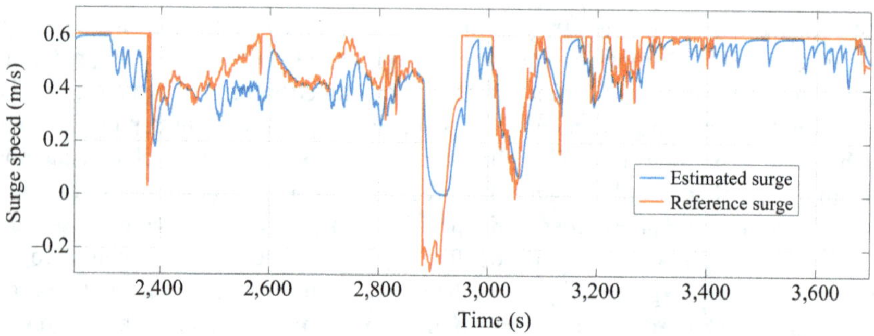

Figure 17.3 Experimental evaluation of the surge speed control

to avoid, due to the presence of the integrator in the controller, that the error is continuously integrated also when the control inputs are saturated. In general, the reference u_{ref} is an input from an outer control loop system that generates velocity references with respect to the goal of the mission. Figure 17.3 shows a comparison between the estimated and desired surge speed during an experimental sea trials.

17.4.3 Heading control

Given the symmetrical shape, and the capability to generate yaw torques using horizontal pumps of the Folaga, the Heading controller can be implemented as a Proportional-Derivative (PD) controller. Given the heading angle tracking error $e_\psi = \psi_{ref} - \psi$ (with ψ_{ref} the desired heading angle and ψ the current heading angle of the vehicle), the continuous time control law is

$$\frac{u_\psi(s)}{e_\psi(s)} = \left(K_{P\psi} + K_{D\psi}s\right)\frac{1}{1 + s\tau_{h\psi}} \tag{17.19}$$

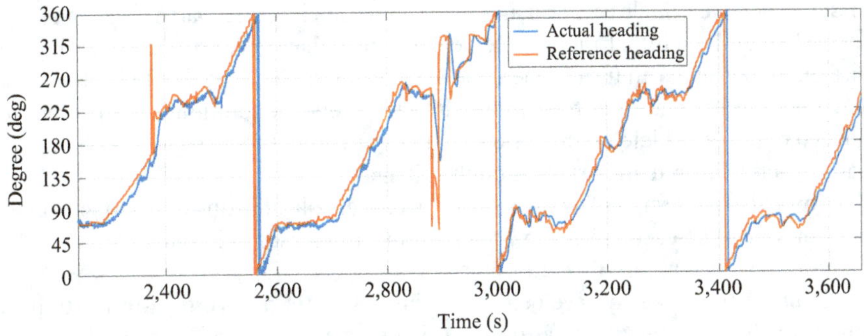

Figure 17.4 Experimental evaluation of the heading control

where the term $\frac{1}{1+s\tau_{h\psi}}$ is needed to make the derivation system causal and to damp the effect of noise. The current heading ψ is estimated by the AUV navigation system as detailed in Section 17.3 and the reference ψ_{ref} is an input from an outer control loop system with respect to the goal of the mission. The continuous time control law is discretized by using the forward Euler method with sampling frequency set to 10 Hz. The command produced by the controller is an input for the control allocation conceptual block that maps the controller output into a torque command for the four horizontal pumps. By considering the pumps if $u_\psi > 0$, then $|u_\psi|$ becomes the actuation command for front-left and rear-right pumps. Instead, if $u_\psi < 0$, $|u_\psi|$ becomes the actuation command for front-right and rear-left pumps. Figure 17.4 shows a comparison between the actual and desired heading angle speed during an experimental sea trial.

17.4.4 Depth control

Depth control can be achieved for the Folaga AUV using two different techniques that we may call pitch-to-dive and vertical-thrust-to-dive. The pitch-to-dive technique, as the name suggests, pitches the vehicle toward the desired vertical direction and exploits horizontal thrust given by the rear propeller to generate a vertical velocity; clearly, this technique cannot be used when the vehicle is not advancing or is cruising at very low speed. It is also quite intuitive that AUV vertical dynamics is a function of the current horizontal velocity.

The vertical-thrust-to-dive technique instead generates vertical motion using the vertical pumps mounted onboard the vehicle. Full effectiveness of this technique requires constant pitch angle and low velocity (close to 0): when water is flowing along the vehicle body, it causes a loss of pressure in front of the pump water inlet, hence a loss of efficiency occurs.

Due to the hull structure and the location of the pumps, the pitch-to-dive technique should be preferred especially for the WiMUST type of missions. This is due to the fact that, in general, the typical geoseismic mission requires changes of depth during the mission when the vehicles are cruising at a certain velocity underwater. Unfortunately, sea trials highlighted that the vehicle velocity may change considerably during the mission, in particular in turns, where the vehicles in the inner part of the curve must slow down to maintain the formation with the others. In addition, the disturbance induced onto the vehicle by the towed streamer turned out to be significantly higher than expected/desired, in particular on the pitch axis.

This led to the design of depth control system capable of fusing the two techniques together: one to be used at very low velocity, one to be used at higher velocity, and a mixture of the two at intermediate velocities.

Depth control at low velocity is achieved with the vertical-thrust-to-dive approach: the necessary vertical thrust u_z is generated using a PD controller from depth error e_d:

$$\frac{u_z(s)}{e_d(s)} = (K_{Pzz} + K_{Dzz}s)\,\frac{1}{1 + s\tau_{hzz}} \tag{17.20}$$

The desired vertical thrust u_z is finally generated allocating equally the control action to the front and rear vertical pumps.

The second technique instead exploits the depth error to generate a vertical speed by pitching the vehicle. Thus, the depth controller, in this case, is implemented using a cascade of two simple controllers: the outer loop controller computes depth tracking error and generates a desired pitch angle; the desired pitch angle becomes an input to the inner-loop controller that regulates vehicle pitch to the desired value. More specifically, given the depth tracking error $e_d = d_{ref} - d$ (with d_{ref} the desired depth and d the current depth of the vehicle), the outer loop controller is a filtered Proportional (P) controller (in continuous time):

$$\frac{\theta_{ref}(s)}{e_d(s)} = \frac{K_{Pd}}{1 + s\tau_{hd}} \tag{17.21}$$

where the pole in $-\frac{1}{\tau_{hd}}$ is employed to limit noise transmission from the depth sensor and to generate a smooth pitch angle reference. The current depth d must be estimated by the AUV navigation system as detailed in Section 17.3 and the reference d_{ref} is an input from an outer control loop system. The output of the first controller becomes an input for the second regulator to steer the pitch angle error to 0. Hence, $e_\theta = \theta_{ref} - \theta$ (with θ_{ref} the output of the outer loop depth controller and θ the current pitch angle of the vehicle) and generates the command u_θ:

$$\frac{u_\theta(s)}{e_\theta(s)} = \left(K_{P\theta} + \frac{K_{I\theta}}{s} + K_{D\theta}s\right)\frac{1}{1 + s\tau_{h\theta}} \tag{17.22}$$

where the pole in $-\frac{1}{\tau_{h\theta}}$ is employed to make the derivative causal and to limit the effect of noise. Detail on how the pitch angle is estimated is presented in Section 17.3. Then, (17.21) and (17.22) are in cascade and the forward Euler method with sampling frequency set to 10 Hz is used to discretize these two controllers. Again, the controller output is mapped, through the control allocation block, producing a pitch torque command. The desired pitch torque u_θ is finally generated allocating symmetrically the control action to the front and rear vertical pumps.

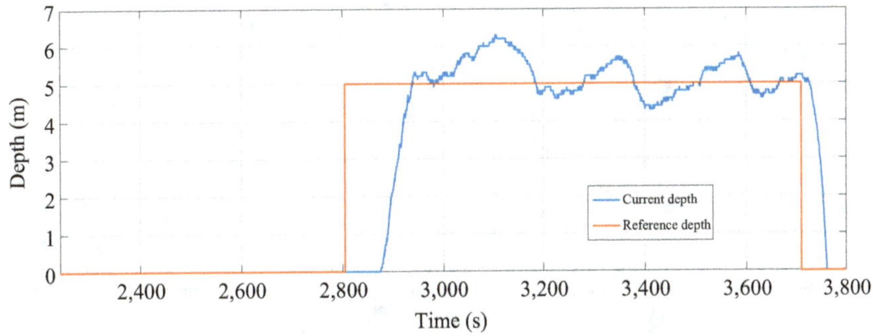

Figure 17.5 Experimental evaluation of the depth control

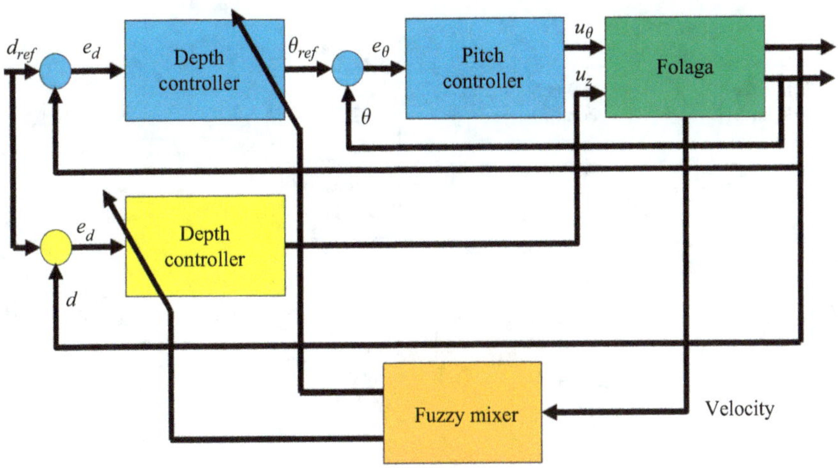

Figure 17.6 Depth control scheme

Figure 17.5 shows a comparison between the estimated and desired depth during an experimental sea trial.

In order to fuse together the two control actions described above, a fuzzy-PID control system was implemented: when the vehicle velocity is very low (below a certain threshold), vertical pumps are used to control depth (with the vehicle almost leveled); when velocity is higher instead (above a certain threshold), the vehicle pitch angle is used as control variable to increase or decrease depth; when the velocity has an intermediate level (in between these two thresholds) both controllers are used at the same time with a control authority proportional to the distance from the respective activation thresholds.

Figure 17.6 shows the conceptual block diagram of the so-called pitch-to-dive and vertical-thrust-to-dive control systems mixed together with a fuzzy decision-maker, based on the vehicle velocity.

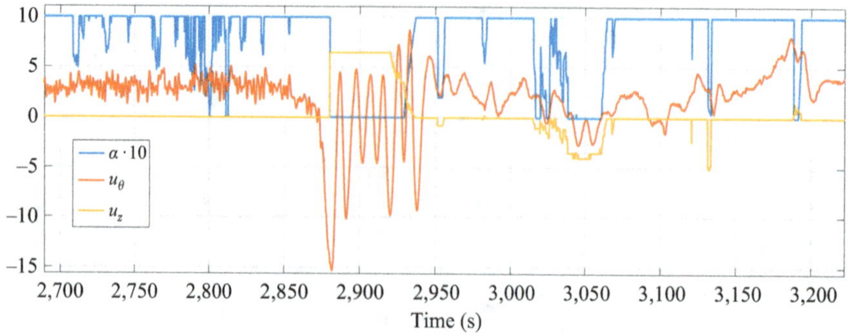

Figure 17.7 Sample

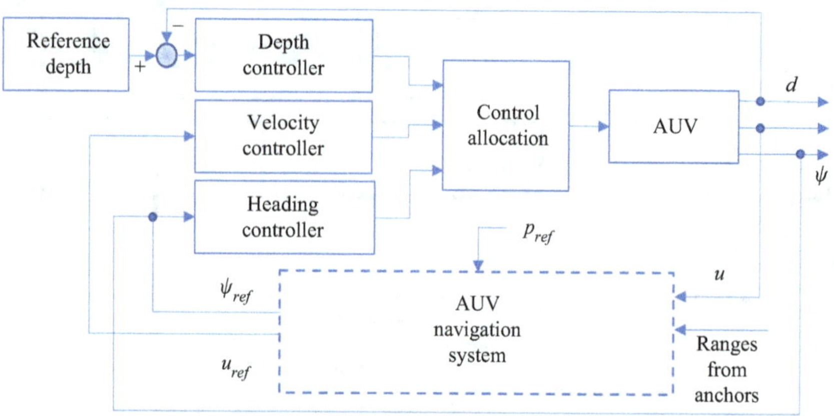

Figure 17.8 Block diagram of the AUV GNC system

Figure 17.7 shows the operation of the fuzzy mixer during another sample sea trial; only a brief portion of the mission is shown for cleanness. The fuzzy scheduling variable α value varies between 1 (when speed is large enough and only the pitch-to-dive controller should be used) and 0 (when only the vertical-thrust-to-dive controller should be used). Also shown are the outputs of the two depth controllers u_θ and u_z. When the fuzzy mixer starts operating, the control action is the result of the mixture of the vertical-trust-to-dive and pitch-to-dive controllers actions; thus, the four vertical pumps are used both differentially to generate a pitch moment, and with same direction to generate the needed vertical thrust.

Figure 17.8 depicts a block diagram of the presented AUV control system where p_{ref} is the desired position, in general the next waypoint to be reached.

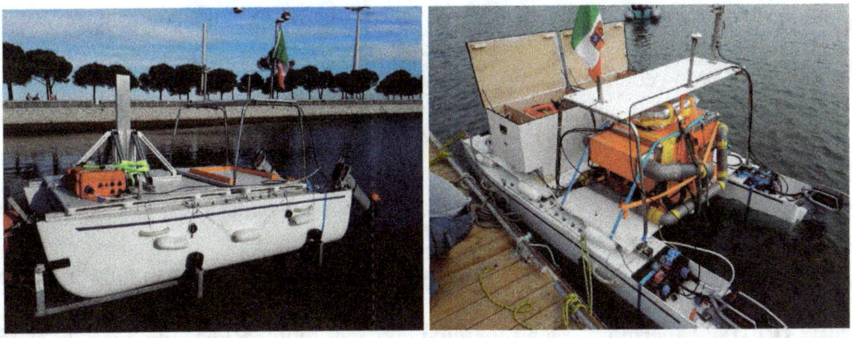

Figure 17.9 Two images of the ULISSE ASV during its use within the WiMUST project

17.5 ASV guidance, navigation, and control

Two ASVs were employed for carrying the sparkers, namely the DELFIM and ULISSE catamarans. The ASVs can also be employed as anchors for the AUVs navigation system, hence the GNC system of the ASVs is relevant for the overall navigation of AUV team. In this section, we will focus on the GNC of the ULISSE ASV.

17.5.1 ULISSE autonomous surface vehicle

The ULISSE ASV, depicted in Figure 17.9, is a 3-m long and 1.8-m wide catamaran, constructed in fiberglass. The two hulls, fixed together thanks to two carbon fiber bars, host the batteries (around 3.2 kWh energy each) and the control electronics including inertial sensors (compass, accelerometers, and gyroscopes). Two Torqeedo Cruise 2R electric motors, with around 1 kW propulsive power each, constitute the actuation system of the catamaran. The robot is connected to other vehicles and to the control station through a 5 GHz wireless antenna. Two aluminum profiles are installed along the length of both hulls, at deck level. They allow installing customized decks depending on the experiment. In particular, in the WiMUST project one deck was manufactured to host sparker's power supply. The second one, installed near the bow, hosted a movable pole to lower down the modem's transducers, an additional waterproof case hosting the modems electronics, and two Honda EU20i portable power generators, necessary to produce the 220 V for the sparker's power supply.

17.5.2 ULISSE GNC architecture

For the execution of the geotechnical survey of the WiMUST project, each ASV must follow a predefined path. Toward this aim, a guidance loop implements a path following algorithm developed by IST-ID as described in [17–19]. Technical details of this solution are here omitted for the sake of brevity. We only recall that the outputs of such an outer loop are two reference signals: a desired heading ψ_{ref} and a desired absolute surge velocity u_{ref} for the ASV. For the ULISSE ASV, the control problem is thus tracking such references through a proper assignment of the motors revolutions per minute (RPM) vector $\mathbf{n}$. The approach adopted is described hereafter.

First, let us recall how the model of the ASV is similar to the one previously developed for an AUV (17.3), except that it is usually limited to the horizontal plane. Hence, $\boldsymbol{\nu}_r = (u_r, v_r, r)^\top$, where u_r is the relative surge velocity, v_r is the relative sway velocity, and r is the yaw rate. Furthermore, let us note that the force vector is a function of the thruster RPMs $\mathbf{u} = \mathbf{f}(\mathbf{n})$.

It is well known that if a constant force vector $\mathbf{u}$ is provided as input in (17.3) a steady-state velocity $\boldsymbol{\nu}_r$ will be reached due to the dissipative effects of the drag matrix $\mathbf{D}(\boldsymbol{\nu}_\mathbf{r})$. Furthermore, even providing constant motor RPMs yield steady-state conditions [20], in accordance with common intuition. With this in mind, it is possible to create a map between the motor RPMs $\mathbf{n}$ and the measured relative surge velocity u_r and yaw rate r, corresponding to the steady-state solution of (17.3). Notice that the relative sway is not an input to the map, and as there are no actuators to control it. The relative sway velocity will simply reach a steady-state value imposed by the desired relative surge and yaw rate values. Therefore, in the following, we will consider a reduced desired relative velocity vector $\bar{\boldsymbol{\nu}}_{r_{ref}} = \left[u_{r_{ref}} \; r_{ref}\right]^\top$.

Once the map has been identified and obtained, it can be used for a steady-state control of the ASV: desired relative velocities $\bar{\boldsymbol{\nu}}_{r_{ref}}$ at steady state can be used as input to the map, obtaining the corresponding reference motor RPMs $\mathbf{n}$. This approach requires that the reference velocity $\bar{\boldsymbol{\nu}}_{r_{ref}}$ varies sufficiently slowly to ensure the correctness of the steady-state assumption. However, given that the survey path that was designed for WiMUST required to follow straight lines and arcs with constant velocity, this condition can be assumed to be satisfied.

The desired relative velocities $\bar{\boldsymbol{\nu}}_{r_{ref}}$ can be generated by PID controllers to regulate the heading ψ and the absolute surge speed u, as in the case of the WiMUST project. Obviously, the gains of such a PID controller must be tuned in such a way that the steady-state hypothesis is still preserved.

The resulting control architecture is depicted in Figure 17.10. The outer loop implements the path following algorithm developed by IST-ID. Two independent PID controllers have been developed to track the desired heading and surge references outputs of the path following algorithm in a similar way as described in Sections 17.4.2 and 17.4.3. In particular,

- a PD (proportional derivative) controller has been used to regulate the heading ψ to the desired value ψ_{ref}, and its output is a yaw rate control signal r_{ref};
- a PI (proportional integrative) controller has been designed to regulate the absolute surge velocity u to the desired value u_{ref}. The main role of this controller is to generate a relative surge control signal $u_{r_{ref}}$ that compensates the effects of currents and wind.

Finally, the thruster allocator block is the one implementing the identified mapping $\bar{\boldsymbol{\nu}}_{r_{ref}} \rightarrow \mathbf{n}$, generating the reference thruster RPMs.

Figure 17.11 shows the actual surge speed and heading compared to the reference ones. The peaks in surge velocity are in correspondence of the abrupt heading

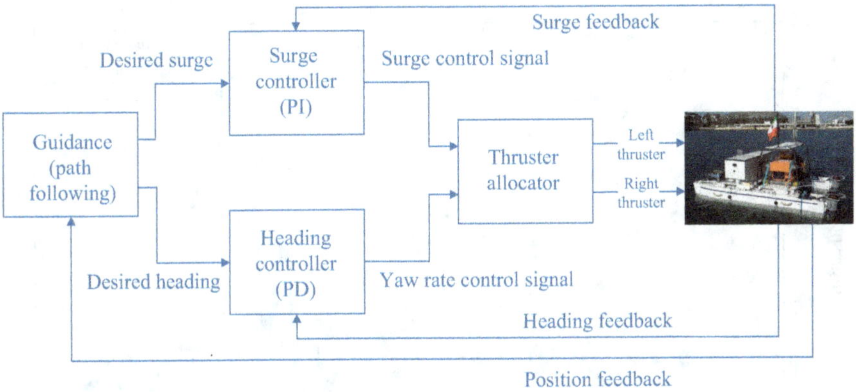

Figure 17.10 The ULISSE ASV GNC architecture

changes, due to the simplifying assumptions about the actuation model $f(\mathbf{n})$. Further, refinements including more realistic propeller model and especially modeling how propeller's thrust changes with the speed of the vessels are outside of the scope of this book.

17.6 Estimation of the streamer shape

The acoustic data acquired by the streamer hydrophones of the WiMUST system (see Figure 17.12) must be processed in order to study the sub-bottom profile. To this aim, beyond knowing the positioning of the acoustic sources (sparkers) and of the AUVs, there is the need to know the positioning of the hydrophones towed by the AUVs during the survey mission. However, the absence of a localization infrastructure requires the development of a proper strategy to estimate the hydrophones' positioning from the available set of information. To the purpose, as reported in [21–23], a localization strategy for the AUV-streamer system was developed in the framework of the WiMUST project that, beyond using the data from AUV's navigation sensors, it also makes use of information gathered from acoustic signals generated for the seismic survey. Indeed, considering that the sparkers and the AUVs of the WiMUST system were equipped with acoustic modems with atomic clocks, allowing precise synchronization, it was possible to use the direct time of arrival of the acoustic signals from the sparkers to the hydrophones to compute range measurements among them. Such range measurements were collected and fused with AUV on-board navigation sensors data to estimate the AUV and hydrophones positioning. In particular, the developed localization strategy was based on a discrete time EKF, where the prediction steps make use of the AUV-streamer dynamic model, and the update steps make use of all the available measurements, namely data from AUV's navigation sensors and range measurements from sparkers to hydrophones.

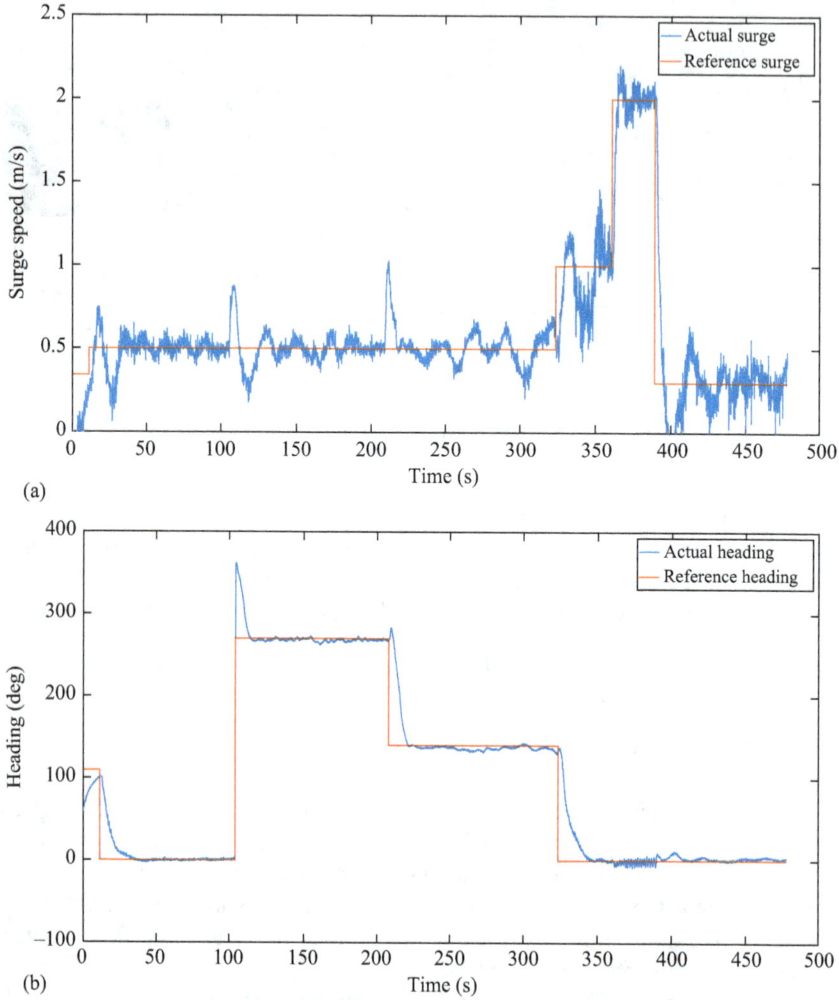

Figure 17.11 Inner-loop step responses: (a) desired and actual surge; (b) desired (wrapped around ±360 degrees) and actual heading (plotted as the desired one plus the heading error)

17.6.1 AUV-streamer modeling

To model the overall AUV-streamer systems, we consider that the streamer is a long and almost neutrally buoyant flexible cable equipped with a set of hydrophones distributed along it. To derive the dynamic model equations, we assumed the AUV as a fully actuated rigid body, and the streamer as a serial chain of rigid links with a spherical unactuated joints in correspondence of each hydrophone. Referring to Figure 17.13,

Figure 17.12 Snapshot of a team of AUVs equipped with streamers with hydrophones

we assume that the AUV has the centroid in C; the streamer is attached to the AUV's hull in the point P; and the ith hydrophone position is denoted as P_i.

Notations: Any frame $\{k\}$ in small letter has the origin at a point K in capital letter; any vector $\mathbf{r}_A^b$ denotes a vector from the origin of the inertial frame to the point A expressed in a frame $\{b\}$; any vector $\mathbf{r}_{AB}^c$ denotes a position vector from a point A to another point B expressed in a frame $\{c\}$; any vector $\mathbf{v}_A^b$ denotes the velocity of the point A with respect to the frame $\{i\}$ expressed in $\{b\}$; any vector $\boldsymbol{\omega}_{ab}^c$ denotes the angular velocity of a frame $\{b\}$ with respect to another frame $\{a\}$ expressed in a frame $\{c\}$; any matrix $\mathbf{R}_a^b$ denotes a rotation matrix of a frame $\{a\}$ with respect to another frame $\{b\}$; any vector $\boldsymbol{\eta}_a^b$ denotes the vector of the Euler angles of frame $\{a\}$ with respect to another frame $\{b\}$; any matrix $\mathbf{T}_{\eta_a^b}$ denotes the angular transformation that transforms the angular velocity to Euler rate, the vector $\boldsymbol{\eta}_a^b$ being the vector of Euler angles; any vector $\mathbf{T}_i$ denotes the tension force applied by the link $i+1$ on the link i; any vector $\mathbf{f}_j^i$ denotes a force expressed in the frame $\{i\}$; any vector $\boldsymbol{\tau}_j^i$ denotes a torque expressed in the frame $\{i\}$.

The underwater dynamics of each rigid body has been modeled following the approach in [24]. In particular, the parameters of the dynamic model of the AUV

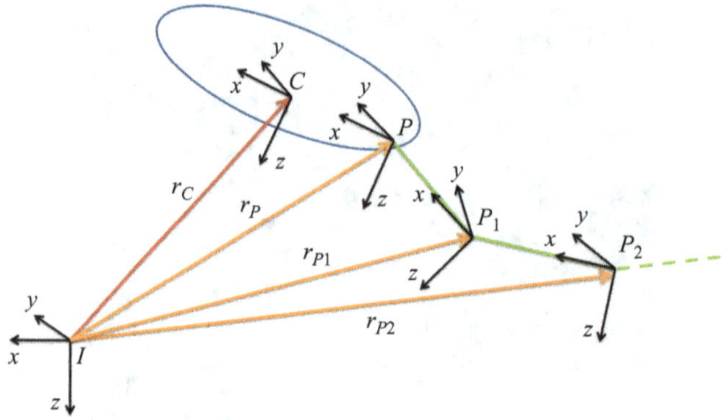

Figure 17.13 Sketch of an AUV towing a streamer of hydrophones

have been defined on the basis of the Folaga vehicles' characteristics, i.e. a 2-m-length torpedo-shaped AUV. Indeed, the Newton–Euler equations for the AUV were expressed as

$$m_{auv}\left(\dot{\mathbf{v}}^i_P + S(\dot{\boldsymbol{\omega}}^i_{ic})\mathbf{r}^i_{PC} + S^2(\boldsymbol{\omega}^i_{ic})\mathbf{r}^i_{PC}\right) = \mathbf{f}_{auv} \tag{17.23}$$

$$\mathbf{f}_{auv} = \mathbf{f}_{act} + \mathbf{f}_{drag} + \mathbf{f}_b + \mathbf{f}_g + \mathbf{T}_0 \tag{17.24}$$

$$m_{auv}S(\mathbf{r}^i_{PC})\dot{\mathbf{v}}^i_P + \mathbf{I}^i_P\dot{\boldsymbol{\omega}}^i_{ip} + S(\boldsymbol{\omega}^i_{ip})\mathbf{I}^i_P\boldsymbol{\omega}^i_{ip} = \boldsymbol{\tau}^i_P \tag{17.25}$$

$$\boldsymbol{\tau}^i_P = \boldsymbol{\tau}^i_{act} + \boldsymbol{\tau}^i_{drag} + \boldsymbol{\tau}^i_0 + \mathbf{r}^i_{PC} \times \mathbf{f}_{act} + \mathbf{r}^i_{PC} \times \mathbf{f}_g + \mathbf{r}^i_{PB} \times \mathbf{f}_b \tag{17.26}$$

where m_{auv} is the AUV mass, $S(\boldsymbol{\omega})$ is the skew-symmetric matrix having for elements those of the angular velocity vector $\boldsymbol{\omega}$, $\mathbf{f}_{auv}$ is the sum of the forces acting on the AUV ($\mathbf{f}_{act}$, $\mathbf{f}_{drag}$, $\mathbf{f}_b$, and $\mathbf{f}_g$ are, respectively, the actuation, drag, buoyancy, and gravity forces, and $\mathbf{T}_0$ is the force applied by the streamer), $\mathbf{I}^i_P$ is the inertia matrix computed with pole in P, and τ_{auv} is the sum of the torques acting on the AUV (τ_{act} and τ_{drag} are the actuation and drag torques, respectively).

The streamer has been modeled as a set of almost neutrally buoyant thin cylindrical elements connected via unactuated spherical joints. Hydrodynamic terms and drag coefficients have been chosen accordingly to the effective geometry of the system, and on the basis of the common velocity regimes of the WiMUST system (i.e. cruise velocity about 1 m/s). The Newton–Euler equations for the ith link can be expressed as

$$m_i(\dot{\mathbf{v}}^i_{P_i} + S(\dot{\boldsymbol{\omega}}^i_{ip_i})\mathbf{r}^i_{P_iC_i} + S^2(\boldsymbol{\omega}^i_{ip_i})\mathbf{r}^i_{P_iC_i}) = \mathbf{f}^i_i \tag{17.27}$$

$$\mathbf{f}^i_i = \mathbf{f}_{i,drag} + \mathbf{f}_{i,b} + \mathbf{f}_{i,g} + \mathbf{T}_i - \mathbf{T}_{i-1} \tag{17.28}$$

$$m_iS(\mathbf{r}^i_{P_iC_i})\dot{\mathbf{v}}^i_{P_i} + \mathbf{I}^i_{P_i}\dot{\boldsymbol{\omega}}^i_{ip_i} + S(\boldsymbol{\omega}^i_{ip_i})\mathbf{I}^i_{P_i}\boldsymbol{\omega}^i_{ip_i} = \boldsymbol{\tau}^i_{P_i} \tag{17.29}$$

$$\begin{aligned}\boldsymbol{\tau}^i_{P_i} = &-\boldsymbol{\tau}^i_{i-1} + \boldsymbol{\tau}^i_i + \boldsymbol{\tau}^i_{i,drag} + \mathbf{r}^i_{P_iB_i} \times \mathbf{f}_{i,b} + \mathbf{r}^i_{P_iP_{i-1}} \\ &\times (-\mathbf{T}_{i-1}) + \mathbf{r}^i_{P_iC_i} \times \mathbf{f}_{i,g} + \cdots + \mathbf{r}^i_{P_iC_i} \times \mathbf{f}_{i,drag}\end{aligned} \tag{17.30}$$

where m_i is the mass of the ith link, $\mathbf{I}^i_{P_i}$ is the inertia matrix computed with pole in P_i, $\mathbf{f}_i$ is the sum of the forces acting on the ith link ($\mathbf{f}_{i,drag}$, $\mathbf{f}_{i,b}$, and $\mathbf{f}_{i,g}$ are the drag, buoyancy, and gravity forces, respectively), and τ_{P_i} is the sum of the torques acting on the ith link.

For both numerical simulation purposes and for the derivation of the EKF prediction equations, the dynamic model of the system has been rewritten in a state space form. Thus, denoting as $\mathbf{q} = [\mathbf{r}^i_C, \eta^i_c, \eta^{p_n}_p]^T$, where $\eta^{p_n}_p = [\eta^{p_1}_p, \eta^{p_2}_{p_1}, ..., \eta^{p_n}_{p_{n-1}}]$, and as $\boldsymbol{\tau} = [\mathbf{f}^i_c, \boldsymbol{\tau}^i_c, \boldsymbol{\tau}^i_{P_i}]^T$, where $\boldsymbol{\tau}^i_{P_i} = [\boldsymbol{\tau}^i_{P_1}, ..., \boldsymbol{\tau}^i_{P_n}]$, the dynamic model can be rewritten as

$$\ddot{\mathbf{q}} = \mathbf{M}(\mathbf{q})^{-1}(\boldsymbol{\tau} - \overline{\boldsymbol{\tau}}) \tag{17.31}$$

$$\overline{\boldsymbol{\tau}} = \mathbf{C}(\mathbf{q}, \dot{\mathbf{q}})\dot{\mathbf{q}} + \mathbf{F}_{drag}(\dot{\mathbf{q}}) + \mathbf{F}_b(\mathbf{q}) + \mathbf{F}_g(\mathbf{q}) \tag{17.32}$$

where $\mathbf{M}$ is the mass matrix; $\boldsymbol{\tau}$ is the actuation forces and torques; $\mathbf{C}(\mathbf{q}, \dot{\mathbf{q}})$ is the centrifugal and Coriolis matrix; $\mathbf{F}_{drag}$ is the drag forces and torques; $\mathbf{F}_b$ is buoyancy forces and torques; and $\mathbf{F}_g$ is the gravitational forces and torques. Defining the state variable $\mathbf{x} = [\mathbf{q}, \dot{\mathbf{q}}]^T$, from (17.31) the model can be reformulated in the form $\dot{\mathbf{x}} = \mathbf{f}(\mathbf{x}, \mathbf{u})$, where the $\mathbf{u}$ is the input term.

Given the large dimension of the overall system state, the direct dynamic function, needed for both numerical simulations of the system dynamics and for the computation of the prediction terms of the EKF, has been developed referring to the Newton–Euler recursive procedure used for industrial manipulators [25].

17.6.2 AUV-streamer localization strategy

To develop the AUV-streamer localization algorithm, we refer to a discrete time version of the EKF. Thus, we started by discretizing the continuous time dynamic model:

$$\mathbf{x}_k = f(\mathbf{x}_{k-1}, \mathbf{u}_{k-1}) + \mathbf{w}_{k-1} \tag{17.33}$$

where $\mathbf{x}_k = [q_k, \dot{q}_k]^T \in \mathbb{R}^{2(6+3\boldsymbol{no\_of\_links})}$ and $f(\mathbf{x}_{k-1}, \mathbf{u}_{k-1})$ are the state transition functions.

The EKF has been implemented following the equations reported next:

- State and covariance prediction

$$\widehat{\mathbf{x}}_{k|k-1} = f(\widehat{\mathbf{x}}_{k-1|k-1}, \mathbf{u}_{k-1}) \tag{17.34}$$

$$\mathbf{P}_{k|k-1} = \mathbf{F}_{k-1}\mathbf{P}_{k-1|k-1}\mathbf{F}^T_{k-1} + \mathbf{Q}_{k-1} \tag{17.35}$$

- State and covariance update

$$\boldsymbol{\beta}_k = \boldsymbol{\alpha}_k - h(\widehat{\mathbf{x}}_{k|k-1}) \tag{17.36}$$

$$\mathbf{S}_k = \mathbf{H}_k\mathbf{P}_{k|k-1}\mathbf{H}^T_k + \mathbf{R}_k \tag{17.37}$$

$$\mathbf{K}_k = \mathbf{P}_{k|k-1}\mathbf{H}^T_k\mathbf{S}^{-1}_k \tag{17.38}$$

$$\widehat{\mathbf{x}}_{k|k} = \widehat{\mathbf{x}}_{k|k-1} + \mathbf{K}_k\boldsymbol{\beta}_k \tag{17.39}$$

$$\mathbf{P}_{k|k} = (\mathbf{I} - \mathbf{K}_k\mathbf{H}_k)\mathbf{P}_{k|k-1} \tag{17.40}$$

where the Jacobians of the state transition and observation functions $f(\mathbf{x}_{k-1}, \mathbf{u}_{k-1})$ and $h(\mathbf{x}_k)$ are calculated, respectively, around the small neighborhood δ and ϵ of the points $\widehat{\mathbf{x}}_{k-1|k-1}, \mathbf{u}_{k-1}$ and $\widehat{\mathbf{x}}_{k|k-1}$, respectively, as given by the following equations:

$$\mathbf{F}_{k-1} = \frac{\partial f}{\partial x}\Big|_{\hat{x}_{k-1|k-1}, u_{k-1}} \text{ and } H_k = \frac{\partial h}{\partial x}\Big|_{\hat{x}_{k|k-1}}$$

For the state and covariance update, we defined the output measurements function as

$$\boldsymbol{\alpha}_k = h(\mathbf{x}_k) + \mathbf{v}_k \tag{17.41}$$

$\boldsymbol{\alpha}_k$ represents the set of output measurements available to the system, $h(\mathbf{x}_k)$ and $\mathbf{v}_k$ are the output function and noise, respectively. To build the output function, we assume that the AUV is equipped with various proprioceptive and relative localization sensors, which can be employed depending on specific operational mode. We define below the relative equations and we assume that all the measurements are affected by zero-mean Gaussian noise. Their formulation will be then used in the development of the EKF output function, and in the computation of the Jacobian matrix used in the covariance correction equation. The considered sensors were the following:

- *Compass:* It gives the actual yaw information of the AUV as described by the following equation:

$$\alpha_{comp} = \psi_k + v_{comp} \tag{17.42}$$

  where $v_{comp} \sim \mathcal{N}(0, \sigma^2_{comp})$ is the noise term.
- *Ultra-short baseline (USBL):* It is an acoustic positioning system for the underwater vehicle navigation and tracking. The measurement equation of this sensor is given as follows:

$$\alpha_{usbl} = [x_k\ y_k\ z_k]^T + \mathbf{v}_{usbl} \tag{17.43}$$

  where $\mathbf{v}_{usbl} \sim \mathcal{N}(0, \sigma^2_{usbl})$ is the noise term.
- *Depth sensor:* It gives the actual depth information of the AUV from the sea surface as described by the following equation:

$$\alpha_{depth} = z_k + v_{depth} \tag{17.44}$$

  where $v_{depth} \sim \mathcal{N}(0, \sigma^2_{depth})$ is the noise term.
- *Gyroscope:* It is used to measure the orientation with respect to the inertial frame of the body to which the sensor is attached to. Generally, the Euler angles, roll, pitch, and yaw are measured using this sensor, and the relative equation is given as

$$\alpha_{gyro} = [\phi_k\ \theta_k\ \psi_k]^T + \mathbf{v}_{gyro} \tag{17.45}$$

  where $v_{gyro} \sim \mathcal{N}(0, \sigma^2_{gyro})$ is the noise term.

- *Doppler velocity log (DVL)/inertial measurement unit (IMU):* It is the couple of sensors that allow to measure linear and angular velocity. In a discrete time formulation, its output, including the effect of the sample time dT, is given by

$$\alpha_{dvl} = \begin{bmatrix} x_k - x_{k-1} \\ y_k - y_{k-1} \\ z_k - z_{k-1} \\ \theta_k - \theta_{k-1} \\ \psi_k - \psi_{k-1} \end{bmatrix} + \mathbf{v}_{dvl} \tag{17.46}$$

where $\mathbf{v}_{dvl} \sim \mathcal{N}(0, R_{v,dvl})$ is the noise term.

- *Range from sparkers:* We assume that the AUV can directly collect information from the hydrophones on the streamer, and from the time of flight of acoustic signals, it can compute the range measurements from each hydrophone X_{P_i} and to a sparker in position $X_S = [x_S, y_S, z_S]^T$. The output equation is

$$\alpha_{range} = \| X_{P_i} - X_R \| + v_{range} \tag{17.47}$$

where $v_{range} \sim \mathcal{N}(0, \sigma^2_{range})$ is noise term.

As reported in [23], the overall localization strategy has been extensively tested in numerical simulations in Matlab environment, considering the case of single and multiple sparkers, and considering the AUV equipped with different set of navigation sensors. Moreover, the localization algorithm has been tested using data gathered from the experimental tests performed in Sines, Portugal (Nov. 2016), for the WiMUST project purposes.

17.7 Conclusions

This chapter has presented the challenges encountered during design and testing of the GNC system of the vehicles involved in the WiMUST project. The results presented here could be adopted in many other multi-vehicle missions where formation control at the surface and underwater, localization with minimal acoustic communications, and towing of streamers, cables, or submerged bodies in general is required by the mission.

Acknowledgments

The results described in this chapter build on the work performed within the WiMUST project. The authors wish to thank all the WiMUST partners.

References

[1] WiMUST project webpage [homepage on the Internet]; 2015-2018 [cited 2018, June 1]. Available from: http://www.wimust.eu.

[2] Alves J, Oliveira P, Oliveira R, *et al.* Vehicle and mission control of the DELFIM autonomous surface craft. In: Control and Automation, 2006. MED'06. 14th Mediterranean Conference on. IEEE; 2006. pp. 1–6.

[3] Abreu PC, Botelho J, Góis P, *et al.* The MEDUSA class of autonomous marine vehicles and their role in EU projects. In: OCEANS 2016-Shanghai. IEEE; 2016. pp. 1–10.

[4] Alvarez A, Caffaz A, Caiti A, *et al.* Folaga: A low-cost autonomous underwater vehicle combining glider and AUV capabilities. Ocean Engineering. 2009;36(1):24–38.

[5] Antonelli G, Arrichiello F, Caiti A, *et al.* ISME activity on the use of autonomous surface and underwater vehicles for acoustic surveys at sea. ACTA IMEKO. 2018;7(2):24–31.

[6] Eustice RM, Singh H, Whitcomb LL. Synchronous-clock, one-way-travel-time acoustic navigation for underwater vehicles. Journal of Field Robotics. 2011;28(1):121–136.

[7] Cardeira B, Pascoal AM, Indiveri G. WiMUST deliverable D4.1 distributed localization and navigation final report. Instituto Superior Técnico for Research and Development, Lisbon; 2017.

[8] De Palma D, Indiveri G. Navigation filters for autonomous underwater vehicles during geotechnical surveying experiments. In: 11th IFAC Conference on Control Applications in Marine Systems, Robotics, and Vehicles (CAMS 2018); 2018.

[9] Fossen TI. Handbook of Marine Craft Hydrodynamics and Motion Control. John Wiley & Sons, Ltd; 2011. Available from: http://dx.doi.org/10.1002/9781119994138.

[10] Canepa A. Guida E Controllo Di Veicoli Sottomarini Autonomi. University of Genova, Informatics Engineering; 2011.

[11] Allotta B, Caiti A, Costanzi R, *et al.* A new AUV navigation system exploiting unscented Kalman filter. Ocean Engineering. 2016;113:121–132.

[12] De Palma D, Arrichiello F, Parlangeli G, *et al.* Underwater localization using single beacon measurements: Observability analysis for a double integrator system. Ocean Engineering. 2017;142:650–665. Available from: https://doi.org/10.1016/j.oceaneng.2017.07.025.

[13] Indiveri G, De Palma D, Parlangeli G. Single range localization in 3-D: Observability and robustness issues. Control Systems Technology, IEEE Transactions on. 2016;24(5):1853–1860. Available from: http://dx.doi.org/10.1109/TCST.2015.2512879.

[14] De Palma D, Indiveri G, Pascoal AM. A null-space-based behavioral approach to single range underwater positioning. In: 10th IFAC Conference on Manoeuvring and Control of Marine Craft MCMC 2015. vol. 48. Copenhagen; 2015. pp. 55–60. Available from: http://dx.doi.org/10.1016/j.ifacol.2015.10.258.

[15] Ruckdeschel P, Spangl B, Pupashenko D. Robust Kalman tracking and smoothing with propagating and non-propagating outliers. Statistical Papers. 2014;55(1):93–123. Available from: http://dx.doi.org/10.1007/s00362-012-0496-4.

[16] De Palma, D., and Indiveri, G., Output outlier robust state estimation. International Journal of Adaptive Control and Signal Processing, Vol. 31, No. 4, pp. 581–607, doi: 10.1002/acs.2673, 2017.

[17] Aguiar AP, Pascoal AM. Coordinated path-following control for nonlinear systems with logic-based communication. In: 2007 46th IEEE Conference on Decision and Control. IEEE; 2007.

[18] Vanni F, Aguiar AP, Pascoal AM. Cooperative path-following of underactuated autonomous marine vehicles with logic-based communication. IFAC Proceedings Volumes. 2008;41(1):107–112.

[19] Fernandes Castro Rego F, Aguar AP, Pascoal AM. A packet loss compliant logic-based communication algorithm for cooperative path-following control. In: The 9th IFAC Conference on Control Applications in Marine Systems 2013; 2013.

[20] Caccia M, Bruzzone G, Bono R. Modelling and identification of the Charlie2005 ASC. In: 14th Mediterranean Conference on Control and Automation (MED). IEEE; 2006. pp. 1–6.

[21] Arrichiello F, Antonelli G, Kelholt E. Shape estimate of a streamer of hydrophones towed by an autonomous underwater vehicle. IFAC-PapersOnLine. 2016;49(23):181–186.

[22] Arrichiello F., Sarkar S., Chiaverini S., Antonelli G., Dynamic Modelling of a Streamer of Hydrophones Towed with an Autonomous Underwater Vehicle. In: Mazal J. (eds) Modelling and Simulation for Autonomous Systems. MESAS 2017. Lecture Notes in Computer Science, vol 10756. Springer, Cham.

[23] Arrichiello F, Sarkar S, Chiaverini S, *et al.* Localization of an array of hydrophones towed by an autonomous underwater vehicle. In: The 26th Mediterranean Conference on Control and Automation; 2018.

[24] Fossen TI. Marine Control Systems: Guidance, Navigation and Control of Ships, Rigs and Underwater Vehicles. Trondheim, Norway: Marine Cybernetics; 2002.

[25] Siciliano B, Sciavicco L, Villani L, et al. Robotics: Modelling, Planning and Control. Springer-Verlag London, 2009.

Chapter 18
Adaptive sampling and energy-efficient navigation in time-varying flows

Tahiya Salam[1], Dhanushka Kularatne[1], Eric Forgoston[2] and M. Ani Hsieh[1]

This chapter presents a strategy to enable a team of mobile robots to adaptively sample and track a dynamic spatiotemporal process. We propose a distributed strategy where robots collect sparse sensor measurements, create a reduced-order model (ROM) of the spatiotemporal process, and use this model to estimate field values for areas without sensor measurements of the dynamic process. The robots then use these estimates of the field, or inferences about the process, to adapt the model and reconfigure their sensing locations. We use this method to obtain an estimate for the underlying flow field and use that to plan optimal energy paths for robots to travel between sensing locations. We show that the errors due to the reduced-order modeling scheme are bounded, and we illustrate the application of the proposed solution in simulation and compare it to centralized and global approaches. We then test our approach with physical marine robots sampling a spatially nonuniform time-varying process in a water tank.

18.1 Introduction

The ocean poses an immensely complex fluid dynamical challenge, involving the interplay of rotation, stratification, complex topography, and variable thermal and atmospheric forcing, not to mention thousands of biological, chemical, and physical inputs. It is the "engine" that drives weather–climate systems worldwide [1], inextricably linking human life to the ocean. Recent years have seen an increase in the pursuit of robotics technology to support a wide range of activities in marine and littoral environments. Examples include commerce (e.g., aquaculture), natural resource gathering (e.g., mining and mineral exploration), construction and repair (e.g., pipeline servicing and inspection), scientific exploration (e.g., health

[1]University of Pennsylvania, Philadelphia, PA 19104, USA
[2]Montclair State University, Montclair, NJ 07043, USA

monitoring of coral reefs), search and rescue efforts (e.g., aviation and maritime accidents), and military operations (e.g., surveillance and mine sweeping).

In all these applications, robots must have the ability to estimate, predict, and track complex and dynamic spatiotemporal processes (e.g., plankton assemblages [2], harmful algae blooms [3–5]), and temperature and salinity profiles that vary across both space and time [6–8]. In general, teams of mobile robots are particularly well suited to monitor these processes since they have the ability to span large physical scales and obtain simultaneous, real-time measurements at distinct locations using their onboard sensors. The gathered data can be used to develop a model of the environment/processes being monitored, and the model can be continuously used and updated to adapt the robots' sensing locations and improve the team's ability to track and predict the process they are monitoring. Nevertheless, complex spatiotemporal processes are often difficult to model and represent. Even when representations are available, the representations are often high-dimensional and computationally burdensome, making it difficult for resource-constrained robots to be executed by themselves. Since the processes being tracked often occur in dynamic and uncertain environments, it is critical that robots do not rely on *centralized* strategies to reduce the impact of limited communication and robot failures. As such, a significant challenge is the development of *distributed* modeling and estimation strategies for teams of robots tracking time-varying spatiotemporal processes.

Existing work on multi-robot coordination for environmental monitoring, mapping, and modeling applications generally fall under two categories: coverage control and Gaussian process (GP)-based information maximizing exploration/search. The seminal work by Cortes *et al.* on optimal sensor placement relies on a weighting function to account for sensing quality and to ensure coverage of the given spatial domain [9]. While the original work assumes the weighting function must be specified *a priori*, the work has been extended in several ways. The stochastic uncertainty associated with modeling the weighting function was incorporated online to optimize the deployment of the sensors in [10]. A strategy for learning the weighting functions online, rather than specifying them *a priori*, was proposed in [11]. A significant advantage of existing coverage control strategies is their ease for distributed implementation. However, these techniques only account for the sensor limitations and ignore the fundamental physics that governs the dynamics of processes of interest. As such, sensing locations provided by these strategies often fail to capture the relevant features needed to obtain a suitable estimate of the field.

In recent years, GPs have been widely used to model various spatiotemporal processes. In [12], the environment is modeled using GPs and the objective is to learn confidence measures on the uncertainty of the model. The uncertainty is then employed to plan minimum risk trajectories for underwater robots. GPs are used to model the quantities of interest being monitored by the robot where the models are then used in a stochastic optimization strategy to minimize regret when collecting measurements/samples [13]. GPs are used to create a map of the environment in [14]. The map is partitioned so robots can determine nearby locations and select future sampling locations that reduces the entropy in the map. The work presented in [15] adapts the model in real time based on observations and optimizes next best sensing locations based on the updated model for a small team of mobile robots. However, similar to coverage control strategies, techniques that rely on using GPs to model the environment/process neglect the physical principles that give rise to the dynamics of the fluid flow. Furthermore, existing work only accounts for the spatial variations of the field and neglects the temporal variations of the field. These limitations are partly because GPs are incapable of capturing the important nonlinearities of the process of interest since they are not suitable for modeling multi-scale processes or processes that are described by functions with varying smoothness.

Similar to [6] and [16], our work focuses on the fusion and control of active sensor networks. We consider the modeling and estimation of spatiotemporal processes and how robots can leverage the process model to adapt their sensing locations. We employ proper orthogonal decomposition (POD) to obtain a model of the process being tracked and use this model to adapt the team's sensing locations. The problem of identifying optimal sensor placements has been addressed in existing work [17,18], even in the context of optimal sensor placement for reduced-order modeling using POD techniques [19–24]. Nevertheless, existing work only addresses the placement of static sensors and does not leverage a robot's mobility to adapt their measurements in response to the changing environmental conditions. These works also assume complete knowledge of the time-series data used to derive the ROM using POD and do not address the assimilation of new data into the existing model.

As robots navigate within the workspace to obtain new measurements and the new data are assimilated into the model, the updated model is then used to improve the selection of next best locations for new measurements. A challenge of operating in a time-varying and stochastic flow environment is that the effects of the surrounding fluid dynamics are tightly coupled with the vehicle dynamics. With enough control authority, these environmental effects can be modeled as external disturbances [25,26] or as a known external flow field [27] and corrected by the vehicle's navigation controller. However, it is often difficult for a power-constrained vehicle to reach the desired goal location using these approaches. Recent work has shown that autonomous underwater and surface vehicle (AUV/ASV) motion planning and adaptive sampling strategies are improved by incorporating either historical ocean flow data [28–30] or multi-layer partial differential equation (PDE) models of the ocean [31–34]. Nevertheless, accessibility to and the overall quality of the flow data and/or numerical models are highly dependent on how well a given region of interest is instrumented. This is because numerical PDE models are often derived through a combination of theoretical and field observations. Ocean current hindcasts, nowcasts, and forecasts provided by Naval Coastal Ocean Model (NCOM) databases * and regional ocean model systems (ROMS) [29] are assimilated from satellite and field observations in conjunction with predictions from numerical PDE models [35,36]. Despite significant advances in ocean modeling and data assimilation, high fidelity *in situ* sensing, and increasing computational power, the predictability of existing models has been limited, particularly at the sub-mesoscale. The thousands of biological, chemical,

*URL: http://cordc.ucsd.edu/projects/mapping/maps

and physical inputs, coupled with limited data and computational power for assimilation, make it very challenging to reconcile spare, uneven observational data with values predicted by the field models. As such, any adaptive sampling must take into consideration vehicle motion and prediction uncertainties.

In this chapter, we describe our efforts in developing an adaptive sampling framework for teams of energy-constrained mobile sensors operating in dynamic and uncertain flows. We present a distributed strategy for mobile robot teams to build an ROM of a spatiotemporal process of interest using sparse sensor measurements. The model is then used to predict and infer field values in regions of the workspace that are not covered by the team. Using their predictions, the robots can then adapt and reconfigure their sensing locations to maximize the effectiveness of the new observations for improving the process model. Furthermore, we show how the model can be leveraged to compute minimum energy trajectories for power-constrained vehicles to enable them to leverage the surrounding flow for navigation to their next sensing locations. The contributions of this work are twofold. First, we propose a framework that uses the dynamics of the process to allow robots to compactly model the environment, infer properties of the environment using sparse sensing data, assimilate these inferences to update the model, and compute energy-optimal paths to new sensing locations. The proposed framework allows for a non-balanced assignment of regions to robots, where robots are able to estimate properties of the environment in regions for which there is no available sensing data. Second, we exploit the structures of the model and inference techniques to synthesize a distributed algorithm for the computation of the process model and the estimation of the field values. Different from related works in this domain, we explicitly use the dominant spatial and temporal characteristics of the dynamic process to determine the team's next best sensing locations, compute energy-efficient trajectories to reach those destinations, and update the model and predictions. We conclude with a discussion on how recent geophysical fluid dynamics (GFD) insights are leading to new strategies for distributed modeling, sensing, estimation, and control strategies of autonomous unmanned systems in dynamic and uncertain environments. By exploiting these new geophysical fluid perspectives, it may be possible to develop a large-scale distributed modeling, sensing, and estimation framework where collectives of autonomous unmanned systems have the ability to maintain and update a global description of the flow dynamics in real time.

18.2 Preliminaries

In this section, we provide some preliminaries and present our assumptions on the vehicle models, the process/flow field, and cost functions employed to compute our energy-optimal trajectories for the mobile robots.

18.2.1 Environment and flow model

We consider tracking a dynamic process in a continuous spatial region in a fluidic environment, which we will define as the workspace $\mathscr{W} \in \mathbb{R}^d$, where $d \in \{2, 3\}$. $\mathscr{W}$ can be discretized into n spatial points where at each of the points, a measurement, such as concentration, temperature, or current velocity can be obtained and provides a representation of the spatiotemporal dynamic process, P. The n spatial points can be grouped into s nonoverlapping regions. The process P can be a scalar field, e.g., temperature or salinity field, or a vector field, e.g., flow velocity field.

When P is a time-varying flow field, it can be described as $\mathbf{V_f} : \mathscr{W}_T \mapsto \mathbb{R}^d$, where $\mathscr{W}_T = \mathscr{W} \times [\underline{t},\ \bar{t}]$ and $[\underline{t},\ \bar{t}] \subset \mathbb{R}_{\geq 0}$ denotes the time interval under consideration. As such, for $\mathbf{x} \in \mathscr{W}$ and $t \in [\underline{t},\ \bar{t}]$, $\mathbf{V_f}(\mathbf{x}, t)$ denotes the flow velocity vector. The coordinates of $\mathbf{x}$ in the inertial frame are denoted x_i, $i = 1, \ldots, d$, and similarly components of $\mathbf{V_f}(\mathbf{x}, t)$ are denoted by V_{fi}, $i = 1, \ldots, d$. The speed of the flow is given by $V_f(\mathbf{x}, t) = \|\mathbf{V_f}(\mathbf{x}, t)\|$ and the maximum flow speed encountered in the domain is given by $V_{fm} = \max_{\mathbf{x} \in \mathbb{W}, t \in [\underline{t},\ \bar{t}]} V_f(\mathbf{x}, t)$. In this work, we assume that the robots are estimating a reduced-order description of $\mathbf{V_f}$ given by $\hat{\mathbf{V}}_\mathbf{f}$. As such, the path-planning problem assumes that the reduced-order description, $\hat{\mathbf{V}}_\mathbf{f}$, is completely known since the robots would simply employ their most up-to-date estimate, $\hat{\mathbf{V}}_\mathbf{f}$, for planning purposes. When forecast uncertainties for $\hat{\mathbf{V}}_\mathbf{f}$ are available, we discuss how to plan trajectories subject to these uncertainties in Section 18.4. Finally, we assume that the positions of static and dynamic obstacles in the environment are available. This information could be encoded by a function $\mathbb{O} : \mathscr{W}_T \mapsto \{true, false\}$, where $\mathbb{O}(\mathbf{x}, t) = true$ indicates an obstacle at $(\mathbf{x}, t)$ and vice versa. Such information could be obtained online by equipping the vehicles with suitable sensors, e.g., sonar range finders, lidars, and RGBD cameras, and by running a parallel routine that computes the positions and possibly velocities of the obstacles.

18.2.2 Vehicle model

Given a team of q robots, we assume each robot is equipped with a sensor that is capable of sensing across each of the s sensing regions such that $q < s$. We assume that the quality and range of the onboard sensors are homogeneous and that the robots can localize and communicate small packets of information, e.g., matrices, with their neighbors. Furthermore, robots can obtain an initial estimate of the dynamic process using either historical data or some forecast model. Since $q < s$, each robot is responsible for monitoring multiple regions. However, since each robot's sensor range is finite, the robot can only obtain measurements for a subset of the regions assigned to it. Thus, we assume each robot will have to rely on a pre-existing model of the process and measurements from other areas to infer the value of the process in the assigned regions not covered by its sensor. In this work, robots are assigned disjoint subsets of the workspace so that each robot only has to maintain a model of the environment for its assigned regions. Each robot is then able to share a compact amount of information about its model with its neighbors and use model information aggregated from its neighbors to determine the next best optimal sensing positions to obtain new data to improve their models. As such, we assume that robots communicate via a broadcast network architecture and that the communication network remains connected at all times. Robots can move and create or break their connections with other robots as long as the network remains connected.

For motion planning purposes, we employ a holonomic kinematic model for the autonomous mobile sensor. This assumption is valid if the dimensions of the autonomous marine vehicle (AMV) are small compared with the dimensions of the flow structures, and if the timescales of the dynamics governing the vehicle are much faster than those of the flow field. A typical AMV operating in the ocean satisfies both these criteria. Using this model, the net velocity of the vehicle with respect to the inertial frame is given by

$$\mathbf{V_{net}}(\mathbf{x}, t) = \mathbf{V_f}(\mathbf{x}, t) + \mathbf{V_{still}}(\mathbf{x}, t) \tag{18.1}$$

where $\mathbf{V_{still}}$ is the velocity of the vehicle with respect to the flow, i.e., $\mathbf{V_{still}}$ is the "thrust" vector of the vehicle. To achieve a given velocity $\mathbf{V_{net}}$, the AMV speed with respect to the flow needs to be

$$V_{\text{still}} = \sqrt{(V_{\text{net}} - V_f \cos\theta)^2 + (V_f \sin\theta)^2} \tag{18.2}$$

where $V_{\text{net}} = \|\mathbf{V_{net}}\|$, $V_f = \|\mathbf{V_f}\|$, $V_{\text{still}} = \|\mathbf{V_{still}}\|$, and θ is the angle between $\mathbf{V_f}$ and $\mathbf{V_{net}}$. We further assume that the actuation capability of the vehicle is limited and that its maximum speed is lower than the speed of the surrounding flow, i.e., $V_{\text{still}}(\mathbf{x}, t) \leq V_{\max} < V_{fm}$.

18.2.3 Proper orthogonal decomposition

In this chapter, we are interested in developing ROMs for complex spatiotemporal processes. While there are many techniques for extracting the dominant dynamics of these infinite-dimensional fields [37–39], we focus on modal analysis techniques in the construction of a low-dimensional approximation of the process dynamics. In particular, we use proper orthogonal decomposition (POD) [40,41] to obtain a representative ROM of the flow field.

In POD analysis, m snapshots of the field are collected, either through experimentation or numerical simulations, such that at each time $t = 1, \ldots, m$, $\boldsymbol{z}(\boldsymbol{t}) = [z_1(t), \ldots, z_n(t)]^\top$, where n is the spatial dimension of some discretization of the flow field. A covariance matrix is constructed as

$$\boldsymbol{K} = \frac{1}{\boldsymbol{m}} \sum_{t=1}^{\boldsymbol{m}} \boldsymbol{z}(\boldsymbol{t})\boldsymbol{z}(\boldsymbol{t})^\top = \frac{1}{\boldsymbol{m}} \boldsymbol{X}\boldsymbol{X}^\top \tag{18.3}$$

where $\boldsymbol{X} \in \mathbb{R}^{n \times m}$ with its columns as $\boldsymbol{z}(\boldsymbol{t})$. The low-dimensional basis is created by solving the symmetric eigenvalue problem

$$\boldsymbol{K}\phi_i = \lambda_i \phi_i \tag{18.4}$$

where $\boldsymbol{K}$ has n eigenvalues such that $\lambda_1 \geq \lambda_2 \geq \cdots \geq \lambda_n \geq 0$ and the eigenvectors $\boldsymbol{\phi}$ are pairwise orthonormal. The original basis is then truncated into a new basis $\boldsymbol{\Phi}$ by choosing k eigenvectors that capture a user-defined fraction, E, of the total variance of the system, such that their eigenvalues satisfy

$$\frac{\sum_{i=1}^{k} \lambda_i}{\sum_{i=1}^{n} \lambda_i} \geq E \tag{18.5}$$

Thus, each term $z(t)$ can be written as

$$z(t) = \boldsymbol{\Phi}\boldsymbol{c}(t) \tag{18.6}$$

where $\boldsymbol{c}(t) = [c_1(t), \ldots, c_k(t)]^\top$ holds time-dependent coefficients and $\boldsymbol{\Phi} \in \mathbb{R}^{n\times k}$ with its columns as $\boldsymbol{\phi}_1,\ldots,\boldsymbol{\phi}_k$. The low-dimensional, orthogonal subspace associated with Φ is an optimal approximation of the data with respect to minimizing least squares error.

18.2.4 Cost function

To plan energy-efficient navigation strategies in marine environments, we focus on techniques that reduce the overall energy utilized by the vehicle as it navigates through the environment. As such we consider a cost function that represents the total energy consumption of the AMV, which is given by $E_{\text{total}} = E_{\text{hotel}} + E_{\text{drag}}$, where E_{hotel} is the energy required to operate the vehicle's computing and sensor systems independent of propulsion [42] and E_{drag} is the energy expended to overcome drag forces exerted by the fluid. Assuming a constant power usage K_h by the computing and sensor systems gives $E_{\text{hotel}} = \int_{t_s}^{t_g} K_h dt$. The drag force F_d encountered by the AMV is given by $F_d(t) = K_d V_{\text{still}}^{\alpha-1}(\mathbf{x}, t)$, where K_d is the drag coefficient and $\alpha \in \{2, 3, \ldots\}$. If $\alpha = 2$, the drag is linear, if $\alpha = 3$ the drag is quadratic, and so on. This leads to $E_{\text{drag}} = \int_{t_s}^{t_g} K_d V_{\text{still}}^{\alpha}(\mathbf{x}, t)dt$. Thus, the total cost of a path is given by

$$C = \int_{t_s}^{t_g} K_h + K_d V_{\text{still}}^{\alpha}(\mathbf{x}, t)dt \tag{18.7}$$

Note that K_h and K_d can also be thought of as weighting parameters between minimum time paths and minimum energy paths. If a minimum time path is required, we could set $K_d = 0$ and proceed, and vice versa. If exact energy minimization is required, actual values for K_h and K_d should be used.

18.3 Adaptive sampling and reduced-order modeling

In this section, we describe a distributed strategy to enable a mobile robot team to adaptively sample and track a dynamic spatiotemporal process [43]. The approach enables the team to create an ROM of the process using sparse sensor measurements and using the model to determine the next best locations for obtaining new data to improve the model. The strategy is distributed in that each robot develops its own ROM using its sensor data and sensor data provided from their teammates. The ROM is then used to infer measurements in regions where no measurements are available and new measurement locations are then determined based on these inferences. The robots navigate to these new locations and obtain new measurements which are then assimilated into the model, and the process is repeated. In this section, we briefly summarize our strategy by presenting the centralized approach and describing a distributed implementation of the centralized strategy. We briefly discuss simulation and experimental results. The interested reader is referred to [43] for the details. Finally, in order to quantify the uncertainty between the dynamic process of interest and the ROM constructed by the robots, an error analysis is presented.

18.3.1 Optimizing robot locations for field reconstruction

Given a low-dimensional representation of the subspace on which the data are located, we leverage the properties of the orthogonal bases obtained via POD to compute the optimal set of robot locations in the workspace for the reconstruction of the field using only sparse data in real time. Consider the problem of reconstructing a field from measurements in q arbitrary sensing regions. Given s total sensing regions, let $S \subset \{1, \ldots, s\}$, where S contains the locations of the q sensing regions. Measurements of the field are collected over the q sensing regions as $\boldsymbol{y}_r(\boldsymbol{t})$ for sensing region $r \in S$, where each $\boldsymbol{y}_r(\boldsymbol{t}) \in \mathbb{R}^{n_r \times 1}$ for n_r points of measurements in region r. Let matrices $\boldsymbol{\Phi}_r \in \mathbb{R}^{n_r \times k}$ such that the rows of $\boldsymbol{\Phi}_r$ are the rows of $\boldsymbol{\Phi}$ corresponding to the locations in sensing region r. Using the gappy POD [19,21,41], the time-dependent coefficients that minimize the distance between $\boldsymbol{y}(\boldsymbol{t})$, sensor values, and $\hat{\boldsymbol{y}}(\boldsymbol{t})$, the projection of sensor values onto the subspace associated with the vectors $\{\boldsymbol{\Phi}_r\}_{r \in S}$ can be found using

$$\hat{\boldsymbol{c}}(\boldsymbol{t}) = \boldsymbol{A}^{-1}\boldsymbol{B}$$

$$\text{for } \boldsymbol{A} = \sum_{r \in S} \boldsymbol{\Phi}_r^{\top}\boldsymbol{\Phi}_r \quad \text{and} \quad \boldsymbol{B} = \sum_{r \in S} \boldsymbol{\Phi}_r^{\top}\boldsymbol{y}_r(t) \tag{18.8}$$

where this time-dependent coefficient $\hat{\boldsymbol{c}}(\boldsymbol{t})$ is then applied to $\boldsymbol{\Phi}$ as in (18.6) to recover the values of the field for which there are no sensor measurements.

To optimize the reconstruction of the full field using only measurements from the q regions, one must determine the q sensing regions from the set of S possible regions. Note that the matrix $\boldsymbol{A} \in \mathbb{R}^{k \times k}$ depends only on the set of S sensing regions and is not time varying. If measurements from all sensing regions were used, the matrix $\boldsymbol{A}$ would be the identity matrix since $\boldsymbol{A} = \Phi^{\top}\boldsymbol{\Phi} = \boldsymbol{I}$ for $\boldsymbol{\Phi}$ containing orthonormal columns and the coefficients $\hat{\boldsymbol{c}}(\boldsymbol{t})$ could be calculated exactly using (18.6). However, since only some and not all sensing regions are being used, the sensing regions should be chosen such that the rows of the eigenvectors corresponding to these sensing regions create a basis that is close to orthogonal.

Additionally, [20] provides a criteria for selecting the optimal set of sensing regions S as

$$\max_{S} \min_{i} \lambda_i(\boldsymbol{A}) \tag{18.9}$$

where maximizing the minimum eigenvalue of $\boldsymbol{A}$ in turn minimizes the maximum angle between the subspace associated with $\boldsymbol{\Phi}$ and P_r, the subspace associated with using only the sensor measurements, as shown in Figure 18.1.

Let a_{ij} represent entries in $\boldsymbol{A}$ and $r_i = \sum_{j \neq i} |a_{ij}|$. The Gershgorin circle theorem [44] states that all eigenvalues of $\boldsymbol{A}$ lie in a circle centered at a_{ii} with radius r_i. Using this property of the eigenvalues of $\boldsymbol{A}$, an estimation of (18.9) is given by

$$\max_{S} \min_{i} a_{ii} \tag{18.10}$$

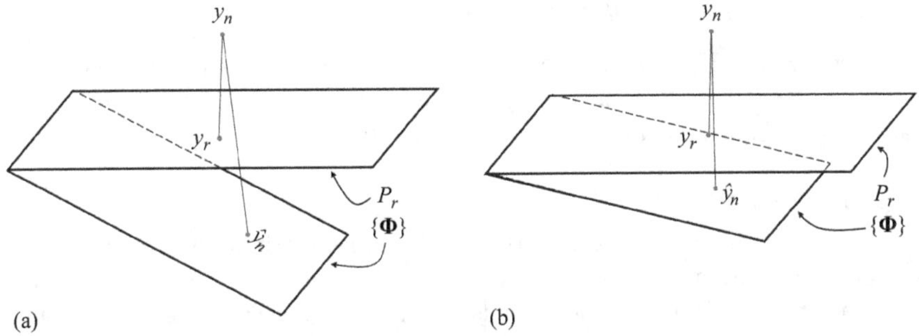

Figure 18.1 Geometric interpretation of maximizing the minimum eigenvalue. The data point y_n containing all the measurements of the field is projected as y_r onto the subspace P_r, where y_r equivalently represents a vector of just the sensor measurements, and is projected as $\hat{y}_n$ onto the low-dimensional subspace $\boldsymbol{\Phi}$. As the angle between P_r and the subspace associated with $\boldsymbol{\Phi}$ decreases, the projection error between y_r and $\hat{y}_n$ also decreases

where maximizing the minimum diagonal element of $\boldsymbol{A}$ seeks the set S that results in the basis used to compute $\boldsymbol{A}$ being both close to orthonormal and minimizing the distance between the subspaces associated with $\boldsymbol{\Phi}$ and $\{\boldsymbol{\Phi}_r\}_{r\in S}$. We employ the algorithm developed in [20] and extended in [17] to find the set S that satisfies criteria (18.10).

18.3.2 Adaptive computation of ROM and robot locations

The techniques described in [17,19–21] rely on computing POD basis vectors using all the available snapshots of data over the process. Instead, we propose a method that dynamically adapts the POD basis vectors using incoming data and reconfigures the robot positions based on the adapted POD.

At the start, i.e., instance 0, POD basis vectors $\boldsymbol{\Phi}(\mathbf{0})$ are computed using T arbitrarily selected snapshots $\{z(t_{0,1}),\dots,z(t_{0,T})\}$, where $z(t_{0,i})\in\mathbb{R}^{n\times 1}$. The T snapshots are gathered from either experiments or numerical simulation based on the equations governing the process of interest. A set of sensing regions $S(0)$ is selected according to the algorithm described above, where robots are then deployed to collect measurements. Estimates of the field are computed using $\boldsymbol{y}_r(\boldsymbol{t})$ for $r\in S_0$ the collected sensing data, $\{\boldsymbol{\Phi}_{r(\mathbf{0})}\}_{r\in S_0}$ the POD basis over the sensing regions, and the relationship (18.8). At instance 1, the new inferences are assimilated into the covariance matrix $\boldsymbol{K}$ as in (18.3) as new snapshots, at which point the POD basis vectors are recomputed as $\boldsymbol{\Phi}(\mathbf{1})$ and a new set of sensing regions $S(1)$ are found. This procedure is repeated for the duration of the mission. This is contrast to other techniques that compute the POD basis requiring all the initial data and do not update the POD basis after new observations.

18.3.3 Distributed implementation

The procedure described above can be implemented in a distributed fashion. In this section, we first show how the procedure can be distributed and conclude with an analysis of the bounds on the error resulting from the ROM.

18.3.3.1 Methodology

A comparison of the centralized and distributed approach is shown in Figure 18.2. The model of the environment is represented as a matrix with columns of eigenvectors, where each row of the matrix corresponds to a spatial point. These can be distributed to different robots, and robots can keep on board the rows corresponding to their assigned regions as shown in Figure 18.3(a). The push-sum algorithm [45] is leveraged to allow for robots to maintain field measurements only over their respective regions, while occasionally exchanging small packets of information with their neighbors to understand the areas of the field without sensor measurements and recompute optimal sensing locations, shown in Figure 18.3(b). Estimating the models in the areas without

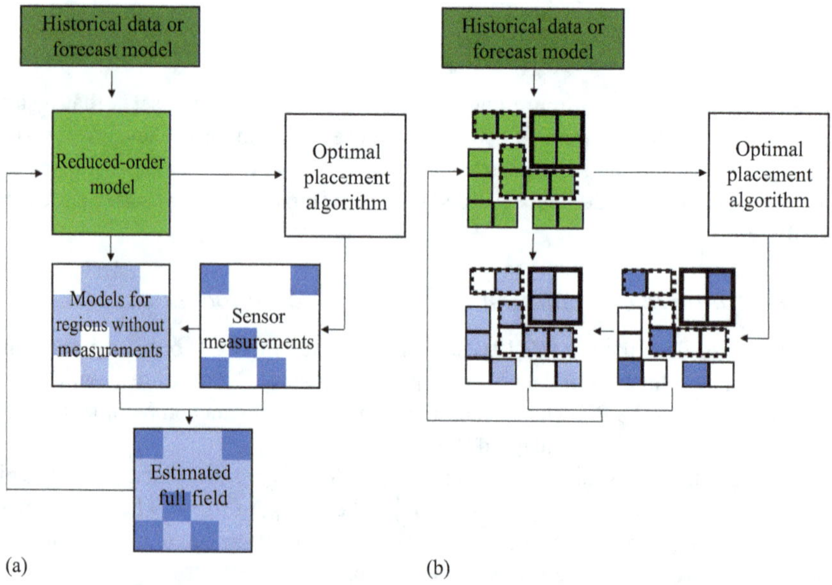

Figure 18.2 Comparison of a centralized framework for model-inference-assimilation scheme and the corresponding distributed framework. (a) The centralized framework keeps a global model which is combined with sensor measurements to estimate the field and update the model. (b) The distributed framework allows robots to take sensing measurements at specific regions and estimate the values of the field using the current model and their neighbors' data. These estimates are used to update the model at the robots' assigned locations

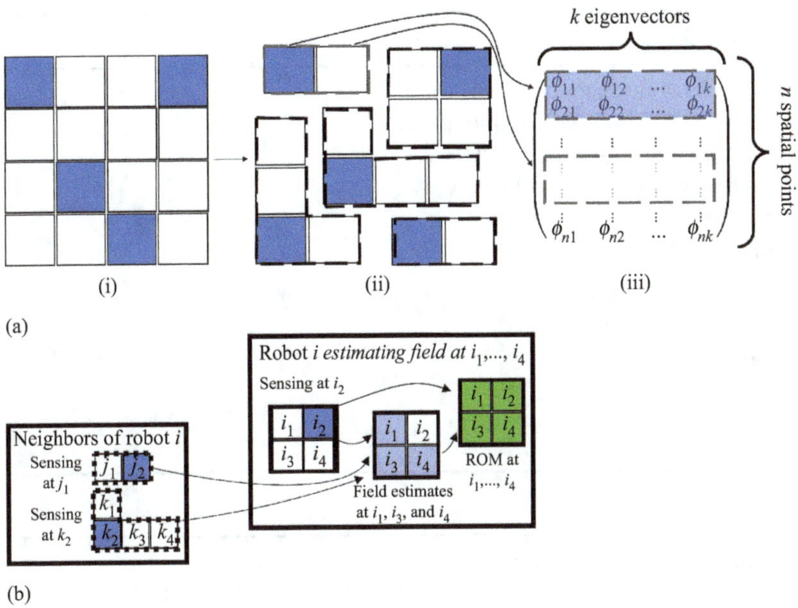

Figure 18.3 (a) Visualization of spatial points corresponding to rows of eigenvectors in the POD basis. In (i), the full field is shown, where each region is a set of points in the field. Blue regions are monitored by the robots and thus are the regions with sensor measurements, while the field in the white regions are inferred using the ROM. In (ii), the dashed lines contain the regions for which each robot either takes measurements or estimates values. The matrix in (iii) shows rows in the POD basis that correspond to a single robot's assigned regions indicated with the gray dashed lines. (b) Regions assigned to a robot for sensing, estimating, and modeling. The robot senses at a location and is assigned to keep track of models of the regions for which no sensor measurement exists. The robot uses its own ROM over its assigned regions, its collected sensor measurements, and sensor measurements from its neighbors to estimate the values of the field for regions without sensor measurements, all of which will then be used to update its ROM

sensor measurements requires aggregating information from previous models over these areas and the new sensing measurements. Instead of having all robots compute the estimates of field measurements for regions without sensor measurements, these regions are assigned arbitrarily to robots, such that the estimates of the values and models for each region are only maintained by one robot.

Algorithm 1: Push-sum algorithm

Input : P_i from each robot
Output: robot i's estimate $\hat{\mathbf{P}}^i$ of matrix $\boldsymbol{P}$
select $\hat{i}$, let $w_{\hat{i}} = 1$, and $w_i = 0\ \forall \hat{i} \neq i$;
for *each robot i in parallel* **do**
  $\hat{P}_i = P_i$;
  for *loop* **do**
    $\hat{P}_i = \sum_{j \in N_i} m_{ij} P_j$;
    $w_i = \sum_{j \in N_i} m_{ij} w_j$;
  end
  return $\hat{P}^i = \frac{\hat{P}_i}{w_i}$
end

Algorithm 2: Orthogonal iteration

set $\boldsymbol{Q} \in \mathbb{R}^{n \times k}$ with random elements;
for *loop* **do**
  $V = \boldsymbol{KQ}$;
  $QR \overset{QR}{=} V$;
end
return columns of $\boldsymbol{Q}$ as eigenvectors;
return diagonals of $\boldsymbol{R}$ as eigenvalues

Details of the push-sum algorithm are now described. Suppose some matrix $\boldsymbol{P} = \sum_i \boldsymbol{P}_i$. Further, there exists agents where each agent i has access to matrix $\boldsymbol{P}_i$ and can communicate with its neighbors N_i. Let $\boldsymbol{M}$ be an arbitrary stochastic matrix such that $m_{ij} = 0$ if agent i is not a neighbor to agent j. A stochastic matrix is used to exploit the equivalence between averaging and Markov chains; we refer the interested reader to [45] for more details. Each agent i can compute $\hat{\boldsymbol{P}}^i$, its own estimate of $\boldsymbol{P}$, as shown in Algorithm 1.

The push-sum algorithm is used for the distributed computations of (a) the covariance matrix from data at sensing regions, (b) the eigenvectors and eigenvalues for the POD basis vectors, and (c) the time-dependent coefficients for estimating the full field. First, we show how to compute the eigenvalues and eigenvectors of a pre-computed covariance matrix in a distributed fashion using existing techniques. Then, we will bypass the need to directly compute the covariance matrix and instead compute the eigenvalues and eigenvectors from onboard data in a distributed setting.

The method of orthogonal iteration allows for the computation of the top k eigenvectors and eigenvalues of a symmetric matrix $\boldsymbol{K} \in \mathbb{R}^{n \times n}$ using Algorithm 2.

The distributed computation of Algorithm 2 rests on the following matrix properties, shown in detail in [45]. However, while [45] assumes a bijection between the rows of the covariance matrix and the robots performing the computation, we show here that this is not strictly necessary, allowing for a non-balanced assignment of rows to robots, where robots can be assigned an arbitrary number of rows. Every row of the covariance matrix $\boldsymbol{K}$ corresponds to a location in the field. Each robot i is assigned the set of rows, L_i, of the matrices $\boldsymbol{K}$ and $\boldsymbol{Q}$ corresponding to the spatial points in its sensing region and some arbitrary subset of the spatial points of the regions not covered by any robot. Let $L = L_i \cup (\cup_{j\in N_i} L_j)$, where N_i is the set of neighbors of robot i so that L is the set that contains all the spatial points assigned to robot i and its neighbors. To start, the rows $\boldsymbol{V}_l$ for $l \in L_i$ can be estimated as a linear combination of the random row vectors $\boldsymbol{Q}_m$ over all $m \in L$ with coefficients a_{lm}. Then each robot can use an estimate of the matrix $\boldsymbol{R}$ to apply to its set of rows $\boldsymbol{V}_l$ for $l \in L_i$ to find the corresponding rows $\boldsymbol{Q}_l$ for the next iteration of orthogonal iteration. An estimate of $\boldsymbol{R}$ is found by leveraging the relation

$$\boldsymbol{W} = \boldsymbol{V}^\top \boldsymbol{V} = \boldsymbol{R}^\top \boldsymbol{Q}^\top \boldsymbol{Q} \boldsymbol{R} \tag{18.11}$$

where $\boldsymbol{Q}^\top \boldsymbol{Q} = \boldsymbol{I}$ since $\boldsymbol{Q}$ is orthonormal and $\boldsymbol{R}$ is a unique upper triangular matrix. Since $\boldsymbol{W} = \sum_{c=1}^{n} \boldsymbol{V}_c^\top \boldsymbol{V}_c$, each agent can compute $\boldsymbol{W}_i$ over its sensing region as $\boldsymbol{W}_i = \sum_{l\in L_i} \boldsymbol{V}_l^\top \boldsymbol{V}_l$. Using the push-sum algorithm, each agent can compute estimates $\hat{\boldsymbol{W}}^i$, perform a Cholesky factorization to compute $\hat{\boldsymbol{W}}^i = \hat{\boldsymbol{R}}_i^\top \hat{\boldsymbol{R}}_i$, and apply $\hat{\boldsymbol{R}}_i^{-1}$ to its rows $\boldsymbol{V}_l$ to compute $\boldsymbol{Q}_l = \boldsymbol{V}_l \hat{\boldsymbol{R}}_i^{-1}$. Then $\boldsymbol{Q}_l$ is used in the next iteration of the orthonormal iteration algorithm. This requires the entries of the covariance matrix $\boldsymbol{K}$ to be known and communication between neighbors to estimate the values $\boldsymbol{V}_l$.

We leverage the following relation presented in [46] to eliminate the centralized computation of $\boldsymbol{K} = \frac{1}{m}\boldsymbol{X}\boldsymbol{X}^\top$ and instead allow for the distributed computation of the eigenvectors and eigenvalues of $\boldsymbol{K}$ directly from $\boldsymbol{X}$ without explicitly constructing $\boldsymbol{K}$. Let the *Diag* operator create a diagonal matrix out of a given vector and the *diag* operator extract the diagonal elements of a given matrix. Each column $\boldsymbol{v}_j$ can be computed as

$$\begin{aligned} \boldsymbol{v}_j &= \frac{1}{m}\boldsymbol{X}\boldsymbol{X}^\top \boldsymbol{q}_j \\ &= \frac{1}{m} diag(\boldsymbol{X}\boldsymbol{X}^\top \boldsymbol{q}_j \boldsymbol{1}^\top) \\ &= \frac{1}{m} diag(\boldsymbol{X}\boldsymbol{X}^\top Diag(\boldsymbol{q}_j)\boldsymbol{1}\boldsymbol{1}^\top) \\ &= \frac{1}{m} diag[\boldsymbol{X}(\boldsymbol{1}\boldsymbol{1}^\top Diag(\boldsymbol{q}_j)\boldsymbol{X})^\top] \end{aligned} \tag{18.12}$$

For $\boldsymbol{q}_j = [q_j(1), \ldots, q_j(n)]$, the lth row of $Diag(\boldsymbol{q}_j)\boldsymbol{X}$ is equal to $q_j(l)\boldsymbol{X}_l$. Furthermore, the quantity $\boldsymbol{1}\boldsymbol{1}^\top Diag(\boldsymbol{q}_j)\boldsymbol{X}$ is a matrix where each row is equal to the sum of all the rows of $Diag(\boldsymbol{q}_j)\boldsymbol{X}$. Thus, only the quantity $\boldsymbol{F} = \sum_{c=1}^{n} \boldsymbol{D}_c$, where $\boldsymbol{D} = Diag(\boldsymbol{q}_j)\boldsymbol{X}$ and $\boldsymbol{D}_c$ denotes the rows of $\boldsymbol{D}$, needs to be computed. Each robot can individually compute the quantity $\boldsymbol{F}_i = \sum_{l\in L_i} q_j(l)\boldsymbol{X}_l$ and then can compute estimates $\hat{\boldsymbol{F}}^i$ using

Algorithm 3: Distributed eigenvectors from data

set $\boldsymbol{Q} \in \mathbb{R}^{n \times k}$ with random elements;
select $\hat{i}$, let $w_{\hat{i}} = 1$, and $w_i = 0 \; \forall \hat{i} \neq i$;
for *loop* **do**
  for *each row r of Q in parallel* **do**
    for *each robot i* **do**
      $\boldsymbol{Z}_i = \sum_{l \in L_i} q_{lr} \boldsymbol{X}_l$;
      Compute $\hat{\boldsymbol{Z}}_i$ with Algorithm 1 (push-sum);
      for $l \in L_i$ **do**
        $v_{lr} = \frac{1}{n} \hat{\boldsymbol{Z}}_i^\top \boldsymbol{X}_l$;
      end
      $\boldsymbol{W}_i = \sum_{l \in L_i} \boldsymbol{V}_l^\top \boldsymbol{V}_l$; Compute $\hat{\boldsymbol{W}}_i$ with Algorithm 1 (push-sum);
      Use Cholesky factorization $\hat{\boldsymbol{W}}^i = \hat{\boldsymbol{R}}_i^\top \hat{\boldsymbol{R}}_i$;
      $\boldsymbol{Q}_l = \boldsymbol{V}_l \hat{\boldsymbol{R}}_i^{-1}$;
    end
  end
end
return rows $\boldsymbol{Q}_l$ as eigenvectors for robot i;
return diagonals of $\hat{\boldsymbol{R}}_i$ as eigenvalues for robot i;

the push-sum algorithm. Then, the lth row of $\boldsymbol{v}_j$ is equal to $\frac{1}{m}\hat{\boldsymbol{F}}^{i\top}\boldsymbol{X}_l$. This is carried out for all k columns of $\boldsymbol{Q}$ and $\boldsymbol{V}$. The full procedure for distributed computation of eigenvectors and eigenvalues is shown in Algorithm 3.

To estimate the time-dependent coefficients, each robot can compute its own $\boldsymbol{A}_i$ and $\boldsymbol{B}_i$ as in (18.8) and use the push-sum algorithm to compute estimates $\hat{\boldsymbol{A}}^i$ and $\hat{\boldsymbol{B}}^i$. Then, robots can compute the estimate $\hat{\boldsymbol{c}}_i = \left(\hat{\boldsymbol{A}}^i\right)^{-1} \hat{\boldsymbol{B}}^i$ and apply coefficients $\hat{\boldsymbol{c}}_i$ to the rows $\boldsymbol{Q}_l$ to estimate the values $\hat{\boldsymbol{y}}_l = \boldsymbol{Q}_l \hat{\boldsymbol{c}}_i$ in the regions $l \in L_i$ for which there are no sensor measurements. We note that, in this framework, data broadcasted by each robot consists of matrices of fixed size corresponding to low-dimensional representations of the evolving process of interest.

18.3.3.2 Error analysis

We begin with a statement of the main result.

Theorem 18.1. *For sufficient mixing time τ, as determined by the criteria for mixing speeds of Markov Chains [45], $\delta = t_{i+1} - t_i$ the time between collected snapshots, $\left\| \frac{df(t)}{dt} \right\| \leq \xi$ for $\boldsymbol{f}$ describing the dynamic process, the bound between $(\boldsymbol{P} + \boldsymbol{P}^\perp)$, the* *projection corresponding to the dynamic process, and* $\boldsymbol{P}_{\hat{Q}_i}$ *the projection onto the POD basis vectors estimated by robots with high probability is defined as*

$$\left\|(\boldsymbol{P}+\boldsymbol{P}^{\perp})-\boldsymbol{P}_{\hat{Q}_i}\right\|_2 \leq e^{\gamma\tau}\left(2\lambda_{k+1}+\frac{\delta^2}{8}\xi\right)+\mathcal{O}\left|\frac{\lambda_{k+1}}{\lambda_k}\right|^{\tau} * n+(C+2)\varepsilon^{4\tau} \tag{18.13}$$

The proof consists of two main components. First, the error between a dynamic process and its centralized POD basis vector estimation is computed. This explicitly takes into account the time between snapshots. Second, the error between the centralized POD basis vector estimation and the decentralized POD basis vector estimation is computed, taking into account the errors incurred by individual robots through the push-sum algorithm. The results are aggregated to derive a final estimation on the error between the dynamic process and the estimated POD basis vectors computed by the robots. The derivation of the bound can be found in Appendix A.

18.3.4 Task allocation

Using the distributed algorithm, individual robots can adaptively calculate their respective eigenvectors and eigenvalues. They can then share the necessary properties of their eigenvectors to their neighbors so each robot can compute the optimal sensing locations. After finding the set of optimal sensing locations, robots are assigned to locations that minimize the total cumulative energy cost incurred by all robots. In the next section, we present how the path costs and the corresponding optimal paths are computed for a given robot and its list of destination points. Once the energy cost matrix between all the robots and the next set of sensing locations is computed, we perform a task assignment using the Hungarian algorithm to reduce the overall energy cost. The flow velocity description, $\mathbf{V_f}$, required for path planning could be obtained using the same reduced-order modeling framework described above. In this case, since the process being observed, i.e., flow velocities are d-dimensional vectors with $d \in \{2, 3\}$, the measurement vector $\mathbf{z(t)}$ would have $d \times n$ entries instead of the n entries for the scalar case.

18.4 Planning in flows

In this section, a graph-based solution method to compute optimal paths in time-varying flows is presented. The method allows for searching across different actuation speeds at each node in order to compute the globally optimum trajectory. It accounts for the time-varying nature of the environment and also considers an adaptive discretization resolution scheme for graph construction. In the context of the previous section, this general method could be used to compute the optimal path cost from the current agent position to the next sensing location. The graph-based nature of the approach makes it possible to compute the optimal energy cost from the current position to each of the next candidate sensing locations in one go.

All the graph-based path-planning strategies mentioned in Section 18.1 are based on a fixed discretization scheme, where the environment is discretized uniformly. The implicit assumption in any graph-based method is that the properties of the system (thrust, flow velocity, etc.) remain constant along edges of the graph. However, in reality, these system properties will vary along edges. Thus, this assumption will give rise to errors in the state-dependent costs (e.g., energy cost) computed for the edges. These errors would be more pronounced if the discretization is coarse and would lead to incorrect results. It is possible to overcome this by allowing system properties to change along edges. However, such a solution is computationally infeasible since the edge cost computations would now require the solution to a two-point boundary value problem at each edge. Alternatively, the discretization resolution could be made very fine, which would again lead to a computationally intractable problem.

In the approach presented in this section, these errors due to discretization resolution are overcome with a novel adaptive discretization scheme. Similar to existing graph-based methods, the environment is represented as a discrete graph and a search algorithm is used to find an optimal path in the graph. The fundamental difference of the presented method lies in the adaptive discretization scheme used in the construction of the graph. In this method, the discretization resolution is selected locally according to the flow characteristics at each point. Specifically, the discretization resolution is selected such that the flow velocity error remains bounded along an edge. Thus, if the flow changes rapidly at a point, the resolution is made finer, and if the flow remains relatively constant, the discretization resolution is made coarser. The result is the first graph-based method that uses adaptive discretization to compute optimal paths in time-varying flows.

We use a discrete graph $\mathscr{G} = (\mathscr{V}, \mathscr{E})$ to represent the workspace $\mathscr{W}_T$, where $\mathscr{V}$ is the vertex set and $\mathscr{E}$ is the edge set. Each $v_i \in \mathscr{V}$ represents a point in $\mathscr{W}_T$ and is identified by the pair $(\mathbf{x_i}, t_i)$. Each $e_{ij} \in \mathscr{E}$ represents a directed edge from v_i to v_j and has an associated traversal cost computed using (18.7). In computing the edge cost, we assume that the flow velocity along the edge remains constant at $\mathbf{V_f}(v_i)$, the flow at the base of the edge. Note that with a little abuse of notation, we sometimes represent the flow velocity at a vertex v_i with $\mathbf{V_f}(v_i)$.

18.4.1 *The need for adaptive discretization*

To illustrate the need for adaptive discretization in graph-based methods for planning optimal paths in time-varying flows, consider three cases of the wind-driven double gyre flow model which is often used to describe large-scale recirculation regions in the ocean [47]. The model is given by

$$V_{f1}(\mathbf{x}, t) = -\pi A \sin(\pi f(x_1, t)) \cos(\pi x_2) - \mu x_1 \tag{18.14a}$$

$$V_{f2}(\mathbf{x}, t) = \pi A \cos(\pi f(x_1, t)) \sin(\pi x_2) \frac{\partial f(x_1, t)}{\partial x_1} - \mu x_2 \tag{18.14b}$$

$$f(x, t) = \varepsilon \sin(\omega t) x^2 + (1 - 2\varepsilon \sin(\omega t)) x \tag{18.14c}$$

where A determines the maximum speed of the flow field, μ is a dissipation constant, and ε and ω, respectively, determine the amplitude and the frequency of the time-varying oscillations of the flow field.

In the three cases considered, the fixed resolution multi-time-step search method described in [48] is used to compute optimal energy paths. In [48], the spatiotemporal workspace is uniformly discretized using fixed intervals Δx and Δt. In the first two cases, the flow field parameters are selected to be $A = 1$, $\omega = 4\pi$ and $\varepsilon = 0.1$. For the third case, a value of $\varepsilon = 0.6$ is selected while leaving the rest of the parameters unchanged. In all three cases, $\mu = 0$. This results in a maximum flow speed of $V_{fm} = 3.74$ m/s and $V_{fm} = 6.73$ m/s for the first two cases and the third case, respectively. However, in both cases the average flow speed is ≈ 2 m/s. Lastly, in all three cases, the maximum vehicle speed is set as $V_m = 2$ m/s.

In Case 1, a discretization resolution of $\Delta x = 0.01$ m and $\Delta t = 0.1$ s was used for the graph search and the results are shown in red in Figure 18.4(a). It can be clearly seen that the resulting path does not match the expected path shown in black. One reason for this discrepancy is due to Δt being set too high with respect to Δx, resulting in an overly slow edge traversal speed (≈ 0.1 m/s). For Case 2, Δt was set to 0.01 s so as to increase the edge traversal speed. The results are shown in Figure 18.4(b), which is much closer to the optimal path. Thus, it is clear that the size of Δx and Δt has to be selected according to the local flow speeds to yield better results. Case 3 further supports this assertion. While the larger ε value in Case 3 (0.1 vs. 0.6) does not affect the average flow speed, it results larger spatiotemporal variations within each region of the flow field. Since the encountered flow speeds are of similar magnitudes for both Cases 2 and 3, one would expect similar results if the same discretization resolutions were used. However, Figure 18.4(c) shows that the resulting path computed for Case 3 does not match the expected path. This mismatch is a result of the large spatiotemporal variations of the flow within the volume of space-time centered at each vertex, which violates the assumption that the flow velocity remains constant along an edge of the graph. As such, appropriate selection of Δx and Δt that ensures that the local flow remains relatively constant is crucial for obtaining good results. This is especially true in highly turbulent regions of the flow field where the discretization resolution has to be made finer to account for the larger changes over the same space-time volume.

18.4.2 Adaptive discretization method

The implicit assumption in any graph-based method for computing optimal paths in time-varying flows is that the flow velocity remains constant along the edges of a graph. Thus, in order to overcome the issues highlighted in the example above, the discretization resolution should be selected to be small enough so that the actual flow velocity variation along an edge in the graph is small.

Consider an edge $e_{ij} = (v_i v_j)$ from node $v_i = (\mathbf{x_i}, t_i)$ to node $v_j = (\mathbf{x_j}, t_j)$ with $\mathbf{x_j} = \mathbf{x_i} + \mathbf{dx}$ and $t_j = t_i + dt$. It is assumed that the flow velocity at v_j can be approximated by a first-order Taylor series expansion at v_i given by

$$\mathbf{V_f}(v_j) = \mathbf{V_f}(v_i) + \tilde{\nabla}\mathbf{V_f}(v_i)\mathbf{dx_t} \tag{18.15}$$

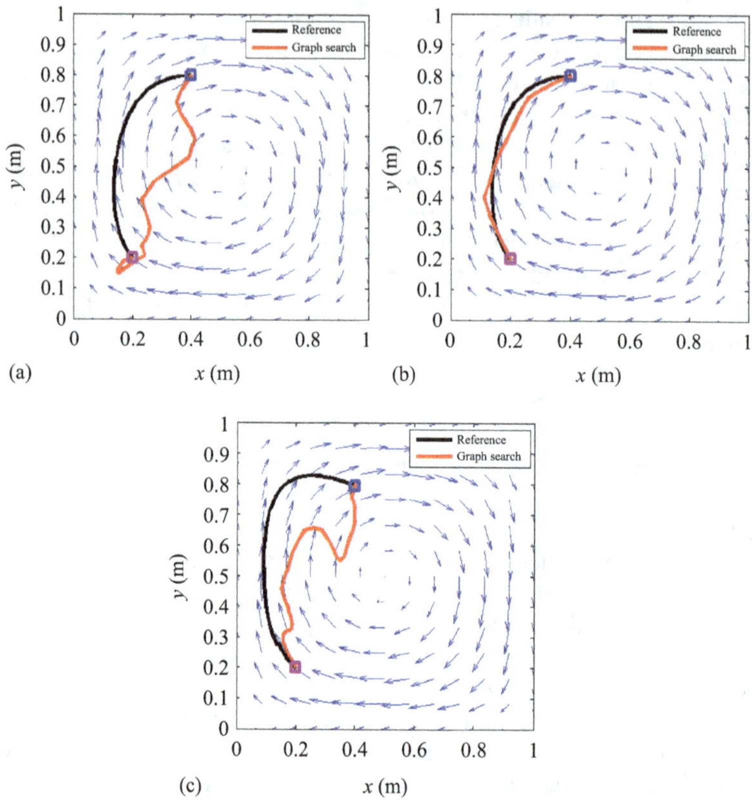

Figure 18.4 *Expected optimal path (in black) vs. computed optimal path (in red). (a) Case 1: $\Delta x = 0.01$ m and $\Delta t = 0.1$ s results in erroneous results as the selected discretization does not match the flow speeds encountered. (b) Case 2: $\Delta x = 0.01$ m and $\Delta t = 0.01$ s results in a better result. (c) Case 3: The flow is changed by setting $\varepsilon = 0.6$, with the same discretization as Case 2. However, the result is incorrect because the discretization does not consider the spatiotemporal variation of the flow*

where $\tilde{\nabla}\mathbf{V_f}(v_i)$ is the gradient of the flow vector at v_i, and $\mathbf{dx_t} = [\mathbf{dx}^T, dt]^T$. Since the flow is assumed to be constant along an edge, the error in the velocity along an edge is given by

$$\mathbf{V_e} = \mathbf{V_f}(v_j) - \mathbf{V_f}(v_i) = \tilde{\nabla}\mathbf{V_f}(v_i)\mathbf{dx_t} \tag{18.16}$$

Thus, the magnitude of the velocity error due to a displacement $\mathbf{dx_t}$ is

$$\|\mathbf{V_e}\| = \sqrt{\mathbf{V_e}^T\mathbf{V_e}} = \sqrt{\mathbf{dx_t}^T\mathbf{H}(v_i)\mathbf{dx_t}}$$

where $\mathbf{H}(v_i) = \tilde{\nabla}\mathbf{V_f}^T(v_i)\tilde{\nabla}\mathbf{V_f}(v_i)$ is a symmetric matrix. As such, it can be shown that

$$\|\mathbf{V_e}\| = \sqrt{\mathbf{dx_t}^T\mathbf{H}(v_i)\mathbf{dx_t}} \leq \lambda_{\max}(v_i)\|\mathbf{dx_t}\| \tag{18.17}$$

where $\lambda_{\max} > 0$ is the maximum eigenvalue of $\mathbf{H}(v_i)$. This maximum error occurs when $\mathbf{dx_t}$ is parallel to $\mathbf{v_{max}}$ which is the direction of the eigenvector corresponding to $\lambda_{\max}$ (see Figure 18.5(a)). Thus, to limit the flow velocity error along the edge, we require

$$\|\mathbf{V_e}\| \leq \lambda_{\max}(v_i)\|\mathbf{dx_t}\| \leq p\|\mathbf{V_f}(v_i)\| \tag{18.18}$$

where $0 < p < 1$ is a ratio that specifies the magnitude of the allowable error in terms of the flow at the base of the edge. To ensure (18.18) is satisfied by any edge emanating from v_i, we require

$$\|\mathbf{dx_t}\| \leq dx_{t_{\max}} = \frac{p\|\mathbf{V_f}(v_i)\|}{\lambda_{\max}(v_i)} \tag{18.19}$$

Since the edge length limit $dx_{t_{\max}}$ in (18.19) is computed using a first-order approximation of the error, it will not be correct for large $\|d\mathbf{x_t}\|$. To address this, Newton's root finding method is used along the $\mathbf{v_{max}}$ direction, with $v_{t_1} = v_i + dx_{t_{\max}}\mathbf{v_{max}}$ as the initial point, to find the point $v_p = (\mathbf{x_p}, t_p)$ along $\mathbf{v_{max}}$, that has a velocity error very close to $p\|\mathbf{V_f}(v_i)\|$ (see Figure 18.5(b)). The maximum allowable spatial and temporal displacements at node v_i, $dx_{\max}$ and $dt_{\max}$ are then, respectively, selected as

$$dx_{\max} = \|\mathbf{x_p} - \mathbf{x_i}\| \tag{18.20}$$

$$dt_{\max} = |t_p - t_i| \tag{18.21}$$

18.4.3 The aSTS method

In Case 1 of the illustrative example, we showed that the spatial and temporal discretization given by Δx and Δt need to be selected according to the speed of the flow. If Δx and Δt were chosen such that the resulting edge traversal speed ($\approx \Delta x/\Delta t$) is either too fast or too slow, the results would be incorrect. Here, we present the adaptive single time-step search (aSTS) method which explicitly considers the local flow speeds, in addition to the limits established in (18.20) when selecting the discretization resolution. Our method ensures that the flow velocities remain relatively constant along edges, and that the resulting edge traversal speeds are commensurate with the underlying flow speeds.

During graph construction, the aSTS method considers the region of space that could be reached from a given node v_i in a single time step Δt, under the influence of both the vehicle actuation and the flow velocity at v_i. This reachable space is demarcated by an n-ball of radius $V_{\max}\Delta t$ centered at $\bar{\mathbf{x}} = \mathbf{x_i} + \mathbf{V_f}(v_i)\Delta t$. In two-dimensional (2D) case, this reachable space is a circle and is discretized using a hexagonal lattice of vertices, centered at $\bar{\mathbf{x}}$ with $2q + 1$ vertices along the main axis ($3q^2 + 3q + 1$ nodes in total; see Figure 18.6(a)). In 3D case, the reachable space is a sphere (see Figure 18.7(a)). In order to discretize this space, results for the "sphere packing problem" are used. It has been shown that densest packing of spheres in three dimensions can be obtained using either the *face centered cubic (FCC)* method or the

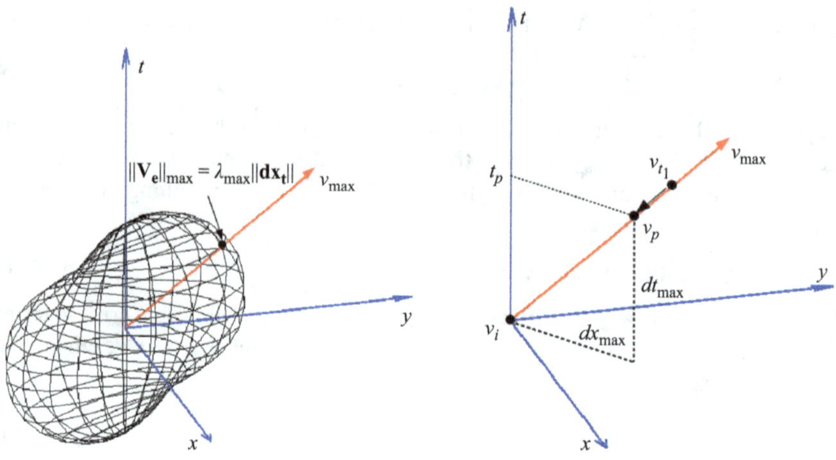

(a) (b)

Figure 18.5 *(a) Velocity error magnitude surface at v_i. The distance of each point on the surface from v_i represents the magnitude of the error in that direction. The maximum velocity error occurs along $\mathbf{v_{max}}$.*
(b) Newton's root finding method is used along the $\mathbf{v_{max}}$ direction to find the point v_p that has a velocity error of exactly $p\|\mathbf{V_f}(v_i)\|$

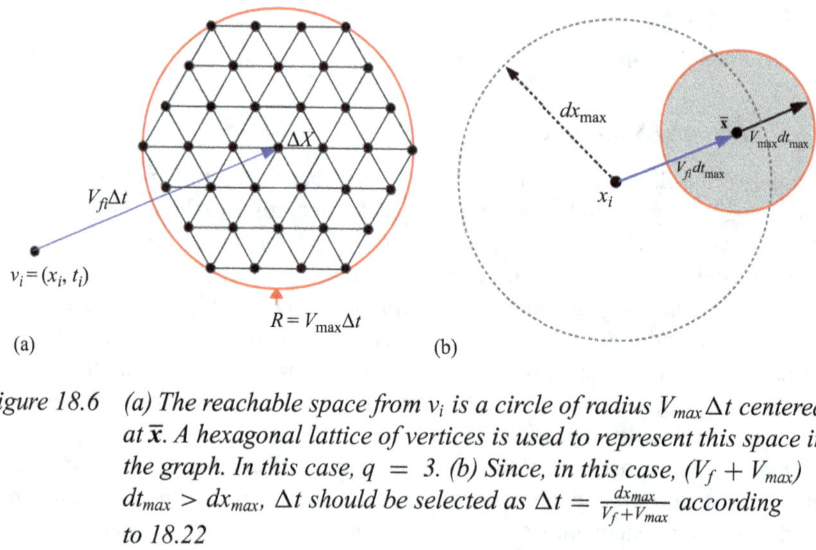

(a) (b)

Figure 18.6 *(a) The reachable space from v_i is a circle of radius $V_{max}\Delta t$ centered at $\bar{\mathbf{x}}$. A hexagonal lattice of vertices is used to represent this space in the graph. In this case, $q = 3$. (b) Since, in this case, $(V_f + V_{max})\,dt_{max} > dx_{max}$, Δt should be selected as $\Delta t = \frac{dx_{max}}{V_f + V_{max}}$ according to 18.22*

hexagonal close packed (HCP) method [49], and in the current application, the HCP method is used. In the HCP method, the base layer (A) consists of a tightly packed planar layer of spheres. The second layer (B) is constructed on top of the A layer, by placing spheres in the hollows created by adjacent spheres in the A layer. HCP packing is achieved by stacking A and B layers alternatively. Figure 18.7(b) shows the reachable space discretized using the HCP method. The center of each sphere is considered to be a node location is the graph constructed for the aSTS method. The number of nodes in the lattice is decided by the number of spheres used to discretize the horizontal base layer through $\bar{\mathbf{x}}$. In the instance shown in Figure 18.7(b), the base layer has 37 spheres (this is equivalent to using a grid with $q = 3$ to discretize the reachable space in the 2D case), which results in 95 nodes in total.

The vertices in the reachable space are added to the neighbor set $\mathcal{N}(v_i)$ of v_i. For each $v_j \in \mathcal{N}(v_i)$, an edge $e_{ij} = (v_i, v_j)$ and a vertex v_j are added to the graph if $\mathbb{O}(v_j) = false$, i.e., if the vertex is not obstructed by an obstacle. All the vertices in $\mathcal{N}(v_i)$ will have the same time coordinate $t_j = t_i + \Delta t$.

The time step Δt at each v_i is selected according to the conditions given in (18.20). The maximum spatial distance between v_i and any $v_j \in \mathcal{N}(v_i)$ is $(V_f + V_{\max})\Delta t$ (at the farthest point on the reachable space). From the discussion in Section 18.4.2, we want $\Delta t \leq dt_{\max}$ and $(V_f + V_{\max})\Delta t \leq dx_{\max}$. Thus, for each v_i, we select Δt according to the following rule:

$$\Delta t = \begin{cases} dt_{\max}, & (V_f + V_{\max})dt_{\max} \leq dx_{\max} \\ \frac{dx_{\max}}{V_f + V_{\max}}, & \text{otherwise} \end{cases} \tag{18.22}$$

Selection of Δt according to (18.22) ensures that the edges added to the graph at each node expansion satisfy the conditions specified in (18.20) (see Figure 18.6(b)). The graph is constructed by repeating this process at each node expansion

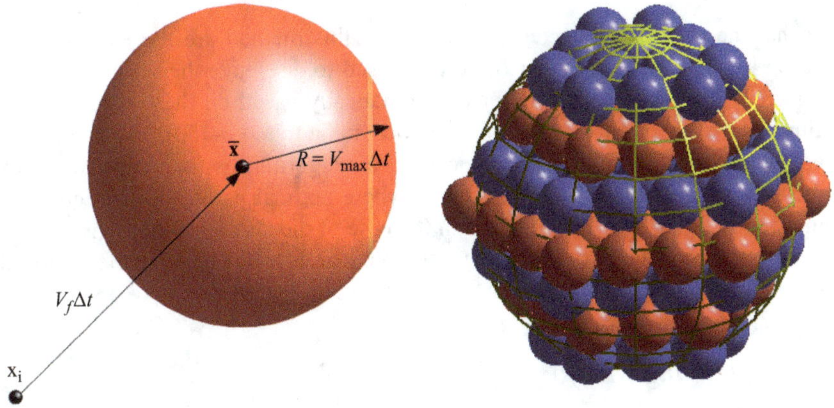

(a) (b)

Figure 18.7 (a) The reachable space in 3D is demarcated by a sphere of radius $V_{max}\Delta t$ centered at $\bar{\mathbf{x}}_i = \mathbf{x}_i + V_f\Delta t$. (b) The HCP method used to discretize the reachable space with spheres. The A layers are in red while the B layers are in blue. For the case shown, there are 95 nodes in total

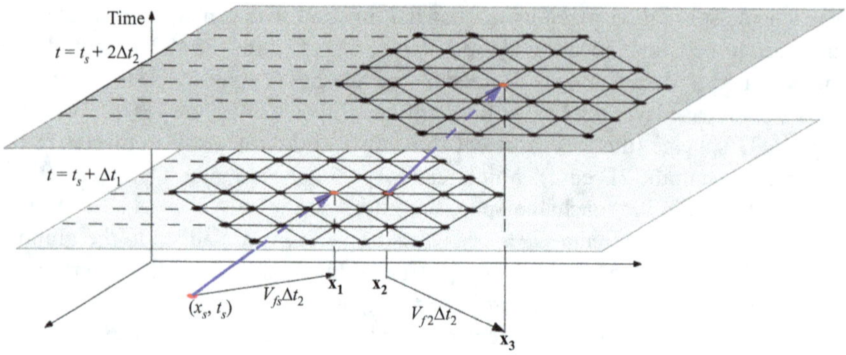

Figure 18.8 Construction of the graph for 2D environments using the aSTS method. All the vertices reachable within a single time step from the base node are considered as neighbors

(see Figure 18.8), and the node expansion is guided by the A* algorithm. The aSTS method guarantees that

1. the maximum velocity error along any edge is less than $pV_f(v_i)$, and
2. the net vehicle speed along any edge is commensurate with the flow velocity at the base of the edge.

Note that the only user-selected parameters in this method are the edge velocity variation limit p and the number of vertices used to discretize the reachable space. All discretization levels are computed automatically based on these values during runtime. More information about this method can be found in [50]. In conventional cases, where there is only one goal location, the algorithm is terminated when the goal has been reached. However, in cases where optimal paths and path costs are required to multiple goal locations, as in the case of the task allocation phase of the reduced-order modeling framework, the graph search could be continued until all goal locations have been reached.

18.4.4 Planning with forecast uncertainties

The reduced-order modeling framework described in Section 18.3 makes it is possible to obtain uncertainty estimates for the flow velocities provided by the ROM. When uncertainty estimates are available, the aSTS method could be used to compute minimum *expected* cost paths. In such cases, the edge cost function used in the aSTS method needs to be modified to be

$$dC_{ij} = (K_h + K_d E[\tilde{V}^{\alpha}_{\text{still}_j}])dT$$

where

$$E[\tilde{V}^{\alpha}_{\text{still}_j}] = \frac{1}{2\pi\sigma_{f1}\cdots\sigma_{fd}} \int\cdots\int \left(\sum_{k=1}^{d} \hat{x}_k^2\right)^{\alpha/2} e^{-\left[\sum_{k=1}^{n} \frac{(\hat{x}_k - V_{\text{still}_{jk}})^2}{2\sigma_{fk}^2}\right]} d\hat{x}_1 \cdots d\hat{x}_n \tag{18.23}$$

In (18.23), $V_{\text{still}_{jk}}$ is the kth component of the thrust vector ($\mathbf{V}_{\mathbf{still}_\mathbf{j}}$) required to reach node v_j from v_i in a deterministic flow, and σ_{fk} is the standard deviation of the flow velocity estimate in the kth direction. More information about this derivation and the associated assumptions can be found in [51].

18.5 Simulations and experiments

In this section, we present simulation and experimental validation for the methodology presented in the previous sections. We first present a systematic validation of the POD-based reduced-order method presented in Section 18.3, and then present simulation-based validation for the path-planning scheme used to navigate the robots from the current sensing location to the next.

18.5.1 Results for the POD-based reduced-order modeling method

Analyses were carried out both in simulation and on physical robots. In simulation, a 1 m × 1 m 2D grid space was modeled using video data from an experimental flow tank at low Reynolds number. The fluid experiment was conducted in a 10 cm × 10 cm tank with glycerol at a depth of 1–2 cm. The tank is equipped with a 4 × 4 array of equally spaced submerged disks. The mechanism creates a cellular flow where two sets of eight disks are separately controlled via independent stepper motors and speed controllers. As such, the resulting flow has both a spatially nonuniform and a temporally complex pattern. The dye was strategically placed at the start of the experiment such that the resulting unsteady flow would stretch the dye along dynamically distinct regions in the flow field. Concentration values of the dye in the tank were estimated for each time step from the grayscale values of the pixels of the images from a grayscale video of the LoRe tank.

In simulation, the grid space was discretized into nine nonoverlapping regions. Four robots were simulated in the field. Concentration values of the field were initially gathered for 100 equally spaced times across the time series, and Gaussian noise was then added to these concentration values. These were then used to construct the initial POD basis for the distributed placement algorithm. Data were collected for another 100 sequential times before adapting the POD basis and recomputing the placement. The distributed placement algorithm was compared to the centralized placement algorithm, where all computations occur on a centralized system and are broadcasted to robots. Additionally, the distributed placement algorithm was compared with radial basis function (RBF) interpolation schemes. Two RBF interpolations were computed using the real data (1) from the regions determined as sensing locations from the distributed method and (2) from randomly selected points across the entire field. All of these methods were compared against the optimal placement that was calculated using noiseless data across the entirety of the time series. Simulations were carried out for various number of robots, discretizations of the spatial domain, and number of snapshots for the initial construction of the POD basis.

The simulation results of the comparison of the various field estimation schemes over the entire time series are shown in Figure 18.9. The Frobenius normwise error between the actual concentration value and the estimated field computed using various algorithms was computed for each time step. The Frobenius normwise error, e, is calculated for field estimate $\hat{z}(t)$ using $e = ||\hat{z}(t) - z(t)||_F / ||z(t)||_F$, where $||z(t)||_F = \sqrt{\sum_i |z_i(t)|^2}$. Both RBF interpolation schemes perform significantly worse than placement algorithms using POD. Even in the case of the RBF with randomly selected sensor points, the field estimation is approximately an order of magnitude

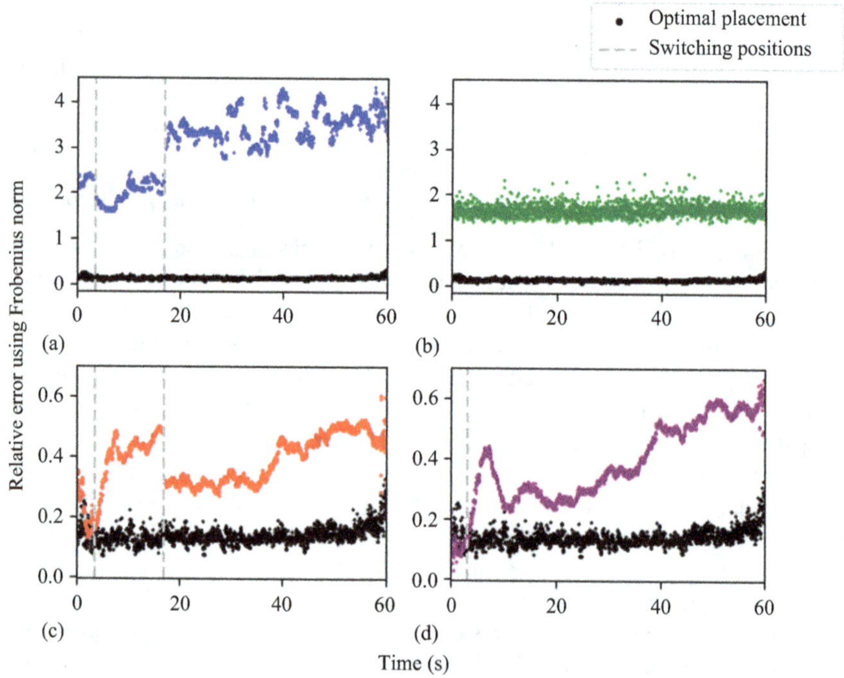

Figure 18.9 Normwise relative error between field simulation and various field estimation algorithms at each time step. Estimations are calculated using (a) RBF using sensing data, (b) RBF using random points, (c) proposed distributed algorithm, and (d) centralized version of proposed algorithm. Black circles represent estimations calculated using the optimal placement determined by using all data to calculate the POD basis. Gray vertical lines indicate robots switching their placement

worse than the three placement algorithms using POD. However, the distributed algorithm and centralized algorithm perform just slightly worse than the optimal placement determined by using all data to calculate the POD basis.

The mean absolute error of the various field estimation schemes is shown in Figure 18.10. The RBF interpolation scheme using the data from the sensor measurements results in high error across the field, as it is unable to adequately estimate values in regions far from the sensing locations. The RBF interpolation scheme using random points fails to capture the interesting features of the process. The distributed algorithm fails in similar areas as compared to the optimal placement algorithm. This can be attributed to little to no data being collected over these regions, which makes it difficult to estimate the concentration values over these areas. Additionally, the distributed algorithm performs slightly worse than the centralized algorithm. This is expected given the fact that distributed algorithm uses only local information in its computation of the field estimate.

The adaptive nature of the algorithm allows robots to rectify tracking errors by recomputing the POD basis and possibly reassigning the sensing locations. This is shown in Figure 18.11, where robots are able to improve field measurements for

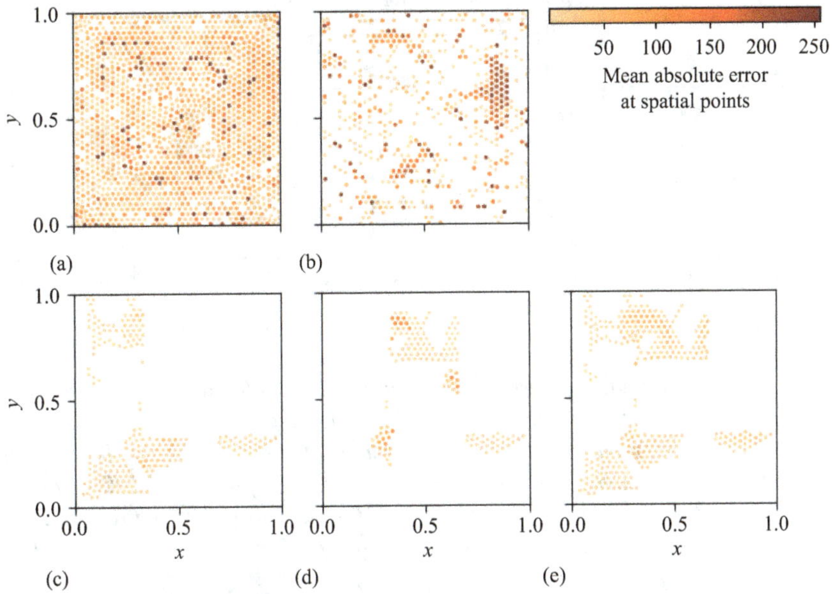

Figure 18.10 Mean absolute error at spatial points calculated over time series for various field estimation algorithms. Concentrations at points are calculated using (a) RBF using sensing data, (b) RBF using random points, (c) optimal placement determined by using all data to calculate the POD basis, (d) centralized version of proposed algorithm, and (e) proposed distributed algorithm

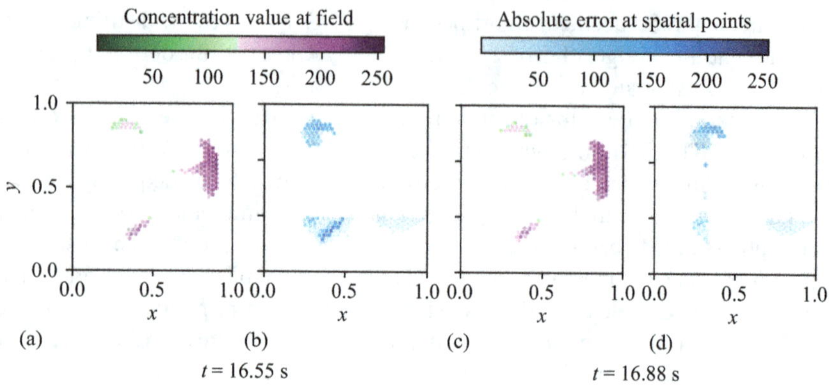

Figure 18.11 Concentration field and absolute error at spatial points before and after robots switch locations using distributed placement algorithm. Concentration field (a) and absolute errors (b) are before the switch; concentration field (c) and absolute errors (d) are after assimilating data and switching positions

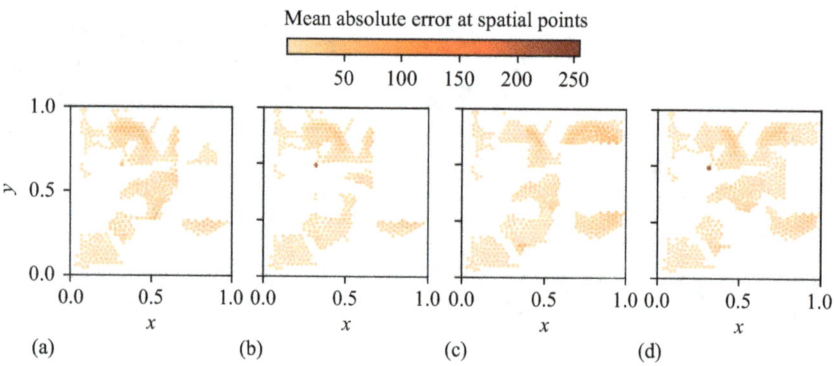

Figure 18.12 Mean absolute error at spatial points calculated over time series between field and distributed algorithm for various discretizations of field, numbers of robots, and snapshots used to compute initial POD basis. Algorithm tested for (a) 4 robots, 9 regions, and 100 snapshots, for (b) 4 robots, 9 regions, and 500 snapshots, for (c) 8 robots, 25 regions, and 100 snapshots, and for (d) 8 robots, 25 regions, and 500 snapshots

areas of high error after reassimilating their collected data to determine new sensing locations and a new POD basis.

The distributed algorithm demonstrates consistent results across various discretizations of the spatial region, various numbers of robots, and various initial models of the dynamic process, as shown in Figures 18.12 and 18.13. Mean absolute errors

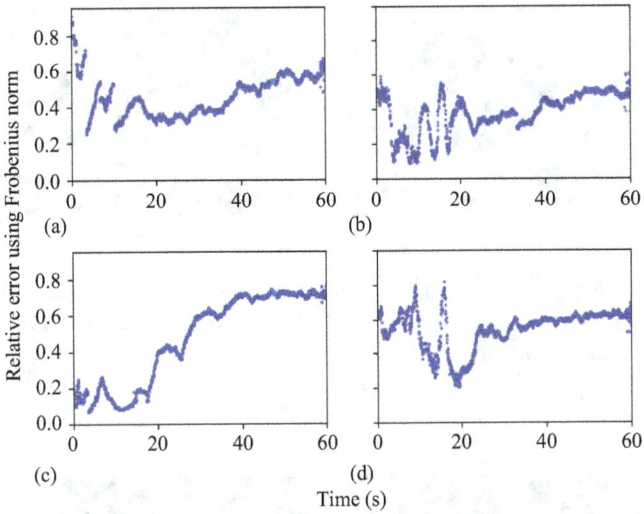

Figure 18.13 Normwise relative error between field and distributed algorithm for various discretizations of field, numbers of robots, and snapshots used to compute original POD basis. Algorithm tested for (a) 4 robots, 9 regions, and 100 snapshots, for (b) 4 robots, 9 regions, and 500 snapshots, for (c) 8 robots, 25 regions, and 100 snapshots, and for (d) 8 robots, 25 regions, and 500 snapshots

between the estimated field and the actual field for 4 robots with 9 total regions in Figure 18.12(a) and for 8 robots with 25 total regions in Figure 18.12(c) perform comparably despite a nearly 15% reduction in the area being sensed by robots. This can be attributed to the robustness of the constructed model. Despite the use of various initial POD bases, the distributed optimal placement eventually results in similar errors estimations as shown by Figure 18.13(a)–(d). This is again due to the adaptive nature of the algorithm.

Experiments were carried out in a 5 m × 3 m water tank using four marine robots, shown in Figure 18.14(a). The concentration field was mapped and projected onto the tank using the video from the LoRe tank, shown in Figure 18.14(b). The robots then tracked the projected concentration field using the distributed algorithm. In the water tank, the robots were able to track the projected dye. The robots collect measurements from their sensing locations and adapt their assigned models. They are able to switch locations to track the process as shown in Figure 18.15.

18.5.2 Simulation results for aSTS method

In this section, the performance of the aSTS method used to compute optimal paths for the sensing agents to move between sensing locations is verified in simulations.

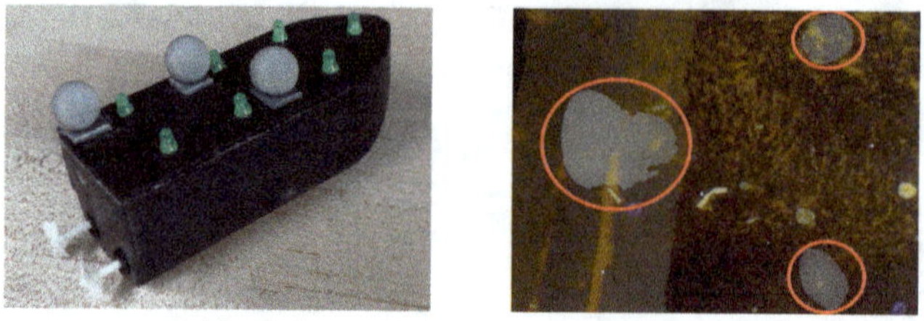

(a) (b)

Figure 18.14 Experimental setup with marine robots. (a) Robot boat equipped with pose information from motion capture system and ability to communicate. (b) Water tank with projection of dynamic process depicted in white and circled in red

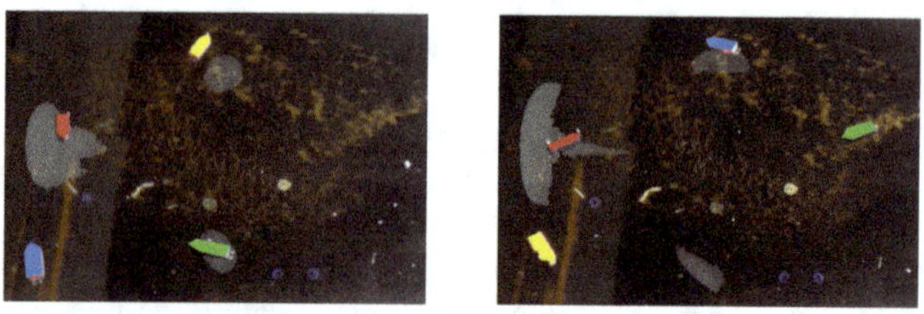

(a) (b)

Figure 18.15 Robotic boats tracking dynamic process in water tank. The dynamic process is shown in white, and robots in blue, red, yellow, and green. (a) Robots assume positions based on the initial POD basis. (b) Robots switch positions after collecting sensor measurements and updating their models

In all simulations, the path parameters were set as $K_h = 0.0005$, $K_d = 1$, and $\alpha = 2$ (linear drag model).

The accuracy of the paths computed by the aSTS method was evaluated by comparing it against a path obtained by solving the corresponding optimal control problem. The optimal control problem involves minimizing the path cost given in (18.7), subject to the kinematic model in (18.1), and the constraint $V_{\text{still}} \leq V_{\max}$. Let $\Gamma^* : [t_s, t_g] \mapsto \mathscr{W}$ be the reference path obtained from the optimal control formulation of the problem, and let $\Gamma : [t_s, t_g] \mapsto \mathscr{W}$ be the path computed by the proposed method. The mean error (mE) between Γ^* and Γ, defined by

$$mE = \int_{t_s}^{t_g} \frac{\|\Gamma^*(t) - \Gamma(t)\|}{t_g - t_s} dt \tag{18.24}$$

was used to evaluate the relative accuracy of computed paths. One could alternatively use the path cost to compare the results. However, the computed path cost is a function of the discretization resolution and thus does not provide a good measure for path quality. For example, a path with only two intermediate nodes might have better cost because it ignores the flow variation at intermediate points along the path.

Simulations using ocean flow data

The aSTS method was also used to compute optimal paths in a 2D ocean environment. The data were obtained from the ROMS for the Santa Barbara Bay area. The Southern California Coastal Ocean Observing System (SCCOOS)† generates these hourly ocean current forecasts everyday and each forecast is for 72 h. The data generated on July 7 and July 8, 2016, were used for the simulations. The ROMS data have a 3 km × 3 km × 1 h spatiotemporal resolution and linear interpolation is used to obtain flow velocities at intermediate coordinates. The maximum flow speed was $V_{fm} = 0.73$ m/s and as such $V_{\max}$ was selected to be 0.5 m/s. A hexagonal lattice with $q = 3$ was used to discretize the reachable space and the error parameter was set as $p = 0.1$. Figure 18.16 shows the comparison between the reference path and the path computed using the aSTS method. The reference path computed using the optimal control formulation had a cost of 3,775 J. It can be seen that the paths match closely with a mean error of $mE = 169$ m. The path cost obtained from the aSTS method was 4,243 J. As before $K_h = 0.0005$ and $K_d = 1$ were used for the simulations, and as a result, more prominence is given to reducing the energy expended to overcome the drag forces. Therefore, the computed path tends to follow the flow as much as possible in order to reduce relative motion between the flow and the vehicle. This leads to the loop structure that can be observed in Figure 18.16. The video for this simulation trial can be found on https://youtu.be/6R0RevaAMmY.

Table 18.1 shows the results of the aSTS method for different p and q values. It shows that increasing q while keeping p constant (Paths 1–3) results in more accurate paths at the expense of running time. Similarly, decreasing p while keeping q constant (Paths 5–3) also results in more accurate paths at the expense of running times. Note that the running time is only a fraction of the total path duration for each case. Table 18.2 shows the results for paths computed using a fixed-grid discretization scheme. In this fixed discretization method, the user has to specify Δx and Δt, the spatial and temporal discretization resolutions, as well as nHx and nHt, the number of spatial and temporal hops considered as neighbors at each expansion. In the results shown in Table 18.2, all paths were computed with $\Delta x = 150$ m and $nHx = 3$ so that it has the same spatial discretization as the mean resolution obtained for Path 3 of the aSTS method. It can be seen that Path 1 has a similar accuracy as Path 3 of the aSTS method, and the computation is faster. However, the results obtained from the fixed discretization method depend heavily on the user-defined discretization resolution. For example, a slight increase in the temporal resolution (Path 2) results in a less accurate path. The results degrade further when the number of temporal neighbors considered are decreased (Path 3). Thus, increased accuracy cannot be guaranteed even if the discretization is made finer with such fixed discretization

†URL: http://www.sccoos.org/data/roms-3km/

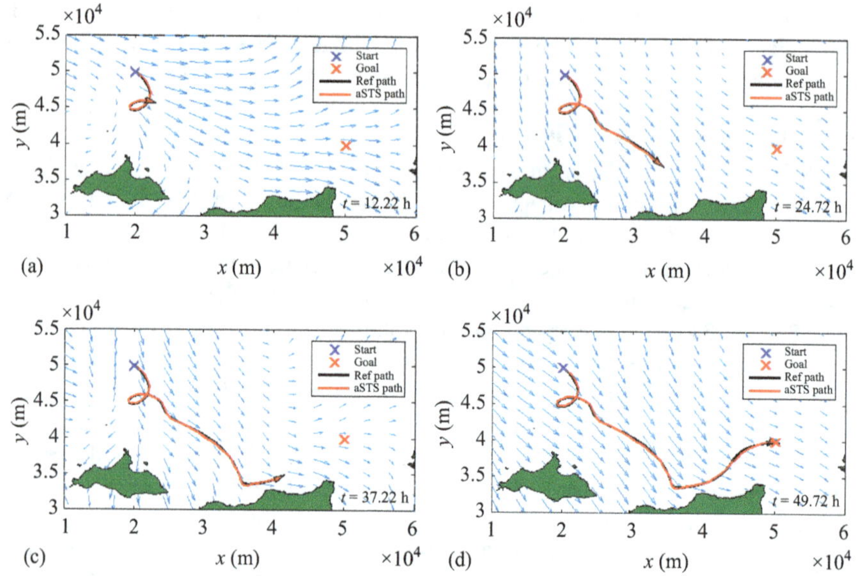

Figure 18.16 Comparison of optimal path computed using the aSTS method (red) with the path computed using the optimal control formulation (black) in an ocean environment. The parameters for the aSTS method were set at $q = 3$, $p = 0.1$. Mean path error for the aSTS path is $mE = 169\,m$

Table 18.1 Results for the aSTS method for different p and n values

Path	**1**	**2**	**3**	**4**	**5**
p	0.1	0.1	0.1	0.2	0.3
q	1	2	3	3	3
mE (m)	6,426	595.3	169.2	656.3	1,095
Cost (J)	9,200	5,284	4,243	4,234	4,179
Running time (s)	385	1,074	2,009	687	364
Path duration (s)	182,800	176,200	1,782,000	179,400	180,600

schemes. However, accuracy is guaranteed to improve in the proposed aSTS method by simply increasing q and decreasing p.

Simulations using the double-gyre flow model

The aSTS method was used to compute optimal energy paths in a 3D environment where the horizontal components $V_{f1}(\mathbf{x}, t)$ and $V_{f2}(\mathbf{x}, t)$ of flow field were given by the double gyre flow in (18.14), and the vertical component V_{f3} was constant. The parameters for the double gyre flow were set as $A = 1$, $\mu = 0$, $\omega = 4\pi$ and $\varepsilon = 0.6$

Table 18.2 Results for the fixed discretization method

Path	1	2	3
Δt	150	100	100
nHt	7	10	5
mE (m)	226.5	326.0	1,256
Cost (J)	4,191	4,236	7,195
Running time (s)	1,399	2,641	816
Path duration (s)	180,000	180,600	176,000

while the constant vertical flow component was set at $V_{f3} = 0.5$ m/s. Figure 18.17 shows snapshots of the path computed by the aSTS method. The computed path cost was 0.0389 J and the path duration was 0.84 s. When the discretization was done with 37 nodes the path cost was increased to 0.0673 J. Thus, as expected, the higher the number of nodes used to discretize the reachable space, the better the computed path is.

18.6 Future outlook

In this chapter, we presented an energy-efficient distributed adaptive sampling strategy for a team of mobile robots to track a dynamic spatiotemporal process. We show how a team of mobile robots can build an ROM of the process using sparse measurements obtained from onboard sensors. The ROM is then used to infer measurements in regions lacking sensor coverage and the inferred data are then used to determine next best locations for new measurements and for planning minimum energy paths to the new locations. In this work, we rely on POD analysis in the derivation of the ROM which identifies the dominant dynamics of the process based on point measurements at specified locations over time. As such, the quality of the ROM is highly dependent of both the spatial and temporal resolution of the measurement data. Consequently, the quality of the minimum energy trajectories computed using the ROM depends on the spatiotemporal resolution of the model. Under these circumstances, the proposed strategies can be significantly improved by incorporating specific GFD insights.

GFDs is the study of natural fluid flows that span large physical scales, such as oceans, the atmosphere, and rivers. GFD flows are naturally stochastic and aperiodic, yet exhibit coherent structure. Recent studies of coherent structures in GFD and other flows have shown that these ubiquitous structures are typically kinematic and play a key role in transport. The Gulf Stream is a prominent example of a coherent jet whose heat transport is a critical component of global weather and climate. Using GFD models, details from flows such as the Gulf Stream can be used to diagnose the underlying GFDs. This, in turn, enables the prediction of various physical, chemical, and biological processes in general geophysical flows. While the nature of a particular flow may originate as the solution to some dynamical (force) balance, the characteristics of the flow are an intrinsic property of its velocity vector field. A number of important

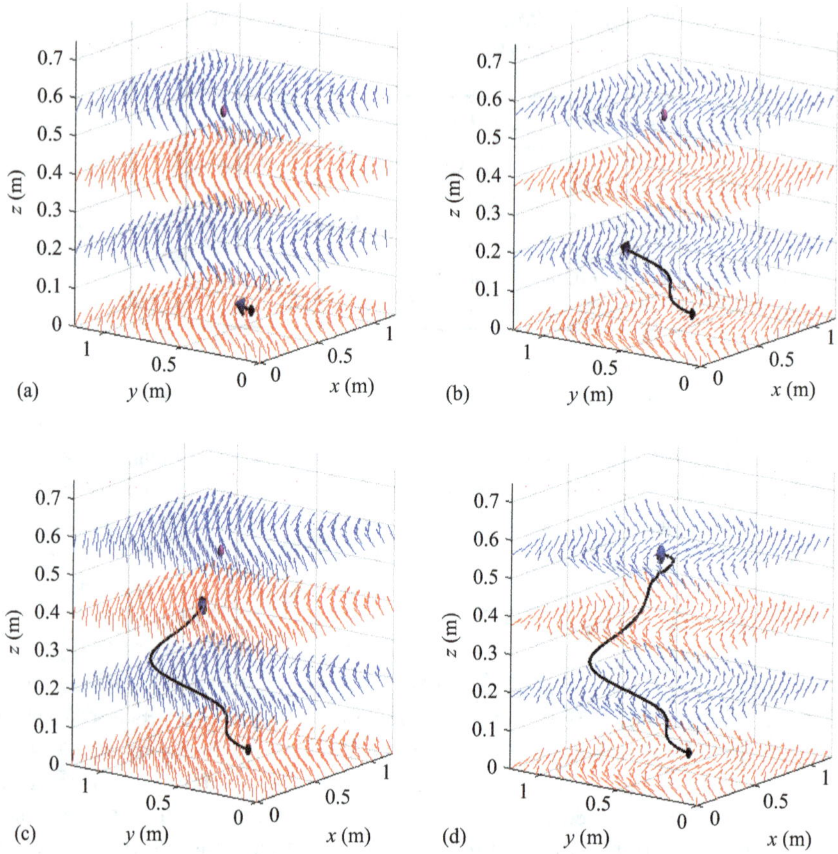

Figure 18.17 Snapshots of the path computed in a 3D environment subject to a time-varying flow field. The blue and red spheres indicate the start and goal positions, respectively

processes, including transport, can be examined by this purely kinematic description of the flow. It is for this reason that we can witness similar transport features across a broad range of flows and that in the control of vehicles, as for tracers, knowledge of coherent structures may be exploited more effectively than knowledge of the detailed predictions offered by the state-of-the-art primitive equation PDE models.

This is further supported by the work of Inanc *et al.* [52] where they showed that minimum energy- and time-optimal paths in the ocean can coincide with a specific class of coherent structures that are important for quantifying transport phenomena in flows called Lagrangian coherent structures (LCS). LCS are the extensions of stable and unstable manifolds to general time-dependent flows [53,54] and act as separatrices that divide the flow into dynamically distinct regions. Since these separatrices are inherently unstable and denote regions of the flow where more escape events may occur [55], knowing the LCS locations is also important for keeping sensors in specific monitoring regions [56–58]. These results strongly suggest that it is possible to improve the overall quality of deployment strategies when incorporating such GFD knowledge in the planning, coordination, and controller synthesis process for AMVs. This makes intuitive sense since coherent structures provide a reduced-order description of the complex fluid environment and enable the estimation of the underlying GFDs. In fact, LCS is one example of a transport controlling feature that can be leveraged to improve the quality ROM of the spatiotemporal process of interest and simultaneously be leveraged to improve vehicle maneuverability when planning minimum energy paths.

In the past few years, we have developed distributed tracking strategies for teams of robots to track these structures relying solely on onboard measurements obtained by the robots [59–63]. An immediate direction for future work is to develop adaptive sampling strategies where the team is tasked to track both the coherent structures within the workspace while simultaneously making measurements in specific regions of the process itself. The idea is to improve the ROM by assimilating information about the locations of the LCS in addition to specific measured quantities of the process itself. Since LCS encodes both the stable and unstable modes of the process dynamics, the incorporation of this information can significantly improve the predictive capabilities of the ROM and thus improve subsequent deployment strategies that employ it.

Another direction for future inquiry is the development of path-planning strategies that leverage the locations of the known LCS in the workspace. Since LCS has been shown to be both persistent [64,65] as well as important for time- and energy-optimal navigation in the ocean [52,66], they could potentially be used as optimal paths in regions for which explicit flow descriptions are not available. Furthermore, if the adaptive sampling strategy includes the explicit tracking of LCS locations, the information can significantly reduce the computational burden of computing an energy-optimal assignment of robots to the next best location for obtaining new measurements. Lastly, such a strategy can further leverage recent stochastic control strategies that leverage LCS knowledge to enable energy-constrained vehicles to efficiently navigate from one LCS-bounded region to another [67]. The ability to develop robust navigation strategies in the presence of model and/or forecast uncertainties is extremely important to ensure prolonged monitoring of dynamic and uncertain environments such as the ocean.

Appendix A

Proof of Theorem 18.2 (see page 525)

Theorem 18.2. *For sufficient mixing time* τ*, as determined by the criteria for mixing speeds of Markov Chains [45],* $\delta = t_{i+1} - t_i$ *the time between collected snapshots,* $\left\| \frac{df(t)}{dt} \right\| \leq \xi$ *for* $\boldsymbol{f}$ *describing the dynamic process, the bound between* $(\boldsymbol{P} + \boldsymbol{P}^{\perp})$*, the* *projection corresponding to the dynamic process, and* $P_{\hat{Q}_i}$ *the projection onto the POD basis vectors estimated by robots with high probability is defined as*

$$\left\|(\boldsymbol{P}+\boldsymbol{P}^{\perp})-\boldsymbol{P}_{\hat{Q}_i}\right\|_2 \leq e^{\gamma\tau}\left(2\lambda_{k+1}+\frac{\delta^2}{8}\xi\right)+\mathcal{O}\left|\frac{\lambda_{k+1}}{\lambda_k}\right|^{\tau} * n+(C+2)\varepsilon^{4\tau} \quad (18.25)$$

Proof. For POD analysis, m snapshots of the field are collected, either through experimentation or numerical simulations, such that at each time $t = 1, \ldots, m$, $z(t) = [z_1(t), \ldots, z_n(t)]^\top$, where n is the spatial dimension of some discretization of the flow field. A covariance matrix is constructed as

$$\boldsymbol{K} = \frac{1}{\boldsymbol{m}}\sum_{t=1}^{\boldsymbol{m}} z(\boldsymbol{t})z(\boldsymbol{t})^\top = \frac{1}{\boldsymbol{m}}\boldsymbol{X}\boldsymbol{X}^\top \quad (A.1)$$

where $\boldsymbol{X} \in \mathbb{R}^{n\times m}$ with its columns as $z(\boldsymbol{t})$.

The low-dimensional basis is created by solving the symmetric eigenvalue problem

$$\boldsymbol{K}\phi_i = \lambda_i\boldsymbol{\phi}_i \quad (A.2)$$

where $\boldsymbol{K}$ has n eigenvalues such that $\lambda_1 \geq \lambda_2 \geq \cdots \geq \lambda_n \geq 0$ and the eigenvectors $\boldsymbol{\phi}$ are pairwise orthonormal.

In practice, $m < n$ and we assume $rank(\boldsymbol{X}) = m$. For $\boldsymbol{X}\boldsymbol{X}^\top \in \mathbb{R}^{n\times n}$ and $\boldsymbol{X}^\top\boldsymbol{X} \in \mathbb{R}^{m\times m}$, $rank(\boldsymbol{X}\boldsymbol{X}^\top) = rank(\boldsymbol{X}^\top\boldsymbol{X}) \leq m$. Let σ_i, for $i = 1, \ldots, m$, be the singular values $\boldsymbol{X}$. The squares of the m largest singular values of $\boldsymbol{X}$ are the m nonzero eigenvalues of $\boldsymbol{X}\boldsymbol{X}^\top$ and $\boldsymbol{X}^\top\boldsymbol{X}$, as in $\sigma_i^2 = \lambda_i$ is an eigenvalue of $\boldsymbol{X}\boldsymbol{X}^\top$ and $\boldsymbol{X}^\top\boldsymbol{X}$.

Additionally, suppose $\boldsymbol{U}$ is the matrix whose columns are the set of orthonormal eigenvectors of $\boldsymbol{X}\boldsymbol{X}^\top$ and $\boldsymbol{V}$ is the matrix whose columns are the set of orthonormal eigenvectors of $\boldsymbol{X}^\top\boldsymbol{X}$. Let Σ be the matrix with diagonal entries σ_i. Then, from singular value decomposition, we have that

$$\boldsymbol{X} = \boldsymbol{U}\Sigma\boldsymbol{V}^\top \quad (A.3)$$

The original basis $\boldsymbol{U}$ is then truncated into a new basis $\boldsymbol{\Phi}$ by choosing k eigenvectors that capture a user-defined fraction, E, of the total variance of the system, such that their eigenvalues satisfy

$$\frac{\sum_{i=1}^{k}\lambda_i}{\sum_{i=1}^{n}\lambda_i} \geq E \quad (A.4)$$

Thus, each term $z(\boldsymbol{t})$ can be written as

$$z(\boldsymbol{t}) = \boldsymbol{\Phi}\boldsymbol{c}(\boldsymbol{t}) \quad (A.5)$$

where $\boldsymbol{c}(\boldsymbol{t}) = [c_1(t), \ldots, c_k(t)]^\top$ holds time-dependent coefficients and $\boldsymbol{\Phi} \in \mathbb{R}^{n\times k}$ with its columns as $\boldsymbol{\phi}_1,\ldots,\boldsymbol{\phi}_k$. The low-dimensional, orthogonal subspace associated with $\boldsymbol{\Phi}$ is an optimal approximation of the data with respect to minimizing least squares error.

Using the singular value decomposition (SVD), the truncated expression for $\boldsymbol{z(t)}$ with just k vectors in the new basis can then be written as

$$\hat{z}(\boldsymbol{t_j}) = \sum_{i=1}^{k} \sigma_i v_{ji} \boldsymbol{\phi}_i \tag{A.6}$$

Define the projection matrix $\boldsymbol{P} = \boldsymbol{\Phi}\boldsymbol{\Phi}^{\top}$. Define $\hat{\boldsymbol{U}}$ such that the columns are the nonzero eigenvectors in $\boldsymbol{U}$ that are not in $\boldsymbol{\Phi}$. Thus, $\boldsymbol{P}^{\perp} = \hat{\boldsymbol{U}}\hat{\boldsymbol{U}}^{\top}$. We know that $\boldsymbol{P}$ is the projection onto $span\{\phi_1, \ldots, \phi_k\}$ and $\boldsymbol{P}^{\perp}$ is the projection onto $span\{\phi_{k+1}, \ldots, \phi_m\}$.

We can derive the following upper bounds on the projections of $\boldsymbol{x}(\boldsymbol{t_j})$ onto the two subspaces, using the representation from (A.6):

$$\begin{aligned}\left\|\boldsymbol{P}z(\boldsymbol{t_j})\right\|_2 &= \left\|\hat{z}(\boldsymbol{t_j})\right\|_2 = \left\|\sum_{i=1}^{k} \sigma_i v_{ji} \boldsymbol{\phi}_i\right\|_2 \\ &= \sqrt{\sum_{i=1}^{k} (\sigma_i v_{ji} \boldsymbol{\phi}_i)^2} = \sqrt{\sum_{i=1}^{k} (\sigma_i v_{ji})^2} \le \sigma_1 \sqrt{\sum_{i=1}^{k} (v_{ji})^2} \le \sigma_1\end{aligned} \tag{A.7}$$

$$\left\|\boldsymbol{P}^{\perp}z(\boldsymbol{t_j})\right\|_2 = \left\|\sum_{i=1}^{m-k} \sigma_{i+k} v_{j(i+k)} \boldsymbol{\phi}_i\right\|_2 = \sqrt{\sum_{i=1}^{m-k} (\sigma_{i+k} v_{j(i+k)} \boldsymbol{\phi}_i)^2} \le \sigma_{k+1} \tag{A.8}$$

In (A.7) and (A.8), σ_i and v_{ji} are constants, and so $(\sigma_i v_{ji} \boldsymbol{\phi}_i)^2 = (\sigma_i^2 v_{ji}^2 \boldsymbol{\phi}_i{}^2)$. Since $\boldsymbol{\phi}_i$ are orthonormal, $\boldsymbol{\phi}_i{}^2 = 1$. $\sigma_1 \ge \sigma_i$ for all i and can thus be factored out of the summation term. $\sqrt{\sum_{i=1}^{k} (v_{j(i+k)})^2} = \boldsymbol{v}_j{}^2 = 1$ from the definition of dot product and since v_j are orthonormal.

Given a function $\boldsymbol{f}$ defining the dynamic process of interest, the POD approximation is defined as

$$\begin{aligned}\dot{\hat{z}}(t) &= \boldsymbol{Pf}(\hat{z}, t) \\ \hat{z}(0) &= \boldsymbol{P}z(0)\end{aligned} \tag{A.9}$$

The exact solution is defined as

$$\begin{aligned}\dot{z}(t) &= (\boldsymbol{P} + \boldsymbol{P}^{\perp})f(z, \boldsymbol{t}) \\ z(0) &= (\boldsymbol{P} + \boldsymbol{P}^{\perp})z(\boldsymbol{0})\end{aligned} \tag{A.10}$$

The error between the exact solution and the POD approximation is defined as

$$\begin{aligned}\dot{\boldsymbol{e}}(t) &= \boldsymbol{P}(\boldsymbol{f}(\boldsymbol{x}, t) - \boldsymbol{f}(\hat{z}, t)) + \boldsymbol{P}^{\perp}\boldsymbol{f}(z, t) \\ \boldsymbol{e}(0) &= \boldsymbol{P}^{\perp}z(0)\end{aligned} \tag{A.11}$$

For $\boldsymbol{f}$ differentiable, $\hat{z}$ is defined by Taylor's Theorem

$$\boldsymbol{f}(z,t)-\boldsymbol{f}(\hat{z},t)=\frac{\partial \boldsymbol{f}}{\partial z}(\hat{z}(t),t)\boldsymbol{e}(t) \tag{A.12}$$

We can define a bound as

$$\begin{aligned} A(t) &= \boldsymbol{P}\frac{\partial \boldsymbol{f}}{\partial z}(\hat{z}(t),t) \\ \|\boldsymbol{A(t)}\|_2 &\le \gamma, \forall t \in [0,T] \end{aligned} \tag{A.13}$$

Then, integrating the error (A.11)

$$\begin{aligned} \boldsymbol{e(t)} &= \boldsymbol{P}^{\perp}z(0)+\int_0^t \boldsymbol{P}(\boldsymbol{f}(z,s)-\boldsymbol{f}(\hat{z},s))+\boldsymbol{P}^{\perp}\boldsymbol{f}(z,s)ds \\ &= \boldsymbol{P}^{\perp}z(0)+\int_0^t A(s)\boldsymbol{e}(s)+\boldsymbol{P}^{\perp}\boldsymbol{f}(z,s)ds \\ &= \boldsymbol{P}^{\perp}z(0)+\int_0^t A(s)\boldsymbol{e}(s)ds+\int_0^t \boldsymbol{P}^{\perp}\boldsymbol{f}(z,s)ds \end{aligned} \tag{A.14}$$

By evaluating $\int_0^t \boldsymbol{P}^{\perp}\boldsymbol{f}(z,s)\boldsymbol{ds}$, we have

$$\begin{aligned} \int_0^t \boldsymbol{P}^{\perp}\boldsymbol{f}(z,s)ds &= \int_0^t \boldsymbol{P}^{\perp}\dot{z}(s)ds = \boldsymbol{P}^{\perp}(z(t)-z(0)) \\ \implies \boldsymbol{e}(t) &= \boldsymbol{P}^{\perp}z(0)+\boldsymbol{P}^{\perp}(z(t)-z(0))+\int_0^t A(s)\boldsymbol{e}(s)ds \\ &= \boldsymbol{P}^{\perp}z(t)+\int_0^t A(s)\boldsymbol{e}(s)ds \end{aligned} \tag{A.15}$$

We can bound the error using the following relations:

$$\begin{aligned} \|\boldsymbol{e}(t)\|_2 &= \left\|\boldsymbol{P}^{\perp}z(t)+\int_0^t A(s)\boldsymbol{e}(s)ds\right\|_2 \\ \|\boldsymbol{e}(t)\|_2 &\le \left\|\boldsymbol{P}^{\perp}z(t)\right\|_2+\left\|\int_0^t A(s)\boldsymbol{e}(s)ds\right\|_2 \\ \|\boldsymbol{e}(t)\|_2 &\le \left\|\boldsymbol{P}^{\perp}z(t)\right\|_2+\int_0^t \|A(s)\boldsymbol{e}(s)ds\|_2 \\ \|\boldsymbol{e}(t)\|_2 &\le \left\|\boldsymbol{P}^{\perp}z(t)\right\|_2+\gamma\int_0^t \|\boldsymbol{e}(s)\|_2\, ds \end{aligned} \tag{A.16}$$

Theorem 18.3. (Gronwall's Theorem) *Let α, β, and u be real-valued and defined on the interval $[a,b]$. Assume β and u are continuous and α_- is integrable on every closed* *and bounded subinterval of* $[a, b]$. *If* $\beta \geq 0$ *and* u *satisfies* $u(t) \leq \alpha(t) + \int_a^t \beta(s)u(s)ds$, *then* $u(t) \leq \alpha(t) + \int_a^t \alpha(s)\beta(s)\exp(\int_s^t \beta(r)dr)ds$, $t \in [a, b]$.

Applying Gronwall's Theorem to (A.16)

$$u(t) = \|\boldsymbol{e}(t)\|_2, \alpha(t) = \left\|\boldsymbol{P}^{\perp}z(t)\right\|_2, \beta(t) = \gamma$$

$$\|\boldsymbol{e}(t)\|_2 \leq \left\|\boldsymbol{P}^{\perp}z(t)\right\|_2 + \int_0^t \left\|\boldsymbol{P}^{\perp}z(t)\right\|_2 * a * e^{\gamma(t-s)} \tag{A.17}$$

$$\|\boldsymbol{e}(t)\|_2 \leq e^{\gamma t} \max_{t\in[0,T]} \left\|\boldsymbol{P}^{\perp}z(t)\right\|_2$$

The above bound does not take into account the construction of POD matrices, namely how the snapshots are collected. Now, we will investigate the use of snapshot matrix $\boldsymbol{X} = [z(t_1), \ldots, z(t_m)]$ with projection matrix $\boldsymbol{P}$.

Using Lagrange interpolation

$$\begin{aligned} z(t) &= z(t_i) * \frac{t - t_{i+1}}{t_i - t_{i+1}} + z(t_{i+1})\frac{t - t_i}{t_{i+1} - t_i} + \boldsymbol{R}(z(t)) \\ \boldsymbol{R}(z(t)) &= \frac{(t - t_i)(t - t_{i+1})}{2}\frac{d^2 z(t)}{dt^2} = \frac{(t - t_i)(t - t_{i+1})}{2}\frac{d\boldsymbol{f}(t)}{dt} \end{aligned} \tag{A.18}$$

Multiplying by $\boldsymbol{P}^{\perp}$

$$\begin{aligned} z(t) &= z(t_i)\frac{t - t_{i+1}}{t_i - t_{i+1}} + z(t_{i+1})\frac{t - t_i}{t_{i+1} - t_i} + \frac{(t - t_i)(t - t_{i+1})}{2}\frac{d\boldsymbol{f}(t)}{dt} \\ \boldsymbol{P}^{\perp}z(t) &= \boldsymbol{P}^{\perp}z(t_i)\frac{t - t_{i+1}}{t_i - t_{i+1}} + \boldsymbol{P}^{\perp}z(t_{i+1})\frac{t - t_i}{t_{i+1} - t_i} \\ &\quad + \boldsymbol{P}^{\perp}\frac{(t - t_i)(t - t_{i+1})}{2}\frac{d\boldsymbol{f}(t)}{dt} \end{aligned} \tag{A.19}$$

For $\delta = t_{i+1} - t_i$, $\max_{t\in[t_i,t_{i+1}]} |(t - t_i)(t - t_{i+1})| = \delta/4$. We assume $\frac{d\boldsymbol{f}(t)}{dt} \in C(O, T)$ and $\left\|\frac{d\boldsymbol{f}(t)}{dt}\right\|_2 \leq \xi$. Taking the norm of both sides,

$$\begin{aligned} \left\|\boldsymbol{P}^{\perp}z(t)\right\|_2 &= \left\|\boldsymbol{P}^{\perp}z(t_i)\frac{t - t_{i+1}}{t_i - t_{i+1}}\right\|_2 + \left\|\boldsymbol{P}^{\perp}z(t_{i+1})\frac{t - t_i}{t_{i+1} - t_i}\right\|_2 \\ &\quad + \left\|\boldsymbol{P}^{\perp}\frac{(t - t_i)(t - t_{i+1})}{2}\frac{d\boldsymbol{f}(t)}{dt}\right\|_2 \end{aligned}$$

$$\left\|\boldsymbol{P}^{\perp}z(t)\right\|_2 \leq \left\|\boldsymbol{P}^{\perp}z(t_i)\right\|_2 + \left\|\boldsymbol{P}^{\perp}z(t_{i+1})\right\|_2 + \frac{\delta^2}{8}\left\|\boldsymbol{P}^{\perp}\frac{d\boldsymbol{f}(t)}{dt}\right\|_2 \tag{A.20}$$

Using (A.8), $\left\|\boldsymbol{P}^{\perp}z(t_i)\right\|_2 \leq \sigma_{k+1}$ and $\left\|\boldsymbol{P}^{\perp}z(t_{i+1})\right\|_2 \leq \sigma_{k+1}$

$$\left\|\boldsymbol{P}^{\perp}z(t)\right\|_2 \leq 2\sigma_{k+1} + \frac{\delta^2}{8}\left\|\boldsymbol{P}^{\perp}\right\|\xi \tag{A.21}$$

Substituting into (A.17), with

$$\|e(t)\|_2 \le e^{\gamma t}\left(2\sigma_{k+1} + \frac{\delta^2}{8}\xi\right) \tag{A.22}$$

In the proposed method, we base our algorithm off of the following centralized relationships for orthogonal iteration used in computing a matrix of eigenvectors $\boldsymbol{Q}$ and eigenvalues contained in the diagonals of $\boldsymbol{R}$. For the design matrix $\boldsymbol{X}$, we have $\boldsymbol{V} = \frac{1}{m}\boldsymbol{X}\boldsymbol{X}^T\boldsymbol{Q} = \boldsymbol{K}\boldsymbol{Q}$, $\boldsymbol{W} = \boldsymbol{V}^T\boldsymbol{V}$, $\boldsymbol{W} \overset{QR}{=} \boldsymbol{R}^T\boldsymbol{R}$, $\boldsymbol{Q}' = \boldsymbol{V}\boldsymbol{R}^{-1}$.

For the decentralized implementation, we start with a matrix $\hat{\boldsymbol{Q}}$, we compute $\hat{\boldsymbol{V}} = \frac{1}{m}\boldsymbol{X}\boldsymbol{X}^T\hat{\boldsymbol{Q}}$. However, instead of $\hat{\boldsymbol{V}}$, for each robot, we have some $\hat{\boldsymbol{V}}_i = \hat{\boldsymbol{V}} + \boldsymbol{E}_i^V$, since $\hat{\boldsymbol{V}}$ is estimated using the push-sum algorithm. Then, we compute $\hat{\boldsymbol{W}} = \hat{\boldsymbol{V}}_i^T\hat{\boldsymbol{V}}_i$. Again, instead of $\hat{\boldsymbol{W}}$, we have $\hat{\boldsymbol{W}}_i = \hat{\boldsymbol{W}} + \boldsymbol{E}_i^W$.

We use the following relationships in the proofs given next. First, we have that $\|\boldsymbol{V}\|_F = \|\boldsymbol{R}\|_F$ and $\|\boldsymbol{V}\|_2 = \|\boldsymbol{R}\|_2$ since $\boldsymbol{Q}$ is orthonormal. For $\boldsymbol{V} = \boldsymbol{K}\boldsymbol{Q}$, we use the submultiplicativity of norms to get $\|\boldsymbol{V}\|_F \le \|\boldsymbol{A}\|_F$ and $\|\boldsymbol{V}\|_2 \le \|\boldsymbol{A}\|_2$. Since $\boldsymbol{W} = \boldsymbol{R}^T\boldsymbol{R}$, $\|\boldsymbol{W}\| \le \|\boldsymbol{R}\|_F^2$, and $\|\boldsymbol{W}\| = \|\boldsymbol{R}\|_2^2$. From $\boldsymbol{V} = \boldsymbol{K}\boldsymbol{Q}$ and $\hat{\boldsymbol{V}}_i = \boldsymbol{K}\hat{\boldsymbol{Q}} + \boldsymbol{E}_i^V$, we have that $\left\|\boldsymbol{V} - \hat{\boldsymbol{V}}_i\right\|_F \le \|\boldsymbol{K}\|_2\left\|\boldsymbol{Q} - \hat{\boldsymbol{Q}} - \boldsymbol{E}_i^V\right\|_F$. In the proof, we assume that the quantity $\left\|\boldsymbol{Q} - \hat{\boldsymbol{Q}} - \boldsymbol{E}_i^V\right\|_F$ is bounded by a constant C.

$$\begin{aligned}
\hat{\boldsymbol{V}}_i &= \frac{1}{m}\boldsymbol{X}\boldsymbol{X}^T\hat{\boldsymbol{Q}} + \boldsymbol{E}_i^V \\
\left\|\boldsymbol{V} - \hat{\boldsymbol{V}}_i\right\|_F &\le \|\boldsymbol{K}\|_2\left\|\boldsymbol{Q} - \hat{\boldsymbol{Q}} - \boldsymbol{E}_i^V\right\|_F \\
\left\|\boldsymbol{V} - \hat{\boldsymbol{V}}_i\right\|_F &\le C\|\boldsymbol{K}\|_2 \\
\left\|\hat{\boldsymbol{V}}_i\right\|_2 - \|\boldsymbol{V}\|_2 &\le C\|\boldsymbol{K}\|_2 \\
\left\|\hat{\boldsymbol{V}}_i\right\|_2 &\le (C+1)\|\boldsymbol{V}\|_2
\end{aligned} \tag{A.23}$$

We can define a bound on error of the matrix $\boldsymbol{W}$ as

$$\begin{aligned}
\left\|\boldsymbol{W} - \hat{\boldsymbol{W}}\right\|_F &= \left\|\boldsymbol{V}^T\boldsymbol{V} - \hat{\boldsymbol{V}}_i^T\hat{\boldsymbol{V}}_i\right\|_F \\
\left\|\boldsymbol{W} - \hat{\boldsymbol{W}}\right\|_F &\le \left\|\boldsymbol{V}^T\boldsymbol{V} - \hat{\boldsymbol{V}}_i^T\boldsymbol{V}\right\|_F + \left\|\hat{\boldsymbol{V}}_i^T\boldsymbol{V} - \hat{\boldsymbol{V}}_i^T\hat{\boldsymbol{V}}_i\right\|_F \\
\left\|\boldsymbol{W} - \hat{\boldsymbol{W}}\right\|_F &\le \|\boldsymbol{V}\|_2\left\|\boldsymbol{V}^T - \hat{\boldsymbol{V}}_i^T\right\|_F + \left\|\hat{\boldsymbol{V}}_i\right\|_2\left\|\boldsymbol{V} - \hat{\boldsymbol{V}}_i\right\|_F \\
\left\|\boldsymbol{W} - \hat{\boldsymbol{W}}\right\|_F &\le (C+2)\|\boldsymbol{V}\|_2\left\|\boldsymbol{V} - \hat{\boldsymbol{V}}_i\right\|_F
\end{aligned} \tag{A.24}$$

Our choice of τ for the number of iterations of the push-sum algorithm, as in [45], to ensure the bound on the error of the projection onto the space spanned by the eigenvectors allows us to establish the bound $\|\boldsymbol{V}\|_F^2 \leq k\|\boldsymbol{K}\|_2^2$, where $\boldsymbol{R} \in \mathbb{R}^{k\times k}$.

Next, we establish the error incurred by estimating $\boldsymbol{W}$ through the decentralized approach.

$$\begin{aligned}
\left\|\boldsymbol{W} - \hat{\boldsymbol{W}}_i\right\|_F &= \left\|\boldsymbol{W} - \hat{\boldsymbol{W}}_i + \hat{\boldsymbol{W}} - \hat{\boldsymbol{W}}\right\|_F \\
&\leq \left\|\boldsymbol{W} - \hat{\boldsymbol{W}}\right\|_F + \|\boldsymbol{E}_i^W\|_F \\
&\leq \left\|\boldsymbol{W} - \hat{\boldsymbol{W}}\right\|_F + (C+1)^2\varepsilon k\|\boldsymbol{K}\|_2^2 \\
&\leq (C+2)\|\boldsymbol{K}\|_2^2\left\|\boldsymbol{Q} - \hat{\boldsymbol{Q}} - \boldsymbol{E}_i^V\right\|_F + (C+1)^2\varepsilon k\|\boldsymbol{K}\|_2^2 \\
&\leq \max\{(C+2),(C+1)^2\}\|\boldsymbol{K}\|_2^2 * \left(\left\|\boldsymbol{Q} - \hat{\boldsymbol{Q}} - \boldsymbol{E}_i^V\right\|_F + \varepsilon k\right)
\end{aligned} \tag{A.25}$$

Using the results presented in [37] and [45], we have that

$$\begin{aligned}
\left\|\boldsymbol{R} - \hat{\boldsymbol{R}}_i\right\|_F &\leq \|\boldsymbol{W}^{-1}\|_2\|\boldsymbol{R}\|_2\left\|\hat{\boldsymbol{W}} - \hat{\boldsymbol{W}}_i\right\|_F \\
&\leq \max\{(C+2),(C+1)^2\}\left\|\boldsymbol{R}^{-1}\right\|_2^2\|\boldsymbol{K}\|_2^3\left(\left\|\boldsymbol{Q} - \hat{\boldsymbol{Q}} - \boldsymbol{E}_i^V\right\|_F + \varepsilon k\right)
\end{aligned} \tag{A.26}$$

Applying Wedin's Theorem [37], we have for invertible matrices $\boldsymbol{R}$, $\hat{\boldsymbol{R}}_i$

$$\left\|\boldsymbol{R}^{-1} - \hat{\boldsymbol{R}}_i^{-1}\right\|_2 \leq \frac{1+\sqrt{5}}{2}\left\|\boldsymbol{R} - \hat{\boldsymbol{R}}_i\right\|_2 \max\left\{\|\boldsymbol{R}^{-1}\|_2^2, \left\|\hat{\boldsymbol{R}}_i^{-1}\right\|_2^2\right\} \tag{A.27}$$

Then, applying (A.26) and (A.27),

$$\begin{aligned}
\left\|\boldsymbol{R}^{-1} - \hat{\boldsymbol{R}}_i^{-1}\right\|_2 &\leq \|\boldsymbol{R}^{-1}\|_2^2\|\boldsymbol{K}\|_2^3\left(\left\|\boldsymbol{Q} - \hat{\boldsymbol{Q}} - \boldsymbol{E}_i^V\right\|_F + \varepsilon k\right) \\
&\leq \max\{(C+2),(C+1)^2\} * (1+\sqrt{5})\|\boldsymbol{K}\|_2^3\left\|\boldsymbol{R}^{-1}\right\|_2^4 \\
&\quad \times \left(\left\|\boldsymbol{Q} - \hat{\boldsymbol{Q}} - \boldsymbol{E}_i^V\right\|_F + \varepsilon k\right)
\end{aligned} \tag{A.28}$$

Using results from [45] related to analyzing the effects of the node on the matrix $\boldsymbol{Q}$, we have that

$$\begin{aligned}
\left\|\boldsymbol{Q}' - \hat{\boldsymbol{Q}}_i'\right\|_F &\leq \left\|\hat{\boldsymbol{V}}_i - \boldsymbol{V}\right\|_F * \max_i\left\|\hat{\boldsymbol{R}}_i\right\|_2^{-1} + \|\boldsymbol{V}\|_F * \max_i\left\|\boldsymbol{R}^{-1} - \hat{\boldsymbol{R}}_i^{-1}\right\|_2 \\
&\leq \|\boldsymbol{K}\|_2\left\|\boldsymbol{Q} - \hat{\boldsymbol{Q}}_i\right\|_F * \sqrt{2}\|\boldsymbol{R}^{-1}\|_2 \\
&\quad + \sqrt{k}\,\|\boldsymbol{K}\|_2 * \max\{(C+2),(C+1)^2\}
\end{aligned}$$

$$*(1+\sqrt{5})\,\|\boldsymbol{K}\|_2^3\|\boldsymbol{R}^{-1}\|_2^4\Big(\Big\|\boldsymbol{Q}-\hat{\boldsymbol{Q}}-\boldsymbol{E}_i^V\Big\|_F+\varepsilon k\Big)$$

$$\leq \|\boldsymbol{K}\|_2\Big\|\boldsymbol{Q}-\hat{\boldsymbol{Q}}_i\Big\|_F*\sqrt{2}\|\boldsymbol{R}^{-1}\|_2 \tag{A.29}$$

$$+\sqrt{k}\|\boldsymbol{K}\|_2*\max\{(C+2),(C+1)^2\}$$

$$*(1+\sqrt{5})\|\boldsymbol{K}\|_2^3\|\boldsymbol{R}^{-1}\|_2^4\Big(\Big\|\boldsymbol{Q}-\hat{\boldsymbol{Q}}_i\Big\|_F+\varepsilon k\Big)$$

$$\leq C^4\sqrt{k}\|\boldsymbol{R}^{-1}\|_2^4\|\boldsymbol{K}\|_2^4\Big(\Big\|\boldsymbol{Q}-\hat{\boldsymbol{Q}}_i\Big\|_F+\varepsilon k\Big) \tag{A.30}$$

This shows that the error is a factor of $\sqrt{k}(C\|\boldsymbol{R}^{-1}\|_2\|\boldsymbol{K}\|_2)^4$. We can calculate the total error, defined as $\|(\boldsymbol{P}+\boldsymbol{P}^{\perp})-\boldsymbol{P}_{\hat{Q}_i}\|$, using (A.29) and the analysis set forth in [45]:

$$\begin{aligned}
\|(\boldsymbol{P}+\boldsymbol{P}^{\perp})-\boldsymbol{P}_{\hat{Q}_i}\|_2 &= \|(\boldsymbol{P}+\boldsymbol{P}^{\perp})-\boldsymbol{P}_{\hat{Q}}+\boldsymbol{P}_{\hat{Q}}-\boldsymbol{P}_{\hat{Q}_i}\|_2\\
&\leq \|(\boldsymbol{P}+\boldsymbol{P}^{\perp})-\boldsymbol{P}_Q\|_2+\|\boldsymbol{P}_Q-\boldsymbol{P}_{\hat{Q}_i}\|_2\\
&= \|(\boldsymbol{P}+\boldsymbol{P}^{\perp})-\boldsymbol{P}_Q\|_2+\Big\|\boldsymbol{Q}\boldsymbol{Q}^T{}_2-\hat{\boldsymbol{Q}}_i\hat{\boldsymbol{Q}}_i^T\Big\|_2\\
&\leq \|(\boldsymbol{P}+\boldsymbol{P}^{\perp})-\boldsymbol{P}_Q\|_2+\Big\|\boldsymbol{Q}\boldsymbol{Q}^T-\boldsymbol{Q}\hat{\boldsymbol{Q}}_i{}^T\Big\|_F+\Big\|\boldsymbol{Q}\hat{\boldsymbol{Q}}_i{}^T-\hat{\boldsymbol{Q}}_i\hat{\boldsymbol{Q}}_i^T\Big\|_F\\
&\leq \|(\boldsymbol{P}+\boldsymbol{P}^{\perp})-\boldsymbol{P}_Q\|_2+\Big(\|\boldsymbol{Q}\|_2+\Big\|\hat{\boldsymbol{Q}}_i\Big\|_2\Big)\Big\|\boldsymbol{Q}-\hat{\boldsymbol{Q}}_i\Big\|_F\\
&\leq e^{\gamma t}\left(2\lambda_{k+1}+\frac{\delta^2}{8}\xi\right)+(C+2)\Big\|\boldsymbol{Q}-\hat{\boldsymbol{Q}}_i\Big\|_F\\
&\leq e^{\gamma t}\left(2\lambda_{k+1}+\frac{\delta^2}{8}\xi\right)+\mathcal{O}\left|\frac{\lambda_{k+1}}{\lambda_k}\right|^{\tau}*n+(C+2)\varepsilon^{4\tau}
\end{aligned} \tag{A.31}$$

□

Acknowledgments

TS, DK, and MAH are supported by ONR Award Nos. N00014-16-1-2216 and N00014-17-1-2690 and National Science Foundation (NSF) grant IIS-1812319. EF is supported by National Science Foundation (NSF) grant DMS-1418956.

References

[1] Delaney JR, Barga RS (2009) A 2020 vision for ocean science. In: Hey T, Tansley S, Tolle K (eds) The Fourth Paradigm: Data-Intensive Scientific Discovery, Microsoft Research, pp. 27–38.

[2] Caron D, Stauffer B, Moorthi S, *et al.* (2008) Macro- to fine-scale spatial and temporal distributions and dynamics of phytoplankton and their environmental driving forces in a small subalpine lake in southern California, USA. Journal of Limnology and Oceanography 53(5):2333–2349.

[3] Zhang F, Fratantoni DM, Paley D, Lund J, Leonard NE (2007) Control of coordinated patterns for ocean sampling. International Journal of Control 80(7):1186–1199.

[4] Chen V, Batalin M, Kaiser W, Sukhatme G (2008) Towards spatial and semantic mapping in aquatic environments. In: IEEE International Conference on Robotics and Automation, Pasadena, CA, pp. 629–636.

[5] Das J, Py F, Maughan T, *et al.* (2012) Coordinated sampling of dynamic oceanographic features with underwater vehicles and drifters. The International Journal of Robotics Research 31(5):626–646, DOI 10.1177/0278364912440736

[6] Lynch KM, Schwartz IB, Yang P, Freeman RA (2008) Decentralized environmental modeling by mobile sensor networks. IEEE Transactions on Robotics and Automation 24(3):710–724.

[7] Wu W, Zhang F (2011) Cooperative exploration of level surfaces of three dimensional scalar fields. Automatica, the IFAC Journal 47(9):2044–2051.

[8] Sydney N, Paley DA (2011) Multi-vehicle control and optimization for spatiotemporal sampling. In: IEEE Conf. Decision and Control, Orlando, FL, pp. 5607–5612.

[9] Cortés J, Martínez S, Karatas T, Bullo F, Member S (2004) Coverage control for mobile sensing networks. IEEE Transactions on Robotics and Automation 20(2):243–255, DOI 10.1109/TRA.2004.824698

[10] Le Ny J, Pappas GJ (2010) Sensor-based robot deployment algorithms. Proceedings of the IEEE Conference on Decision and Control, pp. 5486–5492, DOI 10.1109/CDC.2010.5716979

[11] Schwager M, Vitus MP, Powers S, Rus D, Tomlin CJ (2017) Robust adaptive coverage control for robotic sensor networks. IEEE Transactions on Control of Network Systems 4(3):462–476.

[12] Hollinger GA, Pereira AA, Binney J, Somers T, Sukhatme GS (2016) Learning uncertainty in ocean current predictions for safe and reliable navigation of underwater vehicles. Journal of Field Robotics 33(1):47–66, DOI 10.1002/rob.21613

[13] Das J, Py F, Harvey JB, *et al.* (2015) Data-driven robotic sampling for marine ecosystem monitoring. International Journal of Robotics Research 34(12):1435–1452, DOI 10.1177/0278364915587723

[14] Kemna S, Rogers JG, Nieto-Granda C, Young S, Sukhatme GS (2017) Multi-robot coordination through dynamic Voronoi partitioning for informative adaptive sampling in communication-constrained environments. Proceedings – IEEE International Conference on Robotics and Automation, pp. 2124–2130, DOI 10.1109/ICRA.2017.7989245

[15] Garg S, Ayanian N (2014) Persistent monitoring of stochastic spatio-temporal phenomena with a small team of robots. Proceedings of Robotics: Science and Systems. Retrieved from http://www.roboticsproceedings.org/rss10/p38.html.

[16] Durrant-Whyte HF (1988) Sensor models and multisensor integration. International Journal of Robotics Research 7(6):97–113.

[17] García MR, Vilas C, Banga JR, Alonso AA (2007) Optimal field reconstruction of distributed process systems from partial measurements. Industrial and Engineering Chemistry Research 46(2):530–539, DOI 10.1021/ie0604167

[18] Kumar S, Deshpande A, Sarma SE (2012) Stable arrangements of mobile sensors for sampling physical fields. In: American Control Conference (ACC), 2012, IEEE, pp. 324–331.

[19] Everson R, Sirovich L (1995) Karhunen–Loève procedure for gappy data. Journal of the Optical Society of America 12(8):1657, DOI 10.1364/JOSAA.12.001657

[20] Alonso AA, Kevrekidis IG, Banga JR, Frouzakis CE (2004) Optimal sensor location and reduced order observer design for distributed process systems. Computers and Chemical Engineering 28(1–2):27–35, DOI 10.1016/S0098-1354(03)00175-3

[21] Willcox K (2006) Unsteady flow sensing and estimation via the gappy proper orthogonal decomposition. Computers and Fluids 35(2):208–226, DOI 10.1016/j.compfluid.2004.11.006

[22] Peherstorfer B, Willcox K (2016) Dynamic data-driven model reduction: adapting reduced models from incomplete data. Advanced Modeling and Simulation in Engineering Sciences 3(1), DOI 10.1186/s40323-016-0064-x

[23] Frolov S, Paduan J, Cook M, Bellingham J (2012) Improved statistical prediction of surface currents based on historic HF-radar observations. Ocean Dynamics 62(7):1111–1122, DOI 10.1007/s10236-012-0553-5

[24] Edwards CR, Seim HE (2008) Complex EOF analysis as a method to separate barotropic and baroclinic velocity structure in shallow water. Journal of Atmospheric and Oceanic Technology 25(5):808–821, DOI 10.1175/2007JTECHO562.1

[25] Das J, Py F, Maughan T, *et al.* (2014) Simultaneous Tracking and Sampling of Dynamic Oceanographic Features with Autonomous Underwater Vehicles and Lagrangian Drifters, Springer Berlin Heidelberg, Berlin, Heidelberg, pp. 541–555. DOI 10.1007/978-3-642-28572-1_37, URL https://doi.org/10.1007/978-3-642-28572-1_37

[26] Williams RK, Sukhatme GS (2012) Probabilistic spatial mapping and curve tracking in distributed multi-agent systems. In: 2012 IEEE International Conference on Robotics and Automation, pp. 1125–1130, DOI 10.1109/ICRA.2012.6224689

[27] DeVries L, Paley DA (2012) Multivehicle control in a strong flowfield with application to hurricane sampling. Journal of Guidance, Control, and Dynamics 35(3):794–806, DOI 10.2514/1.55580, URL https://doi.org/10.2514/1.55580

[28] Smith R, Pereira AM, Chao Y, *et al.* (2010) Autonomous underwater vehicle trajectory design coupled with predictive ocean models: A case study. In: Proc. of the 2010 IEEE International Conference on Robotics and Automation, Anchorage, AK, pp. 4770–4777.

[29] Smith RN, Chao Y, Li PP, Caron DA, Jones BH, Sukhatme GS (2010) Planning and implementing trajectories for autonomous underwater vehicles to track evolving ocean processes based on predictions from a regional ocean model. International Journal of Robotics Research 29(12):1475–1497, DOI http://dx.doi.org/10.1177/0278364910377243, URL http://cres.usc.edu/cgi-bin/print_pub_details.pl?pubid=646

[30] Smith R, Kelly J, Sukhatme G (2012) Towards improving mission execution for autonomous gliders with an ocean model and Kalman filter. In: Proc. of the 2012 IEEE International Conference on Robotics and Automation, Minneapolis, MN.

[31] Wang D, Lermusiaux P, Haley P, Eickstedt D, Leslie W, Schmidt H (2009) Acoustically focused adaptive sampling and on-board routing for marine rapid environmental assessment. Journal of Marine Systems 78: 393–407.

[32] Lolla T, Ueckermann MP, Haley P, Lermusiaux PFJ (2012) Path planning in time dependent flow fields using level set methods. In: Proc. IEEE International Conference on Robotics and Automation, Minneapolis, MN, USA.

[33] Kularatne D, Bhattacharya S, Hsieh MA (2016) Time and energy optimal path planning on a flow field. In: Robotics: Science and Systems, Ann Arbor, MI, USA (submitted).

[34] Kularatne D, Bhattacharya S, Hsieh MA (2016) Computing energy optimal paths in time-varying flows. Tech. rep., Drexel University, Philadelphia, USA, DOI:10.17918/D8B66V

[35] Shchepetkin A, McWilliams J (1998) Quasi-monotone advection schemes based on explicit locally adaptive dissipation. Monthly Weather Review 126:1541–1580.

[36] Shchepetkin AF, McWilliams JC (2005) The regional oceanic modeling system (ROMS): A split-explicit, free-surface, topography-following-coordinate oceanic model. Ocean Modeling 9:347–404.

[37] Chen KK, Tu JH, Rowley CW (2012) Variants of dynamic mode decomposition: boundary condition, Koopman, and Fourier analyses. Journal of Nonlinear Science 22(6):887–915.

[38] Schmid PJ (2010) Dynamic mode decomposition of numerical and experimental data. Journal of Fluid Mechanics 656:5–28.

[39] Rowley CW, Mezić I, Bagheri S, Schlatter P, Henningson DS (2009) Spectral analysis of nonlinear flows. Journal of Fluid Mechanics 641:115–127.

[40] Sirovich L (1987) Turbulence and the dynamics of coherent structures. I. Coherent structures. Quarterly of Applied Mathematics 45(3):561–571, DOI 10.1090/qam/910463

[41] Kirby M (2000) Geometric Data Analysis: An Empirical Approach to Dimensionality Reduction and the Study of Patterns. John Wiley & Sons, Inc., New York, NY, USA.

[42] Huynh V, Dunbabin M, Smith R (2015) Predictive motion planning for AUVs subject to strong time-varying currents and forecasting uncertainties. In:

Robotics and Automation (ICRA), 2015 IEEE International Conference on, pp. 1144–1151, DOI 10.1109/ICRA.2015.7139335

[43] Salam T, Hsieh MA (2019) Adaptive sampling and reduced-order modeling of dynamic processes by robot teams. IEEE Robotics and Automation Letters 4(2):477–484, DOI 10.1109/LRA.2019.2891475

[44] Golub GH, Loan CFV (1996) Matrix Computations, 3rd edn. The Johns Hopkins University Press, Baltimore, MD.

[45] Kempe D, McSherry F (2004) A Decentralized Algorithm for Spectral Analysis. Proceedings of the thirty-sixth annual ACM Symposium on Theory of Computing, pp. 561–568, DOI 10.1145/1007352.1007438

[46] Penna F, Stanczak S (2015) Decentralized eigenvalue algorithms for distributed signal detection in cognitive networks. IEEE Transactions on Signal Processing 63(2):427–440.

[47] Veronis G (1966) Wind-driven ocean circulation, Part i and Part ii. Deep-Sea Research and Oceanographic Abstracts. 13(31): 17–29 and 31–55.

[48] Kularatne D, Bhattacharya S, Hsieh MA (2018) Going with the flow: A graph based approach to optimal path planning in general flows. Autonomous Robots 42(7):1369–1387, DOI 10.1007/s10514-018-9741-6, URL https://doi.org/10.1007/s10514-018-9741-6

[49] Conway JH, Sloane NJA (1999) Certain Important Lattices and Their Properties, Springer New York, New York, NY, pp. 94–135. DOI 10.1007/978-1-4757-6568-7_4, URL https://doi.org/10.1007/978-1-4757-6568-7_4

[50] Kularatne D, Bhattacharya S, Hsieh MA (2018) Optimal path planning in time-varying flows using adaptive discretization. IEEE Robotics and Automation Letters 3(1):458–465, DOI 10.1109/LRA.2017.2761939

[51] Kularatne D, Hajieghrary H, Ani Hsieh M (2018) Optimal path planning in time-varying flows with forecasting uncertainties. In: 2018 IEEE International Conference on Robotics and Automation (ICRA), pp. 1–8, DOI 10.1109/ICRA.2018.8460221

[52] Inanc T, Shadden S, Marsden J (2005) Optimal trajectory generation in ocean flows. In: American Control Conference, 2005. Proceedings of the 2005, pp. 674–679, DOI 10.1109/ACC.2005.1470035

[53] Haller G, Yuan G (2000) Lagrangian coherent structures and mixing in two-dimensional turbulence. Physica D 147:352–370.

[54] Haller G (2011) A variational theory of hyperbolic Lagrangian coherent structures. Physica D 240:574–598.

[55] Forgoston E, Billings L, Yecko P, Schwartz IB (2011) Set-based corral control in stochastic dynamical systems: Making almost invariant sets more invariant. Chaos 21(1):013,116.

[56] Mallory K, Hsieh MA, Forgoston E, Schwartz IB (2013) Distributed allocation of mobile sensing swarms in gyre flows. Nonlinear Processes in Geophysics 20(5):657–668.

[57] Hsieh MA, Mallory K, Schwartz IB (2014) Distributed allocation of mobile sensing agents in geophysical flows. In: Proc. of the 2014 American Controls Conference, Portland, OR.

[58] Heckman CR, Schwartz IB, Hsieh MA (2015) Toward efficient navigation in uncertain gyre-like flows. The International Journal of Robotics Research 34(13):1590–1603, DOI 10.1177/0278364915585396

[59] Michini M, Hsieh MA, Forgoston E, Schwartz IB (2014) Robotic tracking of coherent structures in flows. IEEE Transactions on Robotics 30(3):593–603.

[60] Michini M, Rastgoftar H, Hsieh MA, Jayasuriya S (2014) Distributed formation control for collaborative tracking of manifolds in flows. In: Proc. of the 2014 American Control Conference (ACC 2014), Portland, OR.

[61] Michini M, Hsieh MA, Forgoston E, Schwartz IB (2014) Experimental validation of robotic manifold tracking in gyre-like flows. In: Proc. IEEE/RSJ Int. Conf. on Intelligent Robots and Systems (IROS) 2014, Chicago, IL, USA.

[62] Kularatne D, Hsieh MA (2015) Tracking attracting Lagrangian coherent structures in flows. In: Proceedings of Robotics: Science and Systems, Rome, Italy.

[63] Kularatne D, Hsieh MA (2017) Tracking attracting manifolds in flows. Autonomous Robots 41(8):1575–1588, DOI 10.1007/s10514-017-9628-y, URL https://doi.org/10.1007/s10514-017-9628-y

[64] Peacock T, Haller G (2013) Lagrangian coherent structures: The hidden skeleton of fluid flows. Physics Today 66(2):41–47.

[65] Olascoaga MJ, Rypina II, Brown MG, *et al.* (2006) Persistent transport barrier on the West Florida shelf. Geophysical Research Letters 33(22):L22,603.

[66] Senatore C, Ross S (2008) Fuel-efficient navigation in complex flows. In: American Control Conference, 2008, pp. 1244–1248, DOI 10.1109/ACC.2008.4586663

[67] Kularatne D, Forgoston E, Hsieh MA (2018) Exploiting stochasticity for the control of transitions in gyre flows. In: Robotics: Science and Systems, Pittsburgh, PA.

Chapter 19
Summary, conclusions, and future work

Frank Ehlers[1]

In this last chapter, after reading 17 chapters (Chapters 2–18) dealing with various aspects of the design and practice of autonomous underwater vehicles (AUVs), it is worthwhile to summarise the findings, draw conclusions, and take a look into the future. Thereby, we continue the discussion started in the Introduction (Chapter 1). In Section 19.1, we elaborate on the three views on the maritime robotics aspect of AUVs. In Section 19.2 summaries, conclusions, and recommendations for future work from the authors of the chapters are collected. In Section 19.3, the conclusion from the (holistic) viewpoint on autonomy described in Section 1.2 is given.

19.1 Three views on maritime robotics

In the Introduction (Chapter 1), three views on the combination task for robotics and maritime environment have been proposed: the roboticist's view, the statistical signal processing view, and the system-of-systems view. The chapters of the book are ordered guided by the robotics view. Hence, for the generation of conclusions, it is useful to elaborate on the two remaining views.

Important for the statistical signal processing view is the way how uncertainty is handled. The design has to allow for a compensation or absorption of uncertainty, which is inherent in almost all internal and external parameters concerning the AUV operation. From this abstract perspective, several control algorithms presented in this book could be viewed as different methods to minimise, e.g., the energy necessary for the AUV to reach its goal. An abstract conclusion from the chapters dealing with navigation could be that uncertainty is absorbed by the capability to exploit visual or acoustic sensor data. As part of acoustic communications, a 'buffering' can absorb uncertainty about inter-symbol interference, and buffering is also the method that helps the cooperation between taxi-USV and AUVs. Another method to deal with uncertainty is to generate an additional information flow, together with specific behaviours of robot team members. The flexibility of architectures, also to deal with the implementation of fault-tolerant measures, is another design approach. The

[1]Bundeswehr Technical Center for Ships and Naval Weapons, Maritime Technology and Research (WTD 71), Kiel, Germany

'look-through-the-sensor' approach means the exploitation of measurement data via its filtering through additional environmental process modelling, and it is a bridge towards facing more environmental complexity with autonomous capabilities by utilising learning approaches.

From the system-of-systems point of view, the approaches and findings in the chapters suggest the declaration of modules for path planning and control which incorporate additional knowledge on underwater current (and processes related to a more global view on the generation of these currents). Furthermore, they suggest a database module where navigation algorithm could launch service requests regarding data from previous missions. From the systems engineer's view, hardware system design is related to the implementation of flexibility, when, for example, the need for a pressure vessel is no longer given.

The overall design process and, with the possibility to extend this to the (human) learning process related to the results from practicing, the overall future AUV development both depend on the way how we organise our AUV community, in order to progress as fast as possible in bridging the reality gap between the experimentation with and the application of AUVs.

19.2 Chapters' summary, conclusion, and future work

Part I: Control

Chapter 2

Karras *et al.* described the kinematic and dynamic equations of underwater robotic vehicles considering standard assumptions and simplifications that may lead to low-complexity but accurate and operational models. Additionally, the authors presented an RLS method for the online identification of the vehicle dynamic parameters. Moreover, the authors presented a state estimation technique which, depending to the available sensors, can be adopted for either shallow or deep water operation. The proposed state estimation method is based on the Complementary Filters notion; hence low-complexity computational requirements are satisfied. The goal of the chapter was to provide the reader with the appropriate analytical methods, in order to derive a simple yet accurate vehicle model, design a state estimation algorithm that can be easily integrated into the embedded system framework of an underwater robotic vehicle, and perform a fast online dynamic parameter identification.

Chapter 3

Bechlioulis *et al.* presented model-based and model-free motion control strategies for underwater robotic vehicles. Regarding model-based control, the authors presented an NMPC strategy for underwater robotic vehicles operating under various constraints. The purpose of the controller is to guide the vehicle towards specific waypoints while simultaneously satisfying various state and inputs constraints. The proposed scheme incorporates the full dynamics of the vehicle in which the ocean currents are also involved. The controller is designed in order to find the optimal thrusts required for minimising the waypoint tracking error. Furthermore, the control inputs calculated by the proposed approach are formulated in a way that the vehicle will exploit the ocean currents, when they are in favour of the waypoint-tracking mission, which results in reduced energy consumption by the thrusters. Regarding model-free control strategies, we presented position and trajectory tracking control schemes for AUVs. The proposed controllers do not utilise the vehicles' dynamic model parameters and guarantees prescribed transient and steady-state performance despite the presence of external disturbances and kinematic constraints. Moreover, through the appropriate selection of certain performance functions, the proposed control strategies can also guarantee the satisfaction of motion and performance constraints imposed by the desired task.

Chapter 4

Roy *et al.* introduce a new class of ARC tracking control framework for AUV without any knowledge of the system dynamics parameters. Conventional ARC strategies either require complete/partial knowledge of the systems dynamics parameters or presume the overall uncertainty (or its time derivative) is upper bounded by a constant. However, such presumption is restrictive for systems like AUVs which have explicit presence of states of system uncertainty in their structure. Besides, many of the existing ARC-based designs suffer from the over- and underestimation problems of switching gains. In such regards, the proposed ASRC law not only avoids restrictive presumptions of system uncertainty being upper bounded by a constant, it simultaneously alleviates the over- and underestimation problems of switching gain. Further, the effectiveness of the ASRC law is verified in simulation in comparison with the existing ASMC law under the scenario of uncertain system dynamics. An important future work would be to formulate an ARC for AUV considering input/output delay since data acquisition and localisation in underwater scenario is always challenging and a controller needs to take care of such circumstances.

Part II: Navigation

Chapter 5

Ånonsen *et al.* have given an overview of single-vehicle AUV navigation which in all advanced vehicles is based on inertial navigation. They have shown how an inertial navigation system (INS) is able to estimate the AUV's position, velocity, and orientation, and how its accuracy is dependent on manoeuvres and the quality of the sensors. With basic aiding sensors such as pressure sensors and Doppler velocity logs (DVLs), one is able to significantly reduce the inherent drift in the INS, but to facilitate extensive submerged, fully autonomous navigation, additional aiding methods are needed. One possibility is to rely on transponder navigation of some kind (LBL or UTP), but this requires deployment and positioning of the transponders to be used. For the AUV to be completely independent of infrastructure or aiding from other vehicles (manned or unmanned), the only remaining option is to use observations of the underwater environment for navigation purposes, e.g. to use payload sensors (sonars, optical cameras, etc.) to obtain accurate delta position estimates or to use some kind of geophysical navigation. Bathymetric terrain-referenced navigation is by far the most mature geophysical navigation for AUVs today. In areas where no prior DTM exists, terrain- or feature-based techniques can be used as the basis for SLAM algorithms. However, as SLAM by definition is local, the global accuracy obtainable through the loop-closing procedure is bounded above by the global navigation accuracy at the first bathymetry or feature observation involved in the loop closure. To achieve fully autonomous operations with high navigation accuracy, the autonomy system of the AUV must work together with the navigation systems to utilise all possible navigation-aiding methods whenever needed, based on the constraints of the operation. For example, in a non-covert operation in shallow water, the autonomy system could make a decision to pause the mission execution in order to surface for a GNSS fix if the navigation accuracy is unacceptably low. In other scenarios, altering the mission plan in order to visit an area suitable for terrain navigation could be an option. Future research on AUV navigation should therefore focus both on developing new and better aiding methods, as well as optimally using the existing methods autonomously.

Chapter 6

Salavasidis *et al.* present the Autosub Long Range 6000 (ALR6000) and Autosub Long Range 1500 (ALR1500) which are two ultra-endurance AUVs. The long endurance is achieved through the use of high-energy-density lithium primary cells, optimisation of energy consumption, and travelling slowly. Power analysis showed that these vehicles are capable of operating for months in a single deployment, even when unique missions may require power-hungry sensors (e.g. a gyrocompass for operations in high latitudes). However, the localisation problem for AUVs becomes increasingly challenging as vehicle endurance increases. Hence, this type of AUV requires an effective low-power on-board navigation solution. One method to put a bound on the localisation error is the terrain-aided navigation (TAN). A low-power particle-filter-based TAN algorithm was developed and tested using field data from a deep (up to 3,700 m) and long-range (up to 195 km in 77 h) mission performed in the Southern Ocean. To the best of the authors' knowledge, TAN has never been tested in truly deep-water operations and for such long-range missions without periodic surfacing to re-initialise the estimation process via GPS. Due to strong water currents, sparse bathymetric map, and motion along contour lines, this mission is a great test case for evaluating the performance of the TAN algorithm. Following the promising results of using TAN, the authors have investigated the potential of ALR1500 to cross the Arctic Ocean, a science-driven aspirational mission from Svalbard to the Barrow coast in Alaska (about 3,250 km). Simulations of crossing the Arctic Basin have been developed incorporating a low-resolution bathymetric map, models of water currents, and potential errors in heading sensors (magnetic or gyro compasses) due to operations in high latitudes. Despite the challenges of such a mission, simulation results show that it may be possible to cross the Arctic Ocean even using a simple magnetic compass for heading estimation. In terms of future work, further investigation in the performance of heading sensor in high latitudes is required. Moreover, due to the significant power required for using a gyrocompass, a hybrid scheme that alternates between both heading sensors may provide a more energy-wise efficient solution for heading estimation, without exposing the AUV in high risk by using only a magnetic compass. Ongoing results also include improvements in TAN. For example, very risky missions, such as the crossing of the Arctic Ocean, require an enhanced robustness navigation system that is able to ignore poor sensor measurements, deal with inaccurate maps, and, if divergence suspected, able to resort to an alternative on-board solution, avoiding temporarily poor estimates, whilst re-initialising employed filters.

Chapter 7

Sarma builds up from his previous work that the VOG-SLAM method could approximately match the power of conventional SLAM with a fraction of the complexity when the vehicle is performing constant course and speed behaviour – the desired behaviour in long transits. In Chapter 7, he argues the near-optimality under that behaviour from a simpler maximum likelihood perspective. We also suggest a possible sonar signal processing method for rapid data compression and data association for use in all SLAM systems. One observation is that, unlike a DVL-aided INS system, the VOG-SLAM method can utilise a group of features, perhaps deliberately emplaced along a known track, to result in potentially lower error growth rates provided the vehicle attempts constant course and speed behaviour. Specifically, observation and data association of these emplaced contacts can allow the vehicle to more quickly converge to accurate estimates of its own speed along the assumed heading when travelling over a level ocean floor. Although the spatially localised clusters are here determined by accounting for the inherent errors of the sensors, their finite extent and implicit dependence on the DVL-aided INS outputs in their creation must lead to incorrect declarations of persistence and data association. Due to the careful CFAR design of the signal processing, this is expected to have a weak effect when the vehicle obeys the constant course and speed behaviour. However, this can lead to a biased velocity estimate when the vehicle deviates from this behaviour. Note, DVL systems can ultimately also only most reliably provide speed along the current vehicle body axis and not velocity. Note that neither system can provide velocity with respect to a WRF. Thus, VOG-SLAM suffers from a similar drawback to DVL when the vehicle deviates from the constant course and speed behaviour or travels over non-level surfaces. Future work is concerned with a means to analyse and compartmentalise the errors in VOG-SLAM and DVL that most contribute to false data association and incorrect persistence declaration when the vehicle deviates from the preferred behaviour. It is crucial to note that conventional SLAM is not immune and must also reflect a loss every time vehicle behaviour does not match any of the kinematic expectations that govern the assumed Kalman platform state evolution equation. In addition, future work is also concerned with optimal ways to bound the positional error growth rate of VOG-SLAM and DVL systems. This is important as, even after GPS initialisation, the initial estimate of vehicle heading will be biased and no 'velocity' aiding can remove this bias. This bias will lead to positional error growth even assuming error-free outputs from the accelerometers, gyroscopes, and derived estimates of speed either along the body axis when under the constant course and speed behaviour. This growth is akin to a cross-range error.

Part III: Toolset

Chapter 8

Dias *et al.* state that the market offers a diversity of options of unmanned vehicles. A flexible tool is needed that enables the interoperability between assets for the operation of a fleet of vehicles procured from different vendors. LSTS toolchain for autonomous systems has been in the development for several years. Its flexibility is originated by the interplay of several components: IMC (which defines a message protocol independent from the communication means that allows the passing of the information between vehicles and command and control interfaces); DUNE (the on-board control software that fully controls an autonomous vehicle or interfaces with a front seat to control the vehicle); back-seat driver control (that exposes an API from DUNE in order to allow the creation of more complex controllers without compromising the front-seat operation); Neptus Command and Control software (allowing the command of autonomous vehicles, being flexible enough by adding a plug-in framework for easy functionality or protocol extension allowing connection to non-IMC vehicles, also providing a data review and analysis tool with plug-in extensions for data visualisation and export); Ripples (a REST web interface allowing the monitor and command of autonomous systems through a web browser interface, and at the same time exposing functionality and data persistence for command and control interfaces). Several examples of real-world use of the toolchain in scenarios spanning from military to science were shown depicting the flexibility of the toolchain. Additionally, we discussed the integration of the toolchain with other frameworks. The toolchain is used as the main interface, but can also be used to expose IMC vehicles to other frameworks. The development of the toolchain is now in a mature phase, with some adoption not only by other universities, but also from commercial companies that use the full or partial toolchain.

Chapter 9

Alves *et al.* relate, as outlook to future work, their research on underwater communications to new concepts like the Internet of Things (IoT). Today, it is difficult to identify the one dominant application of underwater communications that may drive the market and push the technology. Civilian and military applications alike are taking advantage of the developments in AUV technology and exploring the increased size of their fleets. This trend in AUV employment may well set a future landscape where underwater seamless connectivity will become a need. A subsea world where sensors, instruments, and vehicles are able to connect and build awareness of each other, just as we do now with our 'smart' gadgets, seems to still be a far-fetched proposition. However, under the concept of the 'IoT', one can certainly extrapolate and dream of a world of underwater connectivity. The realisation of underwater IoT is contingent upon the ability to send digital information via acoustic signals. However, the combined effect of platform mobility with multipath variability is still the major limitation for achieving bandwidth-efficient high-speed acoustic communications. Hence, research on advanced receiver structures will not cease until the theoretical capacity limits are achieved. Given the lack of universal UWA channel models, the community is rapidly moving towards machine learning approaches for adaptive modulation. Receiver feedback is essential and future research will manifest the type of channel characteristics that are invariable during the two-way travel time. In addition, future mobile platforms will include transmit arrays to allow spatial-division multiplexing strategies in multi-user networks and prevent energy propagating to an eavesdropper for security purposes. Networking protocols will play an increasingly important role as the numbers of assets increase and the applications explore those numbers to gain efficiency. Standards will be required addressing the format of the messages exchanged in the network and the encoding/decoding of the information, thus enabling interoperability beyond what currently done at the waveform level (i.e. JANUS). The investigation of smarter strategies will require taking full advantage of multimodal and heterogeneous communication and sensing solutions available at the node. This will make possible to truly explore software-defined and cognitive systems. Additionally, a coordinated integration of the communication, sensing localisation, and navigation capabilities will allow mobile nodes to complete the required tasks in a more efficient way and to enhance their cooperation ability. This coordinated integration will make possible to effectively use AUV mobility to cover connectivity holes in the network, increase the quality of the communication links, and avoid the presence of single points of failure, just to name a few. Security for underwater communications will certainly be an important feature in future applications. Novel protocols that will establish cryptographic keys without sharing any data between nodes combined with state-of-the-art cryptography will lead to substantial security gains. Additional security enhancement will be provided by distributed and cross-layer networking solutions that allow cooperation between the nodes of the network.

Chapter 10

Brink and Körner conclude that the biggest advantage of the print-tolerant technology compared to the pressure vessel-based technology is the cost saving by not using pressure housings. The savings potential increases with the depth of use, as the required wall thickness of the pressure vessel increases with depth. Furthermore, there are reduced leakage problems due to lack of pressure difference. The same design and system can be used at most any diving depth. There is a reduction of chemical and galvanic corrosion. Pressure-neutral vehicles are one-half to one-third as heavy as comparable pressure-vessel vehicles, whereby accessibility, mountability, and repair are no worse than for pressure vehicles or oil-compensated systems. There is no danger of loss of buoyancy or complete destruction in the event of a fault or implosion. The authors have described two main types of pressure-tolerant systems, fluid-compensated systems and cast systems. One advantage of fluid-compensated systems is the confidence that fluid will flood every cavity at every pressure cycle. Another advantage of fluid-compensated pressure-tolerant systems is the ability to completely drain the fluid for repair/rework. Fluid-compensated systems also facilitate direct connection of fluid-filled hoses – one can create custom pressure-neutral cables by combining different types of wires in fluid-filled hoses. This has important benefits where high-density or high-current subsea connectors are very expensive or impractical or simply not available. The advantages of pressure-neutral cast systems are that the requirement for the compensator (often spring-pressurised) is added complexity and a potential point of failure in fluid-compensated systems, that they can be repaired, changed, and extended as often as required. They show very good long-term behaviour in continuous underwater operation and allow for the integration of antifouling measures in the casting process. They show good insulation properties even in a water-saturated state provided the appropriate grouting properties are maintained. Due to the transparent encapsulation, important system states can be indicated optically but can also be controlled by optical receivers in the encapsulated assemblies.

Part IV: Cooperation

Chapter 11

Bays and Wettergren have presented a framework for optimising the scheduling of multiple heterogeneous unmanned vehicles in order to perform survey-type missions. The framework utilises mixed-integer linear programming in order to obtain optimal to near-optimal solutions, depending on the complexity of the problem and length of time the algorithm is allowed to search for a result. The framework specifically develops a tasking schedule applicable to multiple UUVs that includes docking, deployment, and movement actions performed throughout the area, and also accounts for scheduling of additional post-survey tasks that can only be done after the conclusion of a survey. A robust scheduling technique was introduced that extends the framework to include probabilistic constraints in order to account for uncertainty in the length of time required to perform tasks. Future work includes extending the framework to specific types of post-survey action tasks. Additionally, the authors will examine pre-processing heuristics to reduce the size and complexity of the MILP problem which leads to faster solve times for efficient solutions.

Chapter 12

Maehle *et al.* have presented the low-cost swarm micro-AUV MONSUN for environmental monitoring and inspection tasks. MONSUN's hardware and software design is robust, modular, scalable, and easily adaptable to a wide variety of applications. The communication among swarm members is based on Wi-Fi on the surface and acoustic modems underwater. The localisation concept assumes that some AUVs stay at the surface using GPS while the other swarm members submerge and localise with the help of acoustic communication and distance measurements provided by the modems. Since no further infrastructure is required, the swarm can be easily deployed from a small boat or the shore. Depending on the application, various swarm behaviours are possible. A robust and fault-tolerant protocol for a V-formation with two surfaced MONSUNs and an arbitrary number of submerged ones has been described in detail. It can, e.g., be used for scanning an area following meander-shaped paths. Also a protocol for general swarms with three surfaces and an arbitrary number of underwater MONSUNs at arbitrary positions is given which is based on triangulation. First field experiments have been executed to prove the MONSUN concept. In the expedition

Clockwork Ocean of Helmholtz Zentrum Geesthacht, two MONSUNs were successfully used for measuring sub-mesoscale eddies in the Baltic Sea. A V-formation with three MONSUNs using our robust protocol designed especially to work with unreliable acoustic communication has been tested under realistic conditions in a lake. Future work will, e.g., consist of using MONSUN micro-AUV swarms for monitoring tasks in ports and harbours in case of water pollution by dangerous chemicals (BMBF Project MoSAIk) or inspection of quay walls and ship hulls (BMVI Project SecurePort). For this, the swarm behaviours are further developed to be tailored to these tasks and additional sensors like oil sensors or imaging sonars will be integrated. Additionally, a Graphical User Interface (GUI) for easy handling of the swarm also by inexperienced users is under development. Further field experiments are planned with a higher number of MONSUNs in different realistic scenarios.

Chapter 13

Djapic elaborates on the details regarding the combination of cooperative AUVs: accurate timing and clock synchronisation, underwater sensors, and acoustic communication for the purposes of detection of various underwater targets for seismic surveys, surveillance, and marine mammals detection applications. Future work will include sensor fusion of passive sensor array of sensors of the same kind or passive sensor array of sensors of different kind (magnetic and acoustic) similar to an underwater monitoring systems by introducing autonomous triggering based on acoustic presence indicators of marine species.

Part V: Methodology

Chapter 14

Redfield *et al.* have introduced Capability Analysis Tables and shown how they can be used to connect behaviours and platforms to create a detailed behaviour specification. Making this connection explicit helps to identify missing or unused capabilities, incompleteness, and inconsistencies in the design. This design method is agnostic to platform and application area, and has been applied to behaviour specification in space robotics and underwater operations.

A further extension of this method is seen by the authors: providing a framework for robotic systems that makes them precise enough for formal methods. Formal methods are a powerful tool, but are not widely accessible to roboticists. The Capability Analysis Table represents a connection between the robotics community and the formal methods community. In the future, the authors expect to be able to link the Capability Analysis Table representation to one or more formal methods representations, enabling formal analysis of arbitrary behaviour designs. Today, it allows roboticists to improve the level of precision in their design without committing to developing the expertise necessary to apply formal methods. Capability analysis does not replace state machine diagrams or similar behaviour-centric representations, just as it does not replace platform design representations. It is a complementary design method that represents information related to both the behaviour design and the platform design to illuminate the facet of the behaviour design problem that focuses on its interaction with the platform.

Chapter 15

Seto and Svendsen state that deliberative AUV fault tolerance has not had as much attention compared to autonomous land or space-based vehicles. This is due to the smaller industry and challenges in communication and control of the vehicles on extended long-range missions. Reactive fault methods for vehicle hardware subsystems have had some attention but there is far less for deliberative fault tolerance. The AUV FDIR currently employed is basic and not adaptable for complex faults. Fault tolerance is a growing field for AUVs which can exploit previous gains in space and other types of unmanned vehicles. There is research into vehicle subsystem fault detection and a growing trend in fault recovery but there is a lack of combined methods for multiple subsystems with in-water testing of more complex methods. Such integrated AUV solutions do not exist yet in the same way as they do for space systems. Promising areas to further AUV FDIR development include learning and probabilistic methods.

Chapter 16

Brink summarises that, in an academic approach, a model is designed at the beginning of a project and validated by means of simulation. The next step is to build this model until an end product is created. Only at this stage are tests for functional capability carried out under real conditions. This approach involves the risk that the real conditions differ from the conditions assumed in the simulation, so that the developed end product is not functional. In a test-driven approach, there is no scientific model at the beginning of the project, but it instead develops simultaneously with the construction of the end product. Through permanent testing, this model or preliminary product is constantly evolving. The experience with the development of the AUV in the Arggonauts project shows that this test-driven approach worked, even if some details could have been solved better in retrospect.

Part VI: Application

Chapter 17

Pollini *et al.* presented the challenges encountered during design and testing of the Guidance Navigation and Control system of the vehicles involved in the WiMUST project. The results presented could be adopted in many other multi-vehicle missions where formation control at the surface and underwater, localisation with minimal acoustic communications and towing of streamers, cables, or submerged bodies in general is required by the mission.

Chapter 18

Salam *et al.* presented an energy-efficient distributed adaptive sampling strategy for a team of mobile robots to track a dynamic spatio-temporal process. They show how a team of mobile robots can build a reduced-order model (ROM) of the process using sparse measurements obtained from on-board sensors. The ROM is then used to infer measurements in regions lacking sensor coverage and the inferred data are then used to determine next best locations for new measurements and for planning minimum energy paths to the new locations. The authors apply the POD analysis scheme for the derivation of the ROM which identifies the dominant dynamics of the process based on point measurements at specified locations over time. As such, the quality of the ROM is highly dependent of both, the spatial and the temporal resolution of the measurement data. Consequently, the quality of the minimum energy trajectories computed using the ROM depends on the spatio-temporal resolution of the model. Under these circumstances, the proposed strategies can be significantly improved by incorporating specific geophysical fluid dynamics (GFD) insights. GFD is the study of natural fluid flows that span large physical scales, such as oceans, the atmosphere, and rivers. GFD flows are naturally stochastic and aperiodic, yet exhibit coherent structure. Recent studies of coherent structures in GFD and other flows have shown that these ubiquitous structures are typically kinematic and play a key role in transport. The Gulf Stream is a prominent example of a coherent jet whose heat transport is a critical component of global weather and climate. Using GFD models, details from flows such as the Gulf Stream can be used to diagnose the underlying GFD. This, in turn, enables the prediction of various physical, chemical, and biological processes in general geophysical flows. While the nature of a particular flow may originate as the solution to some dynamical (force) balance, the characteristics of the flow are an intrinsic property of its velocity vector field. A number of important processes, including transport, can be examined by this purely kinematic description of the flow. It is for this reason that we can witness similar transport features across a broad range of flows and that in the control of vehicles, as for tracers, knowledge of coherent structures may be exploited more effectively than knowledge of the detailed predictions offered by state-of-the-art primitive equation PDE models. The incorporation of knowledge about Lagrangian Coherent Structures will play a key role in future work in the development of new adaptive sampling algorithms and in the development of robust navigation strategies in the presence of model and/or forecast uncertainties, which is extremely important to ensure prolonged monitoring of dynamic and uncertain environments such as the ocean.

19.3 Autonomy in and for AUVs' design and practice

Returning to the set-up described in the Introduction, at the first glance, AUVs belong to the class of unmanned or automated systems, which are removing or remotely locating the operator. This reduces cost of the operation, reduces the size of the platform, and removes risk of life because the maritime operation is a naturally harsh environment. In this sense, AUVs can be viewed as maritime robots. When AUVs are built for carrying sensors only (i.e. without manipulators), they are, in this framework, off-board sensors.

Reading through the chapters of this book, however, we have learnt that AUVs can be (much!) more. They can interactively adapt to their environment, which is a sign of autonomy. On-board processing power will allow them to adapt with machine speed. Sufficient processing power is a prerequisite for implementing learning algorithms on board the AUVs in order to improve with experience, but to apply this learning advances in the organisation of the learning space are needed, as explained in the next paragraph.

At present, we are using AUVs to gather more knowledge about the processes in the maritime environment. Characterising the maritime environment without a certain level of knowledge regarding these processes turns out to end up in data sets which are too 'big' to draw validated conclusions from. In other words, by just measuring and data storing, we cannot disentangle the various processes in the maritime environment and label them according to their relevance given a certain operational task. Therefore, a complete picture of the environment, the operations, and the AUV is needed, including sensors, communications, navigation algorithms, etc., i.e. a holistic view filled with details like those presented in the chapters of this book. Interestingly, the impression from the future work recommendations formulated by the authors of the chapters is that access to information from such complete picture would help to improve the automatic or autonomous capabilities currently implemented in the AUVs. This scenario describes a feedback loop: a better picture of the environment leads to better ways to produce a better picture.

In order to prepare this feedback loop to become available for AUVs with increased autonomous functionality as a tool to cross their reality gap between experimentation and operational application, the relevance of details and their connectivity in the design space need to be explored. To enable this, AUV design and practice has to employ verification and validation mechanisms. The danger there is to make institutional decisions (also called standardisation) too quickly, for the ease of verification and validation. Instead, the community has to generate the data for such institutional decision-making first, by continuing with design and practice of AUVs until the ignition of autonomy embedded in (and thereby solving) the design space exploration task leads to continuously self-adjusting systems. Who knows when, or if at all, this will happen?

Index

www.ingramcontent.com/pod-product-compliance
Lightning Source LLC
LaVergne TN
LVHW011540200926
848484LV00001B/285